Handbook of Research on Artificial Intelligence Applications in the Aviation and Aerospace Industries

Tetiana Shmelova
National Aviation University, Ukraine

Yuliya Sikirda
Flight Academy of National Aviation University, Ukraine

Arnold Sterenharz
EXOLAUNCH GmbH, Germany

A volume in the Advances in Mechatronics and
Mechanical Engineering (AMME) Book Series

Published in the United States of America by
IGI Global
Engineering Science Reference (an imprint of IGI Global)
701 E. Chocolate Avenue
Hershey PA, USA 17033
Tel: 717-533-8845
Fax: 717-533-8661
E-mail: cust@igi-global.com
Web site: http://www.igi-global.com

Library of Congress Cataloging-in-Publication Data

Names: Shmelova, Tetiana, 1961- editor. | Sikirda, Yuliya, 1978- editor. |
  Sterenharz, Arnold, 1953- editor.
Title: Handbook of research on artificial intelligence applications in the
  aviation and aerospace industries / Tetiana Shmelova, Yuliya Sikirda,
  Arnold Sterenharz, editors.
Description: Hershey, PA : Engineering Science Reference, [2020] | Includes
  bibliographical references and index. | Summary: "This book explores
  best practices for AI implementation in aviation to enhance security and
  the ability to learn, improve, and predict. It also examines the
  enhancement of global aviation security as well as the methods of modern
  information systems in the aeronautics industry"-- Provided by
  publisher.
Identifiers: LCCN 2019029233 (print) | LCCN 2019029234 (ebook) | ISBN
  9781799814153 (h/c) | ISBN 9781799814177
  (eISBN)
Subjects: LCSH: Aeronautics--Technological innovations. | Aircraft
  industry--Data processing. | Aerospace industries--Data processing. |
  Artificial intelligence.
Classification: LCC TL553 .H25 2020 (print) | LCC TL553 (ebook) | DDC
  387.70285/63--dc23
LC record available at https://lccn.loc.gov/2019029233
LC ebook record available at https://lccn.loc.gov/2019029234

This book is published in the IGI Global book series Advances in Mechatronics and Mechanical Engineering (AMME) (ISSN: 2328-8205; eISSN: 2328-823X)

British Cataloguing in Publication Data
A Cataloguing in Publication record for this book is available from the British Library.

For electronic access to this publication, please contact: eresources@igi-global.com.

Advances in Mechatronics and Mechanical Engineering (AMME) Book Series

J. Paulo Davim
University of Aveiro, Portugal
ISSN:2328-8205
EISSN:2328-823X

MISSION

With its aid in the creation of smartphones, cars, medical imaging devices, and manufacturing tools, the mechatronics engineering field is in high demand. Mechatronics aims to combine the principles of mechanical, computer, and electrical engineering together to bridge the gap of communication between the different disciplines.

The **Advances in Mechatronics and Mechanical Engineering (AMME) Book Series** provides innovative research and practical developments in the field of mechatronics and mechanical engineering. This series covers a wide variety of application areas in electrical engineering, mechanical engineering, computer and software engineering; essential for academics, practitioners, researchers, and industry leaders.

COVERAGE

- Vibration and acoustics
- Micro and nanomechanics
- Autonomous Systems
- Manufacturing Methodologies
- Control Methodologies
- Sustainable and green manufacturing
- Bioengineering Materials
- Design and Manufacture
- Computational Mechanics
- Intelligent Navigation

IGI Global is currently accepting manuscripts for publication within this series. To submit a proposal for a volume in this series, please contact our Acquisition Editors at Acquisitions@igi-global.com or visit: http://www.igi-global.com/publish/.

Titles in this Series

For a list of additional titles in this series, please visit: www.igi-global.com/book-series

Design and Optimization of Sensors and Antennas for Wearable Devices Emerging Research and Opportunities
Vinod Kumar Singh (S. R. Group of Institutions Jhansi, India) Ratnesh Tiwari (Bhilai Institute of Technology, India) Vikas Dubey (Bhilai Institute of Technology, India) Zakir Ali (IET Bundelkhand University, India) and Ashutosh Kumar Singh (Indian Institute of Information Technology,India)
Engineering Science Reference • ©2020 • 196pp • H/C (ISBN: 9781522596837) • US $215.00

Global Advancements in Connected and Intelligent Mobility Emerging Research and Opporunities
Fatma Outay (Zayed University, UAE) Ansar-Ul-Haque Yasar (Hasselt University, Belgium) and Elhadi Shakshuki (Acadia University, Canada)
Engineering Science Reference • ©2020 • 278pp • H/C (ISBN: 9781522590194) • US $195.00

Automated Systems in the Aviation and Aerospace Industries
Tetiana Shmelova (National Aviation University, Ukraine) Yuliya Sikirda (Kirovograd Flight Academy of the National Aviation University, Ukraine) Nina Rizun (Gdansk University of Technology, Poland) Dmytro Kucherov (National Aviation University, Ukraine) and Konstantin Dergachov (National Aerospace University – Kharkiv Aviation Institute, Ukraine)
Engineering Science Reference • ©2019 • 486pp • H/C (ISBN: 9781522577096) • US $265.00

Design, Development, and Optimization of Bio-Mechatronic Engineering Products
Kaushik Kumar (Birla Institute of Technology, India) and J. Paulo Davim (University of Aveiro, Portugal)
Engineering Science Reference • ©2019 • 312pp • H/C (ISBN: 9781522582359) • US $225.00

Nonlinear Equations for Problem Solving in Mechanical Engineering
Seyed Amir Hossein Kiaeian Moosavi (Babol Noshirvani University of Technology, Iran) and Davood Domairry Ganji (Babol Noshirvani University of Technology, Iran)
Engineering Science Reference • ©2019 • 250pp • H/C (ISBN: 9781522581222) • US $195.00

Handbook of Research on Green Engineering Techniques for Modern Manufacturing
M. Uthayakumar (Kalasalingam University, India) S. Aravind Raj (Vellore Institute of Technology, India) Tae Jo Ko (Yeungnam University, South Korea) S. Thirumalai Kumaran (Kalasalingam University, India) and J. Paulo Davim (University of Aveiro, Portugal)
Engineering Science Reference • ©2019 • 403pp • H/C (ISBN: 9781522554455) • US $275.00

701 East Chocolate Avenue, Hershey, PA 17033, USA
Tel: 717-533-8845 x100 • Fax: 717-533-8661
E-Mail: cust@igi-global.com • www.igi-global.com

List of Contributors

Alexeiev, Oleh / *National Aviation University, Ukraine* 460

Bayramov, Azad Agalar / *Armed Forces War College of the Azerbaijan Republic, Azerbaijan* 193

Boiko, Serhii Mykolaiovych / *Kremenchuk Flight College, Kharkiv National University of
   Internal Affairs, Ukraine* 279

Bondarev, Dmitriy I. / *National Aviation University, Ukraine* 306

Chebukin, Bogdan / *Vilnius Gediminas Technical University, Lithuania* 404

Chorna, Viktoriia / *Kremenchuk Flight College, Kharkiv National University of Internal Affairs,
   Ukraine* 279

Derets, Sergiy / *National Aviation University, Ukraine* 404

Dergachov, Kostiyantin / *Kharkiv Aviation Institute, Ukraine* 36

Dolgikh, Serge / *Solana Networks, Canada* 1

Hashimov, Elshan Giyas / *Armed Forces War College of the Azerbaijan Republic, Azerbaijan* 193

Hlavcheva, Daria / *Kharkiv Polytechnic Institute, National Technical University, Ukraine* 134

Hryshchenko, Yurii / *National Aviation University, Ukraine* 372

Kasatkin, Mykola / *Kharkiv National University of Air Forces named by I. Kozhedub,
   Ukraine* 66

Kataieva, Mariia / *National Aviation University, Ukraine* 352

Kharchenko, Volodimir / *National Aviation University, Ukraine* 460

Kredentsar, Svetlana / *National Aviation University, Ukraine* 438

Kuchuk, Heorhii / *Kharkiv Polytechnic Institute, National Technical University, Ukraine* 134

Kulik, Anatolii / *National Aerospace University, Ukraine* 36

Kuzmenko, Nataliia / *National Aviation University, Ukraine* 180

Kuzmych, Lyudmyla / *National Aviation University, Ukraine* 214

Kvasnikov, Volodymyr / *National Aviation University, Ukraine* 214

Lysenko, Oleksandr / *Igor Sikorsky Kyiv Polytechnic Institute, National Technical University of
   Ukraine, Ukraine* 323

Nozhnova, Marina / *Kremenchuk Flight College, Kharkiv National University of Internal
   Affairs, Ukraine* 279

Ostroumov, Ivan / *National Aviation University, Ukraine* 180

Perederko, Anatoly / *National Aviation University, Ukraine* 263

Pérez, Mario Boyero / *EUROCONTROL, Belgium* 438

Pipa, Daria / *National Aviation University, Ukraine* 372

Podorozhniak, Andrii / *Kharkiv Polytechnic Institute, National Technical University, Ukraine* 134

Polozhevets, Hanna / *National Aviation University, Ukraine* 404

Romanenko, Viktor / *National Aviation University, Ukraine* 372

Rudas, Sergiy I. / *National Aviation University, Ukraine* 306

Sahun, Yelyzaveta Serhiyivna / *Flight Academy, National Aviation University, Ukraine* 419

Shmelev, Yurii / *Kremenchuk Flight College, Kharkiv National University of Internal Affairs, Ukraine* 279

Shmelova, Tetiana / *National Aviation University, Ukraine* 1

Sikirda, Yuliya / *Flight Academy, National Aviation University, Ukraine* 66

Solomentsev, Oleksandr / *National Aviation University, Ukraine* 148

Sterenharz, Arnold / *EXOLAUNCH GmbH, Germany* 1

Sushchenko, Olha / *National Aviation University, Ukraine* 231

Tachinina, Olena / *National Aviation University, Ukraine* 323

Tkachenko, Dmytro / *Ukrainian State Air Traffic Services Enterprise (UkSATSE), Lviv, Ukraine* 66

Vasyliev, Denys / *Ukrainian State Air Traffic Services Enterprise (UkSATSE), Ukraine* 91

Vasyliev, Volodymyr / *National Aviation University, Ukraine* 91

Yaloveha, Vladyslav / *Kharkiv Polytechnic Institute, National Technical University, Ukraine* 134

Yastrub, Maksym / *National Aviation University, Belgium* 438

Yurchuk, Alina / *National Aviation University, Ukraine* 352

Zaliskyi, Maksym / *National Aviation University, Ukraine* 148

Znakovska, Evgeniya A. / *National Aviation University, Ukraine* 306

Zuiev, Oleksii / *National Aviation University, Ukraine* 148

Table of Contents

Foreword xix

Preface xx

Chapter 1
Artificial Intelligence in Aviation Industries: Methodologies, Education, Applications, and
Opportunities 1
    Tetiana Shmelova, National Aviation University, Ukraine
    Arnold Sterenharz, EXOLAUNCH GmbH, Germany
    Serge Dolgikh, Solana Networks, Canada

Chapter 2
Rational Adaptation of Control Systems for the Autonomous Aircraft Motion 36
    Kostiyantin Dergachov, Kharkiv Aviation Institute, Ukraine
    Anatolii Kulik, National Aerospace University, Ukraine

Chapter 3
Intelligent Automated System for Supporting the Collaborative Decision Making by Operators of
the Air Navigation System During Flight Emergencies.. 66
    Yuliya Sikirda, Flight Academy, National Aviation University, Ukraine
    Mykola Kasatkin, Kharkiv National University of Air Forces named by I. Kozhedub, Ukraine
    Dmytro Tkachenko, Ukrainian State Air Traffic Services Enterprise (UkSATSE), Lviv,
        Ukraine

Chapter 4
Cooperative Decision Making Under Air Traffic Conflicts Detection and Resolution........................ 91
    Volodymyr Vasyliev, National Aviation University, Ukraine
    Denys Vasyliev, Ukrainian State Air Traffic Services Enterprise (UkSATSE), Ukraine

Chapter 5
Application of Deep Learning in the Processing of the Aerospace System's Multispectral
Images 134
    Heorhii Kuchuk, Kharkiv Polytechnic Institute, National Technical University, Ukraine
    Andrii Podorozhniak, Kharkiv Polytechnic Institute, National Technical University, Ukraine
    Daria Hlavcheva, Kharkiv Polytechnic Institute, National Technical University, Ukraine
    Vladyslav Yaloveha, Kharkiv Polytechnic Institute, National Technical University, Ukraine

Chapter 6
Intelligence-Based Operation of Aviation Radioelectronic Equipment 148
    Oleksandr Solomentsev, National Aviation University, Ukraine
    Maksym Zaliskyi, National Aviation University, Ukraine
    Oleksii Zuiev, National Aviation University, Ukraine

Chapter 7
Applications of Artificial Intelligence in Flight Management Systems 180
    Ivan Ostroumov, National Aviation University, Ukraine
    Nataliia Kuzmenko, National Aviation University, Ukraine

Chapter 8
Application SMART for Small Unmanned Aircraft System of Systems 193
    Azad Agalar Bayramov, Armed Forces War College of the Azerbaijan Republic, Azerbaijan
    Elshan Giyas Hashimov, Armed Forces War College of the Azerbaijan Republic, Azerbaijan

Chapter 9
The Method of Evaluation of the Resource of Complex Technical Objects, in Particular,
Aviation 214
    Lyudmyla Kuzmych, National Aviation University, Ukraine
    Volodymyr Kvasnikov, National Aviation University, Ukraine

Chapter 10
Computer-Aided Design of Intelligent Control System of Stabilizing Platforms With Airborne
Instrumentation 231
    Olha Sushchenko, National Aviation University, Ukraine

Chapter 11
Correction of the Temperature Component of Error of Piezoelectric Acceleration Sensor 263
    Anatoly Perederko, National Aviation University, Ukraine

Chapter 12
Research of the Reliability of the Electrical Supply System of Airports and Aerodromes Using
Neural Networks 279
    *Serhii Mykolaiovych Boiko, Kremenchuk Flight College, Kharkiv National University of
        Internal Affairs, Ukraine*
    *Yurii Shmelev, Kremenchuk Flight College, Kharkiv National University of Internal Affairs,
        Ukraine*
    *Viktoriia Chorna, Kremenchuk Flight College, Kharkiv National University of Internal
        Affairs, Ukraine*
    *Marina Nozhnova, Kremenchuk Flight College, Kharkiv National University of Internal
        Affairs, Ukraine*

Chapter 13
Artificial Intelligence Methods in Aviation Specialist Training for the Analysis and Transmission
of Operational Meteorological Information 306
Sergiy I. Rudas, National Aviation University, Ukraine
Evgeniya A. Znakovska, National Aviation University, Ukraine
Dmitriy I. Bondarev, National Aviation University, Ukraine

Chapter 14
Methods for the Synthesis of Optimal Control of Deterministic Compound Dynamical Systems
With Branch 323
Olena Tachinina, National Aviation University, Ukraine
Oleksandr Lysenko, Igor Sikorsky Kyiv Polytechnic Institute, National Technical University
of Ukraine, Ukraine

Chapter 15
Intellectual Measuring Complex for Control of Geometrical Parameters of Aviation Details:
Differential-Digital Method of Measurement of Aircraft Parts of Complex Geometric Form 352
Mariia Kataieva, National Aviation University, Ukraine
Alina Yurchuk, National Aviation University, Ukraine

Chapter 16
Methods for Assessing the Glissade Entrance Quality by the Crew 372
Yurii Hryshchenko, National Aviation University, Ukraine
Viktor Romanenko, National Aviation University, Ukraine
Daria Pipa, National Aviation University, Ukraine

Chapter 17
Basic Analytics of Anti-Failure Avionics 404
Hanna Polozhevets, National Aviation University, Ukraine
Sergiy Derets, National Aviation University, Ukraine
Bogdan Chebukin, Vilnius Gediminas Technical University, Lithuania

Chapter 18
Perspective Directions of Artificial Intelligence Systems in Aircraft Load Optimization
Process 419
Yelyzaveta Serhiyivna Sahun, Flight Academy, National Aviation University, Ukraine

Chapter 19
Methods and Techniques of Effective Management of Complexity in Aviation 438
Maksym Yastrub, National Aviation University, Belgium
Mario Boyero Pérez, EUROCONTROL, Belgium
Svetlana Kredentsar, National Aviation University, Ukraine

Chapter 20
Ensuring the Guaranteed Level of Flight Safety: The View of the Future 460
    Volodimir Kharchenko, National Aviation University, Ukraine
    Oleh Alexeiev, National Aviation University, Ukraine

Compilation of References 480

About the Contributors 508

Index 514

Detailed Table of Contents

Foreword xix

Preface xx

Chapter 1
Artificial Intelligence in Aviation Industries: Methodologies, Education, Applications, and
Opportunities 1

Tetiana Shmelova, National Aviation University, Ukraine
Arnold Sterenharz, EXOLAUNCH GmbH, Germany
Serge Dolgikh, Solana Networks, Canada

This chapter presents opportunities to use Artificial Intelligence (AI) in aviation and aerospace industries. The AI used an innovative technology for improving the effectiveness of building aviation systems in each stage of the lifecycle for enhancing the security of aviation systems and the characteristic ability to learn, improve, and predict difficult situations. The AI is presented in Air Navigation Sociotechnical system (ANSTS) because the activity of ANSTS, is accompanied by a high degree of risk of causing catastrophic outcomes. The operator's models of decision making in AI systems are presented such as Expert Systems, Decision Support Systems for pilots of manned and unmanned aircraft, air traffic controllers, engineers, etc. The quality of operator's decisions depends on the development and use of innovative technology of AI and related fields (Big Data, Data Mining, Multicriteria Decision Analysis, Collaboration Decision Making, Blockchain, Artificial Neural Network, etc.).

Chapter 2
Rational Adaptation of Control Systems for the Autonomous Aircraft Motion 36

Kostiyantin Dergachov, Kharkiv Aviation Institute, Ukraine
Anatolii Kulik, National Aerospace University, Ukraine

The possibilities of using an adaptation principle in an application for organizing the close-loop life circuit of autonomous fly vehicles (FV) are discussed in the chapter. The uncertainties arising at each stage of the life cycle of an autonomous FV are considered. To solve a problem the approach with using intelligent, rational objects and using knowledge database tool is proposed. The main theses of rational adaptation control system (CS) are represented. The preliminary designing tools for constructing rational adaptation algorithms for the motion CS are considered. The practical applications of the proposed approach at the stage of preliminary design of control systems for autonomous fly vehicles are presented.

Chapter 3
Intelligent Automated System for Supporting the Collaborative Decision Making by Operators of the Air Navigation System During Flight Emergencies..66
> Yuliya Sikirda, Flight Academy, National Aviation University, Ukraine
> Mykola Kasatkin, Kharkiv National University of Air Forces named by I. Kozhedub, Ukraine
> Dmytro Tkachenko, Ukrainian State Air Traffic Services Enterprise (UkSATSE), Lviv, Ukraine

This chapter researches pilot and air traffic controller collaborative decision making (CDM) during flight emergencies for maximum synchronization of operators' technological procedures. Deterministic models of CDM by the Air Navigation System's human operators were obtained by network planning methods; their adequacy is confirmed by full-scale modeling on a complex flight simulator. For the sequential optimization of the collaborative two-channel network "Air traffic controller-Pilot" to achieve the end-to-end effectiveness of joint solutions, a multi-criteria approach was used: ensuring the minimum time to parry flight emergency with maximum safety/maximum consistency over the time of operators' actions. With the help of the multiplicative function, the influence of organizational risk factors on flight safety in the air traffic control was evaluated. A conceptual model of System for control and forecasting the flight emergency development on the base of Intelligent Automated System for supporting the CDM by operators was developed.

Chapter 4
Cooperative Decision Making Under Air Traffic Conflicts Detection and Resolution..........................91
> Volodymyr Vasyliev, National Aviation University, Ukraine
> Denys Vasyliev, Ukrainian State Air Traffic Services Enterprise (UkSATSE), Ukraine

A number of probabilistic methods for aircraft conflict evaluation are presented. The analytical method is shown. The probability method that enables taking into account the features of stochastic dynamics of motion and the new compositional method of conflict probability evaluation are proposed. The feature of this method is that analytical solution is made using the predicted uncertainty areas of each aircraft position separately, making it possible to manage the position uncertainty and to apply for a wider range of scenarios. The multi-criteria decision-making for conflicts resolution is discussed. Optimality criteria and constraints for conflict resolution are defined. The generic multi-criteria model of conflict-free trajectories selection and the methods of resolution of two- and multi-aircraft conflicts have been developed. These methods provide the synthesis of conflict-free trajectories using different aircraft maneuvers according to criteria of flight regularity, economy, and maneuvering complexity using the multi-criteria dynamic programming.

Chapter 5
Application of Deep Learning in the Processing of the Aerospace System's Multispectral Images 134
> Heorhii Kuchuk, Kharkiv Polytechnic Institute, National Technical University, Ukraine
> Andrii Podorozhniak, Kharkiv Polytechnic Institute, National Technical University, Ukraine
> Daria Hlavcheva, Kharkiv Polytechnic Institute, National Technical University, Ukraine
> Vladyslav Yaloveha, Kharkiv Polytechnic Institute, National Technical University, Ukraine

This chapter uses deep learning neural networks for processing of aerospace system multispectral images. Convolutional and Capsule Neural Network were used for processing multispectral images from satellite Landsat 8, previously processed using spectral indices NDVI, NDWI, PSRI. The authors' approach was applied to wildfire Camp Fire (California, USA). The deep learning neural networks are used to solve the problem of detecting fire hazardous forest areas. Comparison of Convolutional and Capsule Neural Network results was done. The theory of neural networks of deep learning, the theory of recognition of multispectral images, methods of mathematical statistics were used.

Chapter 6

Intelligence-Based Operation of Aviation Radioelectronic Equipment.. 148
Oleksandr Solomentsev, National Aviation University, Ukraine
Maksym Zaliskyi, National Aviation University, Ukraine
Oleksii Zuiev, National Aviation University, Ukraine

This chapter presents the questions of aviation radioelectronic equipment operation. Operation is the main stage in the life cycle of equipment. This stage is the longest in time and very significant in terms of the costs for equipment reliability providing. Control influences are realized in operation system to provide the efficient functioning of the equipment. The operation system is complex in structure intelligence system and is the object of design and modernization. The chapter deals with the structure of the operation system and its elements interconnection and discusses the models of diagnostic variables, reliability parameters for the case of changepoint presence. An important issue during design and improvement of operation systems is the selection, evaluation, and support of the efficiency indicator. This chapter concentrates on the efficiency indicator substantiation taking into account the effectiveness and operational resources costs. Three data processing strategies analysis allowed choosing the most rational option from maximum efficiency point of view.

Chapter 7

Applications of Artificial Intelligence in Flight Management Systems.. 180
Ivan Ostroumov, National Aviation University, Ukraine
Nataliia Kuzmenko, National Aviation University, Ukraine

Flight management system (FMS) is one of the key elements of the modern airplane. It is a computer-based system that helps a pilot with different routine operations. FMS includes numerous algorithms of Artificial Intelligence to support navigation, guidance, and control of aircraft. FMS hosts algorithms of airplane positioning by data from navigational aids and data fusion from multiple sensors. The internal memory of FMS includes global air navigation databases such as runways, airports data, air navigation charts, navigational aids, SIDs, STARs, approaches, and routes for automatic support of airplane operation.

Chapter 8

Application SMART for Small Unmanned Aircraft System of Systems.. 193
Azad Agalar Bayramov, Armed Forces War College of the Azerbaijan Republic, Azerbaijan
Elshan Giyas Hashimov, Armed Forces War College of the Azerbaijan Republic, Azerbaijan

This chapter presents results of SMART for small Unmanned Aerial Vehicles System of Systems complex making, the development of complex control and Infrared communication algorithms, investigations of operation effectiveness of reconnaissance SMART UAVs System-of-System and of the distributed systems of wireless sensors. The development of UAV System-of-System complex and the military tactic tasks and reconnaissance application have been considered. The problems upon providing effective functioning of reconnaissance UAV during analysis and control of highly dynamic scenes are considered. It is shown that during the optimum regime of functioning and taking into account given limiting conditions on total radiation power of radio signals, the reconnaissance UAV should possess the current radiation power inversely proportional to the amount of signal-noise ratio. The technical realization of such optimum regimes is possible by way of development of adaptive control of transmitter power depending on the amount of useful signal detected during reconnaissance activity of UAV.

Chapter 9

The Method of Evaluation of the Resource of Complex Technical Objects, in Particular,
Aviation...214

Lyudmyla Kuzmych, National Aviation University, Ukraine
Volodymyr Kvasnikov, National Aviation University, Ukraine

The method of estimating the resource of complex technical objects, in particular, aviation technique, is proposed, which includes the construction of a mathematical model of the object's functioning to determine its actual technical condition and residual resource for planning, design, production, operation, repairs, and modernization of aviation technique. To determine the actual state and estimate the residual life of structures, it is proposed to simultaneously evaluate several characteristics of the material of the object: the characteristic parameters of the structure of the material, integral parameters of the material, related to strength (for example, hardness), the presence and nature of macro defects, the degree of corrosion wear of the metal. Limit values of the selected diagnostic parameters are determined by available standards or technical conditions for objects. The dynamics of changes in diagnostic parameters can be monitored and modeled on the basis of the data of periodic inspections of the control object.

Chapter 10

Computer-Aided Design of Intelligent Control System of Stabilizing Platforms With Airborne
Instrumentation 231

Olha Sushchenko, National Aviation University, Ukraine

The chapter includes principles of the computer-aided design of the intelligent control system for stabilizing platform with airborne instrumentation. The emphasis is given to the design of the robust controller. The design procedures including mathematical models development, simulation, analysis, and making a decision using such artificial intelligent approaches as soft computing and expert assessments of obtained results are considered. Two basic problems such as the design of new systems and modernization of operating systems are researched. The first problem can be solved by means of the robust parametric optimization, and the second by means of the robust structural synthesis. The basics of algorithmic, software, and method supports are represented. The generalized structural schemes of the interactive design process are given. This approach allows decreasing engineering time and improving the quality of design systems.

Chapter 11

Correction of the Temperature Component of Error of Piezoelectric Acceleration Sensor...............263

Anatoly Perederko, National Aviation University, Ukraine

In the aviation and aerospace industries, systems of artificial intelligence are being intensively implemented. For their effective functioning, it is necessary to provide reliable primary information. Information is collected by primary converters. In various systems of aviation technology, piezoelectric accelerometers are widely used as primary transducers. In particular, they are part of inertial navigation systems. These systems do not need external sources of information and are not affected by external interference. Therefore, they are widely used in aircraft for various purposes. These accelerometers are also used in other aircraft systems. In systems for monitoring the operation of gas turbine engines, systems for registering overloads. However, in conditions of high-temperature differences, shocks, vibrations, and significant accelerations, piezoelectric transducers have inherent flaws that affect the linearity and accuracy in measuring these values. Especially piezoelectric transducers are critical to extreme temperatures.

Chapter 12
Research of the Reliability of the Electrical Supply System of Airports and Aerodromes Using
Neural Networks 279

*Serhii Mykolaiovych Boiko, Kremenchuk Flight College, Kharkiv National University of
Internal Affairs, Ukraine*

*Yurii Shmelev, Kremenchuk Flight College, Kharkiv National University of Internal Affairs,
Ukraine*

*Viktoriia Chorna, Kremenchuk Flight College, Kharkiv National University of Internal
Affairs, Ukraine*

*Marina Nozhnova, Kremenchuk Flight College, Kharkiv National University of Internal
Affairs, Ukraine*

The system of supplying airports and airfields is subject to high requirements for the degree of reliability. This is due to the existence of a large number of factors that affect the work of airports and airfields. In this regard, the control systems for these complexes must, as soon as possible, adopt the most optimal criteria for the reliability and quality of the solution. This complicates the structure of the electricity supply complex quite a lot and necessitates the use of modern, reliable, and high-precision technologies for the management of these complexes. One of them is artificial intelligence, which allows you to make decisions in a non-standard situation, to give recommendations to the operator to perform actions based on analysis of diagnostic data.

Chapter 13
Artificial Intelligence Methods in Aviation Specialist Training for the Analysis and Transmission
of Operational Meteorological Information 306

Sergiy I. Rudas, National Aviation University, Ukraine
Evgeniya A. Znakovska, National Aviation University, Ukraine
Dmitriy I. Bondarev, National Aviation University, Ukraine

The authors present methods for the application of artificial intelligence for operational meteorological information (OPEC). The means of communication for distribution of meteorological data using information technologies are presented. Practical courses for aviation specialists (pilots, air traffic controllers, operators of unmanned aerial vehicles) are considered in which artificial intelligence methods are used: datamining, deep learning, machine learning, using information technologies.

Chapter 14
Methods for the Synthesis of Optimal Control of Deterministic Compound Dynamical Systems
With Branch 323

Olena Tachinina, National Aviation University, Ukraine
*Oleksandr Lysenko, Igor Sikorsky Kyiv Polytechnic Institute, National Technical University
of Ukraine, Ukraine*

This chapter states the result of the development of optimal control methods for deterministic discontinuous systems of optimal control problems for deterministic compound dynamical systems (CDS) with branching paths. The necessary conditions for optimality of the CDS branching paths are formulated in the form convenient for subsequent development of algorithms for the operational synthesis of these paths. The optimality conditions developed by the authors allow both preliminary and in real time (on-line) optimization of the CDS branching paths. The need for an operational synthesis of the CDS branching trajectory is caused by the inaccuracy of prior knowledge of information about the factors affecting CDS movement which are critical for the implementation of the CDS end-use. The developed conditions are universal for solving problems with any finite number of branches of a branching trajectory and are focused on the use of artificially intelligent systems which allow analyzing the structure of optimal control of CDS components as they move along the path branches.

Chapter 15

Intellectual Measuring Complex for Control of Geometrical Parameters of Aviation Details:
Differential-Digital Method of Measurement of Aircraft Parts of Complex Geometric Form 352

Mariia Kataieva, National Aviation University, Ukraine
Alina Yurchuk, National Aviation University, Ukraine

This chapter proposes a new automated method of measuring complex three-dimensional surfaces of aircraft parts in static and dynamic modes. The method allows conducting measurements in closed conditions and at the site of the aircraft disposition. The method consists in the continuous determination of the coordinates of the points of the surface of the detail and their representation in a three-dimensional graphic depiction. New methods of measuring the geometric parameters of parts with the complex spatial surface are suggested. This opens the prospect for the development of new ways of measuring geometric parameters of parts in real-time with high metrological characteristics and computer simulation of the measurement process. The differential-digital method is based on the suggested zero-coordinate principle of the measurement process which involves simultaneous parts availability check, and connects measurement result obtained which provided a reduction in the order of measurement error.

Chapter 16

Methods for Assessing the Glissade Entrance Quality by the Crew 372

Yurii Hryshchenko, National Aviation University, Ukraine
Viktor Romanenko, National Aviation University, Ukraine
Daria Pipa, National Aviation University, Ukraine

This chapter proposes methods for assessing the quality of entry into the glide path using autocorrelation functions and regarding the psycho-physiological state of the human operator. This is due to the possible increase in the psycho-physiological tension of a human operator in special flight situations. The analysis of the pilot's ability to control the trajectory of the aircraft is presented. A method for alerting the crew about the failure in the system for obtaining data on the angle of attack and airspeed to prevent non-coordinated turn is given. The majority of the calculations are based on the method of determining the correlation fields during the flight.

Chapter 17
Basic Analytics of Anti-Failure Avionics 404
    Hanna Polozhevets, National Aviation University, Ukraine
    Sergiy Derets, National Aviation University, Ukraine
    Bogdan Chebukin, Vilnius Gediminas Technical University, Lithuania

The chapter analyzes anti-crash avionics. Classifications are given for complex failures of avionics and onboard aviation equipment. The technological failure of the voice informant in an aircraft crash is investigated using process analysis. New technologies to reduce the risks in the elimination of functional failures of avionics are proposed. Considered are the first "computer" accidents and incidents, as well as the causes of errors. Authors present the chronology of the definition of the category of "failure" of radio-electronic equipment and onboard aviation equipment. The complexity of avionics is estimated using process analysis. The methodology of the analysis of technological processes of flights is proposed.

Chapter 18
Perspective Directions of Artificial Intelligence Systems in Aircraft Load Optimization
Process 419
    Yelyzaveta Serhiyivna Sahun, Flight Academy, National Aviation University, Ukraine

The chapter represents an overview of different approaches towards loading process and load planning. The algorithm and specificities of the current cargo loading process force the scientists to search for new methods of optimizing due to the time, weight, and size constraints of the cargo aircraft and consequently to cut the costs for aircraft load planning and handling procedures. These methods are based on different approaches: mix-integer linear programs, three-dimensional bin packing, knapsack loading algorithms, tabu-search approach, rule-based approach, and heuristics. The perspective direction of aircraft loading process improvement is a combination of multicriteria optimization method and heuristic approach using the expert system.

Chapter 19
Methods and Techniques of Effective Management of Complexity in Aviation 438
    Maksym Yastrub, National Aviation University, Belgium
    Mario Boyero Pérez, EUROCONTROL, Belgium
    Svetlana Kredentsar, National Aviation University, Ukraine

This chapter presents the use of enterprise architecture to manage the growing complexity in aviation. Any aviation organization or air traffic management system can be considered as a complex enterprise which involves different stakeholders and uses various systems to execute its business needs. The complexity of such an enterprise makes it quite challenging to introduce any change since it might have an impact on various stakeholders as well as on different systems inside of the enterprise. That is why there is a need for a technique to manage the enterprise and to anticipate, plan, and support the transformation of the enterprise to execute its strategy. This technique can be provided by enterprise architecture, a relatively new discipline that focuses on describing the enterprise current and future states as well as providing a holistic view of it. The authors describe the modern enterprise architecture frameworks and provide an example of an application of one of them (European ATM Architecture framework) to identify and manage changes in aviation.

Chapter 20
Ensuring the Guaranteed Level of Flight Safety: The View of the Future 460
  Volodimir Kharchenko, National Aviation University, Ukraine
  Oleh Alexeiev, National Aviation University, Ukraine

This chapter presents a methodology for the determination of the guaranteed level of flights safety. Purpose of the methodology consists in association in the only complex of tasks of assessment, providing verification of safety aviation activities as a complex hierarchical structure with independent critical elements and also hardware, program, network, and ergatic component which are both means, and subject to safety. The realization of ensuring the guaranteed result consists in realization of management processes so as not to allow transition of infrastructure or its systems to potentially dangerous state and to provide blocking (exception) of the corresponding technical object in case of threat of transition or upon transition to a dangerous state and minimization of consequences of such transition.

Compilation of References 480

About the Contributors 508

Index 514

Foreword

Artificial intelligence plays a bigger and bigger role in our lives and has been incorporated in many industries. The aviation industry has already successfully implemented the early stages of automation in different areas such as flight operations with flight management systems, in the work of autopilot, in new unmanned and manned aircraft, for procedures of air traffic management. In the book "Artificial Intelligence Applications in the Aviation and Aerospace Industries" the contributing authors of chapters describes questions of artificial intelligence applications on all stage of the lifecycle of the functioning of aviation and aerospace technics.

The target audiences of the book "Artificial Intelligence Applications in the Aviation and Aerospace Industries" are scientists, teachers, pupils, students and postgraduate students, air crash investigators, engineers and managers, aviation operators (pilots, air traffic controllers, operators of unmanned air vehicles, etc.).

This book is a collection of twenty chapters that cover problems about the advantages of using artificial intelligence in aviation and aerospace industries. Scientists from different countries presented their scientific achievements in the field of the use of artificial intelligence such as expert systems and decision support systems in the aeronautics and aerospace industries; using automated systems in aviation personnel training; machine learning; autonomous flying robots and many applications of unmanned air vehicles in civil and military tasks solution; automation on a modern airplane; flight management systems; big data; deep learning in the processing of aerospace system's multispectral images; intelligence based operation of aviation radioelectronic equipment; computer-aided design of intelligent control system of stabilizing platforms with airborne instrumentation; using neural networks for research the reliability of the electrical supply systems of airports and aerodromes, etc.

Much useful and interesting information is presented in all chapters by theoreticians and practitioners: scientists, engineers, pilots, air traffic controllers, etc.

Serhiy Nedilko
Flight Academy, National Aviation University, Ukraine

Serhiy Nedilko, *Doctor of Science, Professor, Rector of the Flight Academy of National Aviation University (Ukraine). The main direction of his scientific work is to ensure the efficiency of air traffic service through the integrated use of modern computer and satellite technologies; studying and analyze methods for ensuring the functional stability of the automated air traffic control systems; main technology of Artificial Intelligence applications in the Aviation and Air navigation systems.*

Preface

Air transport industry plays a major role in the world economic activity and to maintain a safe and efficient operation of aviation enterprises International Civil Aviation Organization (ICAO) in its recent documents extended the existing and defined new approaches to improve the practical and sustainable implementation of preventive aviation security measures based on modern advances in information technology (ICAO, 2018). The first Artificial Intelligence (AI) research was aimed at creating computers with "intelligent" behavior (Nilsson, 1971; Turing, 1950). Scientists sought to obtain artificial systems capable of performing work better than the natural intelligence of humans. Nowadays, the range of AI technologies has expanded considerably with successful applications in many areas, commercial and public. The "White paper" of IATA (International Air Transport Association) presents the advantages of application of the modern technologies of AI such as Machine learning (ML), Natural Language Processing (NLP), Expert Systems, Vision, Speech, Planning, Robotics (IATA, 2018). The effectiveness of aviation systems that manifests in the safe and efficient operation of flights depends primarily upon the reliability of aviation specialists and the quality of human decision making. One of the possible approaches to these challenges is formalization and modeling of operational processes in all stages of the aviation technology lifecycle including human operator state and ability to make decisions in the form of a complex system by applying the methods of system analysis. Addition of AI methods and technologies allows, through the ability to evaluate and process massive amounts of relevant data to significantly enhance the predictive power and versatility of such models that could contribute to practical improvements in all stages of modern aviation and aerospace industry.

In the book *Handbook of Research on Artificial Intelligence Applications in the Aviation and Aerospace Industries*, the contributing authors address key questions, problems and directions of research in applications of methods and technologies of Artificial Intelligence, such as Expert Systems, Collaborative Decision Making, Artificial Neural Network Models and other approaches to air traffic control, design, construction and control of aircraft, safety evaluation and management of manned and unmanned aircraft, avionics and many others. The opportunities for applications of AI in the aviation and aerospace industry is an innovative and rapidly expanding field with strong potential for improving the effectiveness of aviation systems at each stage of the operation.

In Chapter 1, "Artificial Intelligence in Aviation Industries: Methodologies, Education, Applications, and Opportunities", the authors present opportunities to use AI in aviation and aerospace industries. The AI used as an innovative technology for improving the effectiveness of building aviation systems in each stage of the lifecycle for enhancing the security of aviation systems and the characteristic ability to learn, improve, and predict difficult situations. The AI is presented in Air Navigation Sociotechnical system because its activity is accompanied by a high degree of risk of causing catastrophic outcomes.

Preface

The operator's models of decision making in AI systems are presented such as Expert Systems, Decision Support Systems for pilots of manned and unmanned aircraft, air traffic controllers, engineers, etc. The quality of operator's decisions depends on the development and using of innovative technology of AI and related fields (Big Data, Data Mining, Multicriteria Decision Analysis, Collaboration Decision Making, Blockchain, Artificial Neural Network, etc.).

In Chapter 2, "Rational Adaptation of Control Systems for the Autonomous Aircraft Motion", the possibilities of using an adaptation principle in an application for organizing the close-loop life circuit of autonomous fly vehicles (FV) are discussed. The uncertainties arising at each stage of the life cycle of an autonomous FV are considered. To solve a problem the approach with using intelligent, rational objects and using knowledge database tool is proposed. The main theses of rational adaptation control system (CS) are represented. The preliminary designing tools for constructing rational adaptation algorithms for the motion CS are considered. The practical applications of the proposed approach at the stage of the preliminary design of control systems for autonomous fly vehicles are presented.

In Chapter 3, "Intelligent Automated System for Supporting the Collaborative Decision Making by Operators of the Air Navigation System During Flight Emergencies", pilot and air traffic controller collaborative decision making (CDM) during flight emergencies for maximum synchronization of operators' technological procedures was researched. Deterministic models of CDM by the Air Navigation System's human-operators were obtained by network planning methods; their adequacy is confirmed by full-scale modeling on a complex flight simulator. For the sequential optimization of the collaborative two-channel network "Air traffic controller-Pilot" in order to achieve the end-to-end effectiveness of joint solutions, a multi-criteria approach was used: ensuring the minimum time to parry flight emergency with maximum safety/maximum consistency over the time of operators' actions. With the help of the multiplicative function, the influence of organizational risk factors on flight safety in the air traffic control was evaluated. A conceptual model of System for control and forecasting the flight emergency development on the base of Intelligent Automated System for supporting the CDM by operators was developed.

Chapter 4, "Cooperative Decision-Making Under Air Traffic Conflicts Detection and Resolution", explores a number of probabilistic methods for aircraft conflict evaluation. Firstly, the analytical method is shown. Then the probability method is proposed that enables to take into account the features of stochastic dynamics of motion. The new compositional method of conflict probability evaluation is proposed. The feature of this method is that analytical solution is made using the predicted uncertainty areas of each aircraft position separately, that makes it possible to manage the position uncertainty and to apply for a wider range of scenarios. The multi-criteria decision-making for conflicts resolution is discussed. Optimality criteria, constraints and conflict resolution rules are defined. The generic multi-criteria model of conflict-free trajectories selection and the methods of resolution of two- and multi-aircraft conflicts have been developed. These methods provide the synthesis of conflict-free trajectories using different aircraft maneuvers according to criteria of flight regularity, economy and maneuvering complexity using the multi-criteria dynamic programming.

In Chapter 5, "Application of Deep Learning in the Processing of Aerospace System's Multispectral Images", the authors present using deep learning neural networks for processing of aerospace system multispectral images. Convolutional and Capsule Neural Network were used for processing multispectral images from satellite Landsat 8, previously processed using spectral indices NDVI, NDWI, PSRI. Our approach was applied to wildfire Camp Fire (California, USA). In the chapter, the deep learning neural networks are used to solve the problem of detecting fire hazardous forest areas. Comparison of Convolutional and Capsule Neural Network results was done. The theory of neural networks of deep learning, the theory of recognition of multispectral images, methods of mathematical statistics were used.

In Chapter 6, "Intelligence-Based Operation of Aviation Radioelectronic Equipment", the authors present the questions of aviation radioelectronic equipment operation. Operation is the main stage in the life cycle of the equipment. This stage is the longest in time and very significant in terms of the costs for equipment reliability providing. Control influences are realized in the operating system to provide the efficient functioning of the equipment. The operation system is complex in structure intelligence system and is the object of design and modernization. The chapter deals with the structure of the operating system and its elements interconnection and discusses the models of diagnostic variables, reliability parameters for the case of changepoint presence. An important issue during the design and improvement of operating systems is the selection, evaluation, and support of the efficiency indicator. This chapter concentrates on the efficiency indicator substantiation taking into account the effectiveness and operational resources costs. Three data processing strategies analysis allowed choosing the most rational option from a maximum efficiency point of view.

Chapter 7, "Applications of Artificial Intelligence in Flight Management System", deals with using Flight management system (FMS) in aviation as AI system. The Flight management system is one of the key elements of the modern airplane. It is a computer-based system that helps a pilot with different routine operations. FMS includes numerous algorithms of Artificial Intelligence to support navigation, guidance, and control of aircraft. FMS hosts algorithms of airplane positioning by data from navigational aids and data fusion from multiple sensors. The internal memory of FMS includes global air navigation databases such as runways, airports data, air navigation charts, navigational aids, SIDs, STARs, approaches, routes for automatic support of airplane operation.

In Chapter 8, "Application SMART for Small Unmanned Aircraft System of Systems", the authors present the results of SMART (Self-Monitoring Analysis and Reporting Technology) systems for small Unmanned Aerial Vehicles System of Systems complex making, the development of complex control and Infrared communication algorithms, investigations of operational effectiveness of reconnaissance SMART UAVs System-of-System and of the distributed systems of wireless sensors. The development of UAV System-of-System complex and the military tactic tasks and reconnaissance application have been considered. The problems upon providing effective functioning of reconnaissance UAV during analysis and control of highly dynamic scenes are considered. It is shown that during the optimum regime of functioning and taking into account given limiting conditions on total radiation power of radio signals, the reconnaissance UAV should possess the current radiation power inversely proportional to the amount of signal-noise ratio. The technical realization of such optimum regimes is possible by way of development of adaptive control of transmitter power depending on the amount of useful signal detected during reconnaissance activity of UAV.

Chapter 9, "The Method of Evaluation of the Resource of Complex Technical Objects, in Particular, Aviation", presents the method of estimating the resource of complex technical objects, in particular, aviation technique, which includes the construction of a mathematical model of the object's functioning to determine its actual technical condition and residual resource for planning, design, production, operation, repairs, and modernization of aviation technique. To determine the actual state and estimate the residual life of structures, it is proposed to simultaneously evaluate several characteristics of the material of the object: the characteristic parameters of the structure of the material, integral parameters of the material, related to strength (for example, hardness), the presence and nature of macro defects, the degree of corrosion wear of the metal. Limit values of the selected diagnostic parameters are determined by available standards or technical conditions for objects. The dynamics of changes in diagnostic parameters can be monitored and modeled based on the data of periodic inspections of the control object.

Chapter 10, "Computer-Aided Design of Intelligent Control System of Stabilizing Platforms With Airborne Instrumentation", includes principles of the computer-aided design of the intelligent control system for stabilizing platform with airborne instrumentation. The emphasis is given to the design of the robust controller. The design procedures including mathematical models development, simulation, analysis, and making a decision using such artificial intelligent approaches as soft computing and expert assessments of obtained results are considered. Two basic problems such as the design of new systems and modernization of operating systems are researched. The first problem can be solved by means of the robust parametric optimization and the second – by means of the robust structural synthesis. The basics of algorithmic, software and method supports are represented. The generalized structural schemes of the interactive design process are given. This approach allows decreasing engineering time and improving the quality of design systems.

Chapter 11, "Correction of the Temperature Component of Error of Piezoelectric Acceleration Sensor", discusses the using of piezoelectric accelerometers in aviation systems. In the aviation and aerospace industries systems of artificial intelligence are being intensively implemented. For their effective functioning, it is necessary to provide reliable primary information. Information is collected by primary converters. In various systems of aviation technology, piezoelectric accelerometers are widely used as primary transducers. In particular, they are part of inertial navigation systems. These systems do not need external sources of information and are not affected by external interference. Therefore, they are widely used in aircraft for various purposes. These accelerometers are also used in other aircraft systems. In systems for monitoring the operation of gas turbine engines, systems for registering overloads. However, in conditions of high-temperature differences, shocks, vibrations, and significant accelerations, piezoelectric transducers have inherent flaws that affect the linearity and accuracy in measuring these values. Especially piezoelectric transducers are critical to extreme temperatures.

Chapter 12, "Research of the Reliability of the Electrical Supply System of Airports and Aerodromes With Using Neural Networks", is devoted to the use of Neural Networks in airports. The system of supplying airports and airfields is subject to high requirements for the degree of reliability. This is due to the existence of a large number of factors that affect the work of airports and airfields. In this regard, the control systems for these complexes must, as soon as possible, adopt the most optimal criteria for the reliability and quality of the solution. This complicates the structure of the electricity supply complex quite a lot and necessitates the use of modern reliable and high-precision technologies for the management of these complexes. One of them is artificial intelligence, which allows you to make decisions in a non-standard situation, to give recommendations to the operator to perform actions based on analysis of diagnostic data.

In Chapter 13, "Artificial Intelligence Methods Using in the Aviation Specialists Training for the Analysis and Transmission of Operational Meteorological Information", the authors are presenting methods for the application of artificial intelligence for working with operational meteorological information (OPEC). The means of communication for the distribution of meteorological data using information technologies are presented. Practical courses for aviation specialists (pilots, air traffic controllers, operators of unmanned aerial vehicles) are considered in which artificial intelligence methods are used: data mining, deep learning, machine learning, and using information technologies.

In Chapter 14, "Methods for the Synthesis of Optimal Control of Deterministic Compound Dynamical Systems With Branch", the authors state the result of the development of optimal control methods for deterministic discontinuous systems of optimal control problems for deterministic compound dynamical systems (CDS) with branching paths. The necessary conditions for optimality of the CDS branching paths are formulated in the form convenient for subsequent development of algorithms for the operational synthesis of these paths. The optimality conditions developed by the authors allow both preliminary and in real time (on-line), to optimize the CDS branching paths. The need for an operational synthesis of the CDS branching trajectory is caused by the inaccuracy of prior knowledge of information about the factors affecting CDS movement and which are critical for the implementation of the CDS end-use. The developed conditions are universal for solving problems with any finite number of branches of a branching trajectory and are focused on the use of artificially intelligent systems which allow analyzing the structure of optimal control of CDS components as they move along the path branches.

In Chapter 15, "Intellectual Measuring Complex for Control of Geometrical Parameters of Aviation Details: Differential-Digital Method of Measurement of Aircraft Parts of Complex Geometric Form", a new automated method of measuring complex three-dimensional surfaces of aircraft parts in static and dynamic modes is proposed. The method allows conducting measurements in closed conditions and at the site of the aircraft disposition. The method consists in the continuous determination of the coordinates of the points of the surface of the detail and their representation in a three-dimensional graphic depiction. Suggested new methods of measuring the geometric parameters of parts with the complex spatial surface. This opening the prospect for the development of new ways of measuring geometric parameters of parts in real-time with high metrological characteristics and computer simulation of the measurement process. The differential-digital method is based on the suggested zero-coordinate principle of the measurement process which involves simultaneous parts availability check and connects measurement result obtained which provided a reduction in the order of measurement error.

In Chapter 16, "Methods for Assessing of the Glissade Entrance Quality by the Crew", the methods for assessing the quality of entry into the glide path using autocorrelation functions and regarding the psycho-physiological state of the human operator are proposed. This is due to the possible increase in the psycho-physiological tension of a human operator in special flight situations. The analysis of the pilot's ability to control the trajectory of the aircraft is presented. A method for alerting the crew about the failure in the system for obtaining data on the angle of attack and airspeed to prevent non-coordinated turn is given. The majority of the calculations are based on the method of determining the correlation fields during the flight.

Chapter 17, "Basic Analytics of Anti-Failure Avionics", is devoted to the analysis of anti-crash avionics. Classifications are given for complex failures of avionics and onboard aviation equipment. The technological failure of the voice informant in an aircraft crash was investigated using process analysis. Proposed new technologies to reduce the risks in the elimination of functional failures of avionics. Considered the first "computer" accidents and incidents, as well as the causes of errors. Presented the chronology of the definition of the category of "failure" of radio-electronic equipment and onboard aviation equipment. The complexity of avionics is estimated using process analysis. The methodology of the analysis of technological processes of flights is proposed.

Chapter 18, "Perspective Directions of Artificial Intelligence Systems in Aircraft Load Optimization Process", represents an overview of different approaches towards the loading process and load planning. The algorithm and specificities of the current cargo loading process force the scientists to search for new methods of optimizing due to the time, weight and size constraints of the cargo aircraft and consequently to cut the costs for aircraft load planning and handling procedures. These methods are based on different approaches: mix-integer linear programs, three-dimensional bin packing, knapsack loading algorithms, tabu-search approach, rule-based approach, and heuristics. The perspective direction of aircraft loading process improvement is a combination of multicriteria optimization method and heuristic approach using the expert system. The purpose is to analyze these methods in order to find a new objective to solve the optimization problem in the aircraft loading process and to formalize it. Received results can be implemented into a new model of multiple criteria load optimization.

In Chapter 19, "Methods and Techniques of Effective Management of Complexity in Aviation", the authors present the use of enterprise architecture to manage the growing complexity in aviation. Any aviation organization or air traffic management system can be considered as a complex enterprise which involves different stakeholders and uses various systems to execute its business needs. The complexity of such an enterprise makes it quite challenging to introduce any change since it might have an impact on various stakeholders as well as on different systems inside of the enterprise. That is why there is a need for a technique to manage the enterprise and to anticipate, plan and support the transformation of the enterprise to execute its strategy. This technique can be provided by enterprise architecture, a relatively new discipline that focuses on describing the enterprise current and future states as well as providing a holistic view of it. The authors describe the modern enterprise architecture frameworks and provide an example of an application of one of them (European ATM Architecture framework) to identify and manage changes in aviation.

In Chapter 20, "Ensuring the Guaranteed Level of Flights Safety: The View of the Future", the authors present a methodology for the determination of the guaranteed level of flights safety. Purpose of the methodology consists in association in the only complex of tasks of assessment, providing and verification of safety aviation activities as a complex hierarchical structure with independent critical elements and also hardware, program, network and ergatic component which are both means, and subject to safety. The realization of ensuring the guaranteed result consists in realization of management processes so that not to allow transition of infrastructure or its systems to potentially dangerous state and to provide blocking (exception) of the corresponding technical object in case of threat of transition or upon transition to a dangerous state and minimization of consequences of such transition.

The book *Handbook of Research on Artificial Intelligence Applications in the Aviation and Aerospace Industries* is an essential reference source that presents concrete examples and best practices for AI implementation in aviation to enhance security and the ability to learn, improve, and predict. While highlighting topics such as computer-aided design, automated systems, and human factors, this publication explores the progress in global aviation security as well as the methods of modern information systems in the aeronautics industry. This book can be used by a wide audience of professionals in the industry such as: pilots, researchers, managers and project administrators, engineers, aviation operators, air crash investigators, teachers and students seeking current research on the application of AI in the field of aviation as well as anyone with interest in the modern state and promising directions of future developments in the aviation and aerospace industries.

REFERENCES

International Air Transport Association. (2018). *AI in aviation. Exploring the fundamentals, threats and opportunities of artificial intelligence (AI) in the aviation industry.* White Paper. Retrieved from https://www.iata.org/publications/Pages/AI-white-paper.aspx

International Civil Aviation Organization. (2018). Potential of artificial intelligence (AI) in Air Traffic Management (ATM). In *Thirteenth Air Navigation Conference ICAO*. Montréal, Canada: Author.

Nilsson, N. J. (1971). *Problem-solving methods in artificial intelligence.* Menlo Park, CA: Stanford Research Institute.

Turing, A. M. (1950). Computing machinery and intelligence. *Mind. New Series, 59*(236), 433–460. doi:10.1093/mind/LIX.236.433

Chapter 1
Artificial Intelligence in Aviation Industries:
Methodologies, Education, Applications, and Opportunities

Tetiana Shmelova
https://orcid.org/0000-0002-9737-6906
National Aviation University, Ukraine

Arnold Sterenharz
https://orcid.org/0000-0003-3942-1227
EXOLAUNCH GmbH, Germany

Serge Dolgikh
https://orcid.org/0000-0001-5929-8954
Solana Networks, Canada

ABSTRACT

This chapter presents opportunities to use Artificial Intelligence (AI) in aviation and aerospace industries. The AI used an innovative technology for improving the effectiveness of building aviation systems in each stage of the lifecycle for enhancing the security of aviation systems and the characteristic ability to learn, improve, and predict difficult situations. The AI is presented in Air Navigation Sociotechnical system (ANSTS) because the activity of ANSTS, is accompanied by a high degree of risk of causing catastrophic outcomes. The operator's models of decision making in AI systems are presented such as Expert Systems, Decision Support Systems for pilots of manned and unmanned aircraft, air traffic controllers, engineers, etc. The quality of operator's decisions depends on the development and use of innovative technology of AI and related fields (Big Data, Data Mining, Multicriteria Decision Analysis, Collaboration Decision Making, Blockchain, Artificial Neural Network, etc.).

DOI: 10.4018/978-1-7998-1415-3.ch001

BACKGROUND

The air transport industry plays a major role in world economic activity and to maintain the safe and efficient operation of aviation enterprises maximum use needs to be made of the enhanced capabilities provided by technical advances. Nowadays the International Civil Aviation Organization (ICAO) has added in its documents new approaches for achieving the main goals of the organization, which are: to enhance the effectiveness of global aviation security and to improve the practical and sustainable implementation of preventive aviation security measures. These new approaches include the development of security culture and human capability, the improvement of technological resources to improve oversight and quality assurance, and other directions. (ICAO, 2018). The quality of decisions dependences on the development and the use of innovative technologies in today's aviation such including Artificial Intelligence and education of aviation operators using modern information technologies and curriculums such as AI, Data Science, Big Data, Data Mining, Multicriteria Decision Analysis, Collaboration DM (CDM), Blockchain, Neural Network, etc.

The aviation systems should be considered as Sociotechnical systems (STS), which tend to have two common features: advanced technologies and high-risk activities (ICAO, 2004; 2013). In addition, large-scale, advanced technology systems such as Air Navigation System (ANS) can be attributed to the STS, in which the distinguishing feature is the presence of the hazardous kinds of activity as well as the use of the high-level technologies in production (Shmelova, Sikirda, Kharchenko, 2016).

Air Navigation System (ANS) is presented as an STS where Artificial Intelligence (AI) can offer significant benefits. The activity of ANS, as well as STS, is accompanied by a high degree risk of causing catastrophic outcomes. Statistical data show that human errors account for up 80% of all causes of aviation accidents (Leychenko, Malishevskiy, & Mikhalic, 2006; Campbell, Bagshaw, 2017). Since operations in STSs generally involve high-risk / high-hazard activities, the consequences of safety breakdowns are often catastrophic in terms of loss of life and property. The more a human-operator (H-O) is trying to control a production process being aided by high-level technologies, especially in case of distant operation, the more non-transparent becomes the result of the operation of a system (ICAO, 2004; 2013a; 2013a; 2017).

The document "White paper" of IATA (International Air Transport Association) presents the results of IATA research and development activities on AI in collaboration with airlines. The new technologies of AI can be clustered in the following capabilities, such as: Machine learning (ML), Natural Language Processing (NLP), Expert Systems, Vision, Speech, Planning, Robotics (IATA, 2016). So, to increase the productivity of work at each stage of the life cycle (LC) of aviation systems, different AI technologies and methods can be effectively applied (Fig. 1).

One of the possible approaches to the solution of these problems is the formalization and mathematical presentation of the ANS operators' activities in the form of a complex STSs on the base of the systemic analysis and using of AI in each of the LC of modern aerospace and aviation technique: planning, design, production, operation, maintenance / repairs, and modernization of technique. The LC may include all components of AI for improving the effectiveness, safety, and activity of the aviation system and obtains synergetic effect (Fig. 2).

Artificial Intelligence is a system that is able to perform inherent human intellectual activities associated with the perception and processing of knowledge, reasoning and relevant communication (playing chess, making music, poetry, design of complex systems). The development of modern AIS started in the 50's. The AI is a simulation of human intelligence processes by modeling, computer systems, and

Figure 1. The synergetic effect - LC of aviation technique with using AI capability

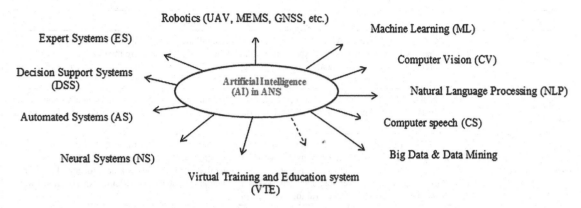

machines. These processes include: learning (the acquisition of information and rules for using the information); reasoning, estimation, and modeling (using rules to reach conclusions (approximate or definite results)); self-correction (estimation of obtained models); particular applications of AI include ES; decision Support system (DSS); automated systems; systems of pattern recognition, speech recognition and machine vision, etc. (ICAO, 2017). These components are described in Table 1.

Many cases of use of AI technology are driven by the emergence, availability, and accessibility. Authors present the synergetic effect using AI for development ANS as STS. Authors used sociotechnical theory for describing Socio-technical systems (Shmelova, Sikirda, Kharchenko, 2016; Shmelova, Sikirda, 2017).

Figure 2. The synergetic effect - LC of aviation technique with using AI capability

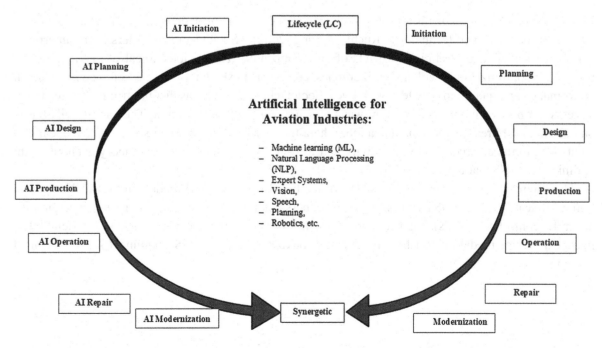

Table 1. Description of AI branches

AI Branches	Component Description
Machine Learning (ML)	ML is a branch within AI. It is a discipline concerned with the implementation of computer software that can learn autonomously. Expert systems and data mining programs are the most common applications for improving algorithms through the use of machine learning. Among the most common approaches are the use of artificial neural networks and genetic algorithms.
Natural Language Processing (NLP)	NLP is a field of computer science, AI concerned with the interactions between computers and human (natural) languages, and, in particular, concerned with programming computers to process large natural language data.
Expert Systems (ES)	In AI, an expert system is a computer system that emulates the decision-making ability of a human expert. Expert systems are designed to solve complex problems by reasoning through bodies of knowledge, represented mainly as if–then rules rather than through conventional procedural code. The first expert systems were created in the 1970s and then proliferated in the 1980s. Expert systems were among the first truly successful forms of AI software
Computer vision (CV)	Computer vision is an interdisciplinary field that deals with how computers can be made for gaining high-level understanding from digital images or videos. From the perspective of engineering, it seeks to automate tasks that the human visual system can do
Computer speech (CS)	Speech recognition is the inter-disciplinary sub-field of computational linguistics that develops methodologies and technologies that enables the recognition and translation of spoken language into text by computers. It is also known as "automatic speech recognition", "computer speech recognition", or just "speech to text". It incorporates knowledge and research in the linguistics, computer science, and electrical engineering fields.
Automated planning and scheduling	Automated planning and scheduling, sometimes denoted as simply AI Planning, is a branch of AI that concerns the realization of strategies or action sequences, typically for execution by intelligent agents, autonomous robots and unmanned vehicles. Unlike classical control and classification problems, the solutions are complex and must be discovered and optimized in multidimensional space
Robotics	Robotics is an interdisciplinary branch of engineering and science that includes mechanical engineering, electrical engineering, computer science, and others. Robotics deals with the design, construction, operation, and use of robots, as well as computer systems for their control, sensory feedback, and information processing

The sociotechnical theory is a theory about the social aspects of people and society and technical aspects of machines and technology (Kuchar &Yang, 2000). "Sociotechnical" refers to the interrelatedness of "social" and "technical" aspects of an organization. The sociotechnical theory is therefore about "joint optimization", with a shared emphasis on the achievement of both excellence in technical performance and quality in people's work lives. Sociotechnical theory, as distinct from socio-technical systems, proposes a number of different ways of achieving joint optimization. They are usually based on designing different kinds of organization, where the relationships between socio and technical elements lead to the emergence of productivity and well-being. The sociotechnical theory is about "joint optimization" (Kuchar &Yang, 2000).

The purposes of the work are: Analysis of the benefits of using AI models in the Air Navigation Sotiotechnical System (ANSTS); training of ANS personnel (pilots of manned and unmanned aircraft, air traffic controllers (ATCs), engineers) in the methods of Deterministic Modeling and Artificial Intelligence including models of collaborative decision making (CDM) by ANS personnel (pilots of manned and unmanned aircraft, ATCs, engineers).

ARTIFICIAL INTELLIGENCE IN AIR NAVIGATION SOCIOTECHNICAL SYSTEMs

Increasing Complexity of Operation and the Role of Human Factor

Most investigations of STS were conducted with the purpose to ensure safety in nuclear power generation, chemical production; hydraulic engineering; chemical and military industries; gas and oil pipelines (Clegg, 2000; Keating, 2001; Flueler, 2006; Carayon, 2006; Bertsch, 2007; Baxter, 2011). Taking into account in the decision-making (DM) by a H-O besides the separate *professional* factors (knowledge, habits, skills, experience) also the factors of *non-professional* nature (individual psychological, psychophysiological and socio-psychological) (Shmelova, Sikirda, Kharchenko, 2016) enables to predict the H-O's actions on the basis of the reflexive theory (Lefebvre, 2008), AI systems (IATA, 2018), new information technologies and current professional education. The effectiveness of aviation systems and assurance of flight safety depend primarily on the reliability of an H-O as well as quality and timeliness of their professional decisions (Shmelova, Sikirda, Kasatkin, 2019).

So, the influence on the DM by H-O of the ANS of the *professional* factors $\overline{F}_p$ (knowledge, skills, abilities, and experience) as well as the factors of *non-professional* nature $\overline{F}_{np}$ (individual-psychological, psycho-physiological and socio-psychological) has been defined using integrated models of DM, reflexive models and AI models (Shmelova, Sikirda, Kharchenko, 2016; Shmelova, Sikirda, Kasatkin, 2019).

The analysis of non-professional factors include the estimation of the influence of social factors on the H-O by defining preferences

$$\overline{F}_{sp} = \left\{ f_{spm}, f_{spe}, f_{sps}, f_{spp}, f_{spl} \right\};$$

diagnostics of individual psychological qualities of the H-O ANS in the situation

$$\overline{F}_{ip} = \left\{ f_{ipt}, f_{ipa}, f_{ipp}, f_{ipth}, f_{ipi}, f_{ipn}, f_{ipw}, f_{iph}, f_{\exp} \right\};$$

monitoring the emotional state of the H-O ANS and the stability of the ANS

$$\overline{F}_{sp} = \left\{ f_{spm}, f_{spe}, f_{sps}, f_{spp}, f_{spl} \right\},$$

and forecasting professional situations development. The H-O ANS preferences systems are influenced by *professional* $\overline{F}_p$ and *non-professional* $\overline{F}_{np}$ factors:

$$\overline{F}_p = \left\{ \overline{F}_{ed}, \overline{F}_{\exp} \right\}; \tag{1}$$

$$\overline{F}_{np} = \left\{ \overline{F}_{ip}, \overline{F}_{pf}, \overline{F}_{sp} \right\}, \tag{2}$$

where $\overline{F}_{ed}$ – are knowledge, skills and abilities, acquired H-O during training; $\overline{F}_{exp}$ – are knowledge, skills and abilities, acquired H-O during professional activity;

$$\overline{F}_{ip} = \left\{ f_{ipt}, f_{ipa}, f_{ipp}, f_{ipth}, f_{ipi}, f_{ipn}, f_{ipw}, f_{iph}, f_{exp} \right\}$$

is set of H-O *individual-psychological* factors (temperament, attention, perception, thinking, imagination, nature, intention, health, experience); $\overline{F}_{pf}$ – is set of H-O psycho-physiological factors (features of the nervous system, emotional types, sociotypes); $\overline{F}_{sp} = \left\{ f_{spm}, f_{spe}, f_{sps}, f_{spp}, f_{spl} \right\}$ – is set of H-O *socio-psychological* factors (moral, economic, social, political, legal factors).

The result of the evaluation of *professional* ($\overline{F}_p$) and *non-professional* (F_{ip}) factors determines the social-psychological impact on DM of H-O by identifying the preferences, diagnosing individual-psychological qualities of H-O ANS in the development of situation and monitoring of the psycho-physiological factors (Figure 3).

The ANS's H-O behavior model can be described by a choice vector $\overline{V}$ (choice towards the positive /negative pole). ANS's H-O behavior model as am example of diagnostics of influencing socio-psychological factors is graphically represented in Figure 4 and has following appearance:

$$\overline{M}_{DM} = \{\overline{F}_{sp\,min}, \overline{F}_{sp}, V_{min}, V, V_{max}, \overline{F}_{sp\,max} \};\qquad(3)$$

$$V_{max} \le V(\overline{F}_{sp}) \le V_{min},$$

Figure 3. Complex influence of professional and non-professional factors activities on DM by ANS's H-O

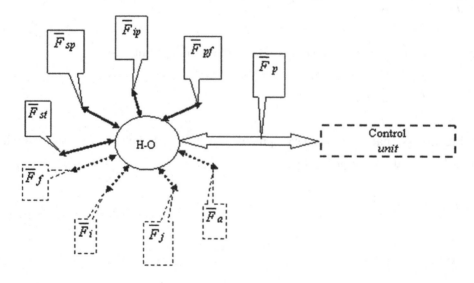

where $\overline{F}_{sp\,min}$, $\overline{F}_{sp\,max}$ – are the minimum (maximum) values of the factor's weighting coefficient; $\overline{F}_{sp}$ – is an actual value of the factor's weighting coefficient; V – is a quantitative measure of choosing; $V(\overline{F}_{sp})$ – is a choosing model; V_{min}, V_{max} – are the minimum (maximum) values of positive choosing.

The influence of the *professional factors* as well as the factors of *non-professional nature* on the DM of H-O is resented on Figure 5, which shows the model of DM of H-O of ANSTS. The scenarios of developing a flight situation in case of selecting either the *positive (A)* or *negative (B)* pole under the

Figure 4. ANS's H-O behaviour model at example of diagnostics of socio-psychological factors influencing

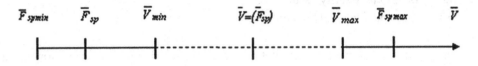

Figure 5. The model of decision making by H-O of ANSTS

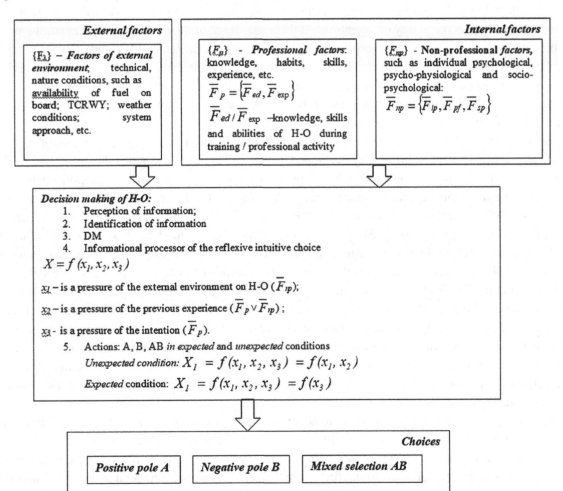

pressure of the external environment (x_1), the preceding experience of an H-O (x_2) and the intentional selection (intention) (x_3) in accordance with the reflexive theory have been obtained (Lefebvre, 2008).

In the recent documents ICAO defined new approaches for achieving principal goals of enhancing the effectiveness of global aviation security, and improving the practical and sustainable implementation of preventive aviation security measure. The Global Aviation Security Plan (GASP) identifies five key outcomes for improving effectiveness, such as (ICAO, 2017):

- Enhancing awareness and response of risk.
- Development of security culture and human capability.
- Improving technological resources and foster innovation.
- Improving oversight and quality assurance.
- Better cooperation and support between states.

The aviation system is a complex system that requires considerable research into how human performance may be affected by its multiple and interrelated components such as technical, political, physical, social, economic, culture, etc. (ICAO, 2007; 2013a; 2013b; Shmelova, Sikirda, Kharchenko, 2016). The Human Factor remains a major cause of aviation accidents (ICAO, 2004). People are the most flexible, adaptable and important element in the aviation system and also most vulnerable in terms of circumstantial impact on quality of operation. At the initial stage of the development of aviation, many problems have been associated with exposure to noise, vibration, heat, and cold and acceleration forces. Similarly, optimization of the human role in complex systems is related to all aspects of human activity, such as DM processes and knowledge; design configuration displays, controls and equipment cockpit, and cabin; maintaining communications and software; preparation of plans and maps; application of AI for aviation systems improving development.

Human Factor (HF) and its effects on safety performance continue evolving. The theory of Human Factor in aviation is being continuously developed, tested and applied in practice. From the end of the previous century and to the present day, ICAO pursues a systematic approach in aviation safety with reference to main establishment of conceptual of safety models. A number of directives, documents and reports of ICAO presented conceptual models of HF and evolution of HF's models (ICAO, 1998, 2002, 2004, 2013a, 2013b, 2014): SHEL, SHELL, SHELL-T, SHCHELL's models; Reason's model of latent conditions; the Threat and Error Management (TEM) model and other models of HF. In the center of the SHEL model is a human individual, the most critical and most flexible component in the system to which other components of the system must be carefully matched if stress and eventual breakdown in the system are to be avoided. In order to achieve this matching, an understanding of the characteristics of this central component is essential. People are subject to considerable variations in performance and suffer many limitations, most of which are now predictable at least in general terms (Campbell, Bagshaw, 2002). ICAO addressed awareness of cultural interfaces and the impact of cross-cultural factors on aviation safety in circular HF Digest (ICAO, 2018) and application for development aviation systems innovation technologies such as AI.

As technology and the market continue to advance, AI technologies will need to play an important role in aviation. With a limited capacity of airspace, airports and supporting operations, the aviation industry is faced with the challenge of accommodating surging air traffic demands. The emergence of new types and areas of flight operations such as unmanned aircraft, commercial space travel and the personal air vehicles of the future are making safe and efficient management of flight operations ever more challenging. The modern aviation systems dynamically adapt and evolve on a regular basis to meet the demands of the society in real time. However, as the system rapidly increases its complexity, it may be facing certain limits that may influence operational safety. Traditional automation cannot handle such "information explosions". Consequently, human cognition needs to be augmented with the support of Artificial Intelligence for fast, confident and error-free processing of large volumes of data. Based on this analysis and conclusion one can identify the five stages of the evolution of the Human Factor models (*Table 2*):

Stage I: "Professional skills of H-O / Interaction of H-O's / Definition of H-O's Errors".
Stage II: "Cooperation in team / Interaction of H-O's in team / Error detection".
Stage III: "Influence of Culture / Safety / Error prevention".
Stage IV: "Safety Management / Safety balance models / Minimization of errors".
Stage V: "Artificial Intelligence Systems / Analyze, detection, prevention, and minimization of errors".

The last stage of evolution, application of AI systems and technologies includes areas as Collaborative Decision Making (CDM), System-Wide Information and Management (SWIM), Flight & Flow Information for a Collaborative Environment (FF-ICE) and other.

Table 2. The evolution of the HF's models

Years	Models	Content of models	Stage
1972	SHEL	Software (procedures) - Hardware (machines) - Environment - Liveware	I stage Professional skills Interaction Errors
1990	Reason's "Swiss Cheese Model"	Active errors - Latent errors - Windows of opportunity - Causation chain	
1993	SHELL	Software (procedures) - Hardware (machines) - Environment - Liveware - Liveware (humans)	
1999	CRM	Crew - Resource - Management	II stage Cooperation in team Error detection
2000	TEM	Threat and Error - Management	
2000	MRM	Maintenance - Resource - Management	
2004	SHELL-T (SHELL-Team)	Software (procedures) - Hardware (machines) - Environment – Liveware - Liveware (humans) - Team	
2004	SCHELL model and CRM	Software (procedures) – Culture - Hardware (machines) - Environment - Liveware - Liveware (humans)	III stage Culture Safety Error prevention
2004	LOSA	Line - Operation - Safety - Audit	
2009	HEAD	Human - Environment - Analysis - Design	
	PBA	Performance-Based Approach	
2010	HFACS	Human Factors - Accident - Classification - System	
2013	SMS Safety Balance Model	Safety Management System	IV stage Safety / Efficiency / Minimization of errors
2016-now	AI CDM FF-ICE	Collaborative Decision Making (CDM) System-Wide Information Sharing and Management (SWIM) Flight & Flow Information for a Collaborative Environment (FF-ICE)	V stage Collaborative DM Artificial Intelligence

To effectively model ANS as a sociotechnical systems one needs to take into account a number of other factors such as the influence of social, cultural environment of H-Os on DM. Culture surrounds people and affects their values, convictions, and behavior, which they share along with other members of different social groups and can introduce its own complex channels of influence on the outcome of the human decision. Culture serves to bind together members of groups and to provide clues as to how to behave in both normal and unusual situations. The psychologist Hofstede suggests that culture is a "collective programming of the mind" (ICAO, 2011). Thus fatal mistakes can be committed by normal, healthy, highly motivated and well-equipped personnel (ICAO, 2011; Leychenko, Malishevskiy, & Mikhalic, 2006). Scientists lately have used the term "deviation of conscience" when they analyzed the causes of aviation events conditioned by the insufficient development of the appropriate cultural values in a person that makes decisions (Leychenko, Malishevskiy, & Mikhalic, 2006). The increasing complexity and expanding range of factors influencing human behavior in critical environment give rise to introduction of advanced decision support and decision making technologies discussed in this chapter.

Artificial Intelligence in Aviation: Opportunities in Education

Effective use of modern and advanced technologies such as AI is possible only by personnel educated, experienced confident in the practice. To this end a number of programs and courses have been developed in Applied AI as part of university course "Informatics of Decision Making" for aviation students and future personal of ANS, as in Table 3.

The differentiating factor of an AI system from a standard software system is the characteristic ability to learn, improve, and predict. Through training, an AI system is able to generate knowledge and apply it to novel situations not encountered before.

Comparison of human and artificial experience is given in Table 4.

Basic properties of AI system:

- **The Ability to Learn**: After receiving input signals the ability to tune by itself, then providing response (output) with the required accuracy.
- **Adaptation**: The system quickly adjusts the parameters under conditions of changing environment.
- **Flexibility**: The system makes generalizations based on incomplete, unclear and inaccurate data.
- **Transparency Interpretation (Explanation)**: The system provides data mined from the knowledge of the kid - friendly way.
- **The Ability to Open New**: The system detects a previously unknown hidden connections and relations in large arrays of numeric and text data.
- The ability of DM in a complex situation and find the optimal solutions for the operators or for robotic mechanisms, etc.

Some of the basic models used in AI:

- DM's deterministic and stochastic models.
- Models of preferences obtained using the method of expert assessments.
- **Neural Networks**: Providing an easy way to model complex nonlinear functions for solving diagnostic and prognostic problems.

Table 3. Applied AI in education course "Informatics of DM"

Content	Pilot	ATC	Engineer	UAV Operator
Aviation man-machine system (MMS)	Analysis and synthesis of aviation using theory of automatic control			
	MMS "pilot – aircraft" Analysis and synthesis of aviation using theory of automatic control	MMS "ATC-aircraft"	MMS	MMS "operator UAV"
Aviation expert system of quantitative estimation	Expert Judgment Method / Multi-criteria decision problems			
	Significance (complexity) of the phases of flight of the aircraft	Controller's work load for aircraft service	Significance of the Landing System (GNSS, ILS, VOR, VOR/DME)	Significance of the UAVs, phases of flight of the UAVs
Decision Support system of H-O in ANS	Models of DSS: deterministic and stochastic models of DM			
	DM of pilot in emergency	DM of ATC in emergency	DM of engineer in service/ emergency	DM of unmanned pilot
Decision making in certainty	Network planning of action of H-O in service/ emergency			
	Graph of procedures of pilot in emergency	Graph of procedures of ATC in emergency	Graph of procedures in service of equipment	Graph of procedures of unmanned pilot in emergency
Design making in risk	Design tree of forecasting of action of H-O in emergency			
	DM in emergency	DM in emergency	DM in service of equipment	DM in emergency
Decision making in uncertainty	Criteria Wald. Laplace, Hurvits, Savidge optimal DM in uncertainty			
	Optimal landing aerodrome in emergency	Optimal landing aerodrome in emergency	Optimal action in emergency	Optimal landing aerodrome/place in emergency
Neural Networks	Forecasting of outcomes in emergencies			
Markov Networks	Neural network admission student to simulator training			
GERT-models	Development and forecasting the emergency situation /Preventing catastrophic situation			
Fuzzy logic	Quantitative estimation of the outcomes /risk in emergencies			
ANSTS	Analysis of ANS as STS and diagnostics, monitoring of the factors (professional and non-professional) that influence on DM by the H-O in STS (individual-psychological, socio-psychological and psychophysiological factors)			

- **Genetic Algorithms:** Are able to approximate complex relationships in incomplete data sets, to find an optimal solution to complex problems.
- **Fuzzy Sets**: Can quantitatively encode qualitative information.
- **Dynamic Structural Models**: Can simulate the basic functions of management.

However, in the forming and modeling of DM, an H-O and an AI have a common property namely the ability to apply different levels of DM complexity depending on the factors that influence the DM. For DM in a difficult situation (S) it is necessary to identify:

- The class of situation (Q).
- Level of Complexity (U).
- Choose the optimal actions (A^*).

Table 4. Comparison of human and artificial experience

		Human Competence	**Artificial Competence**
Advantages		Variable	Constant
		Difficult transmitted	Easily transmitted
		Hard notable	Easy notable
		Unpredictable	Constant
		Expensive	Acceptable cost
Disadvantages		Creative	Programmed
		Adjust	Need hint
		Uses sensory perception	Uses symbolic input
		Broad in scope	Narrow
		Uses public knowledge (common sense)	Uses specialized knowledge

For example, $Q = \{q_j\}$ - the set of consequences of choosing the completion alternative; $U = \{u_j\}$ - vector of the characteristics of the consequences, the results of the choice of the alternative of the completion; $A = \{a_i\}$ is the set of alternative solutions) and choose the optimal actions (A^*).

In Figure 6 a diagram of the process of simplifying a difficult situation presented. This process is necessary to apply for complex systems and AI systems. The AI have high potential in Air Traffic Management (ATM), specifically in areas which involve decision making under uncertainty (e.g. conflict detection and resolution) and prediction with limited information (e.g. trajectory prediction). It is important to create Expert Systems when analyzing the complexity, significance, and responsibility of subsystems before synthase and synchrony of collaborative mathematical models.

The Expert system (ES) being a branch of AI, is a computer system that simulates the DM ability of a human-operators. The ICAO documents recommend developing Intelligent ESs in aviation to support of operators (ICAO, 2011, 2018). Knowledge - characteristics of systems obtained as a result of practice and professional experience of experts. To build an ES, the following Algorithm of the building of Expert Systems is used:

Figure 6. A scheme of process of simplifying a difficult situation

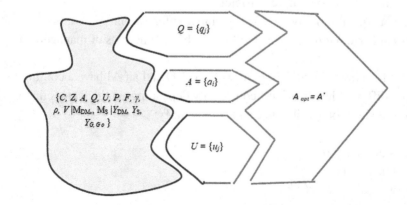

The Algorithm of the Building of Expert Systems

1. Building main components of ES: Users interface; Database; Base Knowledge.
2. System analysis of complex system. Decomposition of complex systems on subsystems:
    a. Definition subsystems for expert estimation of their significance and description of the characteristics of subsystems.
    b. Definition of criteria estimation and description of criteria features.
    c. Estimation of subsystems using EJM by criterion and obtaining weight coefficients of subsystem significance by criterion.
3. Aggregation subsystems in systems.
    a. Additive aggregation of subsystems:

$$W_j = \sum_{i=1}^{n} \omega_i F_{ij}, i = \overline{1,n}, j = \overline{1,m} \tag{4}$$

    b. Multiplicative aggregation of subsystems:

$$W_j^{'} = \prod_{i=1}^{n} F_{ij}^{\omega_i}, i = \overline{1,n}, j = \overline{1,m}, \tag{5}$$

4. Graphical presentation of significance of subsystems in Expert System.

To build an Expert system, it is necessary to determine the significance of the subsystems (parameters, characteristics, values, etc.) in the system, which is investigated with the help of expert knowledge. The main method for building the Knowledge Base in the Expert System is the *Expert Judgment Method (EJM)*. To build an ES, the following Algorithm of *Expert Judgment Method* is used:

Algorithm of Expert Judgment Method (EJM)

1) Questioners for experts, m – is a number of experts, $m \geq 30$

Matrix of individual preferences - determine opinion of the experts and their systems of individual preferences, R_i – is a system of preferences of i-expert, $i = \overline{1,m}$.

2) Matrix of group preferences R_{ij}

3) Determine the experts' group opinion R_{grj} (sample average, arithmetical mean): $R_{grj} = \dfrac{\sum_{i=1}^{m} R_i}{m}$

4) Determine the coordination of experts' opinion:

    a. Dispersion for each factors (procedures, phases of flight of the aircraft,…): $Dj = \dfrac{\sum_{i=1}^{m} \left(R_{grj} - R_i \right)^2}{m-1}$

b. Determine square average deviation (Squared deviations): $\sigma_j = \sqrt{D_j}$

c. Determine coefficient of the variation for each factors (procedure, phases of flight of the aircraft, etc...):

$$v_j = \frac{\sigma_j}{R_{grj}} \bullet 100\%$$

If coefficient of a variation is $v_j < 33\%$ - it means that the opinion of the experts coordinated. The opposite case $v_j > 33\%$ means that the opinion of the experts not coordinated. Therefore calculate the Kendal's coefficient of concordance must be calculated.

5) For evaluation of coordination on all procedures it is necessary to use Kendal's coefficient of concordance or to provide interrogation of the experts again: $W = \dfrac{12S}{m^2(n^3 - n) - m\sum\limits_{j=1}^{m} T_j}$,

If coefficient of concordance is $W > 0,7$ it means that the opinion of the experts coordinated. If coefficient of concordance is $W < 0,7$, the opinion of the experts is not coordinated. So, interrogation of experts is again required.

Systems of preferences R_{gr} and R_i, can be compared by means of the rating correlation coefficient R_s (Spearman's coefficient): $R_{si} = 1 - \dfrac{6\sum\limits_{j=1}^{n}(x_{ij} - y_{ij})^2}{n(n^2 - 1)}$.

6) Significance of the calculations.

a. The significance of the calculations W, criterion-χ^2: $\varsigma_f^2 = \dfrac{S}{\dfrac{1}{2}m(n+1) - \dfrac{1}{12(n-1)}\sum\limits_{j=1}^{m} T_j} > \varsigma_t^2$,

where ς_f^2 - factual value of variable; ς_t^2 - table value of variable.

b. The significance of the calculations R_s (Spirman's coefficient) for using Student's t - criterion:

$$t_{critical} = r_s \sqrt{\frac{n-2}{1-r_s^2}} > t_{st}.$$

7) Weight coefficient w_j: $w_j = \dfrac{C_j}{\sum\limits_{j=1}^{n} C_j}$; $C_j = 1 - \dfrac{R-1}{n}$

8) Graphical presentation of weight coefficients.

For students, training material have been developed that allow you to quickly master complex methods of building AI systems (ES, models of DSS: deterministic and stochastic models of DM, etc.), as demonstrated by examples below.

Training N1 "Expert Judgment Method (EJM) / Multi-criteria decision problems" was building, it includes:

1) Lessons:
    a. Theory. Basic of EJM for ANS (Classification of methods of DM. The algorithm of EJM. The matrix of individual and group preferences. Coordination of experts' opinion. Multi-criteria decision problems.
    b. Practice. Tasks: definition of significance (complexity) of the phases of flight of the aircraft
2) *Individual work*. DM in the Team. Lottery-choice of tasks for the team.
3) *Presentation* of results (Figure 7).

The training N2 "Deterministic models. Network planning. Decision making in an emergency situation (ES)" includes lessons and individual work students. Results of using network planning method for building deterministic models of H-O (air traffic controller) presented on Figure 8.

The training N3 "Stochastic models. Decision making in Risk. Decision making in ES" presented on Figure 9 – example of building the stochastic models for landing in bad weather condition using decision-tree.

The training N4 "Stochastic models. Game Theory. DM in uncertainty. Optimal aerodrome of landing" presented on Figure 10 – example of building the models for landing in bad weather condition using matrix of DM in uncertainty.

Figure 7. The quantitative estimation of the complexity of the phases of flight of the aircraft and obtain the weight coefficients (result - take-off and landing more complex)

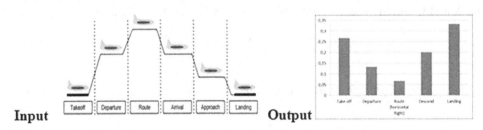

Figure 8. The process of applying solutions for DM in certainty

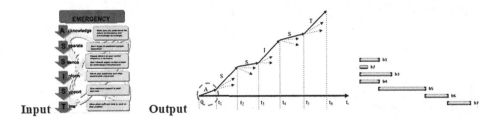

Figure 9. The process of applying solutions for DM in Risk

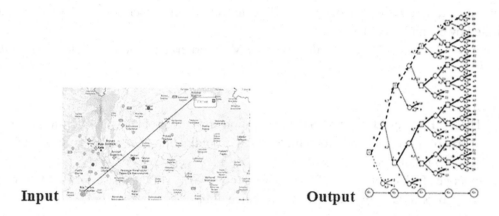

Figure 10. The process of applying solutions for DM in uncertainty

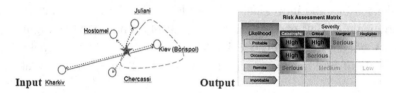

In these examples the decision on selection of the alternative landing/place in the case of an emergency landing (difficult meteorological conditions, etc.) by the method of DM in uncertainty is applied (Wald, Laplace, Savage, Hurwicz).

The training N5 "Dynamic programming (DP). The problem of minimal cost of travel between multiple points, such as application of the method of DP to solve the problem of climbing an aircraft. Grid analysis" presented on Figure 11 provides the examples of solutions for the task of minimal cost between the A and B and the task of climb of the aircraft / take-off of the aircraft.

Figure 11. The process of applying solutions for DP

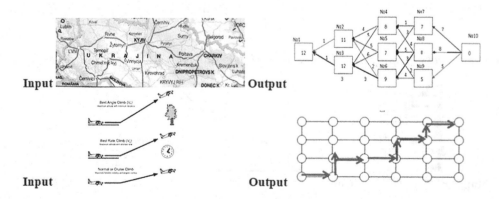

It is important to create highly intelligent joint DM systems for engineers, pilots, air traffic controllers and new airspace users, such as Unmanned Aircraft Systems (UAS). In the existing research DM models for UAV's operators in emergencies (Shmelova, Sterenharz, 2018); the deterministic and stochastic models of DM in ANS; stochastic models type Markov Chains; Stochastic models type GERT's network; Neural Network models; Fuzzy logic models; Reflexive models of bipolar choice; models of diagnostics of emotional state deformation in the professional activity of operators in the ANS; Graphical-Analytical Models of situation Development; Graphical-Analytical Models of DM by H-O etc. (Shmelova, Sikirda, 2017) were investigated and presented.

In the recent documents ICAO defined new approaches - the organization of Collaborative Decision Making (CDM) by all aviation operators using collaborative DM models (CDMM) based on general information on the flight process and ground handling of the UAVs. Models of flight emergencies (FE) development and of DM in Risk and uncertainty by UAV's in FE will allow predicting the operator's actions with the aid of the Informational-analytic and Diagnostics complex for research UAV operator's behavior in extreme situation (ICAO, 2017; ICAO, 2018).

Further research is being directed to finding the optimal solution of practical problems in emergencies with UAV with directions such as:

- Decision Support Systems (DSS) of UAV's operator - forming the optimal actions of UAV's operator in-flight especially in an emergency.
- Board AI systems as Automatic systems or Machine Learning systems - a software creation for UAV distance control in-flight.

The process of analysis and synthesis of DM models of AI in emergency tend to simplify complex models such as stochastic, the neural network, fuzzy, the Markov network, GERT-models, reflexion models to deterministic models. In order to simulate DM under conditions of an emergency, we use the following steps:

- an analysis of an emergency.
- intelligent data processing.
- analysis and identification of the situation using CDMMs.
- decomposition of the situation as a complex situation into subclasses.
- forming of an adapted deterministic models of AI actions.

The models for decision and predicting of emergencies using CDMM technology are presented in Table 5.

Illustrative Example

In this section the examples of model development for expert systems (ES) and DSS are presented. A significant number of ES in ANS has been developed by students of National aviation university (Shmelova, Kharchenko, Sikirda, 2016), such as:

- Quantitative estimation of the complexity of the stages the aircraft flight.
- Quantitative estimation of the complexity of the navigation parameters of flight.
- Quantitative estimation of the significance of the Landing System (GNSS, ILS, VOR, etc.).

Table 5. The models for decision in emergency using CDMM-technology

Models	Describing of Modelling Emergency
	Expert assessment of the complexity of the flight stages (takeoff and climb, echelon, cargo discharge, echelon reverse, descent and landing)
	Neural Network Model to determine potential alternative of the flight completion. Determination of weight coefficients of neural network (probabilities for the model – DM in risk) and effectiveness of flight completion: $\{Y_G ; Y_{Gaer}; Y_{Glf}; W\}$.
	Fuzzy logic to determine quantitative estimates of potential loss - functions of estimation risk R / outcomes U for next models of DM in Risk and Uncertainty-$\{g_r\}$
	DM in Risk. Stochastic models types' tree, GERT's network (Graphical Evaluation and Review Technique) for DM and FE developing. The optimal solution is found by the criterion of an expected value with the principle of risk – A_{dopt}
	DM in certainty using Network Planning method and DM in Risk for each branch. Determined models for an operators / AI with deterministic procedure – t_i; $;T_{cr;} T_{mid;} T_{min;} T_{max}$
	Optimal decision for action in EF (operator / AI model). The authors have developed a computer program for finding optimal solutions.

- Quantitative estimation of the complexity procedures of the operators during working process.
- Quantitative estimation of the HF problem.
- The significance of the procedures performed by the ATC.
- Sources of the projects financing; criteria for assessing the skills.
- The importance of individual-psychological factors influencing the DM.
- The importance of social and psychological factors influencing the decision.
- Definition of the difficult of the procedures for aircraft control by ATC.
- Aviation safety for solving different target tasks (safety, regularity, economic, efficiency), and others.

Example 1. The DSS of Operators of UAV. The synthesis of models for DM in an emergency if is solving logistic problem UAV flight in bad weather condition (emergency - "loss connection") (Figure 12). Steps of analyses and synthesis of the problem:

1. Analyses of the problem as a system, description of the subsystems and criteria of estimation.
2. Emergency: Expert estimation of subsystems using EJM.
3. DSS: DM models in ES: deterministic and stochastic models, etc.

To calculate the efficiency of flight 7 stages of flight UAV are chosen:

1) Takeoff.
2) Climb of UAV.
3) Echelon of UAV.
4) Cargo discharge by UAV.
5) Echelon reverse of UAV before landing.
6) Descent.
7) Landing of UAV.

Figure 12. Solving Logistic task using UAVs flights (1 - take off and climb, 2 - echelon, 3 - cargo discharge, 4 - echelon reverse, 5 - descent and landing)

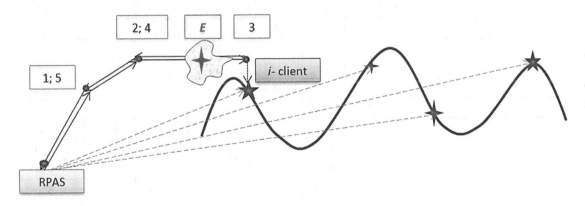

The selected criteria of estimation are as follows:

α_1: Complexity of flight.
α_2: Important of performance of a task.
α_3: Safety of flight.
α_4: Cost of flight, etc.

The results of calculations using EJM determining the coordination of expert opinions for the criterion of estimation "complexity of UAV flight" are shown in Table 6.

The system preferences the group of experts for the criterion of estimation "α_1 - complexity of UAV flight":

$$R_{\alpha 1} = R_7 \succ R_1 \succ R_4 \succ R_2 \succ R_3 \succ R_6 \succ R_5 \tag{6}$$

The graphical presentation of effectiveness (weight coefficients corresponding to priority-ranks R) UAV flight for the criterion of estimation "complexity of UAV flight" is given in Figure 13.

Similar calculations were made for the criteria such as "important of performance of a task (α_2)" and "safety of flight (α_3)" (Figure 14).

Table 6. The results of determining the coordination of experts' opinion

Criteria	Stages of UAV Flight						
α_1 - complexity of flight	1	2	3	4	5	6	7
Experts	takeoff	climb	echelon	cargo discharge	echelon reverse	descent	landing
1	2	4	5	3	6	7	1
2	1,5	4,5	4,5	3	6,5	6,5	1,5
3	2	5	5	2	5	7	2
4	2	4	5	3	6	7	1
5	1,5	4,5	4,5	3	6,5	6,5	1,5
6	2	5	5	2	5	7	2
7	2	4	5	3	6	7	1
8	1,5	4,5	4,5	3	6,5	6,5	1,5
9	2	5	5	2	5	7	2
10	2	4	5	3	6	7	1
Average	1,85	4,45	4,85	2,70	5,85	6,85	1,45
Dispersion	0,06	0,19	0,06	0,23	0,39	0,06	0,19
Squared deviation	0,24	0,44	0,24	0,48	0,63	0,24	0,44
Coefficient of variation	13,06	9,84	4,98	17,89	10,70	3,53	30,19
C	0,88	0,51	0,45	0,76	0,31	0,16	0,94
$w(\alpha_1)$	0,22	0,13	0,11	0,19	0,08	0,04	0,23

Figure 13. Graphical presentation of weight coefficients - results of effectiveness UAV flight for the criterion of estimation "complexity of UAV flight"

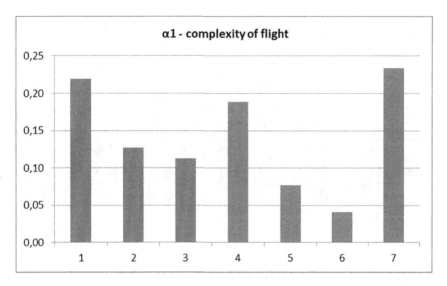

Figure 14a. Graphical presentation of weight coefficients - results of effectiveness UAV flight for the criteria "important of performance of a task (α_2)"

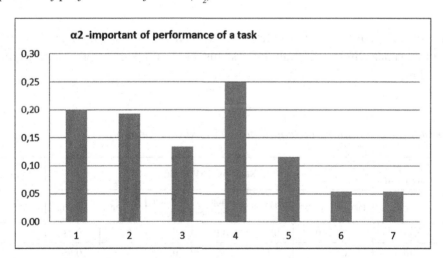

Aggregation of subsystems in systems presented in Table 7 and Figure 15 can be performed following two approaches: additive aggregation (AA) and multiplicative aggregation (MA). If in phase 3 ("echelon of UAV") an emergency ("loss connection") occurred the flight efficiency is additive or multiplicative aggregation of weight coefficients. Additive aggregation of subsystems (criteria / stages of flight):

$$W_j = \sum_{i=1}^{n} \omega_i F_{ij} = 5,08, i = \overline{1,n}, j = \overline{1,m}. \tag{7}$$

Figure 14b. "safety of flight (α_3)"

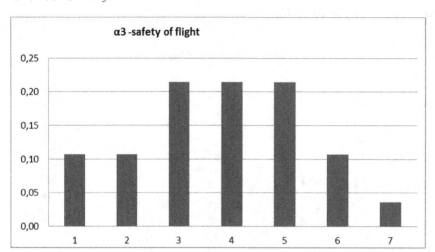

Multiplicative aggregation of subsystems (criteria / stages of flight):

$$W_j' = \prod_{i=1}^{n} F_{ij}^{\omega_i} = 0, \; i = \overline{1,n}, \; j = \overline{1,m}.$$ (8)

Multiplicative aggregation more accurately represents a flight problem. The additive aggregation is mainly used to evaluate students in training systems and in production planning. Multiplicative aggregation is used in real time.

Table 7. The results of determining the coordination of experts' opinion

Criteria	Weight Coefficient	Stages of UAV Flight							Aggregation	
α	w	takeoff	climb	echelon	cargo discharge	echelon reverse	descent	landing	AA	MA
α_1	w (α1)	0,22	0,13	0,11	0,19	0,08	0,04	0,23		
α_2	w (α2)	0,20	0,19	0,13	0,25	0,12	0,05	0,05		
α_3	w (α3)	0,11	0,11	0,21	0,21	0,21	0,11	0,04		
α_1	N=2	0,44	0,25	0,00	0,38	0,15	0,08	0,47	1,78	
α_2	Ne=0	0,40	0,39	0,00	0,50	0,23	0,11	0,11	1,73	
α_3		0,21	0,21	0,00	0,43	0,43	0,21	0,07	1,57	
AA		1,05	0,85	0,00	1,31	0,81	0,40	0,65	5,08	
α_1	N=2	1,16	1,09	0,00	1,14	1,05	1,03	1,18		0,00
α_2	Ne=0	1,15	1,14	0,00	1,19	1,08	1,04	1,04		0,00
α_3		1,08	1,08	0,00	1,16	1,16	1,08	1,03		0,00
MA		1,44	1,34	0,00	1,57	1,33	1,15	1,25		0,00

Figure 15 . Graphical presentation of weight coefficients - results of effectiveness UAV flight for the all criteria

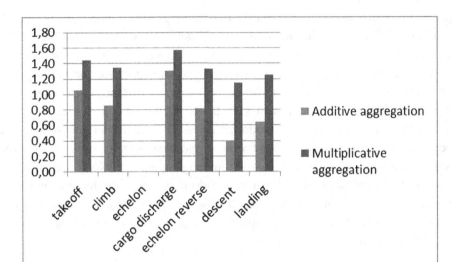

To predict the development of an emergency and determine the optimal decision deterministic and stochastic models were used for the following system: DSS of UAV's operator - forming the optimal actions of UAV's operator in-flight especially in an emergency; board AI models as automatic systems - a software creation for UAV distance control in-flight.

Example 2. Deterministic and Stochastic Models for Remotely Piloted Aircraft (RPA). Within the pre-flight planning, alternate aerodromes/places for return procedure should be defined. These aerodromes/places will be used in the cases of an emergency situation or urgent situation that could be caused by inappropriate meteorological conditions. Stages to find optimal aerodrome/place of landing will be considered:

- Pre-flight preparation piloted aircraft crew is one of the conditions of flight safety provision.

According to regulatory documents, alternate aerodromes are selected by using of following parameters:

- The minimum professional level of the aircraft pilot-in-command.
- Meteorological conditions on alternate aerodrome.
- The amount of fuel on board of aircraft.
- The distance from the alternate aerodrome.

With the aim to optimize the pre-flight preparation the automated systems of pre-flight preparation information and systems of decision making support were developed. The application of the models in these systems depends on the type of flight. An estimation of factors that influence on the selection of optimal landing aerodrome is realized with the help of method of expert estimation, two layers strictly directed network of two layers perceptron.

The usage of RPA should be regulated by the same regulatory documents as for ordinary aircraft. One of the tasks that RPA pilots solve in the process of pre-flight planning is the selection of alternate aerodromes/places for return procedure in the case of an emergency situation that is caused by meteorological conditions. During pre-flight preparation adequate fuel/energy capacity should be predicted. This operation is necessary for a possible excursion from a planned place of landing/return. In this case, RPA can fly to the alternate aerodrome/place safely and perform approach and landing. An RPA pilot should include following factors: adequacy of fuel/energy reserve; distance to the alternate aerodrome/place; reliability of C2 lines for connection with RPA; the possibility of communication with ATC units; meteorological conditions on the alternate aerodrome/place. The multifunctional model of selection of alternate aerodrome/place of RPA is proposed for the economical effectiveness of flight realization of RPA which is used in the remote system of decision making support of UAV's operator.

The decision on the selection of the alternative aerodrome/position in the emergency landing (difficult meteorological conditions, etc.) by the method of decision making under uncertainty by means of the criteria of DM under uncertainty: Wald, Laplace, Savage, Hurwicz (Shmelova & Shulimov, 2017) was implemented with the following output data:

- A calculated route of direction.
- An aerodrome of departure (ADep) and its characteristics.
- An aerodrome of destination (ADest) and his characteristics.
- A list of alternate aerodromes (AA) according to the calculated route.
- A type of RPA and its tactical and technical characteristics (TTC).
- Current flight situation (difficulties in conditions of flight realization, difficult situation).
- Factors that influence on air traffic.
- Adequacy of fuel/energy reserve.
- Distance to the alternate aerodrome/place.
- Reliability of C2 lines for connection with RPA.
- Possibility of communication with ATC units.
- Meteorological conditions on the alternate aerodrome/place.

The solution of finding the optimal landing aerodrome/place (alternate aerodrome/place, (AAP)) for return operation in the case of an emergency situation that is caused by meteorological conditions. Formation of multiplication of alternative decisions $\{A\}$ from ADep, ADest, AAP (AADest – alternate decision about landing ADest; AADep - alternate decision about return to ADep; AAAP - multiplication of alternates APP):

$$\{A\} = \{AADest \ U \ AADep \ U \ \{AAAP\}\} = \{A_1, A_2, ...A_i, ..., A_n\} \tag{9}$$

Formation of multiplication of factors $\{\lambda\}$, that influence on the selection of AAP in the case of H-O decision making in conditions of forced landing of RPA in emergency landing (λ_1 - availability of fuel/energy on board of RPA; λ_2 - distance of RPA from ADep, ADest, AAP; λ_3 - TTC, AAP, ADep, ADest; λ_4 - meteorological conditions on ADep, ADest, AAP; λ_5 – reliability of C2 lines for connection with RPA; λ_6 - possibility of communication with ATC units; λ_7 - subjective factor):

$$\{\lambda\} = \lambda_1, \lambda_2 ..., \lambda_j ..., \lambda_m, \tag{10}$$

Formation of multiplication of possible consequences $\{U\}$ that influence on the selection of AAP in the case of H-O DM in conditions of forced in emergency landing of RPA, where U_{ij} - is defined be means of the method of expert estimation according to estimation scale and data from the regulatory documentation:

$$\{U\} = U_{11}, U_{12}, ..., U_{ij}, ..., U_{nm}, \tag{11}$$

Formation of decision matrix $M = \|M_i\|$ (Table 8) and selection of criteria of DM under uncertainty in the case of an emergency situation that is caused by meteorological conditions: Wald (minmax) criterion – the flight is performed rarely; Laplace criterion – the flight is performed often. The matrix of possible results of DM that relates to the selection of AAP for return operation in the case of an emergency situation or uncertain situation that is caused by meteorological conditions is represented in Table 8. In application of Wald criterion, each action is estimated from the best to the worth states. This criterion uses estimated function that corresponds to the position of extreme caution. According to Wald criterion, the optimal decision is defined for the rule, provide guaranteed result and completely excludes a risk:

$$A^* = \max_{A_i} \left\{ \min_{\lambda_j} u_{ij}(A_i, \lambda_j) \right\} = 3 = A_3 \tag{12}$$

According to Laplace criterion

$$A_i^* = \max O_i = \left\{ \frac{1}{n} \sum_{I=1}^{N} u(a_i, \lambda_j) \right\} = \max \{5, 7.3, 4.3, 4.7, 4.6\} = 7.3 = A_2 \tag{13}$$

Table 13. The matrix of possible results of DM that relates to selection of AAP

Alternate Decisions		Factors that Influence on DM $\{\lambda\}$						Solution	
{A}		Availability of fuel/energy onboard of RPA	Distance of RPA from ADep, ADest, AAP	TTC, AAP, ADep, ADest	Meteorological conditions on ADep, ADest, AAP	Reliability of C2 lines for connection with RPA	Possibility of communication with ATC units	Criterion	
		λ_1	λ_2	λ_3	λ_4	λ_5	λ_6	Wald criterion	Laplace criterion
A1	AADest	1	1	9	6	5	7	1	5
A2	AADep	8	9	9	2	8	8	2	**7.3**
A3	AAAP1	5	6	3	3	6	3	**3**	4.3
A4	AAAP2	6	5	2	5	5	2	2	4.7
A5	AAAP3	5	7	7	4	4	1	1	4.6

Therefore, in pre-flight planning for RPA one of the principal procedures is the selection of the alternative aerodrome/place of landing in the case of an emergency situation that is caused by meteorological conditions. There are set of factors that should be calculated before the flight to perform safe approach and landing such as:

- Adequacy of fuel/energy reserve.
- Distance to the alternate aerodrome/place.
- Reliability of C2 lines for connection with RPA.
- The possibility of communication with ATC units.
- Meteorological conditions on the alternate aerodrome/place.

The models of safe and economical effective landing during an emergency situation that is caused by meteorological conditions presented here were able to achieve good stability due to the fact that four criteria were implemented: Wald, Laplace, Savage, Hurwicz. It is clear from calculations that according to Wald criterion a return to the place of alternative landing is better if flights are performed rarely. In the case of more frequent flights the return to the aerodrome of departure is the better solution according to Laplace criterion. The program is a useful tool to find the optimal solution of any landing problem of described class. Using this computer software one can compute profits and losses of the enterprise but in this project, this software is used for purposes to find an optimal place of landing of RPA. In other words, one will be able to find the best or the worst decision according to the decision of one or a multiple number of persons. Also, one can choose any desired number of assessment criteria. Four different methods of calculations help to optimize the solution.

Example 3. Collaborative Decision Making (CDM). For the last years, the authors have developed computer programs for DSS of the aircraft pilot, air traffic controllers, flight dispatcher, UAV's operator, etc. A recent Masters Diploma of Ya.S. Khaletskyy "Remote expert air traffic management system "decision making in a common environment FF-ICE" presented decentralized-distributed UAVs control system using blockchain technology (Figure 16). There are many advantages of this approach such as enhanced security, security, big data analysis, and record keeping, real-time constant data exchange, etc.

Figure 16. Decentralized-distributed UAVs control system

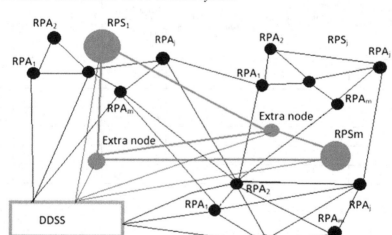

Blockchain technology is ideal as a new infrastructure to secure, share, and verify learning achievements and Collaborative Decision Making (CDM). To ensuring the safety of flights organization of efficient collaborative decision making by all the operational partners – airports, air traffic control services, airlines and ground operators – on the basis of general information on the flight process and ground handling of the aircraft in the airport the key to ensuring the safety of flights.

The Global Operating Concept for Air Traffic Management (ATM) provides for the provision of a joint (pilot – air traffic controller (ATC)) DM air traffic control unit based on a dialogue between them and real-time information evaluation at all stages of the flight. The lives of air passengers in the sky and people on the ground depend on the adequate interaction between the pilot and ATC. According to the statistics of the Aviation Safety Network (ASN), during the second half of the 20th century due to problems in interaction pilot – ATC (language barrier, communicating problems, ATC's interference in the flight crew, wrong ATC instructions / commands, etc.) about 2 000 people died in aviation accidents (ICAO, 2017).

Coherent, clear interaction between pilot and ATC is most important in emergency in flight, which are characterized by a sharp shortage of time in the DM in conditions of incompleteness and uncertainty of information, as well as significant psychophysiological load on the flight crew (emergency). The final decision on the order of the flight in the emergency case is taken by the captain of the aircraft (Capt.), which is fully responsible for the decision.

The ATC is responsible for the correctness and timeliness of the information and advice that given to flight crew, so the ATC in such situations is also given a significant role. The main requirement for the ATC when an emergency case arises is the constant readiness to provide the necessary assistance to the flight crew, depending on the type of situation, taking into account the air situation and meteorological conditions. One of the factors that greatly complicate the interaction between pilot and ATC is the inadequate knowledge of the flight crew procedures performed in emergency. The technology of flight crew and ATC procedures in the emergency case must be in line with the definition of the algorithm prescribed in the normative and regulatory documents, therefore, for the formalization of the actions of the H-O in emergency, it is possible to apply determined models. Since EC is a time-consuming event, when it comes to modeling a collaborative DM pilot and ATC, it is advisable to use network graphs depending on the algorithm of action in EC, which reflects the technological dependence and consistency of operational procedures of operators, ensure their achievement in time, taking into account the cost of resources and the cost of work with the allocation at the same time critical places.

Thus, the problem of optimizing the interaction between pilot and ATC in emergency can be solved by the way of development and synchronization (maximal alignment over time) of deterministic models of H-O collaborative DM, which will minimize the critical time needed to solve emergency, by definition the optimal sequence of execution of technological procedures. The parallel process of simultaneous execution of pilot and ATC technological operations in the emergency case can be represented as a consolidated dual-channel network. For a consistent optimization of such a network in order to achieve the cross-cutting efficacy of joint decisions, it is advisable to use a multi-criteria approach: achieving a minimum time for parity of emergency case with maximum safety / maximum harmonization over the time of H-O actions.

Ways to optimize the network graph for performing procedures by the H-O in the emergency case (by minimizing time with maximum safety) are:

1. Time optimization – by regulating the use of resources minimizing the time of execution of critical paths t_i^k:

$$t_i^{k-1} < t_i^k < t_i^{k+1}, \tag{14}$$

where

$t_i^{k-1} = \max \min \ t_i^k$ – is a minimum time with maximum safety;

$t_i^{k+1} = \min \ \max \ t_i^k$ – is a critical time of the maximum (critical) path;

t_i^k – optimal (minimum) time.

2. Changing the topology of the network due to the multi-varied technology implementation procedures.
3. Introduction of parallel execution of procedures with maximum agreement on time (minimum time for two or more charts), that is, obtaining the optimal consolidated time for the execution of procedures t_j^k

$$t_j^{k-1} < t_j^k < t_j^{k+1}, \tag{15}$$

where

$t_j^{k-1} = \max \min t_j^k$ – is a minimum time with maximum time matching;

$t_j^{k+1} = \min \ \max \ t_j^k$ – is a critical time of the maximum (critical) path;

t_j^k – optimal (minimum) time.

For example, with network planning, synchronized Pilot Flying (PF) and Pilot Monitoring (PM) actions, which determine the time to manage the PF in the event of PM actions in the stages of FP paring (Figure 17)

Stage I: Checking of emergency panels highlighting the failure of generators;
Stage II: Power Check;
Stage III: Checking navigation equipment;
Stage IV: Emergency landing.

Figure 17. A fragment of the network graph for PF and PM procedures in the emergency

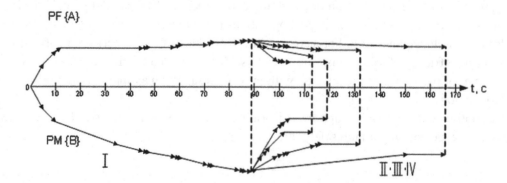

The model of synchronization of operational procedures of pilots PF and PM in the conditions of cross-monitoring in the event of emergency is developed using AI methods.

The formalization of the actions of H-Os (manned, unmanned pilots, ATC, engineers) in the emergency with the help AI, DM methods allows us to determine the optimal sequence and time of execution of procedures for paring *emergency*.

FUTURE RESEARCH DIRECTIONS

In order to simulate DM under conditions of an emergency, the next steps are: deep analysis of an emergency; intelligent data processing; identification of situation; formalization of the situation using integrated models; decomposition of the complex situation into subclasses; synthesis of adapted deterministic models to AI-determined actions. In cases of large and complex data, methods can be integrated into traditional and next-generation hybrid DM systems by processing unsupervised situation data in the deep landscape models (Figure 18), potentially at high data rates and in near real time, producing a structured representation of input data with clusters that correspond to common situation types (Dolgikh, 2018).

In the Figure 18 above, a Deterministic action model is targeted to a specific situation type. Another benefit of these models is a potential ability of such systems to learn to identify relationships between different types of situations, again almost entirely in self-supervised training regime with very limited requirement for ground truth data. Possible applications of such capabilities of machine intelligence models may extend to for example, to learning to detect early signs or symptoms of developing situations via relationships between situation types, and the ability to raise notifications and early alerts that a human operator would be able to attend to proactively before the situation develops.

Figure 18. Hybrid ML-DM Situation Monitoring and Management System (Prototype)

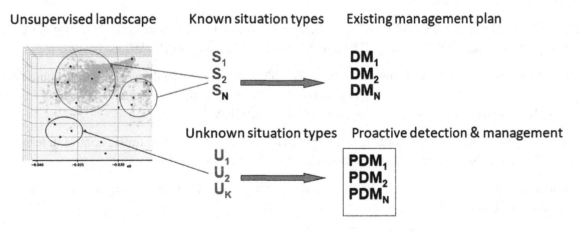

CONCLUSION

In this chapter the authors argued that the aviation systems should be considered as STS, that is, as systems with high technologies and high risk of the operation and reviewed the evolution of the Human Factor models. These concepts have significant and growing influence on safe and efficient operation of the aviation industry. The quality of decisions depends on the development and use of innovative technologies in the aviation of today and not in the least, education of aviation operators in modern information technologies.

Further research can be directed to developing solutions in analysis, prediction and proactive intervention to prevent development of flight situations to emergencies. The models of flight emergency development and of DM by an operator in-flight emergency will be able to predict human operator actions with the aid of the informational-analytic and diagnostics complex for research of operator behavior in extreme situations. It is necessary to develop modern DSSs of Air Navigation System's operator (pilots, air traffic controllers, flight dispatchers, UAV's operators) in-flight emergencies and in other situations, to investigate applied tasks of the DM in Socio-technical System by an operator of aviation systems and other fields using innovative technology such as AI and based on up-to-date education of operators. In development of AI, ES, DSSs new concepts in aviation should be applied (FF-ICE, PBA, SMART, CDM, SWIM, etc.) for different operators and in every stage and process combined with modern information technologies as Artificial Intelligence, Big Data, Data Mining, and others. Incorporating the promising approaches and technological advances discussed in this chapter will allow the aviation industry to evolve according to the increasing demands while offering unsurpassed safety and quality in operation.

REFERENCES

Aviation Accident Statistics. (2017). *National Transportation Safety Board*. Retrieved from www.ntsb.gov/aviation/ aviation.htm

Baxter, G., & Sommerville, I. (2011). Socio-technical systems: From design methods to systems engineering. *Interacting with Computers*, *23*(1), 4–17. doi:10.1016/j.intcom.2010.07.003

Bertsch, V., Treitz, M., Geldermann, J., & Rentz, O. (2007). Sensitivity analyses in multi-attribute decision support for off-site nuclear emergency and recovery management. *International Journal of Energy Sector Management*, *1*(4), 342–365. doi:10.1108/17506220710836075

Bianco, L., Dell'Olmo, P., & Giordani, S. (2006). Scheduling models for air traffic control in terminal areas. *Journal of Scheduling*, *9*(3), 223–253. doi:10.100710951-006-6779-7

Campbell, R. D., & Bagshaw, M. (2002). *Human performance and limitations in aviation* (3rd ed.). UK: Blackwell Science Ltd.

Carayon, P. (2006). Human factors of complex sociotechnical systems. *Applied Ergonomics*, *37*(4), 525–535. doi:10.1016/j.apergo.2006.04.011 PMID:16756937

Civil Aviation Organization. (1998). Human factors training manual (1st. ed.). Doc. ICAO 9683-AN/950. Canada, Montreal: Author.

Clegg, C. W. (2000). Sociotechnical principles for system design. *Applied Ergonomics, 31*(5), 463–477. doi:10.1016/S0003-6870(00)00009-0 PMID:11059460

Dolgikh, S. (2018, November). Spontaneous concept learning with deep autoencoder. *Spontaneous Concept Learning with Deep Autoencoder In International Journal of Computational Intelligence Systems, 12*(1), 1–12. doi:10.2991/ijcis.2018.25905178

Flueler, T. (2006) Decision making for complex socio-technical systems: Robustness from lessons learned in long-term radioactive waste governance. *Environment & Policy*. Springer IATA. Retrieved from https://www.iata.org/publications/Pages/AI-white-paper.aspxInternational

Gianazza, D. (2002). The 21st Digital avionics systems. *Conference Proceedings*. doi:10.1002/9780470774472

International Civil Aviation Organization. (1998). Human factors training manual (1st ed.). Doc. ICAO 9683-AN/950. Montreal, Canada: Author.

International Civil Aviation Organization. (2002). Human factors guidelines for safety audits manual (1st ed.). Doc. ICAO 9806-AN/763. Canada, Montreal: Author.

International Civil Aviation Organization. (2002). Human factors guidelines for safety audits manual (1st ed.). Doc. ICAO 9806-AN/763. Montreal, Canada: Author.

International Civil Aviation Organization. (2002). Global air navigation plan for CNS/ATM Systems 2002 Doc 9750 AN/963.

International Civil Aviation Organization. (2004). *Cross-cultural factors in aviation safety: Human Factors Digest No. 16. Circ. ICAO 302-AN/175*. Montreal, Canada: Author.

International Civil Aviation Organization. (2004). *Cross-cultural factors in aviation safety: Human Factors Digest No. 16. Circ. ICAO 302-AN/175*. Montreal, Canada: Author.

International Civil Aviation Organization. (2005). *Global air traffic management operational concept. Doc. ICAO 9854*. Montreal, Canada: Author.

International Civil Aviation Organization. (2013a). Safety management manual (SMM) (3rd. ed.). Doc. ICAO 9859-AN 474. Canada, Montreal: Author.

International Civil Aviation Organization. (2013a). Safety management manual (SMM) (3rd. ed.). Doc. ICAO 9859-AN 474. Canada, Montreal: Author.

International Civil Aviation Organization. (2013b). *State of global aviation safety*. Montreal, Canada: Author.

International Civil Aviation Organization. (2013b). *State of global aviation safety*. Montreal, Canada: Author.

International Civil Aviation Organization. (2014). *Safety report*. Montreal, Canada: Author.

International Civil Aviation Organization. (2014). *Safety report*. Montreal, Canada: Author.

International Civil Aviation Organization. (2017). Global Aviation Security Plan (GASP). International Civil Aviation Organization, Canada, Montreal (2017) Global Air Navigation Plan (GANP, Doc 9750 International Civil Aviation Organization. (2013). Safety Management Manual (SMM): Doc. ICAO 9859-AN/474 (3rd ed.). International Civil Aviation Organization, Canada, Montreal (2013).

International Civil Aviation Organization. (2018, October). Potential of Artificial Intelligence (AI) in Air Traffic Management (ATM). In *Thirteenth Air Navigation Conference ICAO,* Montréal, Canada, 9-19.

Keating, C., Fernandez, A. A., Jacobs, D. A., & Kauffmann, P. (2001). A methodology for analysis of complex sociotechnical processes. *Business Process Management Journal*, *7*(1), 33–50. doi:10.1108/14637150110383926

Kharchenko, V., Shmelova, T., & Sikirda, Y. (2011). Graphanalytical models of decision-making by air navigation system's human-operator. *Proceedings of the National Aviation University*, *1*, 5–17.

Kharchenko, V., Shmelova, T., & Sikirda, Y. (2011). Methodology for analysis of decision making in air navigation system. *Proceedings of the National Aviation University*, *3*, 85–94.

Kharchenko, V., Shmelova, T., & Sikirda, Y. (2012). *Decision-making of operator in air navigation system*: monograph. Kirovograd, Ukraine: KFA of NAU.

Kharchenko, V., Shmelova, T., & Sikirda, Y. (2012). Modeling of behavioral activity of air navigation system's human-operator in flight. *Proceedings of the National Aviation University*, *2*, 5–17.

Kharchenko, V., Shmelova, T., & Sikirda, Y. (2012). Methodology for analysis of flight situation development using GERT's and Markov's networks: In *Proceedings of the 5th World Congress, Aviation in the XXIst century. Safety in Aviation and Space Technologies*. Kiev, Ukraine.

Kharchenko, V., Shmelova, T., & Sikirda, Y. (2016). *Decision-making in socio-technical systems: Monograph*. Kiev, Ukraine: NAU.

Kharchenko, V., Shmelova, T., & Zubrytskyi, V. (2017) Decision making in air navigation system as a socio-technical system. In *Proceedings of the International Conference, AVIA 2017,* Kiev, Ukraine.

Kuchar, J. K., & Yang, L. C. (2000). *A review of conics detection and resolution modelling methods* (pp. 1388–1397). Cambridge, UK: American Institute of Aeronautics and Astronautics.

Lefebvre, B. (2008, January-June). Intentional-reflective model of the agent. *Reflexion Processes and Control, 1, Vol. 1*, 69–78.

Lefebvre, B., & Adams-Webber, J. (2002, January-June). Functions of fast reflexion in bipolar choice. *Reflexion Processes and Control, 1, Vol. 1*, 29–40.

Lefebvre, V. A. (2001). Algebra of Conscience (2nd enlarged ed.). Dordrecht, The Netherlands: Kluwer Publishers.

Leychenko, S., Malishevskiy, A., & Mikhalic, N. (2006). *Human factors in aviation: monograph in two books. Book 1st.* Kirovograd, Ukraine: YMEKS.

Makarov, R., Nidziy, N., & Shishkin, G. (2000). *Psychological foundations of didactics in flight education: Monograph*. Moscow, Russia: MAPCHAK.

Mescon, M., Albert, M., & Khedouri, F. (2008). *Management. Monograph*. Moscow, Russia: Vilyams.

Mumford, E. (2006). The story of socio-technical design: Reflections on its successes, failures, and potential. *Information Systems Journal*, *16*(4), 317–342. doi:10.1111/j.1365-2575.2006.00221.x

Pasmore, W. (2001). Action research in the workplace: The socio-technical perspective. In *P. Reason & H. Bradbury (2006) Handbook of action research: Concise paperback* (pp. 38–48). London, UK: Sage.

Reason, J. (1990). *Human error*. Cambridge, UK: Cambridge University Press. doi:10.1017/CBO9781139062367

Shmelova, T., Bondarev, D., & Znakovska, Y. (2016). Modelling of the decision making by UAV's operator in emergency situations. In *Proceedings of the IEEE 4th International Conference on Methods and Systems of Navigation and Motion Control (MSNMC-2016)*. Kiev, Ukraine.

Shmelova, T., Danilenko, O., & Boldysheva, T. (2015). Expert estimation of human factor using Reason model's components. In *Proceedings of the XII International conference AVIA -2015*. Kiev, Ukraine.

Shmelova, T., Dolhov, D., & Muliar, P. (2015) Mathematical models of operators' activities in aeronavigation sociotechnical system. In Proceedings of the Conference "Techniques and technology. Science yesterday, today, tomorrow". Warsaw, Poland.

Shmelova, T., & Konovalova, A. (2017). Decision making of ATCO in emergency situation aircraft decompression. In *Proceedings of the International Conference AVIA -2017*. Kiev, Ukraine.

Shmelova, T., Mirskova, M., & Krepak, D. (2017). Decision making in emergency situation of air navigation system's operator. In *Proceedings of the International Conference AVIA -2017*. Kiev, Ukraine.

Shmelova, T., & Sikirda, Y. (2012). Analysis of human-operator's decision-making in air navigation system. *Radioelectronic and Computer Systems*, *7*(59), 319–324.

Shmelova, T., & Sikirda, Y. (2016). Calculation the scenarios of the flight situation development using GERT's and Markov's networks. In *Proceedings of the Conference "Management of high-speed moving objects training operators of complex systems"*. Kirovograd, Ukraine.

Shmelova, T., Sikirda, Y., Rizun, N., Abdel-Badeeh, M., Salem, K. Yu. (2017). Socio-technical decision support in air navigation systems: Emerging research and opportunities. Publisher of Progressive Information Science and Technology Research, Pennsylvania.

Shmelova, T., & Sikirda, Y. (2017). Assessments of the influence of organizational factors on flight safety in air traffic control. *Proceedings of the Kharkov National University of Air Forces*.

Shmelova, T., Sikirda, Y., & Jafarzade, T. (2013). Models of flight situations development while decision-making by air navigation system's human-operator. *Information Processing Systems*, *8*(115), 136–142.

Shmelova, T., Sikirda, Y., & Kasatkin, M. (2019). Applied artificial intelligence for air navigation sociotechnical system development. In *Proceedings of the 15th International Conference on ICT in Education, Research, and Industrial Applications. Integration, Harmonization, and Knowledge Transfer*. Ukraine, pp. 470-475.

Shmelova, T., Sikirda, Y., & Kovalyov, Y. (2013). Models of personality. Behaviour and activity of air navigation system's human-operator: In *Proceedings of the Conference "Management of high-speed moving objects training operators of complex systems "*. Kirovograd, Ukraine.

Shmelova, T., Sikirda, Y., & Sunduchkov, K. (2013). Socio-technical analysis of air navigation system. *Science and technology of the Air Forces of the Armed Forces of Ukraine, 4*(13), 34–39.

Shmelova, T., Sterenharz, A., & Sikirda, Y. (2018). The diagnostics of emotional state deformation in the professional activity of operators in the air navigation system. 8th World Congress. Aviation in the XXIst century. Safety in Aviation and Space Technologies NAU.

Torres, S. (2012). Swarm theory applied to air traffic flow management. *Procedia Computer Science, 12*, 463–470. doi:10.1016/j.procs.2012.09.105

KEY TERMS

AI (Artificial Intelligence) is the simulation of human intelligence processes by modeling: computer systems, and machines.

Air Navigation Socio-technical System: A complex large-scale, high-tech man-machine system, which require complex interactions between their human and technological components; the operations in socio-technical systems generally involve high-risk/high-hazard activities; the consequences of safety breakdowns are often catastrophic in terms of loss of life and property.

Air Navigation System: A complex of organizations, personnel, infrastructure, technical equipment, procedures, rules and information that is used to provide of airspace users of safe, regular and efficient air navigation service.

Expected Aircraft's Operating Conditions: Are the operating conditions, in which pilot's actions are prescribing by Flight Manual.

Expert Systems (ES) -: is a computer system that emulates the decision-making ability of a human expert. Expert systems are designed to solve complex problems by reasoning through bodies of knowledge, represented mainly as if–then rules rather than through conventional procedural code.

GERT-network (Graphical Evaluation and Review Technique): Is an alternative probabilistic method of network planning, applicable in the case when these actions can only start after completion of a prior action including cycles and loops.

Markov Network: This is a graphical model in which a set of random variables possesses the Markov property described by an undirected graph. The Markov network differs from the other graphic model, the Bayesian network, the representation of the dependencies between random variables. It can express some of the dependencies that the Bayesian network cannot express (for example, cyclic dependencies); on the other hand, it cannot express some others.

Non-professional Factors: Are the set of human-operator's individual-psychological, psycho-physiological and socio-psychological factors.

Professional Factors: Are the knowledge, skills and abilities, acquired by human-operator during training and professional activity accordingly.

Reflection: Is the ability of consciousness to focus on itself. The reflection of consciousness is manifested in intuitive thinking.

Socio-technical Systems: A large-scale, high-technology systems, because they require complex interactions between their human and technological components; the operations in socio-technical systems generally involve high-risk/high-hazard activities; the consequences of safety breakdowns are often catastrophic in terms of loss of life and property.

Unexpected Aircraft's Operating Conditions: Are the operating conditions, in which pilot's actions aren't prescribing by Flight Manual.

Chapter 2
Rational Adaptation of Control Systems for the Autonomous Aircraft Motion

Kostiyantin Dergachov
https://orcid.org/0000-0002-6939-3100
Kharkiv Aviation Institute, Ukraine

Anatolii Kulik
https://orcid.org/0000-0001-8253-8784
National Aerospace University, Ukraine

ABSTRACT

The possibilities of using an adaptation principle in an application for organizing the close-loop life circuit of autonomous fly vehicles (FV) are discussed in the chapter. The uncertainties arising at each stage of the life cycle of an autonomous FV are considered. To solve a problem the approach with using intelligent, rational objects and using knowledge database tool is proposed. The main theses of rational adaptation control system (CS) are represented. The preliminary designing tools for constructing rational adaptation algorithms for the motion CS are considered. The practical applications of the proposed approach at the stage of preliminary design of control systems for autonomous fly vehicles are presented.

INTRODUCTION

A Person can live safely only in balance with a Nature. Breaking this balance creates global problems. Global problems are the result of different principles of Nature and human society development. Inartificial Nature functions according to the principle of negative feedback, while human society operates according to the principle of positive feedback. Disrespectful attitude of Man to Nature leads to the extensive use of biosphere resources, which causes a significant deterioration in living conditions on Earth.

The Nobel Prize in Physics Pyotr Kapitsa in 1976 gave a lecture at the University of Stockholm on the topic: «Global Problems and Energy». In this lecture the main cause of the imbalance was formulated.

DOI: 10.4018/978-1-7998-1415-3.ch002

«The cause of global problems is well known: a person differs from an animal mainly in that the animal adapts to nature and the person reworks and adapts it to his needs». How to eliminate this cause and recover the balance, it was suggested that P.L. Kapitsa in the report «Global scientific problems of the near future». He expressed the following idea: «The global crisis associated with the depletion of raw materials, science can prevent by transferring industrial production to so-called «closed processes» as is the case in nature, where nothing is thrown away, since everything is consumed again» (Kapitza, 1987).

Let's consider the possibility of using a device – an adaptation to the organization of a «closed process» in relation to the life cycle of autonomous aircraft. A generalized life cycle diagram is shown in figure 1, reflecting typical features of the desired spiral-like development process of autonomous aircraft. Nine stages of the life cycle together pass – the hull structure bearing the airframe of the aircraft, the power plant and the control system. The control system differs from other components of the airframe and power plant in that it has such a functional element as an on-board computer or a complex. The functionality of modern computing tools allows for all components of autonomous aircraft to have an intelligent interface that provides high-quality adaptation to objectively changing both external and internal operating conditions during the life cycle.

At all stages of the control systems life cycle, there are objectively reasons that destabilize operability. Operability is the ability of the control system to perform specified functions in accordance with the terms of reference. Thus, at the stage of conception, a not well-formulated concept of the control system creating generates problems and intractable tasks for the subsequent stages. At the design stage, the use of inadequate people, design errors and a number of other factors lead to poor-quality design solutions. At the third stage – the manufacture and testing of prototypes of the control system affects both errors and violations accumulated at the previous stages, and deviations from the design documentation, poor quality components and a number of other reasons that complicate the process of testing and refining the control system to ensure its operability. At each of the subsequent stages of the control systems life cycle, both errors and shortcomings of the previous stages as well as their own due to the specificity of production activities and operating conditions are manifested.

Figure 1. Generalized life cycle pattern of typical autonomous aircraft

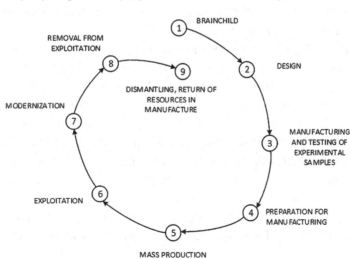

There is a tradition of using different types of models in the practice of designing manufacturing and operating control systems for autonomous aircraft. As a rule at the design stage these are verbal models using quantitative characteristics of the future control system. At the design stage, along with verbal models, graphic in the form of characteristics, functional and structural diagrams, mathematical and machine models, and static graphs, temporal and frequency characteristics are used. At the subsequent stages of the life cycle these are graphical models in the form of principal, assembly, flowcharts of algorithms and other schemes, tabular setup and debugging models and a number of other graphical and text documents corresponding to the life cycle stage.

In accordance with the theory of modeling, each of these models has a certain ability to reflect the process of the control system functioning, its performance. The strongest in terms of the adequate reflection of the efficiency of the control system are mathematical and machine models. These models appear at the stage of preliminary design and are brought to perfection at the stage of detailed design which completes the design stage of autonomous object control systems. The subsequent stages models of the life cycle are significantly less adequate. Such a notion of adequacy leads to an increase in uncertainty in solving the corresponding stages of the tasks as well as to excess costs of resources and reducing the efficiency of the life cycle as a whole.

So at each stage of the autonomous aircraft life cycle there are uncertainties in the operation of control systems due to both external and internal conditions which leads to various deviations from the requirements of the technical specification, i.e. to malfunction.

BACKGROUND

It is necessary to use the principle of adaptation to control the objects in conditions of uncertainties due to incomplete a priori and a posteriori information. The direction of adaptive control is developed in the automatic control theory (Bellman, 1964). The beginning of this direction was laid by the works of a number of famous scientists (Mishkin& Braun, 1961). The adaptive control theory is the intensively developing direction (Tsypkin & Nikolic, 1971).

In the modern representation adaptive control is based on the idea of dualism. On the one hand it is impossible to carry out effective control without knowing the characteristics of the automatic control object, on the other – you can study these characteristics and accumulate knowledge about it in the control process and thereby improve control ensuring its optimality.

The use of artificial intelligence tools appears at the present stage of adaptive control theory development of trend direction (Zadeh, 1963). There are many definitions of artificial intelligence (Russel & Peter, 2003). Artificial intelligence: the science and technology of creating intelligent machines, especially intelligent computer programs; the ability of intelligent systems to perform creative functions that are traditionally considered the prerogative of human. This definition was first given by John McCarthy in 1956 at the conference at Dartmouth University and it is not directly related to understanding human intelligence. According to J. McCarthy the artificial intelligence are methods that are not observed in humans but they are necessary to solve specific problems (Russel & Peter, 2003). Let's use the definition closest to the essence of our research. According to the definition of Andreas Kaplan and Michael Henlein the artificial intelligence is the ability of the system to correctly interpret external data, learn from such data and use the obtained values to achieve specific goals using flexible adaptation (Kaplan & Haenlein, 2019).

Among the many approaches of artificial intelligence to solving the problems of technical objects automation, the most constructive is an agent-oriented approach or an approach based on the use of intelligent (rational) agents (Russel & Peter, 2003). According to this approach intelligence is a computing component of a system that has the ability to achieve goals. Such a component will be the intelligent agent perceiving the world around it with the help of sensors and acting on the control object with the help of executive mechanisms. This approach focuses on those methods and algorithms that help the intelligent control system to survive.

The autonomous aircraft motion control system is one of the most important on-board systems and its role increases with the growth of the functions assigned to it. Instrumental means of generating knowledge about external and internal destabilizing effects into relevant knowledge bases are required to ensure effective adaptation of the motion control system throughout the entire life cycle; about the full diagnosis, registries of operability recovery and means of logical choice of a specific resource for a specific diagnosis.

Tools for the knowledge bases and inference rules formation are developed in the theory of artificial intelligence to apply to static situations and in the absence of restrictions on resource costs. The autonomous aircraft motion control systems are related to the class of dynamic objects with severe restrictions on the time of diagnosis and on the time of recovery, in other words, on the time of adaptation as well as restrictions on the weight and size and energy characteristics. These circumstances determine the search for new approaches to the modification of well-known tools for the formation of intelligent algorithms and effective adaptation programs throughout the entire life cycle of autonomous aircraft motion control systems. Rational (from the Latin. Ratio – mind) adaptation is an adaptation based on logic and reasoning to ensure reasonable operability under conditions of uncertainty in external and internal destabilizing effects.

So the decades-growing contradiction between the growing costs of the entire life cycle of motion control systems and the decreasing efficiency of resource use has led to the emergence of a pressing scientific and technical problem. The problem is to ensure the operability of motion control systems for autonomous aircraft throughout the entire life cycle through rational adaptation to external and internal influences.

THE MAIN PROVISIONS OF RATIONAL CONTROL

1. The control system of the object with uncertain dynamics can be represented as consisting of two interrelated subsystems (Kulik et al., 2016). The first subsystem includes the control object, actuators. This subsystem is a heterogeneous formation since it consists of components that are different in physical principle which determines its vulnerability to various kinds of disturbing influences.

The second subsystem is a controller with the appropriate hardware interface. The perfection of the hardware and software of the subsystem has led to a high level of its reliability and survivability compared with the first subsystem and therefore it can be assumed that it is invulnerable to various disturbing influences.

2. External disturbing effects on the object and internal (faults, malfunctions, failures, breakdowns, noises and interference) acting on the first subsystem are effects that destabilize the operability of the entire control system. Such impacts are destabilizing effects.

3. Destabilizing effects are indefinite events characterized by the appearance at the unknown point in time, in the unknown structural part of the subsystem, with unknown properties. The event uncertainty of destabilizing effects is due to the peculiarities of the life cycle of objects with uncertain dynamics as unique products functioning in changing uncertain conditions.

4. The event uncertainty of destabilizing effects can be reduced by indirectly detecting their occurrence due to a malfunction of the control system. And then the cause that gave rise to the malfunction should be found by the consequences of their effects. Such a procedure for finding the consequences of the cause of the malfunction can only be a deep diagnostics procedure.

5. The procedures of deep diagnostics allow forming by indirect features using the uncertainty sequential removal principle, the diagnosis of a conclusion about a removable cause that impaired the performance of the control system in real time.

6. The procedures for deep diagnosing are formed using formalized and weakly formalized models of the processes of regular and abnormal operation of the control system. Structures of developing dichotomous trees which are a kind of production knowledge bases with a design in nodes «if ... then ... » with logical rules for obtaining a diagnosis of the causes of destabilizing effects are used for the formation of a priori and a posteriori knowledge of the causes of a malfunction.

7. Built-in heterogeneous types of redundant means are created to ensure the ability to maintain operability throughout the entire life cycle for flexible recovery of the operability of the motion control system. The choice of excess mean for the current operability recovery is made according to the diagnosis principle of «stinginess», namely, the least «popular» means are selected from those currently available.

8. The classical adaptation principle is based on the use of information about the deviation of the control object current characteristics from the reference ones. Instrumental design tools for adaptive control, as a rule, are formed on linearized representations of information transformation processes in the system. Linearized models reflect the features of control processes in the «small». In some cases destabilizing effects lead to significant changes in the structure and parameters of the automatic control object which makes it necessary to consider control processes in the «large», following the terminology of A.M. Lyapunov (Lyapunov, 1892). Accidents and disasters in aerospace engineering convince of the need to consider operability as a dynamic state that needs to be controlled «in large» throughout its life cycle in order to stabilize it during destabilizing effects.

9. The rational adaptation to destabilizing effects is based on the use of intelligent procedures for deeply diagnosing the causes of dysfunctional motion control systems and flexible recovery in real-time allowing for productive training to preserve the operability of the motion control system throughout the automatic control system life cycle.

10. The rational adaptation is implemented comprehensively both at the component and system levels on a single methodological basis using the same type of models and tools according to the global criterion of the minimum adaptation time.

11. The further development of the classical adaptation principle is to use throughout the entire life cycle of deeply diagnosing the causes of destabilization and flexible recovery of working capacity. A block diagram for a rational control of an object with an indefinite dynamics is shown in figure 2. The block diagram consists of two interconnected subsystems. The first subsystem is an automatic control object (ACO) consisting of a control object, an actuators unit (AU) and a measurers unit (MU). The subsystem is affected by many uncontrollable destabilizing effects – D which significantly affect the performance of the ACO.

Figure 2. Block diagram for the rational control of the adaptation object of the control system with uncertain dynamics

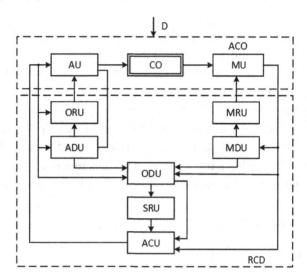

The second subsystem – the rational control device (RCD) that includes a number of specialized units and the corresponding connections. So in order to ensure the operability of the AU the actuators diagnostics unit (ADU) is used. The results of the diagnosis from this unit are delivered to the operability recovery unit (ORU) where control signals are generated to recover the efficiency of the AU. A measurers diagnostics unit (MDU) is used to identifies the cause of the inoperability to diagnose MU and the result of the diagnosis goes to the measurement recovery unit (MRU). At this unit control effects are formed on the MU recovering its measurements.

After the recovering of the AU efficiency and the measurements of the MU are completed the enabling signals go to the object diagnostics unit (ODU) where the signals from the input and output of the ACO are diagnosed. The state diagnosis of the object enters the system recovery unit (SRU) where changes in the structure and parameters of the automatic control unit (ACU) are formed. In this device control effects are formed at the OAU ensuring the recovery and operability stabilization of the entire automatic control system of the motion object.

In the absence of destabilizing effects the diagnostic units: the ADU, MDU and ODU consistently issue corresponding signals that allow the normal operation mode of the ACU ensuring the operability of the closed automatic control system of the motion object.

We need a new technology for conceptual design to implement the rational control of objects with uncertain dynamics. The productivity of the schematic design technology for rational control of objects with uncertain dynamics depends on the used tools. The classical tools that are used in the traditional schematic design technology of motion control systems do not fully correspond to the goals and objectives of rational control. This circumstance led to the development of new schematic design tools.

INSTRUMENTAL MEANS FOR SCHEMATIC DESIGN OF RATIONAL ADAPTATION ALGORITHMS OF MOTION CONTROL SYSTEMS

At the stage of schematic design of control systems for autonomous aircraft (Alekseev et al., 2012) the composition of instrumentation equipment is formed in accordance with the technical specifications and the modeling of functional properties is performed. The dynamics simulation of the motion autonomous aircraft is carried out using the Lagrange equations of the second kind which allow to take into account both the energy aspect of movement and all freedom degrees which makes it possible to formalize the movement features most fully. As a result of such modeling nonlinear differential equations with variable coefficients are obtained. Linearized models are used in the form of linear differential equations with constant coefficients in the Cauchy form represented in the discrete state space in order to analyze and synthesize the characteristics of the designed control system with the help of traditional tools:

$$x\left[(k+1)T_0\right] = Ax(kT_0) + Bu(kT_0) + \xi(kT_0); \ x(kT_0) = x_0;$$
$$y(kT_0) = Cx(kT_0) + Nu(kT_0) + \varsigma(kT_0),$$

(1)

where $x(kT_0)$ – state vector, $x(kT_0) \in X^n$;

$u(kT_0)$ – control vector, $u(kT_0) \in U^r$;

$y(kT_0)$ – measured variable vector, $y(kT_0) \in Y^m$;

$\xi(kT0)$ and $\varsigma(kT0)$ – vectors of modeling errors, measurement noise, disturbance effects $\xi(kT_0) \in \Xi^n$,

$\varsigma(kT_0) \in Z^m$;

A, B, C and N – matrices of corresponding dimensions;

k – discrete number, $k \in K$;

T_0 – quantization period.

Mathematical models reflecting the relationship (influence) of destabilizing effects on the output signals are needed for the synthesis of procedures for deep diagnostics of the object of rational adaptation (ORA) components. Such models have been developed.

Diagnostic Functional Models

This is a relatively new class of mathematical models which was first presented in publications.

In general, the construction of the equations of such models can be represented as follows:

$$\Delta x_i\left[(k+1)T_0\right] = A\Delta x_i(kT_0) + \left[A_i x(kT_0) + B_i u(kT_0)\right]\Delta\lambda_i; \ \Delta x_i(0) = \Delta \dot{x}_{i0};$$
$$\Delta y_i(kT_0) = C\Delta x_i(kT_0) + \left[C_i x(kT_0) + N_i u(kT_0)\right]\Delta\lambda_i.$$

(2)

There $x_i(kT_0)$ – vector of the state deviations of the disturbed motion relative to the reference state; A_i, B_i, C_i, N_i – matrix of sensitivity functions by parameter λ_i; $\Delta\lambda_i$ – deviation of the parameter λ_i.

Parameter λ_i – this is a parameter of a generic destabilizing effect: $\alpha_i \in D$, $\lambda_i \in \{\alpha_i, \beta_i, \gamma_i\}$. Parameter α_i – characterizes the class of destabilizing effects; parameter β_i – characterizes the place of occurrence of destabilizing effects; parameter γ_i – characterizes the presence of destabilizing effects.

The formation of the complete diagnosis of the ORA components is associated with the solution of three main diagnostic tasks. The first task is the detection of destabilizing effects. The second is localization, i.e. search for the place of occurrence of destabilizing effects. The third is the establishment of the class to which the destabilizing effect belongs.

Diagnostic functional models are used to form algorithms for obtaining estimates of direct features of destabilizing effects: α_i, β_i and γ_i with the help of indirect $\Delta x_i(kT_0)$. In other words, these models are necessary for solving inverse problems, since diagnostics tasks are inverse problems in a mathematical sense. The ability to solve inverse problems is established using appropriate criteria.

Diagnostic Criteria

In order to assess the quality of diagnostic functional models which consists the possibility of indirect features $\Delta x_i(kT_0)$ unambiguously determine the estimates of direct features of destabilizing effects: $\widehat{\alpha}_i, \widehat{\beta}_i, \widehat{\gamma}_i$, appropriate tools are required (Salyga et al., 1989). Such tools for assessing the quality of diagnostic functional models are called diagnostic criteria (Shi & Zhou, 2009). The first results on the criteria for diagnosability of linear continuous dynamic systems were published in (Cordier et al., 2006). Subsequently, the design criteria presented in (Tanaka et al., 1987) were formed. So, diagnosability is a property of diagnostic functional models which characterizes the possibility of unambiguously determining the direct features of each destabilizing effect $\alpha_i \in D$ by the indirect features available for observation during a finite time interval (Eriksson et al., 2013).

The criterion of complete structural diagnosability is used for evaluation in diagnostic functional models (DFM) of unambiguous connections between direct and indirect features (Kulik, 1992).

It is necessary and sufficient that the matrices to complete structural diagnosability of the DFM

$$L_i = \begin{bmatrix} A_i & B_i \\ C_i & N_i \end{bmatrix}, \; i = \overline{1, \mu}, \tag{3}$$

are linearly independent in all pairwise combinations.

Diagnosing the components of the ORA is possible only in the mode of operation caused by the supply of working or test control actions, as well as the presence of non-zero conditions. These effects should provide signal diagnosability – this is the property of input signals that allow one to unambiguously obtain estimates of any direct indication from available measurements of indirect signs during a finite time interval. The criterion of complete signal diagnosability is used to assess the presence of this property.

It is necessary and sufficient for the complete signal diagnosability of the DFM and fully structurally diagnosable that the vectors

$$L_i^* \upsilon (kT_0) = \begin{bmatrix} A_i & B_i \\ C_i & N_i \end{bmatrix} \begin{bmatrix} \Delta x_i (kT_0) \\ u(kT_0) \end{bmatrix}, \; i = \overline{1, \mu} \tag{4}$$

are linearly independent in all pairwise combinations.

The use of diagnosability criteria allows not only to assess the diagnostic properties of the structure and the DFM signals but also to purposefully make changes in the DFM that provide complete diagnosability for each $\alpha_i \in D$.

Diagnostic logic models. In order to ensure the shortest possible time of adapt control systems to changing operating conditions, it is required to choose from all indirect features $\Delta y_i(kT_0)$, $i = \overline{1, q}$ a minimum population, processing which can form a balanced diagnostic procedure.

The solution of these problems is based on the use of Boolean variables, the corresponding tables connecting direct features $\Delta \lambda_i$ with indirect ones in the form of Boolean variables z_j and the method of their minimization. Solving these problems allowed us to form a series of diagnostic logic models and a method for their minimization (Katipamula & Brambley, 2005). The simplest diagnostic logic model (DLM) is a table (Kulik et al., 1990) that reflects the connection of direct features with indirect ones using Boolean variables (Table 1).

The table variable takes the value «1» if there is a connection between the direct feature and the indirect feature. Otherwise, it takes the value «0». More complex DLM reflect the association of pairwise combinations of rows in Table 1 with features. In order to minimize various DLM the following recurrent procedure based on the optimality principle and the Bellman method is used (Bellman,1990).

$$M_1 = \{z_{j1}\};$$

$$M_2 = \{\min_{\{z_{j2}\}}[\{z_{j2}\} \wedge M_1]\},$$

(5)

where z_{ji} –features z_j that contains «1» in the i-th row of the table; M_1 – a set of indirect Boolean features that contains «1» in the first row of the table; $\wedge$ – conjunction symbol; M_2 – a set of minimal aggregates in the second step of the procedure.

At the last q-th step of the recurrent procedure all minimal conjunctive forms are obtained using the equation

$$M_q = \{\min_{\{z_{jq}\}}[\{z_{jq}\} \wedge M_{q-1}]\}.$$

(6)

Table 1. The simplest diagnostic logic model

$\Delta \lambda$	Indirect Features				
	z_1	...	z_j	...	z_m
$\Delta \lambda_1$	υ_{11}	...	υ_{1j}	...	υ_{1m}
$\vdots$	$\vdots$	$\vdots\vdots\vdots$	$\vdots$	$\vdots\vdots\vdots$	$\vdots$
$\Delta \lambda_i$	υ_{i1}	...	υ_{ij}	...	υ_{im}
$\vdots$	$\vdots$	$\vdots\vdots\vdots$	$\vdots$	$\vdots\vdots\vdots$	$\vdots$
$\Delta \lambda_q$	υ_{q1}	...	υ_{qj}	...	υ_{qm}

The choice of a rational set of Boolean signs is made on the basis of the conditions of a specific task of diagnosing through the technical and economic analysis of the obtained minimum number of sets M_q and the selection of the suitable rational one. A rational set of minimal Boolean features allows to create a new canonical DLM with which the production knowledge base and the structure of the dichotomic search tree of direct features $\Delta\lambda_i$ using indirect z_j, obtained using two-digit predicate equations from DFM.

Production Diagnosis Model

In order to formalize knowledge in destabilizing effects products based on rules that allow knowledge to be presented in the form of «if ... then ...» sentences are used. Two-digit predicate equations of this type are used to form a set of products:

$$z_i = S_2\left[\Delta\lambda_i\left(kT_0\right)\right] = \begin{cases} 1, & if\ \Delta\lambda_i\left(kT_0\right) \in \Omega_i; \\ 0, & if\ \Delta\lambda_i\left(kT_0\right) \notin \Omega_i, \end{cases} \tag{7}$$

where $S_2\left[\Delta\lambda_i\left(kT_0\right)\right]$ – symbol of a two-digit predicate equation; $\Delta\lambda_i\left(kT_0\right)$ – the current value of the i-th direct destabilizing feature; Ω_i – set of values of the i-th feature. Knowledge of the direct features of destabilizing effects is formed on the basis of the DFM that meets the criteria of structural and signal diagnosability, as well as minimized by the number of indirect features and presented in a convenient canonical form by the following system of equations:

$$\Delta x_i'\left[\left(k+1\right)T_0\right] = A'\Delta x_i'\left(kT_0\right) + \left[A'\Delta x_i'\left(kT_0\right) + B_i'u\left(kT_0\right)\right]\Delta\lambda_i;\ \Delta x_i'\left(0\right) = x_{i0}';$$
$$\Delta y_i'\left(kT_0\right) = C'\Delta x_i\left(kT_0\right) + \left[C_i'x'\left(kT_0\right)\right]\Delta\lambda_i, \tag{8}$$

where symbol «'» means new variables and matrices obtained from system (2) and satisfying the described properties.

In the presented system of algebraic equations, the unknown parameter will be $\Delta\lambda_i$. From all the components of the vector equations of system (2), the simplest in structure scalar equation is selected, from which the algorithm for finding values $\Delta\lambda_i\left(kT_0\right)$ for a number of quantization cycles is formed $k = \overline{1, m}$.

More detailed two-digit predicate equations have the form

$$z_i = S_2\left\{\left|\Delta\lambda_i\left(kT_0\right)\right| \geq \delta_{0i}\right\};\ k = \overline{1, n_{0i}};\ \rho_{0i} = 0,9, \tag{9}$$

where δ_{0i} – tolerance for deviation; n_0 – number of measurements per diagnostic interval; ρ_{0i} – confidence coefficient.

Or:

$$z_t = S_2 \left\{ \Delta\lambda_i - \left| \Delta\lambda_i \left[(k+1)T_0 \right] - \Delta\lambda_i \left(kT_0 \right) \right| \right\}; \quad k = \overline{1, n_{0i}}; \quad \rho_{0i} = 0,85. \tag{11}$$

Other forms of representation of a two-digit predicate equation are possible.

A method of ordering products is used for the formation of correct logical conclusions – two-digit predicate equations using dichotomous trees. A dichotomous tree is a means of a hierarchical organization of knowledge, where each node contains specific knowledge and references are left or right descendants. The node at the very top is called the root. Nodes without descendants – leaves. The organization of knowledge through dichotomous trees can significantly reduce the time of formation of the current diagnosis. The search for knowledge in the linear structures of knowledge is carried out by sequential enumeration of all elements of this structure. Searching the tree does not require brute force search; therefore, it takes much less time, thus ensuring prompt diagnosis.

In order to ensure the efficiency of diagnosis dichotomous trees are balanced according to various criteria which allows to create the rational tree structure that meets the requirements of the speed of diagnostic processes.

The nodes of a dichotomous tree are formed using two-digit predicate equations for the current values of direct features of destabilizing effects. At the root of a dichotomous tree is knowledge about the detection of a destabilizing effect. Next is the knowledge of the place of his appearance. Below are the predicate equations for the class and specific types of destabilizing effects. Leaves of a tree are specific physical types of destabilizing effects. In the predicate equations (9) and (10), instead of a direct feature $\Delta\lambda_i \left(kT_0 \right)$, its expressions are used, derived from the structure of the simplest scalar equation connecting the indirect feature – the vector-accessible component $\Delta y_i' \left(kT_0 \right)$ with the direct feature $\Delta\lambda_i$.

Means of Operability Recovery

The ORA diagnosability property is only one property of an object that is necessary for adaptation. The cause of destabilization identified in the process of diagnosing is required to be parried by using appropriate excess funds to achieve the main goal of rational adaptation - ensuring the operability of the ORA. Recoverability is an ORA property that characterizes the possibility of its transfer from an inoperable state to a working state by means of parrying destabilizing effects $\alpha_i \in D$ on a finite time interval. ORA is considered to be repairable if the recovery means are formed, allowing to compensate for the effect of direct features $\Delta\lambda_i$, $i = \overline{1, q}$ so that $\Delta y_i \left(kT_0 \right) \to 0$ in the final allowable time interval. The selection of redundant recovery tools is based on the current knowledge base. The base of knowledge about the means of operability recovery can be represented using Table 2, where horizontally placed types of destabilizing effects $d_i, i = \overline{1, q}$, and vertically - the means of operability recovery used to parry them $v_j, j = \overline{1, \mu}$.

When designing the OPA, such means are chosen that can parry several species d_i, which is reflected by a variable σ_{ij} taking the value «1» if it is possible to neutralize the type of destabilization d_i with the help of a recovery tool v_j, or the value «0» if it is not possible.

Parameters $l_j, j = \overline{1, \mu}$ numerically equal to the number «1» in the column characterize the recovery rank. The more types of destabilization d_i can parry by using the tool v_j, the higher its rank.

Table 2.The base of knowledge about the means of operability recovery

Kinds	Recovery Means				Level
	V_1	V_2	...	V_μ	
d_1	σ_{11}	σ_{12}	...	$\sigma_{1\mu}$	c_1
d_2	σ_{21}	σ_{22}	...	$\sigma_{2\mu}$	c_2
$\vdots$	$\vdots$	$\vdots$	$\vdots\vdots\vdots$	$\vdots$	$\vdots$
d_i	σ_{q1}	σ_{q2}	...	$\sigma_{q\mu}$	c_q
Rank	l_1	l_2	...	l_μ	

The parameter c_i numerically equal to the sum «1» in the line, characterizes the level of recoverability d_i of the destabilization type. The more c_i, the more money can be recovered. The table is formed from the condition of satisfying the following criteria for recoverability: $\forall c_i \geq c_T, i = \overline{1,q}$ and $\forall l_j \geq l_D, j = \overline{1,\mu}$. Here c_T is the required level of recoverability, and l_D is the allowable rank of the recovery tool.

In the process of evolution of the system of rational adaptation of table. 2 should reflect the current state of the cash recovery, i.e. each time a recovery tool is selected and used v_j the corresponding column is deleted from the table. In essence, the table represents the dynamic structure of the knowledge base of the recovery tools. Production rules in the equivalent format of the production knowledge base on the recovery of working capacity are formed according to the lines of the table like this: «if the diagnosis is d_i, then the means of recovery correspond to it, for which $\sigma_{ij} = 1$». The means of recovery for the current situation is selected by analyzing the ranks of the means of recovery according to the following rule: «if the diagnosis is α_i and among all the means for which $\sigma_{ij} = 1$, l_j has the minimum value, then a means of recovery is chosen v_j».

In order to fully recover the operability of the ORA, a procedure is required to introduce the selected mean in the process of parry α_i the type of stabilization, in order to ensure both the sustainability of the transfer process from the inoperative state to the operational state and its quality.

Signal or parametric adjustment, reconfiguration of algorithms and devices can be acceptable means of recovery for many objects. The transition based on the functions of A.M. Lyapunov is needed for the formation of procedures that ensure the stability and quality of recovery processes. The specificity of the recovery tools generates the corresponding methods for the synthesis of recovery algorithms.

The recovery process is described by the following system of recurrent equations:

$$\Delta y_i \left[(k+1)T_0 \right] = R\Delta y_i \left(kT_0 \right) + h\varphi \left[\sigma \left(kT_0 \right) \right];$$
$$\sigma \left(kT_0 \right) = c^T \Delta y_i \left(kT_0 \right); \Delta y_i \left(0 \right) = \Delta y_{0i},$$

(11)

where $\Delta y_i\left(kT_0\right)$ –deviations vector of the adaptation object caused by destabilizing effects $d_i \in D$; R, h, c – matrices of corresponding dimensions; $\varphi\left[\delta\left(kT_0\right)\right]$ – nonlinear discrete scalar function.

The synthesis problem of recovery algorithms is reduced to the problem of synthesis of equations for nonlinear discrete systems described by equations (11) using the discrete analogue of the direct method A.M. Lyapunov based on the task of a special auxiliary function $V\left[\Delta y_i\left(kT_0\right)\right]$ and the formation by means of equation (11) of a function $V\left[kT_0,\left(k+1\right)T_0\right]$ that satisfies the condition

$$\Delta V\left[kT_0,\left(k+1\right)T_0\right] = V\left[\Delta y\left(k+1\right)T_0\right] - V\left[\Delta y\left(kT_0\right)\right], \tag{12}$$

where $V\left[\Delta y\left(kT_0\right)\right]$ and $V\left[\Delta y\left(k+1\right)T_0\right]$ – definitely positive functions for discrete argument values kT_0 and $(k+1)T_0$.

An algorithm is formed from the providing condition a certain negativity of functions, and it ensures the asymptotic stability of the recovery process.

The described tools allow to formalize the implementation of individual stages of the process of designing a rational adaptation of motion control systems of autonomous aircraft.

ILLUSTRATIVE EXAMPLE

The developed design tools were used in research on the formation of technology that ensures the units rational adaptation of motion control systems for a number of aircraft.

Flying models are used in the process of designing new aircraft. The flying model is a special-purpose unmanned flying vehicle designed for autonomous flight according to a given program and providing the possibility of obtaining and recording flight data. Such a flying model is a research tool for studying the most dangerous flight regimes and phenomena that are associated with the risk of losing the aircraft. On flying models, fundamental decisions are made that are made at all stages of the aircraft life cycle, from the development of the aircraft design concept to the formation of its appearance and to the implementation of the mass-produced aircraft modifications.

Flying models are equipped with on-board equipment and equipment that implements automatic control of aerodynamic surfaces in accordance with the flight program to provide autonomous flight. Information about the angular position of the flying model relative to the center of mass comes from the sensor unit, located along the axes of the associated Cartesian coordinate system. The design of the sensor unit and its installation on the flying model allows to obtain information about the angular evolution of the three channels. Each channel of the block contains one angle sensor and two angular velocity sensors. In total, the unit uses 3 angle sensors and 6 angular velocity. This unit is shown in figure 3. Two types of channel-by-channel redundancies are incorporated in the presented sensor layout scheme. The first one is instrument, as each channel uses two angular velocity sensors. The second is functional, due to the integral relationship between the angle and the angular velocity.

In order to test the technology of rational adaptation of the sensor unit, a research bench was developed for the course channel sensors in accordance with the functional diagram that is shown in figure 4.

Figure 3. Placement of sensors in the unit

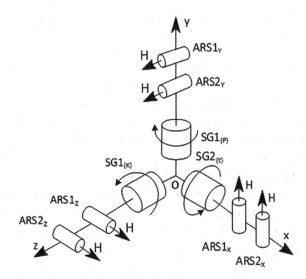

Figure 4. Functional diagram of the course channel

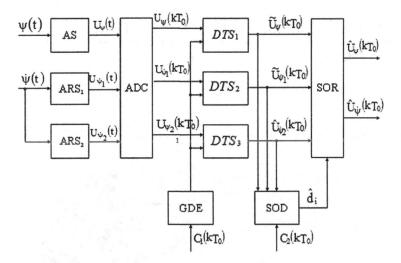

The voltage from the angle and angular velocity sensors, digitized by the ADC, is fed to the corresponding blocks destabilization type simulator (DTS_i), $i = \overline{1,3}$, which perform the function of imitating destabilization types by commands from the destabilizing effects generator (GDE). DTS_i in conjunction with GDE according to certain scenarios set by the team $C_1(kT_0)$, the signals from the sensors are deformed. The deformed signal, corresponding to the effect of the current destabilizing effect on the sensor, enters the sensor operability diagnostic unit (SOD), where, by indirect features, the cause of the malfunction is detected and the diagnosis is formed $\widetilde{U}_i$. SOD is included in teamwork $C_2(kT_0)$. The result of diagnosis – the diagnosis enters the unit that performs the sensor operability recovering function (SOR), where, in accordance with the diagnosis and with the help of available means, the signal deformed by destabilizing effect is recovered and the estimated voltage values are generated $\tilde{U}_\psi\left(kT_0\right)$ and $\tilde{U}_{\dot{\psi}}\left(kT_0\right)$.

Figure 5 presents a general view of the stand for carrying out and testing technological experiments. In the process of research, 9 types of destabilization were simulated for each sensor, i.e. 36 types for all course channel sensors. Specifically: the type of destabilization α_1 is zero disposable positive drift; α_2 – zero drift negative disposable; α_3 – zero drift is both positive and unrecoverable negative; α_4 – reduction of the sensor coefficient is removable; α_5 – reduction of the coefficient of the sensor is unavoidable; α_6 – breakage of the sensor signal wire; α_7 – wire breakage of the negative power supply; α_8 – wire breakage positive power; α_9 – unknown type of destabilization.

The development of diagnostic support for the pitch channel with the help of the diagnostic tools described earlier made it possible to form a structured production base of diagnostic knowledge for the search for a diagnosis. Figure 6 shows the dichotomous tree of searching for types of destabilization in the course channel sensors. In Figure 6, S_1, S_2, S_3 are failure detection rates. Predicate equations ($z_1, z_2, ..., z_{12}$) are used in the nodes of the tree and are obtained by solving the corresponding diagnostic problems. So, for example, the predicate argument for detecting the inoperability of sensors is described by the following equations:

$$\Delta\Psi\left(kT_0\right) = \frac{\tilde{U}_\psi\left[(k+1)T_0\right] - \tilde{U}_\psi\left(kT_0\right)}{\eta_\psi''{}_0} - \frac{1}{2}\left[\frac{U_{\dot{\psi}1}\left(kT_0\right)}{\eta_{\dot{\psi}1}} + \frac{U_{\dot{\psi}2}\left(kT_0\right)}{\eta_{\dot{\psi}2}}\right];$$

Figure 5. General view of the stand

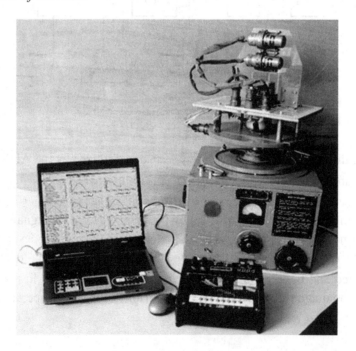

Figure 6. Dichotomous tree of diagnosis

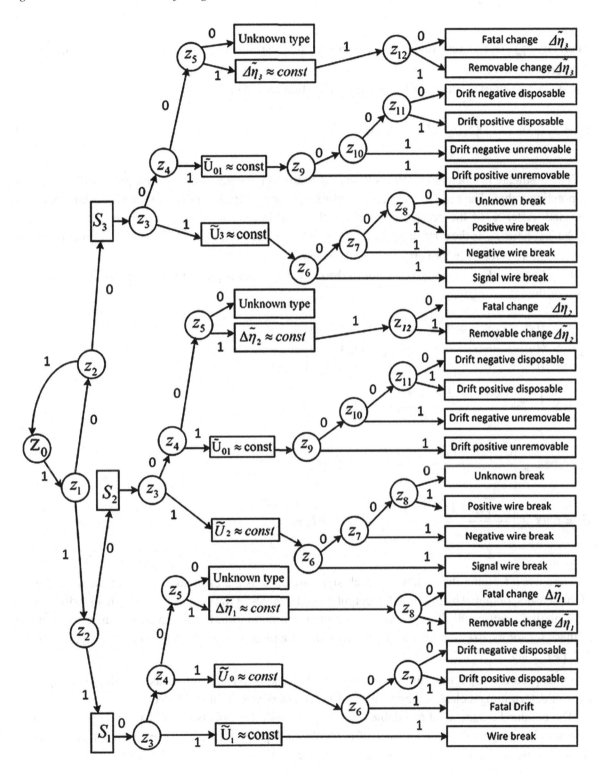

$$\delta_0 = \max\nolimits_{nom} \left| \Delta\Psi\left(kT_0\right) \right|, \tag{13}$$

where $\eta_\psi, \eta_{\dot\psi 1}, \eta_{\dot\psi 2}$ – transmission coefficients of the respective sensors; δ_0 – the maximum permissible error between the readings of the sensors in the nominal operation mode.

Such a two-digit predicate is formed from the obtained relations:

$$z_0 = S_2 \left\{ \left| \Delta\psi\left(kT_0\right) \right| > \delta_0 \right\}; \ k = \overline{1, n}, \ \rho_0 = 0,9. \tag{14}$$

The predicate is calculated n times, and as a result of the calculation at each point will be the corresponding value. The counter counts the number of units – n_0, i.e. points at which destabilization is detected. After n_0 of the comparisons made, the final detection of destabilization is carried out, if $n_0 \geq n\rho_0$, here ρ_0 is the confidence coefficient, which allows to discard «false» measurements, «zero» points, ignore errors, random effects and other factors.

Localization of an inoperative sensor in the course channel is performed using two predicate equations of the following type:

$$z_1 = S_2 \left\{ \left| \frac{\tilde{U}_\psi\left[(k+1)T_0\right]}{\eta_\psi} - \frac{1+T_0}{\eta_{\dot\psi 1}}\tilde{U}_{\dot\psi 1}\left(kT_0\right) < \delta_1 \right| \right\};$$

$$z_2 = S_2 \left\{ \frac{\tilde{U}_\psi\left[(k+1)T_0\right]}{\eta_\psi} - \frac{1+T_0}{\eta_{\dot\psi 2}}\dot\psi_2\tilde{U}_{\dot\psi 2}\left(kT_0\right) < \delta_2 \right\};$$

$$\delta_1 = \max\nolimits_{nom} \left| \frac{\tilde{U}_\psi\left[(k+1)T_0\right]}{\eta_\psi} - \frac{1+T_0}{\eta_{\dot\psi 1}}\tilde{U}_{\dot\psi 1}\left(kT_0\right) \right|;$$

$$\delta_2 = \max\nolimits_{nom} \left| \frac{\tilde{U}_\psi\left[(k+1)T_0\right]}{\dot{\ }_\psi} - \frac{1+T_0}{\dot{\ }_{\dot\psi 2}}\dot\psi_2\tilde{U}_{\dot\psi 2}\left(kT_0\right) \right|. \tag{15}$$

Here when forming the arguments, such signs are used as the estimated values of the input signals of the sensors, obtained by dividing the output discrete values of the signals by the corresponding sensor transmission coefficients. The procedure of numerical integration is used to ensure the comparability of these sensor localization features by the rectangle method for signals of the angular velocity sensors $\tilde{U}_{\dot\psi 1}\left(kT_0\right)$ and $\tilde{U}_{\dot\psi 2}\left(kT_0\right)$.

Other predicate equations for the nodes of a dichotomous tree are formed as a result of solving problems of determining a class and establishing the type of destabilizing influence.

The obtained diagnosis of avoidable causes of destabilizing effects allows us to proceed to the next technological stage - the development of algorithmic support for the recovery process of the course channel sensors. The following redundant means are available for the considered sensor unit fragment:

1) **Instrumentation**: 2 TUS.

2) **Functional**: $\dot{\psi}(t) = \dfrac{d\psi(t)}{dt}$

3) **Algorithmic**: Signal and parametric adjustments.

All types of destabilization (Figure 6) based on the available surplus funds can be divided into the following three groups. The first group will be: unknown type, wire breakage, non-removable drift, fatal change $\Delta\eta_i, i = \overline{1,3}$. The second group will include recoverable drifts, both positive and negative. The third group is a removable change $\Delta\eta_i, i = \overline{1,3}$.

A sample of the table 2 is drawn up from the following table presented in figure 7 for the analysis of the properties of the recoverability of a block fragment.

The table uses the following notation. For AS: $\Delta\lambda_1$ – this is the first group of destabilization; $\Delta\lambda_2$ – the second group and $\Delta\lambda_3$ – the third group. For ARS1: $\Delta\lambda_4$ – the first group of destabilization; $\Delta\lambda_5$ – the second group and $\Delta\lambda_6$ – the third group. For ARS2: $\Delta\lambda_7$ – the first group, $\Delta\lambda_8$ – the second and $\Delta\lambda_9$ – the third group of destabilization. Recovery tools are used as follows: V_1 – it is ARS1; V_2 – ARS2; V_3 – signal adjustment; V_4 – parametric adjustment and V_5 – AS (Figure 7). If enter a valid rank of recoverability $l_d = 2$, then the criterion of the rank of recoverability $\forall l_j \geq l_d$ will be fulfilled for all $j = \overline{1,5}$. If the required level of recoverability is determined as $c_T = 1$, then by the criterion of the level of recoverability $\forall c_i \geq c_T$, the fragment is recovered for each $i = \overline{1,9}$.

Figure 7. The base of knowledge about the means of operability recovery

S		P	V₁	V₂	V₃	V₄	V₅	Level
	A {	$\Delta\lambda_1$	1	1	0	0	0	2
		$\Delta\lambda_2$	0	0	1	0	0	1
		$\Delta\lambda_3$	0	0	0	1	0	1
S1	AR {	$\Delta\lambda_4$	0	1	0	0	1	2
		$\Delta\lambda_5$	0	0	1	0	0	1
	ARS2 {	$\Delta\lambda_6$	0	0	0	1	0	1
		$\Delta\lambda_7$	1	0	0	0	1	2
		$\Delta\lambda_8$	0	0	1	0	0	1
		$\Delta\lambda_9$	0	0	0	1	0	1
		Rank	2	2	3	3	2	

Types of destabilization of the first group indicate that the sensor is completely inoperable and therefore for a heading sensor, for example, it can be recovered to a measurement corresponding to the current heading of a flying model using the functional link between the heading sensor measurements and one of the angular velocity sensors, namely $\psi(t) = \int_{t_0}^{t} \dot{\psi}(t)\,dt$. Using the Euler formula, we represent the relation in digital form:

$$\dot{\psi}(t) \approx \frac{\psi\left[(k+1)T_0\right] - \psi\left(kT_0\right)}{T_0} = \psi_1\left(kT_0\right), \tag{16}$$

then, to obtain the estimated values of the angle using numerical integration, we obtain this ratio:

$$\widehat{\psi}\left[(k+1)T_0\right] = \widehat{\psi}\left(kT_0\right) + T_0\psi_1\left(kT_0\right) \tag{17}$$

Taking advantage of the linear relationship between the output and input signals of the sensors, we get the following recurrence equation:

$$\widetilde{U}_\psi\left[(k+1)T_0\right] = \widetilde{U}_\psi\left(kT_0\right) + T_0\frac{\eta_1}{\eta_2}U_{\dot{\psi}}\left(kT_0\right). \tag{18}$$

If use the output signal of the second ARS, then

$$\widetilde{U}_\psi\left[(k+1)T_0\right] = \widetilde{U}_\psi\left(kT_0\right) + T_0\frac{\eta_1}{\eta_3}U_{\dot{\psi}2}\left(kT_0\right). \tag{19}$$

With the inoperability of one of the ARS, its measurements can be recovered both by connecting to the second ARS that is able to work, and using numerical differentiation of the signal from the heading sensor.

$$\widetilde{U}_{\dot{\psi}1}\left[(k+2)T_0\right] = \frac{\eta_2}{\eta_1}T_0\left[U_\psi\left(k+1\right)T_0 - U_\psi\left(kT_0\right)\right]. \tag{20}$$

The parameters $\Delta\lambda_2$, $\Delta\lambda_5$ and $\Delta\lambda_8$ (Table 3), as a result of the diagnosis, the estimated drift values U_0^i, $i = \overline{1,3}$ and the drift sign «+» or «-» are obtained for the drift group. Measurements recovery of the sensor inoperability $\tilde{U}_i\left(kT_0\right)$ with this type of destabilization is performed using the means of signal adjustment according to the following algorithm:

$$\widetilde{U}_i\left(kT_0\right) = \widetilde{U}_i\left(kT_0\right) \mp \widetilde{U}_0^i, \; i = 1,3. \tag{21}$$

The third group of types of destabilization is a disposable deviation. In table 3 are direct features of the third group: and. The correct measurements of an inoperative sensor may be recovered by using a tool such as a parametric adjustment – V_4 (Figure 7). Consider the features of the formation of the parametric adjustment algorithm for remote control. The estimated value of the voltage of the angle sensor can be obtained, for example, for an event of a decrease in the sensor transmission coefficient by an amount, using this ratio

$$\widetilde{U}_\psi\left(kT_0\right)=\tilde{\eta}_1\psi\left(kT_0\right)+\Delta\tilde{\eta}_1\psi\left(kT_0\right).$$

(22)

If take into account that the first term describes the output signal of an inoperative sensor – $\tilde{U}_1\left(kT_0\right)$ and the angle $\psi\left(kT_0\right)$ in the second term can be expressed as $\dfrac{\tilde{U}_1\left(kT_0\right)}{\tilde{\eta}_1}$, then the estimated value of the output signal is determined by such an expression

$$U_\psi\left(kT_0\right)=\tilde{U}_\psi\left(kT_0\right)+\Delta\eta_1\frac{\tilde{U}_\psi\left(kT_0\right)}{\tilde{\eta}_1}.$$

(23)

Let the coefficient $\tilde{\eta}_1$ are represented as $\tilde{\eta}_1=\eta_{1n}+\Delta\eta_1$ substituting in the expression (23) and making the conversion

$$\widetilde{U}_\psi\left(kT_0\right)=\tilde{U}_\psi\left(kT_0\right)+\Delta\eta_1\frac{\tilde{U}_\psi\left(kT_0\right)}{\eta_{1n}-\Delta\eta_1}=\left(1+\frac{\Delta\eta_1}{\eta_{1n}-\Delta\eta_1}\right)\tilde{U}_\psi\left(kT_0\right)=$$
$$=\left(\frac{\eta_{1n}}{\eta_{1n}-\Delta\eta_1}\right)\tilde{U}_\psi\left(kT_0\right)=\frac{\eta_{1n}}{\eta_{1n}-\Delta\eta_1}\tilde{U}_\psi\left(kT_0\right).$$

(24)

As a result of the transformations, the algorithm for parametric adjustment of the output signal by the inoperability of the angle sensor was obtained in this form:

$$\widetilde{U}_\psi\left(kT_0\right)=\frac{\eta_{1n}}{\eta_{1n}-\Delta\eta_1}\tilde{U}_\psi\left(kT_0\right).$$

(25)

Similarly for ARS1

$$\widetilde{U}_{\dot\psi1}\left(kT_0\right)=\frac{\eta_{2n}}{\eta_{2n}-\Delta\eta_2}\tilde{U}_{\dot\psi1}\left(kT_0\right)$$

(26)

and for ARS2

$$\tilde{U}_{\psi 2}\left(kT_0\right) = \frac{\eta_{3n}}{\eta_{3n} - \Delta\eta_3}\tilde{U}_{\psi 2}\left(kT_0\right). \tag{27}$$

Enlarged basic functions to operability recovery of the sensor unit fragment are shown in figure 8.

As a result of the diagnosis, it is detected when destabilization has occurred, then where the destabilization occurred, in which sensor, after the destabilization class is established and determining the type of destabilization, the process of forming the characteristics of the inoperable sensor state is completed. The task of the «Choice of recovery tools» functional block is to implement the correspondence functions presented in table 3, in other words, in choosing a unit that implements the function of recovering measurements of an inoperative sensor. In table 3 six correspondence functions are single-valued and only three are two-valued. Therefore, for features $\Delta\lambda_1$, $\Delta\lambda_4$ and $\Delta\lambda_7$ the choice of recovery means is more complex – two-stage, depending on the technical condition of the corresponding means of recovering. So, if destabilization appeared in the AS, described by the feature $\Delta\lambda_1$, and the means V_1 has already undergone earlier destabilization and has been recovered, then the means is selected V_2, i.e. ARS2. In a situation where the means V_2 – ARS2 was previously inoperable, and then recovered, it is more expedient to choose the means V_1, i.e. ARS1. If both means were previously workable V_1 and V_1 are equal, then one of them is chosen. Similarly, the choice of means of recovery for the features $\Delta\lambda_4$ and $\Delta\lambda_7$. Thus, the algorithms for choosing the means of operability recovery are formed on the basis of the correspondence functions and the technical state of the used two-level means of operability recovery.

Figure 8. Functional diagram of the operability recovery process of the course channel sensors

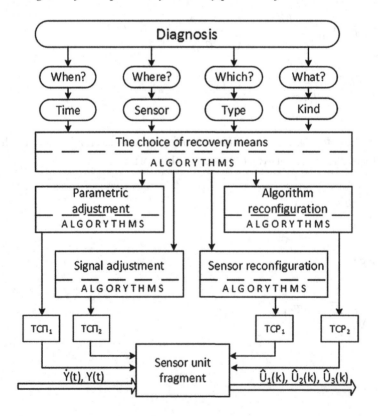

Subsequent blocks of the functional circuit are implemented using the previously described algorithms of the functions of the restoration of working capacity through parametric and signal adjustments, as well as reconfiguration of algorithms and sensors. Built-in technical means of adjustment – TMA_1 and TMA_2, as well as technical means of reconfiguration – TMR_1 and TMR_2 are used to change the functional state of inoperative sensors. In the aggregate, a sensor unit fragment, diagnostic tools and operability recovery capacity represent a new fragment in which a rational adaptation of sensors under conditions of uncertain destabilizing effects is implemented.

At the stand (Figure 5) experimental studies of the developed technology of rational adaptation were carried out. The results of experiments in the normal operation of the sensors in the course channel are shown in figure 9. The left row of graphs reflects the nature of the change in the angle – $\psi(t)$ and the angular velocity – $\dot{\psi}(t)$ of the turntable on which the sensors are installed. The right-hand row of graphs is represented by the estimated values of the angular velocity $\hat{\dot{\psi}}(t)$ obtained by numerical differentia-

Figure 9. Normal operation of sensors

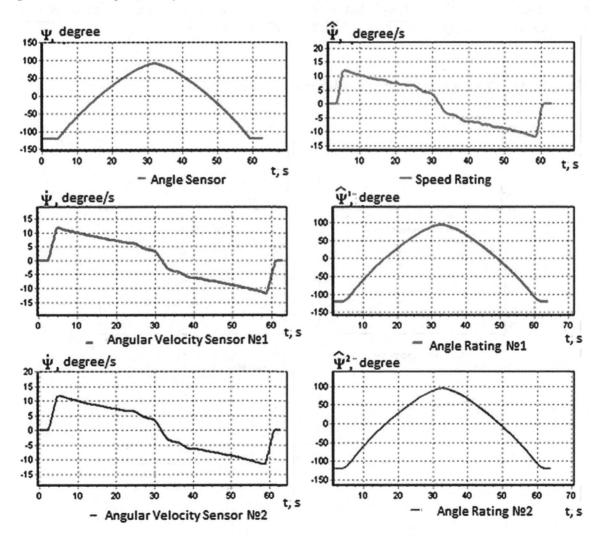

tion of the output signal AS $- U_\psi(t)$. The estimated values of the angle $\widehat{\psi}_1(t)$ and $\widehat{\psi}_2(t)$ obtained as a result of appropriate processing of the output signals $U_{\psi 1}(t)$ and $U_{\psi 2}(t)$. It is obvious that the characters of the input signals and their estimates from the output signals of the sensors indicate their satisfactory coincidence.

The results of destabilization are presented in figure 10. Programmatically, on 15p, a break in the AS of the positive power wire was simulated, as indicated by the graphs $\psi(t)$ and $\widehat{\psi}(t)$. The results of the recovery of the AS measurements are presented in figure 11 using appropriate processing of the output signal with ARS1 $- U_{\psi 1}(t)$ and also presents the results of the speed estimation $\widehat{\psi}(t)$, indicating a satisfactory quality of the recovery of the output signal of the AS when the positive supply wire is broken.

Figure 12 presents the results of destabilization and restoration of measurements of ARS1 in destabilization in the form of unavoidable zero drift by 9 seconds of operation that is presented in figure 13.

The signals and estimates of the input signals of the sensors are shown in figure 14 during destabilization by 5 seconds in ARS2 in the form of a decrease in its transmission coefficient. The bottom row of graphs reflects their significant deformation as compared with the normal mode of operation of ARS2. The process of recovering measurements of ARS2 is reflected in figure 15.

Figure 10. The AS wires positive power break (destabilization)

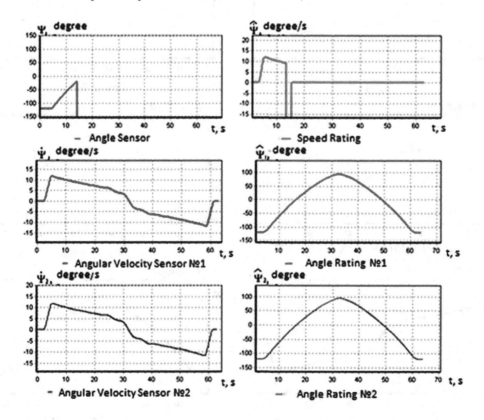

Figure 11. The AS wires positive power break (recovery)

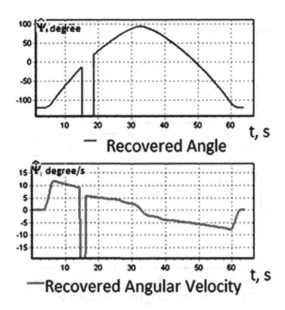

Figure 12. Fatal zero drift in ARS1 (destabilization)

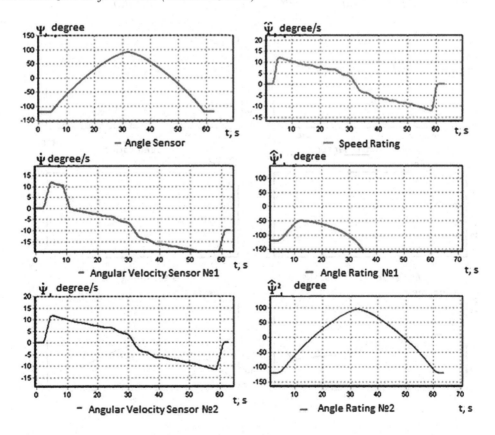

Figure 13. Fatal zero drift in ARS1 (recovery)

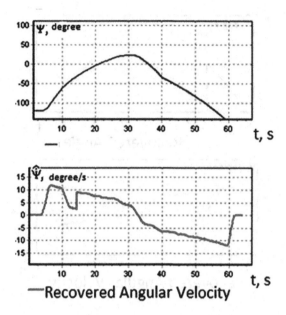

Recovered Angular Velocity

Figure 14. Reduction of the transmission coefficient in ARS2 (destabilization)

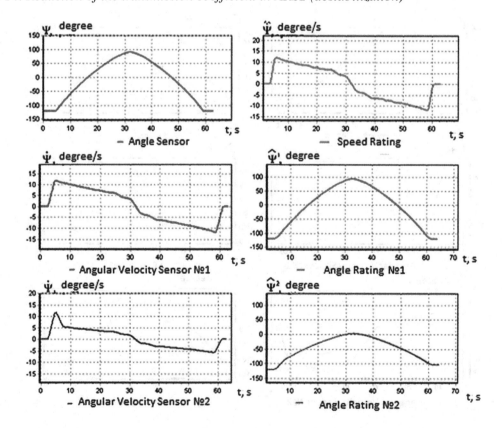

The results of experimental studies indicate the fundamental possibility of the proposed technology of rational adaptation to ensure the operability of the sensor unit of flying models, under the conditions of eventual uncertainty of destabilizing effects.

On flying models, pneumatic actuators are used as drives for controlling ailerons, stabilizers, rudders. Figure 16 presents the functional diagram of such a servo drive. The functional diagram consists of a power amplifier (PA), a steering machine (SM), two potentiometers in feedback P1, P2 and signal adders. Input driver signal – voltage $U_3(t)$ and output signal – rod movement $\mu(t)$. A research stand was developed for the pneumatic formation of the technology of rational adaptation of the servo drive. At the stand, 45 types of destabilizing effects that impair the performance of a pneumatic servo were imitated. A production knowledge base and a logical deduction about the causes of destabilization, containing 43 predicate equations, was developed and investigated. Algorithmic and software have been developed for the recovery process. The results of the bench experiments confirmed the effectiveness of the technology used rational adaptation.

Disruptions in the glider operability of flying models are associated with such destabilizing effects as loss of wing console area, partial loss of flaperones, stabilizers, and a number of other damages. Corresponding diagnostic functional models were created, production knowledge bases and inference bases were built, and algorithms for the recovery of the airframe were developed. Methods of digital modeling produced a study of the capabilities of the developed system of rational adaptation. The performed computational experiments allowed real-time identification of the cause of simulated destabilization and efficiently countering it, ensuring a satisfactory performance of the flying model glider.

Figure 15. Reduction of the transmission coefficient in ARS2 (recovery)

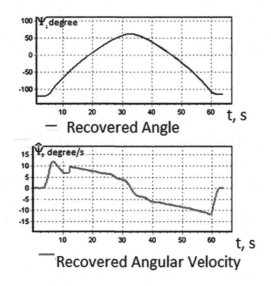

Figure 16. Functional diagram of the pneumatic servo

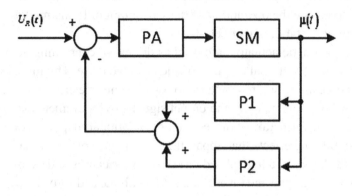

FUTURE RESEARCH DIRECTIONS

The main areas of further research in the field of rational adaptation of control systems for the autonomous aircraft motion are the following:

1. Development of the diagnostic methods for control systems whose operation is described by complex non-linear equations.
2. Development of diagnostic methods for control systems for the autonomous aircraft motion based on the principle of closed-loop diagnostics.
3. In order to ensure high speed and real-time operation it is necessary to perform optimization of the algorithmic support for diagnostic methods.
4. Perform the full-scale test of the diagnostic methods of control systems for the autonomous aircraft motion on real model of technology.

CONCLUSION

This study is focused on the use of the adaptation principle in application for organizing the closed-loop life circuit of autonomous fly vehicles.

The study formulates the main provisions of the rational control of moving objects with uncertain dynamics. On the basis of these provisions tools have been developed that can be applied at the stage of schematic design of algorithms for rational adaptation of motion control systems, including:

- **Diagnostic Functional Models.**
- **Diagnostic Criteria.**
- **Diagnostic Logic Models.**
- **Production Diagnosis Model.**
- **Operability Recovery Tools.**

The expediency of applying the proposed rational adaptation approach and the use of tools at the stage of schematic design is proved by the technological experiments results of rational adaptation of motion control systems units.

Technological experimental studies were carried out on the laboratory bench that is the hardware and software platform for the study of the measuring system control principles according to the diagnosis and methods for recovering measurements.

The obtained results allow to make a conclusion about the effectiveness of the proposed approach.

REFERENCES

Alekseev, Yu. S. Yu. M., Zlatkina, V. S., Krivtsov, A. S., Kulik, V. I., Chumachenko, et al. (2012). Designing control systems for rocket and space technology objects: In 3 volumes. Kharkiv, Ukraine: Nat. Aerospace. un-t "Kharkov. Aviation in-t", SPE Hartron-ARKOS, 2012.

Bellman, R. (1964). *Control processes with adaptation*. Moscow, Russia: Nauka.

Bellman, R. (1966). Dynamic programming. *Science, 153*(3731), 34–37. doi:10.1126cience.153.3731.34 PMID:17730601

Cordier, M. O., Travé-Massuyes, L., & Pucel, X. (2006, June). Comparing diagnosability in continuous and discrete-event systems. In *Proceedings of the 17th international workshop on principles of diagnosis (DX-06)* (pp. 55-60).

Dergachov, K., & Kulik, A. (2019). Ensuring the safety of UAV flights by means of intellectualization of control systems. In Cases on Modern Computer Systems in Aviation (pp. 287-310). Hershey, PA: IGI Global. doi:10.4018/978-1-5225-7588-7.ch012

Eriksson, D., Frisk, E., & Krysander, M. (2013). A method for quantitative fault diagnosability analysis of stochastic linear descriptor models. *Automatica, 49*(6), 1591–1600. doi:10.1016/j.automatica.2013.02.045

Kapitza, P. L. (1987). Experiment, theory, practice. In *Experiment, Theory, Practice*. Moscow, Russia: Nauka.

Kaplan, A., & Haenlein, M. (2019). Siri, Siri in my hand, who's the fairest in the land? On the Interpretations, Illustrations, and Implication of Artificial Intelligence. *Business Horizons, 62*(1), 15-26.

Katipamula, S., & Brambley, M. R. (2005). Methods for fault detection, diagnostics, and prognostics for building systems—A review, part I. *HVAC & R Research, 11*(1), 3–25. doi:10.1080/10789669.200 5.10391123

Kulik, A. (1992). Evaluation of the diagnosability of linear dynamic systems. *Avtomatika i Telemekhanika,* (1), 184-187.

Kulik, A. (2000). Signal-parametric diagnostics of control systems. Kharkiv, Ukraine: National Aerospace University «KhAI»

Kulik, A., ... (2016). Elements of the theory of rational management of objects. Kharkiv, Ukraine. National Aerospace University «KhAI»

Kulik, A., ... (2018). Rational control of objects: theory and applications. Kharkiv, Ukraine. National Aerospace University «KhAI»

Kulik, A., Sirodzha, I. B., & Shevchenko, A. N. (1990). Construction of diagnostic models in the development of diagnostic software for dynamic systems.

Kulik, A., Sirodzha, I. B., & Shevchenko, A. N. (1990). *Construction of diagnostic models in the development of diagnostic software for dynamic systems*. Kharkiv, Ukraine: Part II.

Kulik, A. S. (1991). Fault diagnosis in dynamic systems via signal-parametric approach. In *IFAC/IMACS Symp. Baden-Baden* (Vol. 1, pp. 157-162).

Lee, T. H., Adams, G. E., & Gaines, W. M. (1968). *Computer process control: modeling and optimization*. New York.

Lyapunov, A. M. (1892). The general problem of motion stability. *Annals of Mathematics Studies*, 17.

Mishkin, E., & Braun, L. (Eds.). (1961). *Adaptive control systems*. McGraw-Hill.

Rusell, S., & Norvig, P. (2003). *Artificial intelligence: A modern approach* (2nd ed.). Upper Saddle River, NJ: Prentice Hall.

Salyga, V. I., Sirodga, I. B., Kulik, A. S., & Obruchev, V. L. (1989). Synthesis of fault-tolerant dynamic control systems with fault identification. PROBL. CONTROL INF. *THEORY*, *18*(1), 43–54.

Shi, J., & Zhou, S. (2009). Quality control and improvement for multistage systems: A survey. *IIE Transactions*, *41*(9), 744–753. doi:10.1080/07408170902966344

Tanaka, S., Okita, T., & Müller, P. C. (1987). Failure detection in linear discrete dynamical systems and its detectionability. *IFAC Proceedings, 20*(5), 75-80. 10.1016/S1474-6670(17)55355-0

Tsypkin, Y. Z., & Nikolic, Z. J. (1971). *Adaptation and learning in automatic systems* (Vol. 73). New York: Academic Press.

Zadeh, L. A. (1963). On the definition of adaptivity. *Proceedings of the IEEE*, *51*(3), 469–470. doi:10.1109/PROC.1963.1852

ADDITIONAL READING

Blanke, M., Kinnaert, M., Lunze, J., Staroswiecki, M., & Schröder, J. (2006). Diagnosis and fault-tolerant control (Vol. 2). Berlin: springer.

Dergachov, K., Kulik, A., & Zymovin, A. (2019). Environments Diagnosis by Means of Computer Vision System of Autonomous Flying Robots. In Automated Systems in the Aviation and Aerospace Industries (pp. 115-137). IGI Global. doi:10.4018/978-1-5225-7709-6.ch004

Ducard, G. J. (2009). *Fault-tolerant flight control and guidance systems: Practical methods for small unmanned aerial vehicles*. Springer Science & Business Media. doi:10.1007/978-1-84882-561-1

Kulik, A., & Dergachev, K. (2016). Intelligent transport systems in aerospace engineering. In *Intelligent Transportation Systems–Problems and Perspectives* (pp. 243–303). Cham: Springer. doi:10.1007/978-3-319-19150-8_8

Pelliccione, P. (2007). *Software engineering of fault tolerant systems* (Vol. 19). World Scientific. doi:10.1142/6362

Stengel, R. F. (1991). Intelligent failure-tolerant control. *IEEE Control Systems Magazine, 11*(4), 14–23. doi:10.1109/37.88586

KEY TERMS

Autonomous Fly Vehicles: An autonomous aircraft that is capable of performing its basic functions under destabilizing effects and has a vitality.

Destabilizing Effects: Impacts that interfere with a performance of automatic control system.

Diagnosability: A property of automatic control object that allows to unambiguously establish the causes of destabilizing effects by indirectly features that available for measurements in a finite time.

Operating State of the Automatic Control System: Such a state in which the values of all parameters characterizing the ability to perform specified functions correspond to the requirements of the technical specification.

Rational Control Object: An automatic control object that has properties of controllability and observability, as well as diagnosability and recoverability.

Recoverability: A property of automatic control object that allows to use excess funds in its composition to recover an operability in a finite time.

Chapter 3

Intelligent Automated System for Supporting the Collaborative Decision Making by Operators of the Air Navigation System During Flight Emergencies

Yuliya Sikirda

ⓘ https://orcid.org/0000-0002-7303-0441
Flight Academy, National Aviation University, Ukraine

Mykola Kasatkin

Kharkiv National University of Air Forces named by I. Kozhedub, Ukraine

Dmytro Tkachenko

ⓘ https://orcid.org/0000-0001-8033-9992
Ukrainian State Air Traffic Services Enterprise (UkSATSE), Lviv, Ukraine

ABSTRACT

This chapter researches pilot and air traffic controller collaborative decision making (CDM) during flight emergencies for maximum synchronization of operators' technological procedures. Deterministic models of CDM by the Air Navigation System's human operators were obtained by network planning methods; their adequacy is confirmed by full-scale modeling on a complex flight simulator. For the sequential optimization of the collaborative two-channel network "Air traffic controller-Pilot" to achieve the end-to-end effectiveness of joint solutions, a multi-criteria approach was used: ensuring the minimum time to parry flight emergency with maximum safety/maximum consistency over the time of operators' actions. With the help of the multiplicative function, the influence of organizational risk factors on flight safety in the air traffic control was evaluated. A conceptual model of System for control and forecasting the flight emergency development on the base of Intelligent Automated System for supporting the CDM by operators was developed.

DOI: 10.4018/978-1-7998-1415-3.ch003

INTRODUCTION

It is considered that aviation is the most fail-safe type of transfer. In as little as century, aviation, in the sphere of flight safety, rose through the ranks from an unstable system to the first "ultra-safe" system in the history of transport, it means system in which the number of catastrophic failures, in the sphere of safety, make up less than one per one million of production cycles (ICAO, 2013). The year 2018 was one of the safest years ever for commercial aviation, Aviation Safety Network data show (Aviation Safety Network, 2019b). Yet, 2018 year was worse than the five-year average. Over the year 2018, the Aviation Safety Network recorded a total of 15 fatal airliner accidents, resulting in 556 fatalities. This makes 2018 the third safest year ever by the number of fatal accidents and the ninth safest in terms of fatalities. The safest year in aviation history was 2017 with 10 accidents and 44 lives lost (Aviation Safety Network, 2018). Looking at that five-year average of 14 accidents and 480 fatalities, 2018 year was worse on both accounts. Twelve accidents involved passenger flights, three were cargo flights. Three out of 15 accident airplanes were operated by airlines on the E.U. "blacklist", up by two compared to 2017. Given the estimated worldwide air traffic of about 37.800.000 flights, the accident rate is one fatal accident per 2.520.000 flights. Since 1997 the average number of aircraft accidents has shown a steady and persistent decline due to the continuing flight safety-driven efforts by international aviation organizations.

Despite the improvements in aircrafts control systems and air traffic control systems, the human factors still have a significant impact on flight safety – nearly 80% of aviation events are due to the fault of people (Friedman, Carterette, Wiener, & Nagel, 2014). The theory of human factor is gradually developing, tested and institutionalized. The evolution of the aviation system in the direction of a complex socio-technical system with gradual changes and additions to the well-known model of the human factor SHEL (1972) to date is given in documents of International Civil Aviation Organization (International Civil Aviation Organization [ICAO], 2002, 2003, 2009, 2012, 2013, 2014).

The authors distinguish five stages of the evolution of human factors models in aviation, related to the emergence of new components of the aviation system and to improve the diagnosis of Air Navigation System's (ANS's) human-operators (H-O) errors (Table 1):

Stage 1: Professional Skills / Interaction / Errors.
Stage 2: Cooperation in team / Interaction in team / Error detection.
Stage 3: Culture / Safety / Errors prevention.
Stage 4: Safety / Efficiency / Minimization of errors.
Stage 5: Artificial Intelligence Systems / Analyze, detection, prevention, and minimization of errors.

For today, the key to ensuring the safety of flights is the problem of the organization of Collaborative (joint) Decision Making (CDM) by all the operational partners – airports, air traffic control services, airlines and ground operators – on the basis of general information on the flight process and ground handling of the aircraft in the airport (ICAO, 2014).

The Global Operating Concept for Air Traffic Management (ATM) (ICAO, 2005) provides for the provision of a joint (pilot – air traffic controller (ATCO)) decision making (DM) air traffic control unit based on a dialogue between them and real-time information evaluation at all stages of the flight.

The lives of air passengers in the sky and people on the ground depending on the adequate interaction between the pilot and ATCO. According to the statistics of the Aviation Safety Network (Aviation Safety Network, 2019a), during the second half of the 20th century due to problems in interaction pilot

Table 1. Evolution of human factors models

Year	Model	Content of Model	Content of Stage of Evolution of Human Factors Model	Number of Stages
1972	SHEL	Software (procedures) - Hardware (machines) - Environment - Liveware	Professional skills Interaction Errors	I
1990	Reason's "Swiss Cheese Model"	Active errors - Latent errors - Windows of opportunity - Causation chain		
1993	SHELL	Software (procedures) - Hardware (machines) - Environment - Liveware - Liveware (humans)		
1999	CRM	Crew - Resource - Management	Cooperation in team Interaction in team Error detection	II
2000	TEM	Threat and Error - Management		
2000	MRM	Maintenance - Resource - Management		
2004	SHELL-T (SHELL-Team)	Software (procedures) - Hardware (machines) - Environment - Liveware - Liveware (humans) - Team		
2004	SCHELL model and CRM	Software (procedures) – Culture - Hardware (machines) - Environment - Liveware - Liveware (humans)	Culture Safety Error prevention	III
2004	LOSA	Line - Operation - Safety - Audit		
2009	HEAD	Human - Environment - Analysis - Design		
2010	HFACS	Human Factors - Accident - Classification - System		
2012	SMS	Safety Management System Safety Balance Models Collaborative Decision Making Artificial Intelligence	Safety Efficiency Minimization of errors	IV
2016	AI CDM SWIM FF-ICE	Artificial Intelligence Collaborative Decision Making System-Wide Information Sharing and Management Flight & Flow Information for a Collaborative Environment	Artificial Intelligence Systems Analyze, detection, prevention, and minimization of errors	V

– ATCO (language barrier, communicating problems, ATCO's interference in the flight crew, wrong ATCO instructions / commands, etc.) killed about 2 000 people in aviation accidents.

Coherent, clear interaction between pilot and ATCO is most important in flight emergency (FLEM) in flight, which is characterized by a sharp shortage of time in the DM in conditions of incompleteness and uncertainty of information, as well as the significant psychophysiological load on the flight crew (FC). The final decision on the order of the flight in the FLEM is taken by the captain of the aircraft (Capt.), which is fully responsible for the decision.

The ATCO is responsible for the correctness and timeliness of the information and advice that given to flight crew, so the ATCO in such situations is also given a significant role (Ministry of Civil Aviation of the USSR, 1985; ICAO, 2007).

The main requirement for the ATCO when FLEM arises is the constant readiness to provide the necessary assistance to the flight crew, depending on the type of situation, taking into account the air situation and meteorological conditions. One of the factors that greatly complicate the interaction between pilot and ATCO is the inadequate knowledge of the flight crew procedures performed in FLEM (Golovnin, 2016).

The technology of flight crew and ATCO procedures in the FLEM must be in line with the definition of the algorithm prescribed in the normative and regulatory documents, therefore, for the formalization of the actions of the H-O in FLEM, it is possible to apply determined models (Kharchenko, Shmelova, & Sikirda, 2016; Shmelova & Sikirda, 2018). Since FLEM is a time-consuming event, when it comes to modeling a collaborative DM pilot and ATCO, it is advisable to use network graphs depending on the algorithm of action in FLEM, which reflects the technological dependence and consistency of operational procedures of operators, ensure their achievement in time, taking into account the cost of resources and the cost of work with the allocation at the same time critical places.

Thus, the problem of optimizing the interaction between pilot and ATCO in FLEM can be solved by the way of development and synchronization (maximal alignment over time) of deterministic models of H-O collaborative DM, which will minimize the critical time needed to solve FLEM, by definition the optimal sequence of execution of technological procedures.

BACKGROUND

In 1979, KLM developed the first flight training program for effective methods of interaction and information exchange, known as Cockpit Resource Management (CRM) (Muñoz-Marrón, 2018; İnan, 2018). Gradually the emphasis was shifted, the decoding of the first letter "C" into the abbreviation changed three times - from Cockpit (cabin) to Crew (crew), and finally - when the concept "crew" consisted of consecutive cabin attendants, ATCO's, technical and managerial personnel, and eventually all the airline - "C" became known as Company, and the name of the discipline – "Company resource management". From that time, the awareness of security became a systemic quality, corporate culture (ICAO, 2013).

Optimization of the flight crew and ATCO interactions as small groups (Malyshevskiy, 2010; Sikirda & Shmelova, 2018) is based on the socionics and sociometrics methods (Blutner & Hochnadel, 2010; Cillessen, 2009).

ATM's global operational concept presents the ICAO vision for an integrated, harmonized and globally interoperable ATM system. Its goal is to move towards the implementation of holistic, cooperative and joint decision-making processes, in which the expectations of the entities of the ATM system will be balanced and aimed at achieving optimal results on the basis of equality and ensuring the access of all participants (ICAO, 2005).

The introduction of new technological solutions, in particular, CDM (Collaborative Decision Making), requires the use of a modern information environment based on the concept of system-wide information management (SWIM) and the concept of information on air and traffic flows for the joint use of air Space (FF-ICE - Flight & Flow Information for a Collaborative Environment) (ICAO, 2012, 2014).

As part of the implementation of the Global Air Navigation Plan (ICAO, 2016), ICAO is working on a step-by-step improvement of the civil aviation system, and now the development of "network center systems" (the SWIM concept) has begun, as the current way of data-sharing "point-to-point" has ceased to keep pace with the increase in airspace transport and be effective. SWIM is a kind of internet for aviation: the network is based on providing information when interacting with different aviation systems.

The concept of FF-ICE is limited to the exchange of information about the flight between the subjects of the ATM system (ICAO, 2012). It begins with the timely submission by the user of the airspace of flight information to the ATM system and ends with the archiving of relevant information after the flight. FF-ICE supports all components of the ATM operational concept requiring flight information: Demand

and Capacity Balancing (DCB), Conflict Management (CM), Service Delivery Management (SDM), Airspace Organization and Management (AOM), Aerodrome Operations (AO), Traffic Synchronization (TS), Airspace User Operations (AUO), and clarifies the Global ATM operational concept for flight information management. It creates the necessary foundation for the most up-to-date ATM systems and develops a 4D-trajectory management mechanism.

CDM involves an uninterrupted process of presenting information and individual DM to different interacting parties, as well as ensuring the synchronization of decisions taken by participants and the exchange of information between them. It is important to ensure the possibility of adopting a joint, integrated solution at an acceptable level of efficiency. One possible approach is the preliminary joint development of procedures to be applied in FLEM (ICAO, 2014). This requires the creation of a database of models of possible scenarios of the flight situations development, based on models of the joint pilot and ATCO DM in flight emergencies.

In works (Kharchenko, Shmelova, & Sikirda, 2016; Shmelova & Sikirda, 2018) is presented using the methods of network planning of the deterministic models of DM by the ANS's H-O (pilot, ATCO) in the conditions of normalized algorithms of professional activity with deterministic and probabilistic time for the implementation of technological procedures. The authors outline the critical course and time for the pilot and ATCO (separate) operations in flight emergencies and the main stages of the DM according to the crew's operating manuals, flight guidance for different types of aircraft, ASSIST guidance (Acknowledge, Separate, Silence, Inform, Support, Time) for "Typical Air Traffic Controller Checklist in Flight Emergencies" European Organization for the Safety of Air Navigation [Eurocontrol], 2003, 2004; Federal Aviation Administration [FAA], 2008), issues related to the synchronization of Pilot Flying (a pilot that performs piloting operations) and Pilot Monitoring (pilot performing communication functions) procedures under cross-monitoring in FLEM (Shmelova, Shyshakov, & Shostak, 2015; Shmelova & Shyshakov, 2016), but a problem of CDM by pilot and ATCO in flight emergencies has not been investigated.

The research tasks are:

- Conducting a detailed analysis of CDM by pilot and ATCO in FLEM (for example, the failure of one engine and other engine fires on the one side during take-off on a multi-engine aircraft) with network planning methods.
- Synchronization of ANS's H-Os procedures with an optimal sequence of actions and minimum flight completion time.
- Obtaining a multiplicative function for evaluating the influence of organizational risk factors on the flight safety in the air traffic control.
- Working-out the Intelligent Automated System (IAS) for supporting the CDM by ANS's H-Os in flight emergencies.

THE DETERMINISTIC MODEL OF COLLABORATIVE DECISION MAKING BY AIR NAVIGATION SYSTEM'S HUMAN-OPERATORS DURING FLIGHT EMERGENCIES

One of the DM methods recommended by aviation guidance in the FLEM is FADEC (European Aviation Safety Agency, 2012).

- **Fly the Aircraft:** Remember the limitations for the aircraft and, if conditions permit, use all available automatic systems, such as autopilot.
- **Assess the Situation (Risk & Time):** Spending more time to assess a situation can lead to a better result. Try to avoid instant / quick decisions if the time is not too limited.
- **Decide on a Workable Option and Refer to Abnormal or Emergency Checklist:** Situation in which the "human resources-equipment-software" system, that is, the natural environment, will have to operate.
- **Evaluate:** Continue to evaluate the situation and actions as the situation develops (feedback lines).
- **Communicate:** Keep in touch with air traffic control authorities to make joint decisions, as well as with other personnel as needed.

The parallel process of simultaneous execution of pilot and ATCO technological operations in the FLEM can be represented as a consolidated dual-channel network. For a consistent optimization of such a network in order to achieve the cross-cutting efficacy of joint decisions, it is advisable to use a multi-criteria approach: achieving a minimum time for parity of FLEM with maximum safety / maximum harmonization over the time of H-O actions.

Ways to optimize the network graph for performing procedures by the H-O in the FLEM (by minimizing time with maximum safety) are:

1. Time optimization – by regulating the use of resources minimizing the time of execution of critical paths t_i^k (1):

$$t_i^{k-1} < t_i^k < t_i^{k+1},\qquad(1)$$

where $t_i^{k-1} = \max \min t_i^k$ – is a minimum time with maximum safety;

$t_i^{k+1} = \min\max t_i^k$ – is a critical time of the maximum (critical) path;

t_i^k – is an optimal time.

2. Changing the topology of the network due to the multi-varied technology implementation procedures.
3. Introduction of parallel execution of procedures with maximum agreement on time (minimum time for two or more charts), that is, obtaining the optimal consolidated time for the execution of procedures t_j^k (2):

$$t_j^{k-1} < t_j^k < t_j^{k+1},\qquad(2)$$

where $t_j^{k-1} = \max \min t_j^k$ – is a minimum time with maximum time matching;

$t_j^{k+1} = \min \max t_j^k$ – is a critical time of the maximum (critical) path;

t_j^k – is an optimal time.

Illustrative Example

To investigate the interaction between the flight crew and ATCO in the FLEM, consider the incident on November 28, 2010, with the aircraft IL-76TD of the Georgian private airline Sun Ways Airlines, which performed a flight from Karachi to Khartum (Pakistan) with a cargo weighing 31 tons (Aviation Safety Network, 2010). Immediately after take-off, engine number four failed, then engine number three was on fire. The flame of the engine was noticed from the ground, about which the ATCO informed the captain. The flight crew tried to make an emergency landing.

At 1:48 local time (UTC + 5), four minutes after the take-off, the aircraft fell to the open ground (6 km from the end of the runway). All flight crew members (seven Ukrainians and Russians) and four men on the ground died. During the accident investigation, it was discovered that at the time of the aircraft fall, two of the four engines did not work.

On the flight simulator KTS-32 (aircraft IL-76TD), the simulation of the flight crew and ATCO procedures was carried out in case of one engine failure and other engine fires on the one side during the take-off. Two possible scenarios for the development of events were investigated, when the captain decides to land at the departure airport with direct or reverse heading. Different meteorological conditions were created, the weight of the load and the centering of the aircraft changed, airport charges, and so on.

Based on the results obtained on the simulator KTS-32, a deterministic model of the flight crew and ATCO procedures was developed in case of one engine failure and other engine fires on the one side during the take-off. In Table 2 is shown the structure-time table for the execution of flight crew and ATCO procedures in case of one engine failure and other engine fires on the one side during the take-off.

The time required to perform procedures aimed at paring flight emergency was measured during the Ukraine flight crews and ATCO's simulator training, Ukraine Air Force pilots and ATCO's simulator training and several foreign airlines flight crews' simulator training.

With network planning, flight crew and ATCO procedures were synchronized, resulting in a determined action time by the operators at the stages of parity FLEM, namely:

Stage I: Engine failure.
Stage II: Another engine fire on one side.
Stage III: Approach.
Stage IV: Emergency landing.

The obtained data are statistically processed, their statistical characteristics are within the permissible limits: the standard deviation does not exceed 0.5 sec.; the coefficient of variation does not exceed 19%. Therefore, the average results can be considered reliable. It was also evaluated the competence of experts who participated in the study, with analysis of their professional activities, open-mindedness, and general erudition; the coefficient of competence is received.

A network diagram (Figure 1) of flight crew and ATCO procedures in the FLEM (one engine failure and another engine fire on one side during take-off) allows to determine the critical time depending on the decision taken by the captain (to make a forced landing at the departure airport with direct or reverse heading), which makes $T_{crit\,dir}$ = 6 min. 02 sec. and $T_{crit\,rev}$ = 4 min. 10 sec. Thus, depending on the conditions and circumstances in case of failures quickly perform aircraft landing with a reverse course. So, this is the best variant for completing the flight.

Table 2. The structured-time table for the execution of flight crew and ATCO procedures in case of one engine failure and other engine fires on the one side during take-off

Stage	Procedure	Flight Crew Procedure Description	Relies on Procedure	Execution Time, t, s	Procedure	ATCO Procedure Description	Relies on Procedure	Execution Time, t, s
I	a_1	Flight engineer (FE) find engine failure	-	2	-	-	-	-
	a_2	FE report Capt. about engine failure	a_1	2				
	a_3	Capt. give FE order to shut down the engine, radio-operator (RO) order to switch off the generator	a_2	4				
	a_4	Capt. give RO order to report ATCO about engine failure	a_3	2	b_1	Receive from Capt. report about the engine failure	-	5
	a_5	Capt. give FE order to retract landing gear at height 5m	a_4	2				
	a_6	Capt. reduce the rate of climb, continue taking off	a_5	4				

continues on following page

Figure 1. Network graph of ATCO and flight crew procedure in case of one engine failure and other fires from one side during take-off

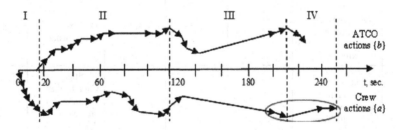

In this context, the use of flight simulators during ATCO professional training is relevant. They will help ATCO to get acquainted with the situation in the flight crew cabin and the parameters of the aircraft's devices during the FLEM. At the same time the ATCO:

- Will receive the experience of the crew members during the FLEM.
- Will pay attention to how the intervention of the dispatcher can disrupt crew members.
- Will complete exercises on the use of radio during the FLEM.
- Will complete the checklist in the FLEM.
- Will participate in captain decision making during the FLEM.
- Will observe the features of the go-around procedure.

Table 2. Continued

Stage	Procedure	Flight Crew Procedure Description	Relies on Procedure	Execution Time, t, s	Procedure	ATCO Procedure Description	Relies on Procedure	Execution Time, t, s
II	a_7	Voice annunciator «Fire», red lamp «Fire» on	a_6	2	b_2	Receive from Capt. report about the engine fire	b_1	8
	a_8	FE check engine fire, report Capt.	a_7	3				
	a_9	Capt. give RO order to report ATCO about the engine fire	a_8	3				
	a_{10}	Capt. set horizontal flight for increase airspeed	a_9	30				
	a_{11}	At height 120 m. and speed for flaps Capt. give FE order to retract flaps	a_{10}	4	b_3	Inform Capt. about external signs of failures, fixes the time	b_2	10
	a_{12}	At speed for slats Capt. give FE order to retract slats	a_{11}	5	b_4	Check Capt. setting emergency squawk 7700	b_3	5
	a_{13}	FE report Capt. flaps and slats retracted	a_{12}	15	b_5	Report supervisor about flight emergency	b_4	5
	a_{14}	Capt. set emergency squawk 7700	a_{13}	4	b_6	Provide clear air space in close proximity to aircraft	b_5	15
	a_{15}	Capt. give FE order to shut down the engine, close fuel valve, switch on the fire extinguisher	a_{14}	8	b_7	If necessary set radio silence	b_6	4
	a_{16}	FE check fire, switch on the second and third bottle fire extinguisher	a_{15}	30	b_8	Clarifies Capt. further intentions for landing at the departure aerodrome.	b_7	10
	a_{17}	FE report Capt. about fire extinguished or not	a_{16}	2	b_9	Facilitates the decision implementation	b_8	37
					b_{10}	Displays emergency board information	b_9	5
					b_{11}	Ask weather forecast	b_{10}	5

continues on following page

Table 2. Continued

Stage	Procedure	Flight Crew Procedure Description	Relies on Procedure	Execution Time, t, s	Procedure	ATCO Procedure Description	Relies on Procedure	Execution Time, t, s
III	a_{18}	Capt. report ATCO that they manage to extinguish the fire or not, the decision to land	a_{17}	10	b_{12}	Clarifies whether the engine fire was eliminated	b_{11}	10
					b_{13}	Provides extraordinary landing	b_{12}	4
	a_{19}	Capt. make an approach, give FE order to extend landing gear, flaps, and slats	a_{18}	77	b_{14}	Gives Capt. directions for approach, reports wind direction and speed	b_{13}	8
					b_{15}	Controls the aircraft movement, informs Capt. about the deviation from heading and the glide path	b_{14}	64
	a_{20}	Capt. give FE order to switch on hydraulic pump station on the failed engines	a_{19}	3	b_{16}	Passes the plane to the Tower ATCO	b_{15}	4
IV	a_{21}	Capt. proceed landing	a_{20}	30	b_{17}	Clear runway according to local instructions	b_{16}	10
	a_{22}	After stopping on the runway, if fire not extinguished, Capt. turn aircraft direction to the wind	a_{21}	10	b_{18}	By supervisor order set on the readiness of rescue equipment	b_{17}	5

In the flight emergency, ATCO is advised to use a checklist that will help to handle incidents in order to establish optimal actions to achieve better cooperation between pilot and ATCO. A supervisor who works with ATCO, using a checklist, can provide better support as it will more clearly understand the ATCO in the FLEM.

EVALUATION OF INFLUENCE OF ORGANIZATIONAL RISK FACTORS ON THE FLIGHT SAFETY IN THE AIR TRAFFIC CONTROL

The analysis of flight safety in the air traffic control (ATC) (State Aviation Administration of Ukraine, 2017) indicates the lack of attention of air navigation service providers (ANSP) to the organizational aspects, in particular, the division of responsibility, the coordination of operational question, the development and improvement of procedures, control over documentation in the implementation of quality management systems and flight safety. Issue of organization of activities remains the most critical element and requires careful attention.

Elimination of accidents remains the key point for all kinds of aviation activity. But it is impossible for aviation systems to be completely free of hazardous factors and associative risks. Neither human activity nor human designed systems are completely free of operational errors and its consequences (ICAO, 2013). Flight safety is a dynamical parameter of the aviation system. Thus risk factors should continuously mitigate. It is important to note that the adoption of indicators for the effectiveness of safety of flights is often influenced by internal and international standards, as well as cultural features (ICAO, 2004). When the risk factors and operational errors are reasonably monitored, flight safety can be managed (ICAO, 2013). Risk is an integral part of any aviation activity. A key function of safety management systems is risk management, which is carried out to ensure acceptable levels of safety. Risk management is performed in relation to the identified hazards and includes: identifying factors that threaten the safety of flights; analysis of the revealed factors; evaluating the magnitude and acceptability of the risks associated with these factors; development of means for reducing risks to admissible levels; control of residual risks during operation (State Aviation Administration of Ukraine, 2005; Ministry of Transport and Communications of Ukraine, 2010).

In order to determine the degree of influence of organizational risk factors on flight safety in ATC, an expert questioning has conducted on 30 area air traffic controllers at ANSP Lviv Regional Branch of Ukrainian State Air Traffic Services Enterprise (UkSATSE) (Sikirda, Shmelova, & Tkachenko, 2018). The questionnaire has formed in accordance with the "Swiss-Cheese" model by Professor James Reason on the base of the structural elements of the "organizational accidents" (ICAO, 2013). It has consisted of eight selected main groups of organizational factors: operational environment, procedures and manuals, engineering procedures and maintenance, cooperation between ATC sectors, ATC systems and equipment, infrastructure, airspace structure, company management and structure (Table 3).

Each expert has filed the matrix of individual preferences. With the help of the pairwise comparison method and ranking (Velychko, Kolomiyets, Hordiyenko, Karpenko, & Haber, 2015), the significance rank of each group of factors according to individual expert's priorities has determined. Next step was to form the group preferences matrix and to obtain the average index of the group of experts concerning each group of organizational factors R_{gri} and rank of each group R'_{gri}. The competences of experts have considered being equal. Two stages of questioning have performed in order to achieve the agreement among professionals on the level of influence on safety by each group (Kurtov, Polikashyn, Potikhenskyy, & Aleksandrov, 2017). Level of significance of each group of organizational risk factors has described with weight coefficients ω_i. The obtained weight coefficients of groups of organizational risk factors that affect the safety of flights in the ATC system are presented in Table 4.

The graphics interpretation of the results of an expert questioning in the form of a histogram is shown in Figure 2.

As a result of performed expert questioning, the following order of groups of organizational risk factors depending on the degree of their influence on flight safety has achieved (from the most significant to the least significant one):

1. ATC systems and equipment ($R_{gr5} = 2,63$; $R'_{gr5} = 1$; $\omega_5 = 0,22$).
2. Cooperation between ATC sectors ($R_{gr4} = 3,12$; $R'_{gr4} = 2$; $\omega_4 = 0,19$).
3. Procedures and manuals ($R_{gr2} = 3,5$; $R'_{gr2} = 3$; $\omega_2 = 0,17$).

Table 3. Groups of organizational factors that influence on flight safety in ATC

Group Number	Group Name	Group Description
1	Operational environment	Factors, connected with the physical environment (temperature, air circulation, illumination, level and timing of noise, atmospheric pressure, etc.).
2	Procedures and manuals	Factors, associated with adequacy/inadequacy of procedures/manuals, failure to comply or possibility/impossibility of compliance with them (working instruction, job description, technology of work, Air Code, Regulations on the Use of Airspace, Flight Rules and Air Traffic Services in the Classified Airspace, etc.).
3	Engineering procedures and maintenance	Factors, connected with work of engineers, including routing technical checks and equipment maintenance after the failure. In addition, procedures of design, installation, and implementation of new equipment.
4	Cooperation between ATC sectors	Factors, relating to the technical aspects of system operations between air traffic control sectors as well as between adjacent air traffic control systems (compatibility of coordination procedures, Letters of Agreement, acceptance of information from other sources).
5	ATC systems and equipment	Factors, associated with work of the hardware, software and its compatibility.
6	Infrastructure	Factors, connected with aerodrome (physical parameters, the configuration of maneuvering areas, restriction zones) and environmental layout.
7	Airspace structure	Factors, relating to classification of airspace structure, route network, capacity, the configuration of sectors.
8	Company management and structure	Factors, connected with the style of company management at all levels, corporate culture, SMS.

Table 4. Calculation of weight coefficients of organizational risk factors' groups

Group Number	Group Rank, R'_{gri}	Intermediate Assessment, C_i	Weight Coefficient, ω_i
1	7	0,250	0,06
2	3	0,750	0,17
3	6	0,375	0,08
4	2	0,875	0,19
5	1	1,000	0,22
6	5	0,500	0,11
7	4	0,625	0,14
8	8	0,125	0,03
Σ		4,5	1

4. Airspace structure ($R_{gr7} = 3,98$; $R'_{gr7} = 4$; $\omega_7 = 0,14$).
5. Infrastructure ($R_{gr6} = 4,38$; $R'_{gr6} = 5$; $\omega_6 = 0,11$).
6. Engineering procedures and maintenance ($R_{gr3} = 5,27$; $R'_{gr3} = 6$; $\omega_3 = 0,08$).
7. Operational environment ($R_{gr1} = 6,37$; $R'_{gr1} = 7$; $\omega_1 = 0,06$).
8. Company management and structure ($R_{gr8} = 6,75$; $R'_{gr8} = 8$; $\omega_8 = 0,03$).

The results of the expert questioning have presented as a system of advantages (3):

Figure 2. Graphics interpretation of expert answers

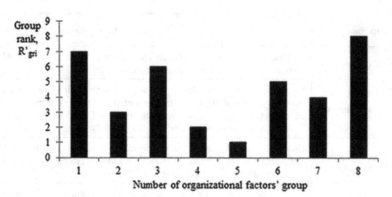

$$R'_{gr5} \succ R'_{gr4} \succ R'_{gr2} \succ R'_{gr7} \succ R'_{gr6} \succ R'_{gr3} \succ R'_{gr1} \succ R'_{gr8},\qquad (3)$$

where R'_{gri} – is the rank of i-group of organizational risk factors.

It is clear that out of all groups of organizational risk factors "Air traffic control systems and equipment" group has the most significant impact on flight safety in ATC, and "Company management and structure" – the least one.

In accordance with the matrix of the risk index (ICAO, 2013), which takes into account the probability and severity of possible consequences, the scale of acceptability (admissibility) of organizational risk factors has constructed on the basis of the fuzzy sets theory with the use of linguistic variables (Zimmermann, 2006): extreme risk (100 points), high risk (80 points) moderate risk (60 points), low risk (35 points) and scarce risk (10 points) (Kharchenko, Shmelova, & Sikirda, 2016). To ensure a sufficient level of flight safety, the risk indicators must be no more than 60 points taken at the maximum permissible level of hazard. The actual significance of the level of hazard to the groups of organizational factors has been determined by questioning the air traffic controllers at Lviv Regional Branch of UkSATSE and statistical processing of the results, which confirmed the consistency of expert opinions.

The results of flight safety evaluation at Lviv Regional Branch of UkSATSE on the basis of the analysis of organizational risk factors are presented in Table 5.

Convolution of multiparametric indicator of the flight safety status in ATC on the basis of the analysis of the organizational risk factors to the scalar indicator has carried out in a multiplicative way (4):

$$W = \prod_{i=1}^{n} L_i^{\omega_i} = \prod_{i=1}^{n} P_V,\qquad (4)$$

where L_i – is a level of hazard of i-group of risk organizational factors;

ω_i – is a weight coefficient which taking into account the probability and severity of i-group of risk organizational factors' consequence;

P_i – is a parameter of the hazardous level of i-group of risk organizational factors.

The results of the expert questioning are presented graphically using a spider diagram (Figure 3).

Table 5. Results of the flight safety evaluation in ATC at Lviv Regional Branch of UkSATSE

Group of Organizational Risk Factors	Operational Environment	Procedures and Manuals	Engineering Procedures and Maintenance	Cooperation between ATC Sectors	ATC Systems and Equipment	Infrastructure	Airspace Structure	Company Management and Structure
Weight coefficient (degree of influence), w_i	0,06	0,17	0,08	0,19	0,22	0,11	0,14	0,03
Maximum allowable level of hazard, L_{alli}	60	60	60	60	60	60	60	60
Actual level of hazard, L_{acti}	20	35	25	35	20	25	30	35
Parameter of maximum allowable hazardous level, P_{alli}	1,28	2,01	1,39	2,18	2,46	1,57	1,77	1,13
Parameter of expertise of actual hazardous level, P_{acti}	1,20	1,83	1,29	1,97	1,93	1,42	1,61	1,11
The difference between maximum allowable and actual parameters of hazardous level, $\Delta P = P_{alli} - P_{acti}$	0,08	0,18	0,10	0,21	0,53	0,15	0,16	0,02

Figure 3. Graphics interpretation of the expert questioning results for the flight safety in ATC at Lviv Regional Branch of UKSATSE on the basis of the analysis of the organizational risk factors

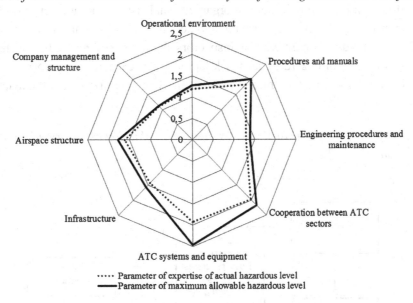

With the help of data from Table 5 the authors have received maximum allowable W_{al} and actual W_{act} value of a multiplicative function for evaluation of flight safety level in ATC at Lviv Regional Branch of UkSATSE on the basis of organizational factors:

$$W_{all} = 60^{0,06} \cdot 60^{0,17} \cdot 60^{0,08} \cdot 60^{0,19} \cdot 60^{0,22} \cdot 60^{0,11} \cdot 60^{0,14} \cdot 60^{0,03} = 60,99;$$

$$W_{act} = 20^{0,06} \cdot 35^{0,17} \cdot 25^{0,08} \cdot 35^{0,19} \cdot 20^{0,22} \cdot 25^{0,11} \cdot 30^{0,14} \cdot 35^{0,03} = 27,33.$$

Comparison of flight safety results at Lviv Regional Branch of UkSATSE has been performed on the basis of the maximum allowable level of hazard and expertise of actual hazardous level of organizational risk factors' groups $\Delta W = W_{all} - W_{act} = 60,99 - 27,33 = 33,66$.

An example of the expertise results, presented in Figure 3, shows the correspondence of the values of all groups of organizational risk factors to the maximum permissible level of hazard, which indicates a high flight safety index in air traffic control at this Regional Branch.

Automated System for Evaluation of the Organizational Risk Factors Influence on Flight Safety in Air Traffic Control

The conceptual model of the automated system for evaluation of the organizational risk factors influence on flight safety in ATC is shown in Figure 4.

The automated system for evaluation of the organizational risk factors influence on flight safety in ATC will create digital safety checklists for conducting inspections of ANSP (Table 6).

The automated system for evaluation of the organizational risk factors influence on flight safety in ATC will allow determining maximum allowable and actual levels of the hazard of organizational risk factors' groups for making decision by aviation inspectors about the issuance of the certificate to ANSP. It is possible to see the level of hazard of each risk factor that is exceeding the norm or is approaching the allowable limit in order to carry out measures in a timely manner to prevent the aviation accident.

Figure 4. The conceptual model of the automated system for evaluation of the organizational risk factors influence on flight safety in ATC

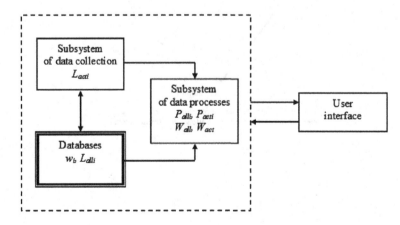

Table 6. Digital safety checklist for conducting inspections of ANSP

Group of Organizational Risk Factors	Allowable Value	Actual Value	Norm / Out of the Norm
Operational environment	1,28	1,20	V
Procedures and manuals	2,01	1,83	V
Engineering procedures and maintenance	1,39	1,29	V
Cooperation between ATC sectors	2,18	1,97	V
ATC systems and equipment	2,46	1,93	V
Infrastructure	1,57	1,42	V
Airspace structure	1,77	1,61	V
Company management and structure	1,13	1,11	V
Complex assessment	60,99	27,33	V

BUILDING AND DEVELOPMENT OF THE INTELLIGENT AUTOMATED SYSTEM FOR SUPPORTING THE COLLABORATIVE DECISION MAKING BY AIR NAVIGATION SYSTEM'S HUMAN-OPERATORS IN THE FLIGHT EMERGENCY

The conceptual model of System for control and forecasting the FLEM development that using DM models on the base of Intelligent Automated System (IAS) / Decision Support System (DSS) / Expert System (ES) was obtained (Figure 5) (Shmelova, Sikirda, Scarponi, & Chialastri, 2018), where $\overline{F}_p = \left\{ \overline{F}_{ed}, \overline{F}_{\exp} \right\}$ – are the professional factors; $\overline{F}_{np} = \left\{ \overline{F}_{ip}, \overline{F}_{pf}, \overline{F}_{sp} \right\}$ – are the non-professional factors; $\overline{F}_{ed}$ – are the knowledge, skills and abilities, acquired H-O during training; $\overline{F}_{\exp}$ – are the knowledge, skills and abilities, acquired H-O during professional activity; $\overline{F}_{ip} = \left\{ f_{ipt}, f_{ipa}, f_{ipp}, f_{ipth}, f_{ipi}, f_{ipn}, f_{ipw}, f_{iph}, f_{\exp} \right\}$ – is a set of H-O individual-psychological factors (temperament, attention, perception, thinking, imagination, nature, intention, health, experience); $\overline{F}_{pf}$ – is a set of H-O psycho-physiological factors (features of the nervous system, emotional types, socio-types); $\overline{F}_{sp} = \left\{ f_{spm}, f_{spe}, f_{sps}, f_{spp}, f_{spl} \right\}$ – is a set of H-O socio-psychological factors (moral, economic, social, political, legal factors).

The analysis of social-physiological factors conducted by the authors allowed to make a conclusion that the activities of pilots are influenced by the own image, the image of the corporation as well as by interests of a family. At the same time, respondents – ATCOs pay special attention to the interests of their families, their own economic status and professional promotion (Kharchenko, Shmelova, & Sikirda, 2016; Rizun & Shmelova, 2016).

Deterministic and stochastic models for ANS's H-O (pilot, ATCO) were obtained in accordance with the flight manual of aircraft or the adopted technologies of controller's work ASSIST (Acknowledge, Separate, Silence, Inform, Support, Time) in FLEM. Deterministic and stochastic models for ATCO are presented in Figures 6-7, where *{A}* – is the set of the operations which are carried out by the controller in accordance with ASSIST; *{T}* – is the time of decision making; *{P}* – is the set of the probabilities of *j*-factor influence during *i*-alternative solution choice; *{U}* – is the set of the losses associated with choosing *i*-alternative solution during *j*-factor influence; *{R}* – is the set of the risks associated with choosing *i*-alternative solution during *j*-factor influence; *{λ}* – is the set of the factors influencing DM.

Figure 5. The conceptual model of System for control and forecasting the FLEM development

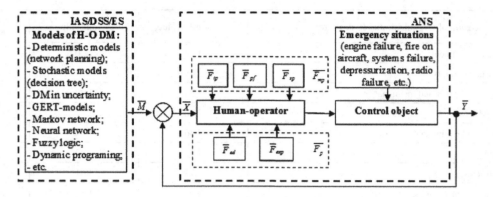

With using neural network models, the values of probabilities (p_n) (Kharchenko, Shmelova, & Sikirda, 2016; Shmelova, Sikirda, Scarponi, & Chialastri, 2018), expected outcomes (r_k) and additional inputs – factors ($_{\xi k}$) (Figure 8) of FLEM development were received.

The network has additional inputs, called the Bias (offset) that takes into account additional restrictions on calculating parameters (5):

$$\sum_{i=1}^{n} p_i u_i - \xi_k \geq 0 .\tag{5}$$

where p_i – are the weight coefficients;
u_i – are the neural network inputs;
$_{\xi k}$ – is a Bias (shift) under influencing factors of uncertainty (Table 7).

Figure 6. Deterministic model of ANS's H-O DM

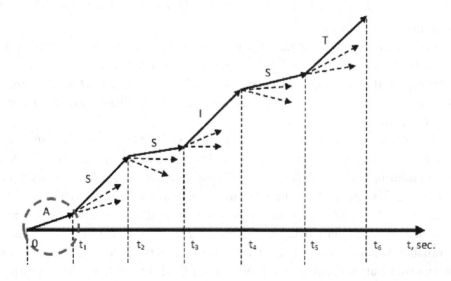

Figure 7. Stochastic model of ANS's H-O DM

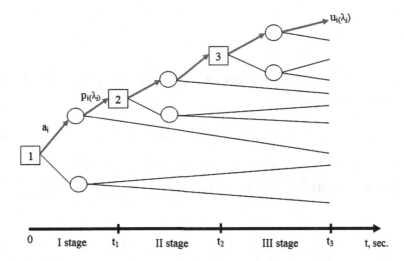

Figure 8. The neural network model of FLEM development with additional inputs of influencing DM factors

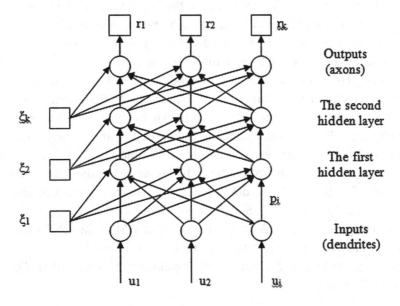

Table 7. The matrix of Bias identification

Alternative Decisions	Factors that Influencing on the ANS's H-O DM					
	λ_1	λ_2	...	λ_j	...	λ_m
A_1	ξ_{11}	ξ_{12}	...	ξ_{1j}	...	ξ_{1m}
A_2	ξ_{21}	ξ_{22}	...	ξ_{2j}	...	ξ_{2m}
...	...	...	...	...	...	...
A_i	ξ_{i1}	ξ_{i2}	...	ξ_{ij}	...	ξ_{im}
...	...	...	...	...	...	...
A_n	ξ_{n1}	ξ_{n2}	...	ξ_{nj}	...	ξ_{nm}

The outcomes of the neural network are (6):

$$\overline{R} = f(\overline{net} - \overline{\xi}),$$ (6)

where f – is a non-linear function (active function) that takes into account the time of decision making t_i; $\overline{net}$ – is a weighted sum of inputs.

The optimal solution is found by the criterion of expected value with the Savage criterion (7):

$$A_{opt} = \min \max \{R\} = \min \max \left\{ t_i (\sum_{i=1}^{n} p_i u_i - \xi_k) \right\}.$$ (7)

The critical time of the flight crew actions in case of an engine failure on take-off and approach to land in the bad weather conditions was obtained (Kharchenko, Shmelova, & Sikirda, 2016). The selection in the direction of the negative pole leads to the maximum expected risk $R=1028$. The choice in the direction of the positive pole when the FLEM occurs at the first stage of DM by H-O ANS (for example, a flight to alternative aerodrome) has a risk which is 60,5 times lesser: $R=17$.

In the stochastic network of the flight situation development of GERT type, the tops are represented by stages of the situation (normal, complicated, difficult, emergency or catastrophic), and the arcs are represented by a process of transition between stages of the situation. The algorithm of stochastic network analysis was developed (Kharchenko, Shmelova, & Sikirda, 2016; Shmelova & Sikirda, 2018). Thus according to results of stochastic network analysis of the flight situation development from normal to catastrophic the following values obtained: the mathematical expectation of flight situation development time $t_{ij} - M[t_{ij}]$; the variance of flight situation development time $t_{ij} - \delta^2 [t_{ij}]$; the probability of flight situation development $p_{ij} - p_{ij} \cdot p_{ji}, p_{ii}$. Based on the W-functions of positive and negative of H-O choice the Markov's network of flight situations' development from normal to catastrophic was constructed (Kharchenko, Shmelova, & Sikirda, 2016; Shmelova & Sikirda, 2018).

In addition, with using reflexive model the risks R_A, R_B of DM in the ANS under the influence of the external environment x_1, the previous H-O's experience x_2 and the intentional choice of H-O x_3 have obtained (Kharchenko, Shmelova, & Sikirda, 2016; Shmelova & Sikirda, 2018). The expected risk in the process of DM of H-O is equal (8):

$$R_{DM} = \begin{cases} R_A = \min \{R_{ij}\} \\ R_B = \{\gamma, \rho\} \\ R_{AB} = \{X(x_1, x_2, x_3), \gamma, \rho\} \end{cases},$$ (8)

where R_A – is an expected risk of the DM for H-O with taking into account the criterion of the expected value minimization:

R_B: Is an expected risk of the DM for H-O with taking into account his model of preferences.

R_{ij}: Is an expected risk for making A_{ij}-decision;.

γ: Is a concept of a rational individual's behavior.

ρ: Is a system of an individual's preferences in a concrete situation of the choice.

R_{AB}: Is a mixed choice made by the H-O.

For example, if the pilot, the ATCO and the society have a choice in the direction of the negative pole B, the preferences model can form the plane of the disaster K (Kharchenko, Shmelova, & Sikirda, 2016; Shmelova & Sikirda, 2018).

The methodology of research and training in ANS as a socio-technical system has developed (Kharchenko, Shmelova, & Sikirda, 2018). Let's consider the individual works of aviation students and post-graduate students in education (course "Basic of DM in ANS" in National Aviation University, Kyiv) after Master class of DM in ANS. Research has shown that the choice of the optimal variant of the forced flight completion in emergencies requires the operator to analyze the significant amount of diverse information. The following conceptual models of DSS in ANS have obtained, such as DSS for ATCO in emergencies, for example, "Aircraft Decompression", "Low oil pressure", "Engine failure", etc.; DSS for flight dispatcher for support of the DM regarding aircraft landing in emergencies to choice alternative landing aerodrome; DSS for operator of Unmanned Aerial Vehicles (UAV) in emergencies situation, for example in losing of communication with UAV and choosing optimal landing place, etc. developed (Kharchenko, Shmelova, & Sikirda, 2016, 2018). DSS contain common sets of components, such as data related components, algorithm related components, user interface, and display related components.

FUTURE RESEARCH DIRECTIONS

The direction of further research is the development of deterministic and non-deterministic network models of CDM by ANS's H-O with probabilistic time for the implementation of technological procedures and identification of appropriate risks.

The developed deterministic models will allow supplementing the database of flight scenarios development in the IAS / DSS / ES of the pilot / ATCO in the FLEM for optimization of CDM and can be used in the future both in the ANS's H-O training process and in real conditions. The operation of the aircraft is based on the use of SWIM and FF-ICE concepts.

What's Next?

On the next steps we are planning to optimize CDM between the ATCOs and the pilots, between the controllers of the adjacent points, between the ATCOs and the service staff with the help of the System for control and forecasting the development of FLEM on the base of IAS / DSS / ES that taking into account the influence of the professional factors (knowledge, habits, skills, experience) as well as the factors of non-professional nature (individual-psychological, psycho-physiological and socio-psychological) on the decision making process by H-O of ANS.

CONCLUSION

With the help of network planning the procedures of ATCO and flight crew in case of one engine failure and other fires from one side during take-off with the optimal sequence of actions and the minimum completion time of the flight, which is 4 min. 10 sec with forced landing with a reverse heading, were synchronized. The deterministic models of the CDM by pilot and ATCO for performing operational procedures by the H-O in the FLEM were obtained. For the sequential optimization of the collaborative two-channel network "Air traffic controller-Pilot" in order to achieve the end-to-end effectiveness of joint solutions, a multi-criteria approach was used: ensuring the minimum time to parry FLEM with maximum safety/maximum consistency over the time of operators' actions.

Organizational factors that influence on the flight safety in ATC have been systematized in eight groups: operational environment, procedures and manuals, engineering procedures and maintenance, cooperation between air traffic control sectors, ATC systems and equipment, infrastructure, airspace structure, company management and structure. The hazard degree of organizational factors in ATC has been determined by the method of expert assessments; the scale of acceptability (admissibility) of the risk factors of organizational nature has been constructed on the basis of the fuzzy sets theory with the use of linguistic variables. It has been revealed that the most significant influence on flight safety has "ATC systems and equipment" group of organizational risk factors and the least – "company management and structure". The multiplicative function for evaluation of the flight safety in ATC has been obtained, which allows checking the conformity of the values of organizational risk factors to the maximum permissible level of hazard. The conceptual model of the automated system for evaluation of the organizational risk factors influence on flight safety in ATC has been designed; the digital safety checklist for conducting inspections of ANSP has been presented. The proposed methodology for evaluation of the organizational risk factors influence on flight safety in ATC will allow developing the safety passports that can be applied by the aviation authorities during the certification inspections of ANSP to compare the normative and actual indicators of their activities.

The conceptual model of System for control and forecasting the FLEM development on the base of IAS / DSS / ES that taking into account the influence on CDM by ANS's H-O of the professional factors (knowledge, habits, skills, experience) as well as the factors of non-professional nature (individual-psychological, psycho-physiological and socio-psychological) was presented. Deterministic and stochastic models for ANS's H-O (pilot, ATCO) were obtained in accordance with the flight manual of aircraft or the adopted technologies of ATCO work ASSIST. With using the neural network model, the values of probabilities of FLEM development were received. The optimal solution was found by the criterion of expected risk minimization.

Perspectives for using estimates of risk organizational factors in ATC in the System for control and forecasting the FLEM development:

- **Operational Level**: The development of an optimal co-operative solution of FLEM paring for ANS's H-Os.
- **Tactical Level**: Optimization of aircraft's flight plans and routes.
- **Strategic Level**: Inspection of ANSP, rationalization of airspace structure, improvement of ATC operations and procedures, determination of the number of staff, etc.

Designing and calculating scenarios of the development of flight situations, forecasting possible collaborative actions of H-Os in FLEM will allow preventing the negative development of the emergency situation toward the catastrophic in a timely manner.

REFERENCES

Aviation Safety Network. (2010). *Final investigation report.* Accident of Sunway Air Carrier (Georgia) IL-76TD Aircraft REG # 4L-GNI at Karachi, Pakistan on Nov. 27, 2010. Retrieved from https://www.reports.aviation-safety.net/2010/20101128-0_IL76_4L-GNI.pdf

Aviation Safety Network. (2018). *ASN data show 2017 was safest year in aviation history.* Retrieved from https://news.aviation-safety.net/2017/12/30/preliminary-asn-data-show-2017-safest-year-aviation-history/

Aviation Safety Network. (2019a). *ASN Wikibase.* Retrieved from https://www.aviation-safety.net/wikibase/

Aviation Safety Network. (2019b). *Aviation Safety Network releases 2018 airliner accident statistics.* Retrieved from https://news.aviation-safety.net/2019/01/01/aviation-safety-network-releases-2018-airliner-accident-statistics/

Blutner, R., & Hochnadel, E. (2010). Two qubits for C.G. Jung's theory of personality. *Cognitive Systems Research, 11*(3), 243–259. doi:10.1016/j.cogsys.2009.12.002

Cillessen, A. S. N. (2009). Sociometric methods. In W. M. Bukowski, B. Laursen, & K. H. Rubin (Eds.), *Handbook of peer interactions, relationships, and group* (pp. 82–98). London, UK: The Guilford Press.

European Aviation Safety Agency. (2012). *Decision making for single-pilot helicopter operations.* Koln, Germany: EASA Publ.

European Organization for the Safety of Air Navigation. (2003). *Guidelines for controller training in the handling of unusual/emergency situations* (2nd ed.). Brussels, Belgium: Author.

European Organization for the Safety of Air Navigation. (2004). *ATM training progression and concepts.* Brussels, Belgium: Author.

Federal Aviation Administration. (2008). *Aviation instructor's handbook.* Washington: Author.

Friedman, M., Carterette, E., Wiener, E., & Nagel, D. (2014). *Human factors in aviation* (1st ed.). Massachusetts: Academic Press Publ.

Golovnin, S. M. (2016). *Training pilots and controllers in the virtual environment of piloting and air traffic control.* Retrieved from www.researchgate.net/publication/303859941_Aviation_Training_Pilots_Air_Traffic_Controller

İnan, T. T. (2018). The evolution of crew resource management concept in civil aviation. *Journal of Aviation, 2*(1), 45–55. doi:10.30518/jav.409931

International Civil Aviation Organization. (2002). Human factors guidelines for safety audits manual (1st ed.). Doc. ICAO 9806-AN/763. Canada, Montreal: Author.

International Civil Aviation Organization. (2004). *Cross-cultural factors in aviation safety: Human Factors Digest No. 16. Cir. ICAO 302-AN/175*. Montreal, Canada: Author.

International Civil Aviation Organization. (2005). *Global air traffic management operational concept. Doc. ICAO 9854-AN/458*. Montreal, Canada: Author.

International Civil Aviation Organization. (2007). Air traffic management (15th ed.). Doc. ICAO 4444-ATM/501. Canada, Montreal: Author.

International Civil Aviation Organization. (2009). Global performance of the air navigation system (1st ed.). Doc. ICAO 9883. Canada, Montreal: Author.

International Civil Aviation Organization. (2012). Manual on flight and flow information for a collaborative environment (FF-ICE) (1st ed.). Doc. ICAO 9965. Canada, Montreal: Author.

International Civil Aviation Organization. (2013). Safety management manual (SMM) (3rd ed.). Doc. ICAO 9859-AN 474. Canada, Montreal: Author.

International Civil Aviation Organization. (2014). Manual on collaborative decision-making (CDM) (2nd ed.). Doc. ICAO 9971. Canada, Montreal: Author.

International Civil Aviation Organization. (2016). Global Air Navigation Plan 2016-2030 (5th ed.). Doc. ICAO 9750. Canada, Montreal: Author.

Kharchenko, V., Shmelova, T., & Sikirda, Y. (2016). *Decision-making in socio-technical systems. Monograph*. Kiev, Russia: National Aviation University.

Kharchenko, V., Shmelova, T., & Sikirda, Yu. (2018). Methodology of research and training in air navigation socio-technical system. *Proceedings of the National Aviation University, 1*, 8–23. doi:10.18372/2306-1472.74.12277

Kurtov, A., Polikashyn, O., Potikhenskyy, A., & Aleksandrov, V. (2017). Expert assessments. "Delphy" method as technology of administrative decision making. *Collections of scientific works of the Kharkiv National University of Air Force, No. 1*(50), 118–122.

Malyshevskiy, A. V. (2010). Improving the management and planning in the field of air transport by means of socionics selection of aviation personnel. *Scientific Bulletin of the Moscow State Technical University of Civil Aviation. Ser. Aeromechanics and Strength, no. 1(151)*, 150–157.

Ministry of Civil Aviation of the USSR. (1985). *Manual on the production of flights in the civil aviation of the USSR (with changes and additions)*. Moscow, Russia: Air Transport Publ.

Muñoz-Marrón, D. (2018). Human factors in aviation: CRM (Crew Resource Management). *Papeles del Psicólogo / Psychologist Papers, vol. 39(3)*, 191-199. doi:. doi:10.23923/pap.psicol2018.2870

Rizun, N., & Shmelova, T. (2016). Decision-making models of the human-operator as an element of the socio-technical systems. In R. Batko & A. Szopa (Eds.), *Strategic imperatives and core competencies in the era of robotics and artificial intelligence. Manuscript* (pp. 167–204). Hershey, PA: IGI Global. doi:10.4018/978-1-5225-1656-9.ch009

Shmelova, T., & Shyshakov, V. (2016). System analysis of internal system and external system factors of the non-deterministic subsystem "crew-aircraft". *Bulletin of the Engineering Academy of Ukraine, no. 4*, 48-54.

Shmelova, T., Shyshakov, V., & Shostak, O. (2015). Deterministic models of the operations of the aircraft crew in the case of an unusual situation in flight. *Collegiate of Scientific Works of the Kharkiv National University of the Air Forces, no. 2(19)*, 33-37.

Shmelova, T., & Sikirda, Yu. (2018). Models of decision-making operators of socio-technical system. In T. Shmelova, Yu. Sikirda, N. Rizun, A.-B. M. Salem, & Yu. Kovalyov (Eds.), *Socio-technical decision support in air navigation systems: emerging research and opportunities. Manuscript* (pp. 21–48). Hershey, PA: IGI Global. doi:10.4018/978-1-5225-3108-1.ch002

Shmelova, T., Sikirda, Yu., Scarponi, C., & Chialastri, A. (2018). Deterministic and stochastic models of decision making in air navigation socio-technical system. In *Proceedings of the 14th International Conference "ICT in Education, Research and Industrial Applications. Integration, Harmonization and Knowledge Transfer" (ICTERI 2018)*, vol. 2: Workshops, part 3: 4th international workshop on theory of reliability and Markov modelling for information technologies (TheRMIT 2018). Kiev, Russia. Taras Shevchenko National University of Kyiv.

Sikirda, Yu., & Shmelova, T. (2018). Socionics and sociometry diagnosting of air navigation system's operator. In T. Shmelova, Yu. Sikirda, N. Rizun, A.-B. M. Salem, & Yu. Kovalyov (Eds.), *Socio-technical decision support in air navigation systems: emerging research and opportunities: Manuscript* (pp. 70–90). Hershey, PA: IGI Global. doi:10.4018/978-1-5225-3108-1.ch004

Sikirda, Yu., Shmelova, T., & Tkachenko, D. (2018). Evaluation of influence of the organizational risk factors on flight safety in air traffic control. *Proceedings of the National Aviation University, no. 2(75)*, 26-34. 10.1109/MSNMC.2018.8576317

State Aviation Administration of Ukraine. (2005). *Regulations on the flight safety management system in aviation transport*. Kiev, Russia: Author.

State Aviation Administration of Ukraine. (2017). *Report of the implementation of flight safety oversight functions in the ATMS in Ukraine 2016*. Kiev, Russia: Author.

Velychko, O., Kolomiyets, L., Hordiyenko, T., Karpenko, S., & Haber, A. (2015). *Group expert assessment and experts' competence: monograph*. Odesa, Ukraine: Aprel Publ.

Zimmermann, H. J. (2006). *Fuzzy set theory & its application* (4th ed.). USA: Softcove, Kluwer Academic Press Publ.

KEY TERMS AND DEFINITIONS

Air Navigation Service Provider: (ANSP): Is a public or a private legal entity providing Air Navigation Services. It manages air traffic on behalf of a company, region or country. Depending on the specific mandate an ANSP provides one or more of the following services to airspace users: Air Traffic Management (ATM), Communications, Navigation and Surveillance Systems (CNS), Meteorological

Service for Air Navigation (MET), Search And Rescue (SAR), Aeronautical Information Services/ Aeronautical Information Management (AIS/AIM).

Air Navigation System (ANS): Is a complex of organizations, personnel, infrastructure, technical equipment, procedures, rules and information that is used to provide of airspace users of safe, regular and efficient air navigation service.

Air Traffic Control (ATC): A service provided for the purpose of preventing collisions between aircrafts on the maneuvering area between aircraft and obstructions; and expediting and maintaining an orderly flow of air traffic.

Air Traffic Controllers (ATCOs): Are the coordinators of the movement of aircraft to maintain safe distances between them. Air traffic controllers typically do the following: monitor and direct the movement of aircraft on the ground and in the air, control all ground traffic at airport runways and taxiways, issue landing and takeoff instructions to pilots, transfer control of departing flights to other traffic control centers and accept control of arriving flights, inform pilots about weather, runway closures and other critical information, alert airport response staff in the event of an aircraft emergency.

Air Traffic Management (ATM): The dynamic, integrated management of air traffic and airspace including air traffic services, airspace management, and air traffic flow management — safely, economically and efficiently – through the provision of facilities and seamless services in collaboration with all parties and involving airborne and ground-based functions.

Collaborative Decision Making (CDM): Is a joint government/industry initiative aimed at improving air traffic flow management through increased information exchange among aviation community stakeholders.

Decision Making (DM): Is the cognitive process resulting in the selection of a belief or a course of action among several alternative possibilities.

Decision Support System (DSS): Is the interactive computer system intended to support different types of activity during the decision making including poorly-structured and unstructured problems.

Expert System (ES): Is a computer system that emulates the decision making ability of a human expert.

Flight Emergency (FLEM): Is one in which the safety of the aircraft or of persons on board or on the ground is endangered for any reason.

Flight & Flow Information for a Collaborative Environment (FF-ICE): A product of the ICAO Global ATM Concept, that defines information requirements for flight planning, flow management, and trajectory management and aims to be a cornerstone of the performance-based air navigation system.

Human-operator (H-O): Is a person who interacts with a complex technique through information processes.

Intelligent Automated System (IAS): Streamline decision making typically use tools for aggregating, extracting, and analyzing information - often, complex information such as human speech or unstructured text. Artificial Intelligence was a distributed computing project undertaken by Intelligence Realm, Inc. with the long-term goal of simulating the human brain in real time, complete with artificial consciousness and artificial general intelligence.

System-Wide Information Sharing and Management (SWIM): Is a Federal Aviation Administration advanced technology program designed to facilitate greater sharing of Air Traffic Management system information, such as airport operational status, weather information, flight data, the status of special use airspace, and National Airspace System restrictions.

Chapter 4
Cooperative Decision Making Under Air Traffic Conflicts Detection and Resolution

Volodymyr Vasyliev
National Aviation University, Ukraine

Denys Vasyliev
Ukrainian State Air Traffic Services Enterprise (UkSATSE), Ukraine

ABSTRACT

A number of probabilistic methods for aircraft conflict evaluation are presented. The analytical method is shown. The probability method that enables taking into account the features of stochastic dynamics of motion and the new compositional method of conflict probability evaluation are proposed. The feature of this method is that analytical solution is made using the predicted uncertainty areas of each aircraft position separately, making it possible to manage the position uncertainty and to apply for a wider range of scenarios. The multi-criteria decision-making for conflicts resolution is discussed. Optimality criteria and constraints for conflict resolution are defined. The generic multi-criteria model of conflict-free trajectories selection and the methods of resolution of two- and multi-aircraft conflicts have been developed. These methods provide the synthesis of conflict-free trajectories using different aircraft maneuvers according to criteria of flight regularity, economy, and maneuvering complexity using the multi-criteria dynamic programming.

BACKGROUND

Regardless of steady increasing of the automation level in the air traffic management (ATM) systems, the final decision-making remains with the air traffic controller (ATCO). Being in the flight control loop, ATCO significantly affects the integral characteristics of the air navigation services system. The human factor is revealed to a large extent with increasing the intensity of air traffic, when the number of potentially conflict situations also increases, that lead to significant delays in air traffic, an increase in the probability of dangerous approaching and aircraft collisions in the air.

DOI: 10.4018/978-1-7998-1415-3.ch004

The main task of the air traffic control (ATC) is to avoid mid-air collisions and collisions with obstacles in the maneuvering area, as well as to regulate air traffic taking into account meteorological conditions and flight restrictions. Air traffic safety is ensured by complying with established safe separation standards (minima).

The solution of the problem of preventing potential conflict situations is carried out in two stages. At first, the threat of a conflict between the aircraft must be detected, and its danger must be evaluated. Then some optimal actions are taken to eliminate the threat.

According to the new concepts, the development of ATM systems involves using of more advanced decision-support tools, including the artificial intelligence components, where the central place is given to the detection and prevention of conflict situations.

The development of ATM systems in the area of integrating of digital data transmission systems, improving surveillance methods, artificial intelligence tools and automation enables to apply the strategy of cooperative decision-making under air traffic conflicts detection and resolution. Cooperative decision-making requires the availability of all necessary relevant information that should be available to the parties involved in the decision-making process, i.e. to the controllers and the pilots. The new qualities of the cooperative decision-making provide a more reliable assessment of air traffic situation and control.

METHODS OF CONFLICT DETECTION AND CONFLICT PROBABILITY EVALUATION

It is considered (Babak et al., 2006; Bakker et al., 2001; Blin et al., 2000; Blom et al., 2001; Kuchar & Yang, 2000; Paielli & Erzberger, 1997; Prandini et al., 2000) that probabilistic methods of conflicts evaluation are more advanced and promising compared to geometric methods. They enable to take into account the probabilistic nature of the flight process because of the influence of numerous disturbances, and at the same time take into account the controllability factor and the features of the applied navigation modes.

However, the well-known probabilistic approaches and methods (Blom et al., 2001; Paielli & Erzberger, 1997; Prandini et al., 2000) have significant limitations. These methods are usually quite complicated, and their algorithmization and computer realization require substantial simplifications.

The methods of conflict detection and evaluation considered in this section relate to a group of methods based on stochastic uncertainty prediction of aircraft location due to their deviation from the flight plan. The proposed methods enable to predict stochastic uncertainties of aircraft location not only on straight flight paths, but also in maneuvering areas, as well as to use the information about stochastic dynamics of motion.

General Statement of Conflict Detection and Conflict Evaluation Problem

In air traffic management systems the evaluation of relative position of the aircraft is performed in a unified coordinate system. Let us consider Cartesian coordinate system $Oxyh$. The axis Ox and Oy are located in a horizontal plane, the axis Oy is directed to the north, and the axis Oh is directed vertically (Figure 1).

Figure 1. Relative position of two aircraft flying at the intersecting routes

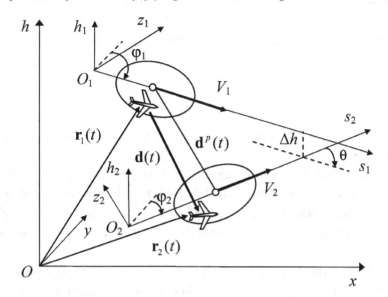

In order to take into account the features of navigation, piloting and control the description of each *j*-th aircraft motion is performed in a rectangular coordinate system $O_j s_j z_j h_j$ (Figure 1) associated with the planned flight route. The axes $O_j s_j$ and $O_j z_j$ are located in the horizontal plane. The axis $O_j s_j$ coincides with the planned route and is aligned with the assigned track angle φ_j. The axis $O_j z_j$ is perpendicular to the axis $O_j s_j$. The third axis $O_j h_j$ is directed vertically. This coordinate system enables to describe the movement of the aircraft in deviations from the planned flight path. The planned relative position of two aircraft at the given time *t* is defined by the vector $\mathbf{d}^p(t)$ at path-crossing angle θ and difference of the planned altitudes Δh.

It is assumed that trajectory parameters of each aircraft motion (coordinates of aircraft position, speed, heading, etc.) at the current time t_0 (initial moment of the prediction time interval) are determined with random errors due to the surveillance system errors and the trajectory estimation errors in ATM system. The main factors affecting the aircraft deviation from the flight plan are navigation errors, flight control errors, and wind.

Thus, for each aircraft at the initial moment of time, there is some area of aircraft position uncertainty. When predicting the aircraft position, the dimensions of uncertainty area increases due to the potential aircraft deviations from the flight plan in the longitudinal, lateral and vertical directions of motion.

The potential conflict between aircraft is determined by their relative position, i.e. by the length of the distance vector $\mathbf{d}(t)$ between the aircraft and the proximity of the areas of predicted aircraft position uncertainty.

Mathematical Statement of Conflict Probability Evaluation Problem

Solution of the conflict detection and evaluation problem is based on the evaluation of probability of separation minima violation between the aircraft.

If the probability density function of deviation distribution from the planned separation of the aircraft d^p in three-dimensional space x, y, h is known as $p(x, y, h)$ than the probability of a conflict for a given time t is defined as the probability of hitting the conflict zone $\Omega(t)$:

$$P_c(t) = \iiint_{\Omega(t)} p(x, y, h) dx \, dy \, dh. \tag{1}$$

The mathematical statement of conflict probability evaluation problem for a pair of aircraft in a general form is formulated as follows.

Let a position of the first aircraft in air space at a given moment of time t be determined by random vector $\mathbf{r}_1$ (Figure 1), which has Gaussian distribution $N(\mathbf{M}_1, \mathbf{D}_1)$, the position of the other aircraft is determined by random vector $\mathbf{r}_2$ with distribution $N(\mathbf{M}_2, \mathbf{D}_2)$. Mean vectors $\mathbf{M}_1, \mathbf{M}_2 \in \mathfrak{R}^3$. The matrices $\mathbf{D}_1, \mathbf{D}_2$ are the positively defined covariance matrices. It is assumed that vector $\mathbf{r}_1$ and vector $\mathbf{r}_2$ are mutually independent.

Under these assumptions the distance vector $\mathbf{d} = \mathbf{r}_2 - \mathbf{r}_1$ between the aircraft has Gaussian distribution $N(\mathbf{M}, \mathbf{D}^2)$, and this vector can be represented as a multidimensional random variable

$$\mathbf{d} = \mathbf{M} + \mathbf{D}\gamma, \tag{2}$$

where $\mathbf{M} = \mathbf{M}_2 - \mathbf{M}_1$; $\mathbf{D} = \sqrt{\mathbf{D}_1 + \mathbf{D}_2}$ is such positive definite matrix that $\mathbf{D}^2 = \mathbf{D}_1 + \mathbf{D}_2$; γ is a vector of random variables with zero mean values and unit variances, that has Gaussian distribution $N(0, \mathbf{I3})$; $\mathbf{I3}$ is unit 3×3 matrix.

The problem of conflict probability evaluation for a given time t is to calculate the probability that the distance between the aircraft will be less than given separation minimum (distance) d_{min}, that is

$$P_c(t) = P\{\|\mathbf{d}(t)\| \le d_{min}\}, \tag{3}$$

where $\|\mathbf{d}(t)\| = d(t)$, $\|\cdot\|$ denotes the Euclidean norm in $\mathfrak{R}^3$.

Analytical Solution of Conflict Probability Evaluation Problem

Let's reduce the vector of the distance $\mathbf{d}$ between aircraft to a form with independent components. This will provide an analytical solution of the problem in closed-form.

Taking into account (2), the expression (3) can be rewritten in the form

$$P_c = P\{\|\mathbf{M} + \mathbf{D}\gamma\|^2 \le d_{min}^2\}. \tag{4}$$

Let $\mathbf{e}_1, \mathbf{e}_2, \mathbf{e}_3$ be the orthonormal eigen-basis of the matrix $\mathbf{D}$ in the vector $\mathbf{d}$ representation of the form (2). The mean value vector $\mathbf{M}$ is expanded in this basis

$$\mathbf{M} = q_1 \mathbf{e}_1 + q_2 \mathbf{e}_2 + q_3 \mathbf{e}_3, \tag{5}$$

where q_i is determined by the scalar product $q_i=(\mathbf{M},\mathbf{e}_i)$, $i=\overline{1,3}$.

Let's represent $\mathbf{D}\boldsymbol{\gamma}$ in the basis of $\mathbf{e1}\,\mathbf{e2}\,\mathbf{e3}$.

$$\mathbf{D}\boldsymbol{\gamma} = [\lambda 1_\gamma 1\ \lambda 2_\gamma 2\ \lambda 3_\gamma 3]^T \tag{6}$$

where $\boldsymbol{\gamma} = [\gamma 1\,\gamma 2\,\gamma 3]^T$ is a vector of independent standard Gaussian random variables $\gamma 1\,\gamma 2\,\gamma 3$ with characteristics $N(0,\mathbf{I3})$; $\lambda 1\,\lambda 2\,\lambda 3$ are the eigen-values of the matrix D.

Then the random vector of distance **d** between aircraft in the basis of $\mathbf{e}_1,\mathbf{e}_2,\mathbf{e}_3$ based on (5), (6) has a distribution

$$N([q_1, q_2, q_3]^T, diag[\lambda_1^2, \lambda_2^2, \lambda_3^2]) \tag{7}$$

with independent components and can be written as

$$\mathbf{d} = \mathbf{M} + \mathbf{D}\boldsymbol{\gamma} = \begin{bmatrix} q_1 + \lambda_1\gamma_1 \\ q_2 + \lambda_2\gamma_2 \\ q_3 + \lambda_3\gamma_3 \end{bmatrix}. \tag{8}$$

After the expansion of distance vector with the distribution (7) in orthonormalized basis its components in expression (8) become independent that simplifies the solution of the problem. In this case the density function of the distance vector in some coordinate system of x_1, x_2, x_3 can be written as

$$p(x_1,x_2,x_3) = p(x_1)p(x_2)p(x_3) = \frac{1}{2\pi\lambda_1\lambda_2\lambda_3}\exp\left(-\frac{1}{2}\left(\frac{(x_1-q_1)^2}{\lambda_1^2}+\frac{(x_2-q_2)^2}{\lambda_2^2}+\frac{(x_3-q_3)^2}{\lambda_3^2}\right)\right). \tag{9}$$

For two-dimensional case when two aircraft are flying at the same flight level, there are two components in equation (9), and distance vector dispersion is represented in figure 2 by the ellipse relative to a point that is determined by the mean values q_1,q_2 of the variables x_1 and x_2.

The conflict probability P_c is defined as the probability that the random distance vector **d** with the characteristics $N([q_1, q_2]^T, diag[\lambda_1^2, \lambda_2^2])$ is hitting into the forbidden area Ω bounded by a circle with radius d_{min} centered at the origin of coordinates.

Taking into account the independence between random variables x_1 and x_2 the expression of conflict probability (1) for two aircraft can be written as

$$P_c = P\{(x_1,x_2)\in\Omega\} = \iint_\Omega \frac{1}{\lambda_1\sqrt{2\pi}}e^{-\frac{(x_1-q_1)^2}{2\lambda_1^2}}\frac{1}{\lambda_2\sqrt{2\pi}}e^{-\frac{(x_2-q_2)^2}{2\lambda_2^2}}\,dx_1dx_2. \tag{10}$$

Figure 2. The explanation of conflict detection and probability evaluation

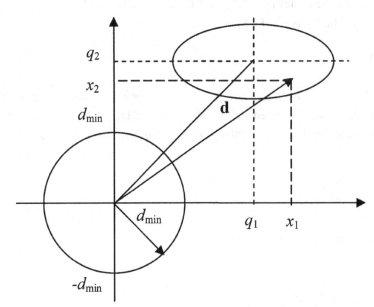

After integrating the external integral of (10) within the boundaries from $-d_{min}$ up to d_{min}, and the internal integral using the equation of a circle boundary $x_1^2 + x_2^2 = d_{min}^2$ from $-\sqrt{d_{min}^2 - x_2^2}$ up to $\sqrt{d_{min}^2 - x_2^2}$, the result is written as

$$
\begin{aligned}
P_c &= \int_{-d_{min}}^{d_{min}} \left(\frac{1}{\lambda_2 \sqrt{2\pi}} e^{-\frac{(x_2 - q_2)^2}{2\lambda_2^2}} \frac{1}{\lambda_1 \sqrt{2\pi}} \int_{-\sqrt{d_{min}^2 - x_2^2}}^{\sqrt{d_{min}^2 - x_2^2}} e^{-\frac{(x_1 - q_1)^2}{2\lambda_1^2}} \, dx_1 \right) dx_2 \\
&= \frac{1}{\lambda_2 \sqrt{2\pi}} \int_{-d_{min}}^{d_{min}} \left(e^{-\frac{(x_2 - q_2)^2}{2\lambda_2^2}} \left(\Phi\left(\frac{\sqrt{d_{min}^2 - x_2^2} - q_1}{\lambda_1} \right) - \Phi\left(\frac{-\sqrt{d_{min}^2 - x_2^2} - q_1}{\lambda_1} \right) \right) \right) dx_2.
\end{aligned}
$$

(11)

where $\Phi(x) = \frac{1}{\sqrt{2\pi}} \int_{-\infty}^{x} e^{-\frac{z^2}{2}} dz$.

The expression (11) determines the conflict probability for a fixed moment of time t. Calculation the conflict probability for successive fixed moments of time t enables to identify dynamics and to estimate the rate of conflict probability changes.

An important indicator of aircraft collision threat is the maximum value of the conflict probability $P_{max} = \max\{P_c(t)\}$.

The total conflict probability can be evaluated by averaging the values of the probabilities obtained for fixed moments of time t over a time of closest approach $t \in (t_0, t_k]$. In this case, for each fixed moment t the values of $q_1(t)$, $q_2(t)$ and $\lambda_1(t)$, $\lambda_2(t)$ are calculated, thus $P_c = P_c(t)$.

The total conflict probability can be approximated by the expression

$$P_k = \frac{1}{t_k - t_0} \sum_{i=0}^{k} P_c(t_i) \Delta t \,,$$

where t_0 is the moment when the violation of separation minimum begins; t_k is the moment when violation ends; Δt is sampling step.

Priori Information and Procedure of Analytical Method Implementation

Let define the procedure of preparation and implementation the conflict probability estimation method being considered when two aircraft are flying at the same altitude level.

For each j-th aircraft the mean value of its position $\mathbf{M}_j$ is the planned flight trajectory, thus is known.

Assuming that aircraft is flying at a constant speed at the predicted time interval in accordance with the plan (Figure 3), we can write down the changing of each aircraft planned position vector as

$$\mathbf{r}_j^p(t) = \mathbf{r}_j^p(0) + \mathbf{V}_j t \,, \tag{12}$$

where $\mathbf{r}_j^p(0)$ is the vector of initial position corresponding to the starting point $s_j(0)$ of the flight route; $\mathbf{V}_j$ is the vector of flight speed.

The ends of vectors $\mathbf{r}_j^p(t)$ in (12) are the centers of the aircraft position distribution $s_j(t)$ and moves along the corresponding path lines (Figure 3).

Figure 3. Relative motion of two aircraft according to flight plan

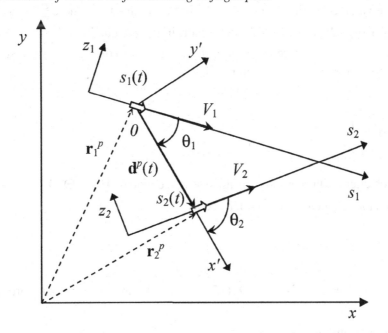

The vector of planned separation distance between the aircraft $\mathbf{d}^P = \mathbf{r}_2^P - \mathbf{r}_1^P$ (Figure 3) for a fixed point at time t is calculated as

$$\mathbf{d}^P(t) = \overrightarrow{s_1 s_2}(t) = \mathbf{r}_2^P(t) - \mathbf{r}_1^P(t) = \mathbf{r}_2^P(0) - \mathbf{r}_1^P(0) + t(\mathbf{V}_2 - \mathbf{V}_1).$$

Let's denote

$$M_d = M_d(t) = \| \overrightarrow{s_1 s_2}(t) \| = \| \mathbf{r}_2^P(0) - \mathbf{r}_1^P(0) + t(\mathbf{V}_2 - \mathbf{V}_1) \|. \tag{13}$$

To make next transformations, let introduce a Cartesian coordinate system $x'Oy'$ with the origin connected to the center of distribution $s_1(t)$ of one of the aircraft and oriented along the line connecting the centers of aircraft distribution $s_1(t)$ and $s_2(t)$, thus along the vector $\overrightarrow{s_1 s_2}$ (Figure 3).

Since it is assumed that in accordance with the plan aircraft must flying at a constant speed, the coordinate system $x'Oy'$ moves without changing its orientation. The positions of the second aircraft in the introduced coordinate system relative to the first one are described by a normal random vector $N(M_d \mathbf{e}_1, \mathbf{D}_d)$, where M_d is determined by the expression (13), $\mathbf{e}_1 = [1,0]^T$, and $\mathbf{D}_d$ is a covariance matrix of the distance vector, which can be determined as follows.

In some studies (Blin et al., 2000; Paielli & Erzberger, 1997) it is assumed that the deviation from a given flight trajectory occurs independently in the longitudinal and lateral movement, then the covariance matrix of aircraft deviations is diagonal $\mathbf{D}(t) = diag[\sigma_s^2(t), \sigma_z^2(t)]$, where σ_s, σ_z are corresponding standard deviations. So, it is assumed that the variance of lateral deviation is constant $\sigma_z^2 = const$ and is determined mainly by navigational error of lateral position measurements. For longitudinal motion it is assumed that the variance of the deviation at the initial moment of time σ_{s0}^2 is equal to the variance of the positioning error and increases with time according to the quadratic law $\sigma_s^2 = t^2 \sigma_{s0}^2$.

Let reduce the covariance matrices $\mathbf{D}_j(t)$ of each j-th aircraft to the coordinate system oriented along the relative distance vector, using a rotation matrix

$$\Theta_j = \begin{bmatrix} \cos\theta_j & -\sin\theta_j \\ \sin\theta_j & \cos\theta_j \end{bmatrix}. \tag{14}$$

As a result, each aircraft has corresponding covariance matrix $\mathbf{D}_j' = \Theta_j \mathbf{D}_j \Theta_j^T$, and covariance matrix for distance vector between aircraft is defined as

$$\mathbf{D}_d = \mathbf{D}_1' + \mathbf{D}_2'. \tag{15}$$

Let $\mathbf{e}_1, \mathbf{e}_2$ be the eigenvectors of matrix $\mathbf{D}_d$ (15), and λ_1^2, λ_2^2 are corresponding proper numbers $\lambda_i \geq 0$, $i=1,2$.

The expansion in eigen-basis $(\mathbf{e}_1, \mathbf{e}_2)$ gives

$$\mathbf{M} = q_1\mathbf{e}_1 + q_2\mathbf{e}_2; \quad \mathbf{D}_d = \lambda_1^2\mathbf{e}_1\mathbf{e}_1^T + \lambda_2^2\mathbf{e}_2\mathbf{e}_2^T. \tag{16}$$

Thus, in the basis of $(\mathbf{e}_1,\mathbf{e}_2)$ a random distance vector $\mathbf{d}$ between aircraft for a fixed time has a distribution at (16) as $N([q_1, q_2]^T, diag[\lambda_1^2, \lambda_2^2])$ and can be presented as

$$\mathbf{d} = \begin{bmatrix} q_1 \\ q_2 \end{bmatrix} + \begin{bmatrix} \lambda_1\gamma_1 \\ \lambda_2\gamma_2 \end{bmatrix}, \tag{17}$$

where γ_1,γ_2 are independent standard Gaussian random variables.

The probability of separation minimum violation for two aircraft under accepted conditions (17) is defined for (4) as

$$P_c = P\{\| \mathbf{d} \|^2 \le d_{min}^2\} = P\{(q_1 + \lambda_1\gamma_1)^2 + (q_2 + \lambda_2\gamma_2)^2 \le d_{min}^2\}. $$

After these preparatory actions have been completed, the estimation of conflict probability can be determined by the expression (11).

Conflict Probability Evaluation Considering Stochastic Flight Dynamics

Let's write down the mathematical model of the process of aircraft deviation from the planned trajectory, based on next assumptions (Figure 1):

- The motion of one aircraft does not depend on the other.
- For each aircraft the motion in the s, z, h directions are independent of each other.
- Aircraft motion is affected by random perturbations.

For each aircraft the system of stochastic dynamic models describing its deviations in the longitudinal, lateral, and vertical directions respectively can be determined in common case as:

$$\dot{\mathbf{S}} = f_s(\mathbf{S},t) + \mathbf{G}_s w_s;$$

$$\dot{\mathbf{Z}} = f_z(\mathbf{Z},t) + \mathbf{G}_z w_z; \tag{18}$$

$$\dot{\mathbf{H}} = f_h(\mathbf{H},t) + \mathbf{G}_h w_h,$$

where $\mathbf{S},\mathbf{Z},\mathbf{H}$ are the state vectors, each of vector includes deviations of position, speed, etc. for the corresponding coordinate *s, z,* and *h*; w_s, w_z, w_h are the Gaussian processes of "white noise" with zero mean values and unit intensities, simulating random perturbations of the flight process; $\mathbf{G}_s, \mathbf{G}_z, \mathbf{G}_h$ are corresponding matrices.

Let's define the system of linear dynamic models describing each aircraft deviations in a local coordinate system for a combined state vector $\mathbf{X}=[\mathbf{S},\mathbf{Z},\mathbf{H}]^T$ as

$$\begin{bmatrix} \dot{\mathbf{S}} \\ \dot{\mathbf{Z}} \\ \dot{\mathbf{H}} \end{bmatrix} = \begin{bmatrix} \mathbf{F}_s & 0 & 0 \\ 0 & \mathbf{F}_z & 0 \\ 0 & 0 & \mathbf{F}_h \end{bmatrix} \begin{bmatrix} \mathbf{S} \\ \mathbf{Z} \\ \mathbf{H} \end{bmatrix} + \begin{bmatrix} \mathbf{G}_s & 0 & 0 \\ 0 & \mathbf{G}_z & 0 \\ 0 & 0 & \mathbf{G}_h \end{bmatrix} \begin{bmatrix} w_s \\ w_z \\ w_h \end{bmatrix} \text{ or } \dot{\mathbf{X}} = \mathbf{F}\mathbf{X} + \mathbf{G}\mathbf{W}, \tag{19}$$

where $\mathbf{F}=diag[\mathbf{F}_s,\mathbf{F}_z,\mathbf{F}_h]$; $\mathbf{W}=[w_s,w_z,w_h]^T$; $\mathbf{G}=diag[\mathbf{G}_s,\mathbf{G}_z,\mathbf{G}_h]$.

Regardless of which mathematical model in (19) is used, the first component of the vectors $\mathbf{S},\mathbf{Z},\mathbf{H}$ is the corresponding aircraft position, it is s, z or h respectively. By combining these components the vectors $\mathbf{X}_j$ that include coordinates of aircraft position are defined:

- For the first aircraft $\mathbf{X}_1=[s_1, z_1, h_1]^T$
- For the second aircraft $\mathbf{X}_2=[s_2, z_2, h_2]^T$

The statistical characteristics of the vector $\mathbf{X}$ are defined by mean value vector $\mathbf{M}=E\{\mathbf{X}\}$ and covariance matrix $\mathbf{P}(t) = E\{[\mathbf{X}(t) - \mathbf{M}(t)][\mathbf{X}(t) - \mathbf{M}(t)]^T\}$.

The covariance matrix $\mathbf{P}(t)$ is symmetric and positive definite matrix which satisfies the following vector-matrix differential equation

$$\frac{d\mathbf{P}(t)}{dt} = \mathbf{F}\mathbf{P}(t) + \mathbf{P}(t)\mathbf{F}^T + \mathbf{G}\mathbf{G}^T, \text{ with } \mathbf{P}(0)=\mathbf{P}_0. \tag{20}$$

For the steady state condition, when $\lim_{t\to\infty}\dfrac{d\mathbf{P}(t)}{dt} = 0$, the matrix $\mathbf{P}' = \mathbf{P}(\infty)$ in (20) is a constant symmetric positive definite matrix and can be obtained by solving the algebraic Riccati equation

$$\mathbf{F}\mathbf{P}' + \mathbf{P}'\mathbf{F}^T + \mathbf{G}\mathbf{G}^T = 0. \tag{21}$$

The solution $\mathbf{P}' = diag[\mathbf{P}_s, \quad \mathbf{P}_z, \quad \mathbf{P}_h]$ includes the covariance matrices $\mathbf{P}_s,\mathbf{P}_z,\mathbf{P}_h$ of the vectors $\mathbf{S},\mathbf{Z},\mathbf{H}$ where the first component corresponds to the coordinates s,z,h.

Thus, the first components of the covariance matrices $\mathbf{P}_s,\mathbf{P}_z,\mathbf{P}_h$ (21) are the variances of the deviations in the longitudinal p_s, lateral p_z and vertical p_h directions respectively. By combining these components a diagonal matrix $\mathbf{P}_j = diag[p_s^j, p_z^j, p_h^j]$ for each j-th aircraft is formed, and this matrix determines the variance of the deviations of coordinates in the local coordinate system.

The characteristics of the random vector variable of the relative position of the aircraft are determined in a uniform coordinate system, which coincides with the local coordinate system for the first aircraft (Figure 1).

As a result, the position vector $\mathbf{X}_2'$ for the second aircraft in coordinate system $O_1 s_1 z_1 h_1$ is obtained

$$\mathbf{X}_2' = \mathbf{\Theta}' \mathbf{X}_2 + \mathbf{\Delta}, \tag{22}$$

where $\mathbf{\Theta}' = \begin{bmatrix} \mathbf{\Theta} & 0 \\ 0 & 1 \end{bmatrix}$; $\mathbf{\Theta}$ is the rotation matrix (14) for $\theta = \phi_1 - \phi_2$; $\mathbf{\Delta}$ is a vector of displacement the system $O2_s2_z2_h2$ relative to $O1_s1_z1_h1$.

Let's determine the process of relative position between aircraft in the coordinate system $O_1 s_1 z_1 h_1$ as

$$\mathbf{d}(t) = \mathbf{X}_2'(t) - \mathbf{X}_1(t) = [(s_2' - s_1), \quad (z_2' - z_1), \quad (h_2' - h_1)]^T. \tag{23}$$

The characteristics of the process (23) are determined as follows

$$\mathbf{M}_d(t) = E\{\mathbf{d}(t)\} = \mathbf{M}_2'(t) - \mathbf{M}_1(t), \tag{24}$$

where $\mathbf{M}_2' = E\{\mathbf{X}_2'\} = \mathbf{\Theta}' \mathbf{M}_2 + \mathbf{\Delta}$;

$$\mathbf{P}_d(t) = E\{[\mathbf{d}(t) - \mathbf{M}_d(t)][\mathbf{d}(t) - \mathbf{M}_d(t)]^T\}. \tag{25}$$

Since the aircraft movements are independent of each other, then (25) is written as

$$\mathbf{P}_d(t) = E\left\{[\mathbf{X}_2'(t) - \mathbf{M}_2'(t)][\mathbf{X}_2'(t) - \mathbf{M}_2'(t)]^T\right\} + E\left\{[\mathbf{X}_1(t) - \mathbf{M}_1(t)][\mathbf{X}_1(t) - \mathbf{M}_1(t)]^T\right\} = \mathbf{P}_2'(t) + \mathbf{P}_1(t).$$

Taking into account the rotation of the coordinate system $O_2 s_2 z_2 h_2$ covariance matrix is written as

$$\mathbf{P}_d(t) = \mathbf{\Theta}' \mathbf{P}_2(t) \mathbf{\Theta}'^T + \mathbf{P}_1(t), \tag{26}$$

where $\mathbf{P}_2(t) = E\left\{[\mathbf{X}_2(t) - \mathbf{M}_2(t)][\mathbf{X}_2(t) - \mathbf{M}_2(t)]^T\right\}$.

Vectors $\mathbf{M}_1, \mathbf{M}_2$ are planned aircraft positions. The matrices $\mathbf{P}_1$ and $\mathbf{P}_2$ are diagonal matrices. The components of the matrices are the variances of deviations in the longitudinal, lateral and vertical positions respectively for the first and second aircraft. In turn, these components are the first elements of the covariance matrices $\mathbf{P}_s, \mathbf{P}_z, \mathbf{P}_h$ that are the solution of equation (21).

Obtained above data is a priori information for estimating the conflict probability. On the basis of (24), (25) relative position vector can be represented as a multidimensional random variable $\mathbf{d} = \mathbf{M}_d + \mathbf{D}_s \boldsymbol{\gamma}$, where $\mathbf{D}_s = \sqrt{\mathbf{P}_d}$; $\boldsymbol{\gamma}$ is vector of standard Gaussian variables. Then the task of the conflict probability evaluation is reduced to an analytical method (11).

To apply the method, it is necessary to specify the mathematical models describing deviations from the programmed trajectory in the state space. For example, the laws of real trajectory control from Flight Management System (FMS) can be used, or models can be derived from the known correlation functions of deviation process for each coordinate.

Note that the proposed method makes it possible to solve the problem by synthesizing models of motion (18), (19) taking into account the laws of trajectory control, navigation errors and disturbances.

Case Study: Conflict Probability Evaluation Considering Stochastic Flight Dynamics for Two Aircraft Flying at the Same Level

Let's consider the solution of the problem when two aircraft are flying at the same level (Figure 4).

In this case the position vector for each aircraft consists of two components $\mathbf{X}=[s,z]^{\mathrm{T}}$.

To transform position vector of the second aircraft to the coordinate system of the first aircraft the displacement vector in (22) became as $\mathbf{\Delta}=\mathbf{\Theta}_1\mathbf{\Delta}_{xy}$, where $\mathbf{\Theta}_1$ is the rotation matrix (14) at the angle of $(\varphi_1-90°)$; $\Delta_{xy}=[\Delta x,\Delta y]^{\mathrm{T}}$; $\Delta x=x_{02}-x_{01}$; $\Delta y=y_{02}-y_{01}$.

The mean value vector of relative aircraft position (24) in this case is defined as

$$\mathbf{M}_d = (\mathbf{\Theta M2} + \mathbf{\Delta}) - \mathrm{M1}, \tag{27}$$

where $\mathrm{M1}\,\mathrm{M2}$ are aircraft planned positions in their local coordinate systems.

The second elements of these vectors according to the flight plan must be equal to zero, then

$$\mathbf{M}_1(t) = \begin{bmatrix} s_1^p(t) \\ 0 \end{bmatrix}; \; \mathbf{M}_2(t) = \begin{bmatrix} s_2^p(t) \\ 0 \end{bmatrix}.$$

Figure 4. Geometric explanation for the case study

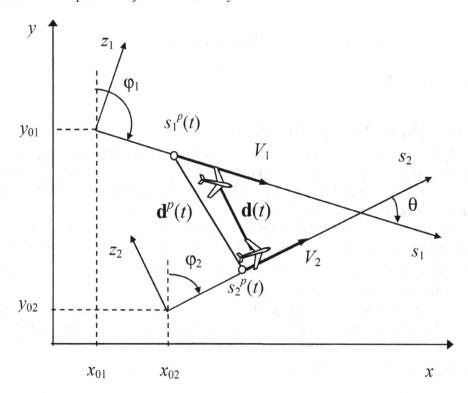

The covariance matrix (26) for the relative aircraft position in the horizontal plane is

$$\mathbf{P}_d(t) = \mathbf{P}_2'(t) + \mathbf{P}_1(t) = \mathbf{\Theta}\mathbf{P}_2(t)\mathbf{\Theta}^{\mathrm{T}} + \mathbf{P}_1(t), \tag{28}$$

where $\mathbf{P}_1$ and $\mathbf{P}_2$ are the diagonal matrices with components that are determined by solving the equation (21) for the variances of the deviations along the coordinates s_j, z_j.

The obtained statistical characteristics of the relative aircraft position process (27), (28) enable to represent the vector of relative position as a random two-dimensional variable $\mathbf{d}=\mathbf{M}_d+\mathbf{D}_s\boldsymbol{\gamma}$, where $\mathbf{D}_s = \sqrt{\mathbf{P}_d}$; $\boldsymbol{\gamma}$ is the vector of random unit variables $N(0, \mathbf{I2})$.

After performed transformations the evaluation of conflict probability is reduced to the procedure discussed above, which after orthonormalizaton leads to the expression (11).

Compositional Method of Conflict Probability Evaluation

Now the new method of conflict probability estimation is proposed which is called compositional method. This method uses a new approach to solve the problem of conflict probability estimation for a wider range of air traffic situations. Thus, it becomes possible to manage the uncertainty of the aircraft predicted position and to estimate the probabilities of conflict in the maneuvering areas.

The essence of the method is – the uncertainty of the future position of each aircraft is taken into account separately, and conflict probability is estimated as a result of a composition of two probabilities:

1. The probability of conflict due to deviations from the flight plan of one of the aircraft for a fixed position of the other aircraft.
2. The probability that this other aircraft will be in this fixed position.

Let us determine the conflict probability P_{co} between two aircraft at the predicted time t as the probability that the distance between one of the aircraft (for example, the first), which will be located at fixed position O with a certain probability, and the predicted uncertain position of the other aircraft will be less than separation minimum.

This definition is mathematically written as follows

$$P_{co} = P_{c/o}P_o, \tag{30}$$

where $P_{c/o}$ is the probability of separation minimum violation when one of the aircraft is located at a given fixed point O; P_o is the probability that this aircraft is located at this point O.

Let's derive the analytical expression for conflict probability evaluation and define the procedure for calculating the conflict probability with corresponding illustrations. For better clarity, let's consider the solution of the problem when two aircraft are flying at the same level.

The characteristics of predicted position uncertainty areas for the aircraft relative to their planned positions O_1, O_2 are calculated for a given prediction time (Figure 5).

Firstly, let's fix the predicted position of the first aircraft at the point O with the coordinates s_1, z_1 and determine the probability of separation minimum violation when aircraft is located at this fixed point. To do this, the auxiliary rectangular coordinate system xOy with the origin at the expected fixed posi-

Figure 5. Probabilities composition for conflict evaluation

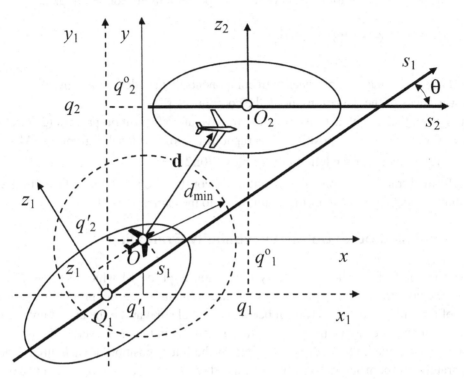

tion point of the first aircraft is introduced (Figure 5). The axis Ox of this coordinate system is oriented in parallel to the track line of the second aircraft.

Note that such orientation is always possible, and this enables to preserve the independence of random longitudinal and lateral deviations of the second aircraft in new coordinate system.

Under these conditions, the desired probability can be derived analytically using the results obtained earlier. The probability $P_{c/o}$ of separation minimum violation for the fixed point at time t is determined as the probability that the end of random distance vector $\mathbf{d}$ with coordinates x,y that are defined by the mean vector $[q_1^o, q_2^o]^T$ (the coordinates of the planned position of the second aircraft in the coordinate system xOy) and by the diagonal covariance matrix $diag[\lambda_{s2}^2, \lambda_{z2}^2]$ (variances of the longitudinal and lateral motion respectively) is hitting into the forbidden area Ω bounded by a circle with radius d_{min} centered at the origin of coordinates.

Taking into account the independence of random variables x and y the expression for the conflict probability can be written as

$$P_{c/o} = \iint_{\Omega} \frac{1}{\lambda_{s2}\sqrt{2\pi}} e^{-\frac{(x-q_1^o)^2}{2\lambda_{s2}^2}} \frac{1}{\lambda_{z2}\sqrt{2\pi}} e^{-\frac{(y-q_2^o)^2}{2\lambda_{z2}^2}} \, dxdy, \tag{31}$$

and integrating the external integral (31) within the boundaries from $-d_{min}$ up to d_{min}, and the internal integral using the equation of a circle boundary $x^2 + y^2 = d_{min}^2$ from $-\sqrt{d_{min}^2 - x_2^2}$ up to $\sqrt{d_{min}^2 - y^2}$ (Figure 5).

The result is written as

$$P_{c/o} = \frac{1}{\lambda_{z2}\sqrt{2\pi}} \int\limits_{-d_{min}}^{d_{min}} \left(e^{-\frac{(y-q_2^o)^2}{2\lambda_{z2}^2}} \left(\Phi\left(\frac{\sqrt{d_{min}^2 - y^2} - q_1^o}{\lambda_{s2}} \right) - \Phi\left(\frac{-\sqrt{d_{min}^2 - y^2} - q_1^o}{\lambda_{s2}} \right) \right) \right) dy .$$

(32)

The form of expression (32) is the same as (11), but in represented method the parameters $q_1, q_2, \lambda_{s2}, \lambda_{z2}$ are determined not as a result of orthonormalization, but using parallel orientation of axes of auxiliary coordinate system and a priori data. Thus it greatly simplifies the solution.

Obviously, for any other possible position of the first aircraft (Figure 5) in a point O, which is random, calculations of the probability of separation minimum violation is performed using the same expressions (31), (32), but after shifting the origin of xOy system to the new possible location O of first aircraft. This shifting is performed by replacing the mean values vector by $[(q_1 - q_1'), \quad (q_2 - q_2')]^T$, where q_1, q_2 are the coordinates of the planned position of the second aircraft relative to the planned position of the first aircraft in the $x_1 O_1 y_1$ system; q_1', q_2' are the deviations of the first aircraft from the planned position.

The probability of first aircraft position at a point with coordinates q_1', q_2' can be determined on the basis of a priori information about the variance of the deviations $\lambda_{s1}^2, \lambda_{z1}^2$ in longitudinal and lateral motions.

Under the condition of lateral and longitudinal deviations independences, the density of the deviations distribution is written as

$$p(s_1, z_1) = \frac{1}{2\pi\lambda_{s1}\lambda_{z1}} e^{-\frac{1}{2}\left(\frac{s_1^2}{\lambda_{s1}^2} + \frac{z_1^2}{\lambda_{z1}^2}\right)} .$$

(33)

Ultimately, according definition (30) the conflict probability, taking into account all possible first aircraft location, is found using the expressions (32), (33) as a result of their composition

$$P_c = \int\limits_{-\infty}^{\infty} \int\limits_{-\infty}^{\infty} p(s_1, z_1) P(q_1 - q_1'(s_1, z_1), q_2 - q_2'(s_1, z_1)) ds_1 dz_1 ,$$

(34)

where

$$P(q_1 - q_1'(s_1, z_1), q_2 - q_2'(s_1, z_1)) = \frac{1}{\lambda_{z2}\sqrt{2\pi}}$$

$$\int\limits_{-d_{min}}^{d_{min}} \left(e^{-\frac{(y-q_2-q_2'(s_1,z_1))^2}{2\lambda_{z2}^2}} \times \left(\Phi\left(\frac{\sqrt{d_{min}^2 - y^2} - q_1 - q_1'(s_1,z_1)}{\lambda_{s2}} \right) - \Phi\left(\frac{-\sqrt{d_{min}^2 - y^2} - q_1 - q_1'(s_1,z_1)}{\lambda_{s2}} \right) \right) \right) dy$$

is the probability of separation minimum violation (32) taking into account changing of first aircraft position; $q_1'(s_1, z_1) = s_1 \cos\theta - z_1 \sin\theta$, and $q_2'(s_1, z_1) = s_1 \sin\theta + z_1 \cos\theta$, are the deviations of the first aircraft from planned position, transformed from coordinate system s_1, z_1 to $x_1 O_1 y_1$; θ is the path-crossing angle.

Validation of Compositional Method of Conflict Probability Evaluation

To validate the proposed method and the correctness of performed analytical transformations, computer simulation was performed.

Computer simulation was carried out according the following scenario. It was assumed that each of aircraft is flying with planed speed of 480 knots; the angle of tracks intersection is equal to 90 degrees; the mean square deviation in lateral position is equal to 1 nautical mile; the mean square deviation in longitudinal position starts from zero and increases with a speed of 15 knots; safe separation minimum is equal to 5 nautical miles. The initial position of the aircraft was set at an equal distances of 80 miles from the point of tracks intersection.

The results of conflict probability calculation (34) for the given prediction time are shown in the table 1 in the row I. Calculated predicted closest approach time is equal to $t=10.0$ minutes. For comparison (the row II) the conflict probability was calculated using the analytical method by equation (11) for previously founded statistical characteristics of the random distance vector between the aircraft. The results of both methods are the same. That gives grounds to conclude that analytical transformations carried out when deriving the compositional method, and the approaches to evaluating the conflict probability are correct.

DECISION-MAKING UNDER AIR TRAFFIC CONFLICTS RESOLUTION

Most of well-known methods of air traffic conflicts resolution have a number of significant limitations. For example, the main disadvantage of the force fields methods (Eby, 1994; Eby & Kelly, 1999; Kosecka, Tomlin, Pappas, & Sastry, 1997) and of their developments is the complexity of synthesized conflict-free trajectories for realization. The disadvantage of the optimization methods (Bicchi, & Pallottino, 2000; Cafieri, & Durand, 2014; Cetek, 2009; Hu, Prandini, & Sastry, 2000, 2002) is the application of a single optimality criterion for trajectories selection. The common disadvantage is that existing methods do not use a combination of heading, speed and altitude change maneuvers and are not used in ATC nowadays.

New methods should provide the synthesis of realistic conflict-free trajectories using combined maneuvers in three-dimensional space according to different optimality criteria.

Table 1. Computer simulation results

Prediction Time t, Minutes		9.0	9.2	9.4	9.6	9.8	10.0	10.2	10.4	10.6	10.8	11.0
Conflict Probability	I	0.003	0.036	0.177	0.457	0.725	0.822	0.707	0.443	0.188	0.052	0.009
	II	0.003	0.036	0.177	0.457	0.725	0.822	0.707	0.443	0.188	0.052	0.009

Multi-Criteria Decision-Making for Air Traffic Conflicts Resolution

Multi-Criteria Problem of Conflict-Free Trajectories Selection

The multi-criteria problem of conflict resolution is the determination of conflict-free flight trajectories taking into account several optimality criteria and constraints. In order to perform the multi-criteria selection the alternatives, optimality criteria and constraints are defined.

Alternatives are either a set of trajectories of a single aircraft, or combinations of trajectories of multiple aircraft that simultaneously make maneuvers to resolve the conflict. Maneuvering is considered as a change of heading, airspeed and vertical speed.

A set of possible trajectories is formed taking into account the various constraints that are: aircraft performance, aircraft priorities, rules of airspace use, weather conditions. Such constraints define the set of possible trajectories (combinations of trajectories) $\mathbf{D}_T$:

$$\mathbf{D}_T = \{\mathbf{T}_1, \mathbf{T}_2, \ldots, \mathbf{T}_k\} = \{\mathbf{T}_j\}, \, j = \overline{1, k}.$$

It is necessary to choose the conflict-free and optimal trajectory (combination of trajectories) from the set $\mathbf{D}_T$ according to the defined optimality criteria.

An unconditional requirement is the flight safety that is ensured by maintaining the separation minima.

Optimization is performed according to the flight regularity criterion, the flight economy criterion and the complexity of maneuvers criterion. These criteria should be taken into account simultaneously, because total deviations from a flight plan, fuel consumption and the number of flight profile changes do not have a direct relationship between themselves when combined maneuvers are applied.

Trajectories or their combinations are estimated according to defined optimality criteria. The following numerical estimations are used: c_1 – estimation of regularity based on deviations from the flight plan; c_2 – estimation of economy based on the fuel consumption; c_3 – estimation of complexity of maneuvers based on the number of flight profile changes. Numerical estimations constitute the vector:

$$\mathbf{C} = \{c_i\}, \, i = \overline{1, 3}.$$

To solve the problem of conflict resolution is to select the flight trajectory (combination of trajectories) $\mathbf{T}^*$ which is conflict-free (ensures the elimination of the conflict) and complies with the flight regularity criterion (minimal deviations from a flight plan), the flight efficiency criterion (minimal fuel consumption), and the criterion of the complexity of maneuvering (minimal number of flight profile changes).

The multi-criteria problem of conflict-free trajectories selection is written as follows:

$$\mathbf{T}^* = \arg \min \mathbf{C}(\mathbf{T}), \mathbf{T} \in \mathbf{D}_T, \mathbf{T} \notin \Omega,$$

where Ω – the space of a conflict where the separation minima are violated.

Generic Multi-Criteria Model of Conflict-Free Trajectories Selection

The following model of optimal conflict-free trajectory (combination of trajectories) selection is used.

The first step is the determination of the set of conflict-free trajectories (combination of trajectories) $\mathbf{D}_{\mathbf{T}}^{S}$ from the set of possible trajectories $\mathbf{D}_{\mathbf{T}}$. A trajectory (combination of trajectories) belongs to the conflict-free set if there are no conflicts during the flight:

$$\mathbf{D}_{\mathbf{T}}^{S} = \left\{ \mathbf{T} \in \mathbf{D}_{\mathbf{T}} \mid \mathbf{T} \notin \Omega \right\}.$$

The second step is the determination of the set of Pareto-optimal trajectories (combination of trajectories) Π from the set $\mathbf{D}_{\mathbf{T}}^{S}$:

$$\Pi = \left\{ \mathbf{T}_{p} \in \mathbf{D}_{\mathbf{T}}^{S} \mid \neg\exists\, \mathbf{T} \in \mathbf{D}_{\mathbf{T}}^{S} : \mathbf{C}(\mathbf{T}) \leq \mathbf{C}(\mathbf{T}_{p}), \mathbf{T}_{p} \neq \mathbf{T} \right\}.$$

If the set Π contains only one trajectory (combination of trajectories), the problem is solved. In the opposite case, a narrowing of the set Π is performed.

The numerical estimations c of trajectories from the set Π are transformed to normalized values $\overline{c}$ with the domain of allowable values $\mathbf{D}_{\overline{c}} = \left\{ \overline{c} \mid \overline{c} \in [0,1] \right\}$ using a linear transformation:

$$\overline{c}(\mathbf{T}) = \frac{c(\mathbf{T}) - \min\limits_{\mathbf{T} \in \Pi} c(\mathbf{T})}{\max\limits_{\mathbf{T} \in \Pi} c(\mathbf{T}) - \min\limits_{\mathbf{T} \in \Pi} c(\mathbf{T})}.$$

The normalized numerical estimations constitute the vector:

$$\overline{\mathbf{C}} = \left\{ \overline{c}_{i} \right\}, i = \overline{1,3}.$$

Selection of optimal conflict-free trajectory (combination of trajectories) $\mathbf{T}^{*}$ from the set of Pareto-optimal Π is performed with use of linear convolution of optimality criteria by solving the optimization problem:

$$\mathbf{T}^{*} = \arg\min\limits_{\mathbf{T} \in \Pi} F\left(\overline{\mathbf{C}}(\mathbf{T}), \mathbf{W} \right) = \arg\min\limits_{\mathbf{T} \in \Pi} \sum_{i=1}^{3} w_{i} \overline{c}_{i}(\mathbf{T}), \tag{35}$$

where w – weighting coefficients reflecting the relative importance of criteria and forming a vector $\mathbf{W} = \left\{ w_{i} \right\}, i = \overline{1,3}$ with the domain of allowable values $\mathbf{D}_{w} = \left\{ \mathbf{W} \mid \sum_{i=1}^{3} w_{i} = 1; w_{i} > 0, i = \overline{1,3} \right\}$.

It is impractical to use methods of weighting coefficients determination that are based on establishing the strict advantages of optimality criteria given in numerical form. This is due to the fact that the relative importance of criteria depends on the type and characteristics of the particular conflict situation.

Selection of optimal trajectory (combination of trajectories) $\mathbf{T}^*$ is considered as a decision-making problem under uncertainty where alternatives are trajectories (combination of trajectories) $\mathbf{T} \in \Pi$ and uncertain states are possible values of the vector of weighting coefficients $\mathbf{W} \in \mathbf{D}_w$.

It is proposed to determine the weighting coefficients taking into account qualitative and quantitative information about the set of optimality criteria. Flight regularity is considered as the priority indicator in comparison with economy. This is due to the fact that significant deviations from the planned flight trajectory can cause new conflict situations, overloads of airspace elements, etc. Economy, in turn, has a priority over the complexity of maneuvers because it is assumed that the aircraft fly in automatic mode.

Then the weighting coefficients ordered as follows: $w_1 \geq w_2 \geq w_3$. The domain of allowable values of weighting coefficients is defined as:

$$\mathbf{D}_w = \left\{ \mathbf{W} \middle| \sum_{i=1}^{3} w_i = 1; \ w_i \geq w_{i+1}, \ i = \overline{1,2}; \ w_3 \geq w_0 > 0 \right\}, \tag{36}$$

where w_0 – minimal weighting coefficient.

It is proposed to determine the numerical values of the weighting coefficients for each alternative trajectory (combination of trajectories) $\mathbf{T} \in \Pi$, taking into account their dependence on the values of the normalized estimations $\overline{c}(\mathbf{T})$.

Applying a cautious decision-making strategy, selection of the optimal trajectory is the selection of the best trajectory from the worst (the minimax rule is applied).

Taking into account the domain of allowable values $\mathbf{D}_w$ (36) the vector of weighting coefficients $\mathbf{W}^*$ for each trajectory (combination of trajectories) $\mathbf{T}$ can be determined as a result of solving the linear programming problem, where the function $f(\mathbf{W})$ is maximized, weighting coefficients w are variables and normalized estimations $\overline{c}(\mathbf{T})$ are constants:

$$f\left(\mathbf{W}^*\right) = \max_{\mathbf{W}} f\left(\mathbf{W}\right) = \max_{\mathbf{W}} \sum_{i=1}^{3} w_i \overline{c}_i \left(\mathbf{T}\right), \tag{37}$$

$$\begin{cases} \sum_{i=1}^{3} w_i = 1, \\ w_i - w_{i+1} \geq 0, \ i = \overline{1,2}, \\ w_3 \geq w_0 > 0. \end{cases}$$

The coefficient w_0 characterizes the minimum allowable importance of the complexity of maneuvering criterion and can be selected in accordance with the type and characteristics of the conflict situation. The allowable values of the coefficient w_0 for the convolution of the three criteria are in the interval

$$w_0 \in \left(0, \frac{1}{3} \right).$$

Thus, the optimization problem (35) taking into account the proposed method for determination of weighting coefficients (37) is reduced to the following:

$$\mathbf{T}^* = \arg \min_{\mathbf{T} \in \Pi} \max_{\mathbf{W} \in \mathbf{D}_w} F\left(\overline{\mathbf{C}}(\mathbf{T}), \mathbf{W} \right) = \arg \min_{\mathbf{T} \in \Pi} \max_{\mathbf{W} \in \mathbf{D}_w} \sum_{i=1}^{3} w_i \overline{c}_i (\mathbf{T}). \tag{38}$$

Method of Multi-Criteria Resolution of Two-Aircraft Conflict

Problem Statement

The problem of multi-criteria resolution of potential conflict between two aircraft in three-dimensional space is considered.

It is assumed that one aircraft changes heading, airspeed and vertical speed to avoid the conflict, the other aircraft flies according to planned trajectory.

The process is observed in the time interval $[t_0, t_k]$ where t_0 is the moment of detection of a potential conflict, t_k is the planned time of crossing the boundary of an area of air traffic service (ATS) airspace.

Controlled motion of the maneuvering aircraft is described using the vector differential equation:

$$\dot{\mathbf{X}}(t) = f\left(\mathbf{X}(t), \mathbf{U}(t), t \right), \ \mathbf{X}(t_0) = \mathbf{X}_0, \tag{39}$$

where $\mathbf{X} = [x \ y \ h \ V \ V_h \ \varphi]^{\mathrm{T}}$ – state vector; x, y – horizontal coordinates; h – altitude; V – true airspeed; V_h – vertical speed; φ – heading; $\mathbf{U} = [\beta_a \ V_a \ V_{ha}]^{\mathrm{T}}$ – vector of controls; β_a – assigned bank angle; V_a – assigned true airspeed; V_{ha} – assigned vertical speed.

The space of a conflict $\Omega(t)$ is a space of states where the separation minima are violated:

$$\Omega(t) = \left\{ \mathbf{X}(t) \big| \left(d\left(\mathbf{X}(t), \mathbf{X}_{ref}(t) \right) < d_S \right) \wedge \left(\Delta h\left(\mathbf{X}(t), \mathbf{X}_{ref}(t) \right) < h_S \right) \right\},$$

where $d(\mathbf{X}(t), \mathbf{X}_{ref}(t))$, $\Delta h(\mathbf{X}(t), \mathbf{X}_{ref}(t))$ – horizontal and vertical distance between aircraft; d_s, h_s – lateral (horizontal) and vertical separation minimum; $\mathbf{X}_{ref}(t)$ – state of the second aircraft.

The state $\mathbf{X}(t)$ belongs to the set of conflict-free states $\mathbf{D}_{\mathbf{X}}(t)$ if the separation minima are not violated $\mathbf{X}(t) \in \mathbf{D}_{\mathbf{X}}(t) \big| \mathbf{X}(t) \notin \Omega(t)$. The initial state is conflict-free $\mathbf{X}(t_0) \notin \Omega(t_0)$.

Controls are limited according to the aircraft performances. Limitations are dependent on states and time $\mathbf{U}(t) \in \mathbf{D}_{\mathbf{U}}(\mathbf{X}(t), t)$, where $\mathbf{D}_{\mathbf{U}}(\mathbf{X}(t), t)$ – a set of possible controls $\mathbf{U}(t)$.

The numerical estimations of trajectories are: $J_1(\mathbf{X}(t), \mathbf{U}(t), t)$ – absolute deviation from planned flight time; $J_2(\mathbf{X}(t), \mathbf{U}(t), t)$ – absolute deviation from planned altitude; $J_3(\mathbf{X}(t), \mathbf{U}(t), t)$ – fuel consumption; $J_4(\mathbf{X}(t), \mathbf{U}(t), t)$ – number of flight profile changes. The numerical estimations are defined as follows:

$$J_1 = \Lambda_1(\mathbf{X}(t_k), t_k), \tag{40}$$

$$J_2 = \Lambda_2(\mathbf{X}(t_k), t_k), \tag{41}$$

$$J_3 = \int_{t_0}^{t_k} \vartheta_3\left(\mathbf{X}(t), \mathbf{U}(t), t\right) dt + \Lambda_3\left(\mathbf{X}(t_k), t_k\right), \tag{42}$$

$$J_4 = \int_{t_0}^{t_k} \vartheta_4\left(\mathbf{X}(t), \mathbf{U}(t), t\right) dt + \Lambda_4\left(\mathbf{X}(t_k), t_k\right), \tag{43}$$

where Λ_1, Λ_2– absolute deviation from planned flight time and absolute deviation from planned altitude at the time moment t_k; ϑ_3 – instantaneous fuel consumption; ϑ_4 – rate of flight profile changes; Λ_3, Λ_4 – estimation of fuel consumption and estimation of flight profile changes for real crossing the boundary of ATS area relatively to actual position at the time moment t_k. Estimations constitute the vector

$$\mathbf{J}\left(\mathbf{X}(t), \mathbf{U}(t), t\right) = \left\{ J_i\left(\mathbf{X}(t), \mathbf{U}(t), t\right) \right\}, i = \overline{1,4}.$$

As a result the problem of multi-criteria conflict resolution is determined as follows:

$$\min_{\mathbf{U}(t) \in \mathbf{D}_U\left(\mathbf{X}(t), t\right)} \mathbf{J}\left(\mathbf{X}(t), \mathbf{U}(t), t\right), \mathbf{X}(t) \in \mathbf{D}_X(t), t \in [t_0, t_k]. \tag{44}$$

Synthesis of Conflict-Free Trajectories

The application of the Bellman's optimality principle (Bellman, 1957) defines the equation of multi-criteria dynamic programming for determination of a set $\mathbf{E}(\mathbf{X}(t),t)$ of Pareto-optimal estimations of conflict-free trajectories for a state $\mathbf{X}(t)$ (Vasyliev, 2016):

$$\mathbf{E}\left(\mathbf{X}(t), t\right) = \text{eff} \bigcup_{\mathbf{U}(t) \in \mathbf{D}_U\left(\mathbf{X}(t), t\right)} \left(\left\{ 0, 0, \int_t^{t+\tau} \vartheta_3 dt, \int_t^{t+\tau} \vartheta_4 dt \right\} \oplus \mathbf{E}\left(\mathbf{X}(t+\tau), t+\tau\right) \right), \ \mathbf{X}(t) \in \mathbf{D}_X(t),$$

$$\mathbf{X}(t+\tau) \in \mathbf{D}_X(t+\tau) \tag{45}$$

with boundary condition:

$$\mathbf{E}\left(\mathbf{X}^m(t_k), t_k\right) = \left\{ \Lambda_1\left(\mathbf{X}(t_k), t_k\right), \Lambda_2\left(\mathbf{X}(t_k), t_k\right), \Lambda_3\left(\mathbf{X}(t_k), t_k\right), \Lambda_4\left(\mathbf{X}(t_k), t_k\right) \right\},$$

where eff – the operator of determination of Pareto-optimal estimations ; τ – small value; $\oplus$ – direct sum.

The set of Pareto-optimal conflict-free flight trajectories Π is determined by the set of estimations $\mathbf{E}(\mathbf{X}_0, t_0)$ at the moment of conflict detection t_0.

The conflict resolution process is decomposed in time into k stages. It is assumed that aircraft maneuvers for conflict avoidance during stages $j = \overline{1, k-1}$ and returns to the planned flight trajectory during last stage k. Each stage corresponds to a time interval $[t_{j-1}, t_j]$.

Generally, it is considered that maneuvering aircraft can transit into the state $\mathbf{X}(j)$ at the stage j from several states $\mathbf{X}(j\text{-}1)$ at the previous stage $(j\text{-}1)$:

$$\mathbf{X}(j) = f(\mathbf{X}(j\text{-}1), \mathbf{U}(j\text{-}1)),$$

where $f(\bullet)$– is the transition function defined in a model of aircraft motion.

The final state $\mathbf{X}_k = \mathbf{X}(k)$ is specified only by the horizontal coordinates of the point at which an aircraft crosses the boundary of ATS area (control point). Aircraft can transit into the final state from all the states of the previous stage.

The sets $\mathbf{D}_U(\mathbf{X}(j\text{-}1))$ are defined at each stage j. Determination of sets of conflict-free controls $\mathbf{D}_U^S\left(\mathbf{X}(j-1)\right) \in \mathbf{D}_U\left(\mathbf{X}(j-1)\right)$ is based on the prediction of separation minima violations when aircraft transits from states $\mathbf{X}(j\text{-}1)$ under the action of controls $\mathbf{U}(j\text{-}1) \in \mathbf{D}_U(\mathbf{X}(j\text{-}1))$. The prediction within a time interval $[t_{j-1}, t_j]$ is performed using the geometrical method described in (Vasyliev, 2016). If the separation minima are not violated under control $\mathbf{U}(j\text{-}1) \in \mathbf{D}_U(\mathbf{X}(j\text{-}1))$, this control belongs to a set of conflict-free: $\mathbf{U}\left(j-1\right) \in \mathbf{D}_U^S\left(\mathbf{X}(j-1)\right)$.

Then the simulation of trajectories is performed, a set of conflict-free states $\mathbf{D}_X(j)$ and estimations $\Delta J_i(\mathbf{X}(j\text{-}1), \mathbf{U}(j\text{-}1))$ when aircraft transits from states $\mathbf{X}(j\text{-}1)$ under the action of conflict-free controls $\mathbf{U}\left(j-1\right) \in \mathbf{D}_U^S\left(\mathbf{X}(j-1)\right)$ are defined. As a result, the set of Pareto-optimal estimations $\mathbf{E}(\mathbf{X}(j) \in \mathbf{D}_X(j))$ is determined.

The estimations ΔJ_i are defied using following expressions:

$$\Delta J_1\left(\mathbf{X}(j-1), \mathbf{U}(j-1)\right) = \begin{cases} 0, & j \neq k, \\ \left| t_k - t_f \right|, & j = k, \end{cases} \tag{46}$$

$$\Delta J_2\left(\mathbf{X}(j-1), \mathbf{U}(j-1)\right) = \begin{cases} 0, & j \neq k, \\ \left| h_k - h_f \right|, & j = k, \end{cases} \tag{47}$$

$$\Delta J_3(\mathbf{X}(j\text{-}1), \mathbf{U}(j\text{-}1)) = Q, \tag{48}$$

$$\Delta J_4(\mathbf{X}(j\text{-}1), \mathbf{U}(j\text{-}1)) = \vartheta_{41} + \vartheta_{42} + \vartheta_{43} \tag{49}$$

$$\vartheta_{41} = \begin{cases} 1, & \left| \phi(j) - \phi(j-1) \right| > \Delta\phi \\ 0, \end{cases}, \quad \vartheta_{42} = \begin{cases} 1, & V_a(j-1) \neq V(j-1) \\ 0, \end{cases}, \quad \vartheta_{43} = \begin{cases} 1, & V_{ha}(j-1) \neq V_{h0} \\ 0, \end{cases}$$

where t_f – actual time of reaching the final state $\mathbf{X}_k$; h_k, h_f – planned and actual altitude of the control point overflight; Q – fuel consumption; $\Delta\varphi$ – parameter that takes into account the small heading changes; V_{h0} – planned vertical speed.

The set of Pareto-optimal estimations $\mathbf{E}(\mathbf{X}(j))$ is determined using the equation of multi-criteria dynamic programming in a recursive form (Vasyliev, 2016):

$$\mathbf{E}\left(\mathbf{X}\left(j\right)\right) = \text{eff} \bigcup_{\mathbf{X}(j-1)\in\blacklozenge(\mathbf{X}(j))} \left[\mathbf{E}\left(\mathbf{X}\left(j-1\right)\right) \oplus \left\{\Delta J_i\left(\mathbf{X}\left(j-1\right), \mathbf{U}'\left(j-1\right)\right)\right\}\right], \tag{50}$$

where $\boldsymbol{\Gamma}(\mathbf{X}(j))$ – the set of states at the stage (j-1) from which the transition into the state $\mathbf{X}(j)$ is possible; $\mathbf{U}'\left(j-1\right) \in \mathbf{D}_{\mathbf{U}}^s\left(\mathbf{X}\left(j-1\right)\right)$ – controls which allow an aircraft to transit from the state $\mathbf{X}(j-1)\in\boldsymbol{\Gamma}(\mathbf{X}(j))$.

The solution of equation (11) also defines the set $\mathbf{D}_{\mathbf{U}}^{\mathbf{E}}\left(\mathbf{X}\left(j\right)\right)$ of Pareto-optimal controls that allow an aircraft to transit into the state $\mathbf{X}(j)$ and the corresponding set of states $\boldsymbol{\Gamma}_{\mathbf{E}}(\mathbf{X}(j))\in\boldsymbol{\Gamma}(\mathbf{X}(j))$.

The set of Pareto-optimal conflict-free trajectories is defined as:

$$\Pi = \{\mathbf{T}\in\mathbf{K}|\mathbf{J}(\mathbf{T})\in\mathbf{E}(\mathbf{X}_k)\},$$

where $\mathbf{K}$ – the set of full trajectories $\mathbf{T}=\{\mathbf{X}_0,\mathbf{X}(1),\ldots,\mathbf{X}_k\}$ by which an aircraft transits from the initial state $\mathbf{X}_0$ into the final state $\mathbf{X}_k$. Each Pareto-optimal conflict-free trajectory $\mathbf{T}\in\Pi$ corresponds to the Pareto-optimal controls program $\mathbf{T}_{\mathbf{U}}$.

Selection of optimal conflict-free trajectory $\mathbf{T}^*$ from the set of Pareto-optimal trajectories Π is performed by solving the optimization problem (38), where the vector of normalized estimations $\overline{\mathbf{C}} = \left\{\overline{c}_1, \overline{c}_2, \overline{c}_3\right\}$ is defined from the vector of estimations $\mathbf{J}=\{J_1,J_2,J_3,J_4\}$ using the following procedure:

- Normalized estimation of economy $\overline{c}_2$ and normalized estimation of complexity of maneuvers $\overline{c}_3$ are calculated using linear transformation:

$$\overline{c}_i\left(\mathbf{T}\right) = \frac{J_{i+1}\left(\mathbf{T}\right) - \min\limits_{\mathbf{T}\in\mathbf{P}} J_{i+1}\left(\mathbf{T}\right)}{\max\limits_{\mathbf{T}\in\mathbf{P}} J_{i+1}\left(\mathbf{T}\right) - \min\limits_{\mathbf{T}\in\mathbf{P}} J_{i+1}\left(\mathbf{T}\right)}, \; i = \overline{2,3}.$$

- Normalized estimation of flight regularity $\overline{c}_1$ is the linear convolution of deviation from planned flight time J_1 and deviation from planned altitude J_2 that are equally important:

$$\overline{c}_1\left(\mathbf{T}\right) = \frac{J_{12}\left(\mathbf{T}\right) - \min\limits_{\mathbf{T}\in\mathbf{P}} J_{12}\left(\mathbf{T}\right)}{\max\limits_{\mathbf{T}\in\mathbf{P}} J_{12}\left(\mathbf{T}\right) - \min\limits_{\mathbf{T}\in\mathbf{P}} J_{12}\left(\mathbf{T}\right)},$$

$$J_{12}\left(\mathbf{T}\right) = \frac{J_1\left(\mathbf{T}\right) - \min\limits_{\mathbf{T}\in\mathbf{P}} J_1\left(\mathbf{T}\right)}{\max\limits_{\mathbf{T}\in\mathbf{P}} J_1\left(\mathbf{T}\right) - \min\limits_{\mathbf{T}\in\mathbf{P}} J_1\left(\mathbf{T}\right)} + \frac{J_2\left(\mathbf{T}\right) - \min\limits_{\mathbf{T}\in\mathbf{P}} J_2\left(\mathbf{T}\right)}{\max\limits_{\mathbf{T}\in\mathbf{P}} J_2\left(\mathbf{T}\right) - \min\limits_{\mathbf{T}\in\mathbf{P}} J_2\left(\mathbf{T}\right)}.$$

It means that the normalized estimation $\bar{c}_1$ in the problem (38) is the embedded linear convolution of the estimations J_1 and J_2 (Fig. 6).

Model of Aircraft Motion

The simulation of trajectories is performed using the kinematics-energy model of the controlled aircraft motion (Vasyliev, 2013). The functional scheme of the kinematics-energy model is presented in Fig. 7.

The kinematic model provides determination of aircraft coordinates, true airspeed, heading and course. The simplified dynamic model is the BADA Operations Performance Model (OPM) (Eurocontrol Experimental Centre, 2011) that uses real performance characteristics of various aircraft types from the database and allows calculating the fuel consumption. Trajectory control is provided by the formation of the assigned values of engine thrust, bank and pitch in the model of controls. Stabilized parameters, depending on the navigation method, are linear lateral deviation from the track, heading or course, as well as true airspeed and altitude.

The input parameters for BADA OPM are: true airspeed V, vertical speed V_h and altitude h from the kinematic model; engine thrust Th and assigned bank angle β_a from the model of controls. The output parameters of BADA OPM are acceleration a and fuel consumption Q. The input parameters for kinematic model are: acceleration a from BADA OPM; assigned bank angle β_a and assigned pitch angle θ_a from the model of controls. The output parameters of kinematic model are coordinates x, y, altitude h, heading φ, course ψ, true airspeed V, ground speed W and vertical speed V_h. All parameters defined by the kinematic model, and acceleration from BADA OPM are transmitted to the model of controls.

Figure 6. Transformation of estimations

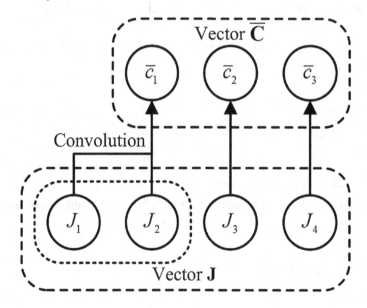

Figure 7. Functional scheme of the kinematics-energy model

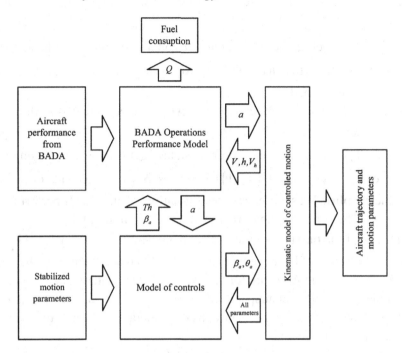

Discretization Aspects

Discretization step Δt is defined taking into account values of possible controls. The main requirement is the stabilization of assigned airspeed during time interval Δt. The number of stages is defined using following expression:

$$k = \left[\frac{t_k + \left(t_{los} + t_{end} \right)/2 - 2t_0}{2\Delta t} \right],$$

where t_{los} – time of potential conflict start; t_{end} – time of potential conflict end; $[\bullet]$– rounding operator.

Time interval $[t_{j-1}, t_j]$, $t_j = t_{j-1} + \Delta t$ corresponds to each stage j, except for the last one. The time interval of the last stage $j=k$ is different because of the different time of reaching the fixed final state when transiting from the states at the previous stage (k-1).

It is assumed that each aircraft can make a left/right turn with absolute value of bank angle $|\beta|$ or to do not change the heading; to increase/decrease the airspeed on ΔV or to do not change it; to increase/decrease the vertical speed on ΔV_h or to do not change it.

When applying controls $\mathbf{U}(j-1)$, an aircraft transits into different states $\mathbf{X}'(j)$ with efficiency estimations $\Delta J_i'\big(\mathbf{X}(j-1), \mathbf{U}(j-1)\big)$ that are defined using expressions (46)-(49). However, the using of discrete dynamic programming requires the ability of aircraft to transit into the state $\mathbf{X}(j)$ from several states $\mathbf{X}(j$-1).

It is proposed to introduce the rule for formation of new states $\mathbf{X}(j)$ which combine states $\mathbf{X}'(j)$. At the stage $j=1$ all states are new. At the stages $j = \overline{2, k-1}$ the new states $\mathbf{X}(j)$ are formed as a combination of states $\mathbf{X}'(j)$ in which altitudes are equal, horizontal coordinates and headings have proximate values. It means that an aircraft can transit into the state $\mathbf{X}(j)$ under the action of several controls $\mathbf{U}'(j-1)$. As a result the set $\Gamma(\mathbf{X}(j))$ is defined.

Coordinates and heading of an aircraft in new state $\mathbf{X}(j)$ are determined as arithmetic mean of these parameters for the states $\mathbf{X}'(j)$ which are combined in this new state $\mathbf{X}(j)$. Estimations $\Delta J_i\left(\mathbf{X}(j-1), \mathbf{U}'(j-1)\right)$ when transiting into new states $\mathbf{X}(j)$ is determined using nearest-neighbor interpolation of values $\Delta J_i'\left(\mathbf{X}(j-1), \mathbf{U}(j-1)\right)$ for states $\mathbf{X}'(j)$ which are combined.

Let's consider the simplified case of new states formation when aircraft performs maneuvers with constant vertical speed. The corresponding graphs are represented in Fig. 8.

There are 9 states $\mathbf{X}'(1)$ at the stage $j=1$ and they all are new states $\mathbf{X}(1) = \mathbf{X}'(1)$. These states are grouped according to the values of bank angle that were used to reach them from the initial state $\mathbf{X}_0$. As a result 3 groups $\mathbf{A}_1(1)$, $\mathbf{A}_2(1)$ and $\mathbf{A}_3(1)$ are formed. Group $\mathbf{A}_1(1)$ contains the states that were reached during left (bank angle -β) on different airspeed. Group $\mathbf{A}_2(1)$ contains the states that were reached during straight flight (β=0) on different airspeed. Group $\mathbf{A}_3(1)$ contains the states that were reached during right turn (bank angle β) on different airspeed.

Groups $\mathbf{A}(1)$ are considered separately at the stage $j=2$. Generally, aircraft transits into 9 states $\mathbf{X}'(2)$ from each state $\mathbf{X}(1)$ under the action of controls. Particularly, there are 27 states $\mathbf{X}'(2)$ that are reached when transiting from all the states of a group $\mathbf{A}(1)$. Such states are grouped according to the values of bank angle that were used to reach them. As a result 3 groups $\mathbf{A}'(2)$ are formed, each group contains 9 states $\mathbf{X}'(2)$. Then each group $\mathbf{A}'(2)$ is analyzed separately. States from $\mathbf{A}'(2)$ are grouped according to values of airspeed. As a result 5 groups $\mathbf{A}''(2)$ are formed, each group contains from 1 to 3 states with equal airspeeds. States from each group $\mathbf{A}''(2)$ are combined into new state $\mathbf{X}(2)$ using techniques described above. Five new states $\mathbf{X}(2)$ that correspond to group $\mathbf{A}'(2)$ constitute a group $\mathbf{A}(2)$. Overall there are 45 new states $\mathbf{X}(2)$ and 9 group $\mathbf{A}(2)$ at the stage $j=2$.

For the stages $j = \overline{3, k-1}$ the procedure is similar. It is necessary to note that controls are applied only if they are conflict-free.

Proposed approach significantly increases the computation efficiency of the discrete dynamic programming algorithm.

Method of Multi-Criteria Resolution of Multi-Aircraft Conflict

Problem Statement

The problem of multi-criteria resolution of potential conflict between $n \geq 2$ aircraft in three-dimensional space is considered.

It is assumed that all aircraft change heading, airspeed and vertical speed to avoid the conflict. Controlled motion of each aircraft is described using the vector differential equation (39).

Figure 8. Graphs of states

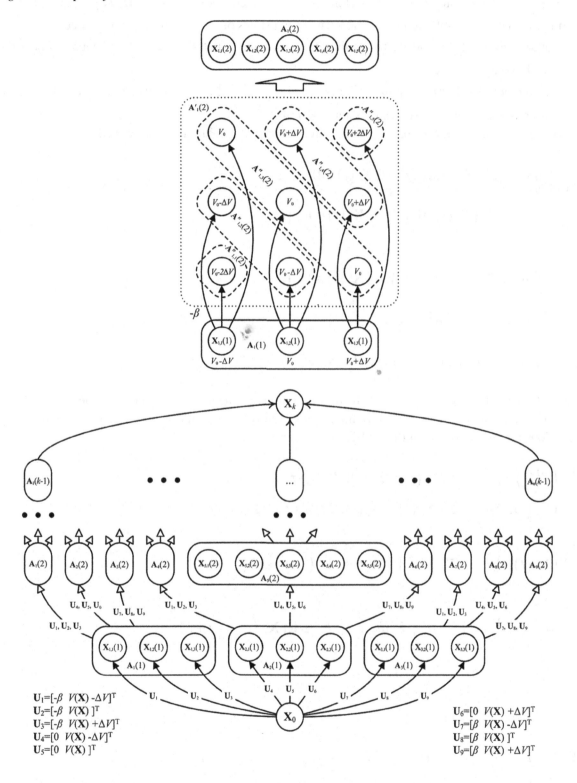

The state $\mathbf{X}^m(t)$ of aircraft $m = \overline{1,n}$ belongs to the set of conflict-free states $\mathbf{D}_{\mathbf{X}}^m(t)$ if the separation minima with other aircraft are not violated. The initial states of all aircraft are conflict-free.

Controls are limited according to the aircraft performances. Let $\mathbf{D}_{\mathbf{U}}^m\left(\mathbf{X}^m(t),t\right)$ be a set of possible controls $\mathbf{U}^m(t)$ in a state $\mathbf{X}^m(t)$.

The vectors of numerical estimations of trajectories $\mathbf{J}^m = \left\{J_i^m\right\}$, $i = \overline{1,4}$ at time interval $[t_0, t_k]$ for each aircraft m are defined using expressions (40)-(43).

As a result the problem of optimal multi-aircraft conflict resolution is determined as follows:

$$
\begin{cases}
\displaystyle\min_{\mathbf{U}^1(t)\in\mathbf{D}_{\mathbf{U}}^1\left(\mathbf{X}^1(t),t\right)} \mathbf{J}^1\left(\mathbf{X}^1(t),\mathbf{U}^1(t),t\right), \mathbf{X}^1(t)\in\mathbf{D}_{\mathbf{X}}^1(t), \\[3mm]
\displaystyle\min_{\mathbf{U}^2(t)\in\mathbf{D}_{\mathbf{U}}^2\left(\mathbf{X}^2(t),t\right)} \mathbf{J}^2\left(\mathbf{X}^2(t),\mathbf{U}^2(t),t\right), \mathbf{X}^2(t)\in\mathbf{D}_{\mathbf{X}}^2(t), \\[3mm]
\cdots \\[3mm]
\displaystyle\min_{\mathbf{U}^n(t)\in\mathbf{D}_{\mathbf{U}}^n\left(\mathbf{X}^n(t),t\right)} \mathbf{J}^n\left(\mathbf{X}^n(t),\mathbf{U}^n(t),t\right), \mathbf{X}^n(t)\in\mathbf{D}_{\mathbf{X}}^n(t),
\end{cases}
\tag{51}
$$

$$
t \in [t_0, t_k]
$$

Synthesis of Conflict-Free Trajectories

Let $\mathbf{E}_{\Xi}(t)$ be a set of Pareto-optimal estimations of combinations of the conflict-free trajectories. Based on Bellman's principle of optimality (Bellman, 1957) the system of multi-criteria dynamic programming equations for determination of the set $\mathbf{E}\Xi_{(t)}$ is written as follows:

$$
\begin{cases}
\mathbf{E}\left(\mathbf{X}^1(t),t\right) = \operatorname{eff}\mathfrak{I}^1, \mathbf{X}^1(t)\in\mathbf{D}_{\mathbf{X}}^1(t), \mathbf{X}^1(t+\tau)\in\mathbf{D}_{\mathbf{X}}^1(t+\tau), \\[2mm]
\mathbf{E}\left(\mathbf{X}^2(t),t\right) = \operatorname{eff}\mathfrak{I}^2, \mathbf{X}^2(t)\in\mathbf{D}_{\mathbf{X}}^2(t), \mathbf{X}^2(t+\tau)\in\mathbf{D}_{\mathbf{X}}^2(t+\tau), \\[2mm]
\cdots \\[2mm]
\mathbf{E}\left(\mathbf{X}^n(t),t\right) = \operatorname{eff}\mathfrak{I}^n, \mathbf{X}^n(t)\in\mathbf{D}_{\mathbf{X}}^n(t), \mathbf{X}^n(t+\tau)\in\mathbf{D}_{\mathbf{X}}^n(t+\tau),
\end{cases}
$$

$$
\mathfrak{I}^m = \bigcup_{\mathbf{U}^m(t)\in\mathbf{D}_{\mathbf{U}}^m\left(\mathbf{X}^m(t),t\right)} \left(\left\{0,0,\int_t^{t+\tau}\lambda_3^m dt, \int_t^{t+\tau}\lambda_4^m dt\right\} \oplus \mathbf{E}\left(\mathbf{X}^m(t+\tau),t+\tau\right)\right),
$$

$$
m = \overline{1,n},
$$

with boundary conditions:

$$
\mathbf{E}\left(\mathbf{X}^m(t_k),t_k\right) = \left\{\Lambda_1^m\mathbf{X}^m(t_k),t_k, \Lambda_2^m\mathbf{X}^m(t_k),t_k, \Lambda_3^m\mathbf{X}^m(t_k),t_k, \Lambda_4^m\mathbf{X}^m(t_k),t_k\right\}.
$$

The set Π_{Ξ} of Pareto-optimal combinations of conflict-free flight trajectories is determined by the set of estimations $E\Xi_{(t}0)$ at the moment of conflict detection t0.

The conflict resolution process is discretized in time and in state space. Generally, discretization procedure is similar to the described procedure for the method of two-aircraft conflict resolution. Some discretization aspects will be described below.

The flight of each aircraft is considered separately, taking into account the overall limitation on maintaining the separation minima. States $\mathbf{X}^m(j)$ of an aircraft m form the set $\mathbf{D}^m(j)$ of possible states.

The separation minima violations are monitored in defined states at each stage. The set of conditionally conflict-free states $\widehat{\mathbf{D}}_{\mathbf{X}}^m(j)$ of aircraft m includes the states $\mathbf{X}^m(j) \in \mathbf{D}^m(j)$ in which the violations of separation minima with other aircraft may be present for some of their states $\mathbf{X}^{ref}(j)$, or the violations of separation minima are absent:

$$\mathbf{X}^m(j) \in \widehat{\mathbf{D}}_{\mathbf{X}}^m(j) : \exists \mathbf{X}^{ref}(j), \left((d \geq d_s) \wedge (\Delta h \geq h_s) \right) \vee \left((d \geq d_s) \wedge (\Delta h < h_s) \right) \vee \left((d < d_s) \wedge (\Delta h \geq h_s) \right),$$

$$\mathbf{X}^{ref}(j) \in \mathbf{D}^{ref}(j), ref = \overline{1, n}, m \neq ref,$$

where d, Δh – horizontal distance and vertical interval between aircraft in states $\mathbf{X}^m(j)$ and $\mathbf{X}^{ref}(j)$ respectively.

The sets $\mathbf{D}_{\mathbf{U}}^m\left(\mathbf{X}^m(j-1)\right)$ are defined at each stage j. Then the simulation of trajectories is performed, sets of conditionally conflict-free states $\widehat{\mathbf{D}}_{\mathbf{X}}^m(j)$ and estimations $\Delta J_i^m\left(\mathbf{X}^m(j-1), \mathbf{U}^m(j-1)\right)$ when aircraft transits from states $\mathbf{X}^m(j-1)$ under the action of controls $\mathbf{U}^m(j-1)$ are defined.

The trajectories simulation is performed using the described kinematics-energy model of the controlled aircraft motion. The estimations ΔJ_i^m for aircraft m are defied using expressions (46)-(49).

The set $\mathbf{E}(\mathbf{X}^m(j))$ of Pareto-optimal estimations for aircraft m is determined using the recursive equation of multi-criteria dynamic programming (50) taking into account the conditions:

$$\mathbf{X}^m(j) \in \widehat{\mathbf{D}}_{\mathbf{X}}^m(j), \ \mathbf{X}^m(j-1) \in \widehat{\mathbf{D}}_{\mathbf{X}}^m(j-1).$$

At each stage, except the last one, sets of Pareto-optimal estimations $\mathbf{E}(\mathbf{X}^m(j))$, corresponding sets $\mathbf{D}_{\mathbf{U}}^E\left(\mathbf{X}^m(j)\right)$ of Pareto-optimal controls and sets $\Gamma E_{(}\mathbf{X}m(j)) \in \Gamma(\mathbf{X}m(j))$ of Pareto-optimal states are determined.

Then the conflict-free combinations I(j) of states are determined. Each combination I(j) determines the combinations of Pareto-optimal trajectories, which transfer all the aircraft to the states of this combination. All conflict-free combinations at the stage j create the set DI(j). Let denote a conflict-free combination as follows:

$$\mathbf{I}(j) = \left\{ \mathbf{X}^1(j), \mathbf{X}^2(j), \ldots, \mathbf{X}^n(j) \right\} = \left\{ \mathbf{X}^m(j) \right\}, \mathbf{X}^m(j) \in \widehat{\mathbf{D}}_{\mathbf{X}}^m(j).$$

Each combination contains the states which meet the following conditions:

Condition 1: There are no separation minima violations between all aircraft:

$$\forall a,b : \left(\left(A \geq d_S \right) \wedge \left(B \geq h_S \right) \right) \vee \left(\left(A \geq d_S \right) \wedge \left(B < h_S \right) \right) \vee \left(\left(A < d_S \right) \wedge \left(B \geq h_S \right) \right),$$

$A = d(\mathbf{X}^a(j), \mathbf{X}^b(j))$, $B = \Delta h(\mathbf{X}^a(j), \mathbf{X}^b(j))$, $\mathbf{X}(j) \in \mathbf{I}(j)$, $a = \overline{1,n}$, $b = \overline{1,n}$, $a \neq b$.

Condition 2: All aircraft transit into states $\mathbf{X}^m(j) \in \mathbf{I}(j)$ under the action of Pareto-optimal controls $\mathbf{U}^m \left(j - 1 \right) \in \mathbf{D}_{\mathbf{U}}^{\mathbf{E}} \left(\mathbf{X}^m \left(j \right) \right)$ from the states $\mathbf{X}^m(j{-}1) \in \Gamma_{\mathbf{E}}(\mathbf{X}^m(j))$, which combinations where conflict-free at the previous stage:

$$\left\{ \mathbf{X}^m \left(j - 1 \right) \right\} \in \mathbf{D}_{\mathbf{I}} \left(j - 1 \right), \mathbf{X}^m \left(j - 1 \right) \in \Gamma_{\mathbf{E}} \left(\mathbf{X}^m \left(j \right) \right), \mathbf{X} \left(j \right) \in \mathbf{I} \left(j \right).$$

Condition 3: One of the following sub-conditions must be met for aircraft pairs in states $\mathbf{X}^a(j)$ and $\mathbf{X}^b(j)$:

- If $(A \geq d_S) \wedge (B \geq h_S)$ than aircraft must transit into the states $\mathbf{X}^a(j)$, $\mathbf{X}^b(j)$ from any states, which meet the Condition 2.
- If $(A \geq d_S) \wedge (B < h_S)$ than aircraft must transit into the states $\mathbf{X}^a(j)$, $\mathbf{X}^b(j)$ from states, which meet the Condition 2 and in which the horizontal and vertical separation or only horizontal separation between this aircraft is ensured.
- If $(A < d_S) \wedge (B \geq h_S)$ than aircraft must transit into the states $\mathbf{X}^a(j)$, $\mathbf{X}^b(j)$ from states, which meet the Condition 2 and in which the horizontal and vertical separation or only vertical separation between this aircraft is ensured.

At the last stage k for each aircraft m the set ξ^m of full trajectories $\mathbf{T}^m = \left\{ \mathbf{X}_0^m, \mathbf{X}^m \left(1 \right), \dots \mathbf{X}_k^m \right\}$, which transfer the aircraft to the final state $\mathbf{X}_k^m$ from the states $\mathbf{X}^m \left(k - 1 \right) \in \widehat{\mathbf{D}}_{\mathbf{X}}^m \left(k - 1 \right)$, is determined. Then the combinations $\mathbf{Z}$ of full conflict-free trajectories of all aircraft are determined. Combination Ξ is a combination of trajectories, which transfer the all aircraft from states of combination I(k-1) to the final states $\mathbf{X}_k^m$:

$$\Xi = \left\{ \mathbf{T}^1, \mathbf{T}^2, \dots, \mathbf{T}^n \right\} = \left\{ \mathbf{T}^m \right\},$$
$$\mathbf{T}^m \in \xi^m : \mathbf{X}^m \left(k - 1 \right) \in \mathbf{T}^m, \mathbf{X}^m \left(k - 1 \right) \in \mathbf{I} \left(k - 1 \right).$$

Each possible combination Ξ is characterized by extended vector of estimations JΣ(Ξ):

$$\mathbf{J}_\Sigma \left(\Xi \right) = \left\{ \mathbf{J}^m \left(\mathbf{T}^m \right) \right\}, \mathbf{J}^m \left(\mathbf{T}^m \right) \in \mathbf{E} \left(\mathbf{X}_k^m \right), m = \overline{1,n}.$$

Then the set Π_Ξ of Pareto-optimal combinations of conflict-free trajectories is determined:

$$\Pi_\Xi = \left\{ \Xi_p \middle| \neg \exists \, \Xi : \mathbf{J}_\Sigma \left(\Xi \right) \leq \mathbf{J}_\Sigma \left(\Xi_p \right), \Xi_p \neq \Xi \right\}.$$

Selection of optimal combination of conflict-free trajectories Ξ^* from the set of Pareto-optimal trajectories $\Pi\Xi$ is performed by solving the optimization problem (38), where the vector of normalized estimations $\overline{\mathbf{C}} = \left\{\overline{c}_1, \overline{c}_2, \overline{c}_3\right\}$ is defined from the vectors of estimations $\mathbf{J}^m = \left\{J_1^m, J_2^m, J_3^m, J_4^m\right\}$ using the following procedure:

- Normalized estimation of flight regularity $\overline{c}_1$ is the linear convolution of the total deviation from planned flight time and total deviation from planned altitude of all aircraft:

$$\overline{c}_1\left(\Xi\right) = \frac{J_{12}\left(\Xi\right) - \min\limits_{\Xi\in\Pi_\Xi} J_{12}\left(\Xi\right)}{\max\limits_{\Xi\in\Pi_\Xi} J_{12}\left(\Xi\right) - \min\limits_{\Xi\in\Pi_\Xi} J_{12}\left(\Xi\right)},$$

$$J_{12}\left(\Xi\right) = \frac{\sum\limits_{m=1}^{n} J_1^m\left(\mathbf{T}^m\right) - \min\limits_{\Xi\in\Pi_\Xi}\sum\limits_{m=1}^{n} J_1^m\left(\mathbf{T}^m\right)}{\max\limits_{\Xi\in\Pi_\Xi}\sum\limits_{m=1}^{n} J_1^m\left(\mathbf{T}^m\right) - \min\limits_{\Xi\in\Pi_\Xi}\sum\limits_{m=1}^{n} J_1^m\left(\mathbf{T}^m\right)} + \frac{\sum\limits_{m=1}^{n} J_2^m\left(\mathbf{T}^m\right) - \min\limits_{\Xi\in\Pi_\Xi}\sum\limits_{m=1}^{n} J_2^m\left(\mathbf{T}^m\right)}{\max\limits_{\Xi\in\Pi_\Xi}\sum\limits_{m=1}^{n} J_2^m\left(\mathbf{T}^m\right) - \min\limits_{\Xi\in\Pi_\Xi}\sum\limits_{m=1}^{n} J_2^m\left(\mathbf{T}^m\right)}, \quad \mathbf{T}^m \in \Xi.$$

- Normalized estimation of economy $\overline{c}_2$ and normalized estimation of complexity of maneuvers $\overline{c}_3$ are calculated using linear transformation of corresponding total values:

$$\overline{c}_i\left(\Xi\right) = \frac{\sum\limits_{m=1}^{n} J_{i+1}^m\left(\mathbf{T}^m\right) - \min\limits_{\Xi\in\Pi_\Xi}\sum\limits_{m=1}^{n} J_{i+1}^m\left(\mathbf{T}^m\right)}{\max\limits_{\Xi\in\Pi_\Xi}\sum\limits_{m=1}^{n} J_{i+1}^m\left(\mathbf{T}^m\right) - \min\limits_{\Xi\in\Pi_\Xi}\sum\limits_{m=1}^{n} J_{i+1}^m\left(\mathbf{T}^m\right)}, \quad \mathbf{T}^m \in \Xi, \, i = \overline{2,3}.$$

The normalized estimations $\overline{c}_1$, $\overline{c}_2$ and $\overline{c}_3$ in the problem (38) are the embedded linear convolutions (Fig. 9).

Discretization Aspects

To ensure separation minima between aircraft while they transiting between the states it is necessary to check the violations at fixed time moments in limited time periods during which an unobservable violations of separation cannot occur.

When the changes of heading, airspeed speed and vertical speed are used to avoid a conflict with crossing angle of initial tracks $\leq 90°$ the discretization step Δt is determined according to inequalities system:

$$\begin{cases} \Delta t < d_s/V_{\max}, \\ \Delta t < \Delta h_s/V_{h\max}, \end{cases}$$

where $V_{\max}$ – maximum ground speed of all aircraft; $V_{h\max}$ – maximum vertical speed of all aircraft.

Figure 9. Transformation of vectors of estimations

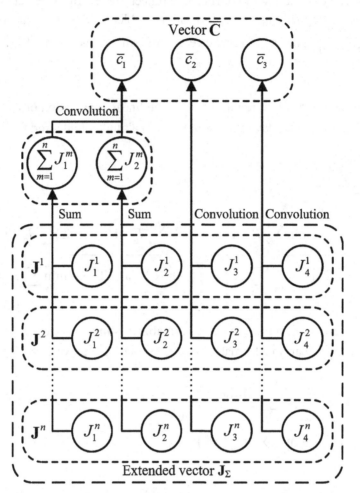

The number of stages is defined using following expression:

$$k = \left\lceil \frac{\sum\limits_{m=1}^{n}\left(t_m - t_0\right)}{2\Delta t \cdot n} + \frac{\sum\limits_{p=1}^{q}\left(t_{conf}^{pair} - t_0\right)}{2\Delta t \cdot l} \right\rceil,$$

where t_m –planned time of crossing the boundary of ATS area for aircraft $m = \overline{1, n}$; t_{conf}^{pair} – center of the time interval when the conflict between a pair of aircraft $pair = \overline{1, l}$ is existed; l – number of aircraft pairs.

The discretization of states and controls is performed using procedure described for the method of two-aircraft conflict resolution.

Computer Simulation

The analysis of the proposed methods is performed using computer simulation. Three conflict situations were simulated. The value of the lateral separation minimum is equal to d_s=18.5 km (10 nautical miles) and the value of vertical separation minima is equal to Δh_s=300 m (1000 feet).

Conflict situations N°1 and N°2 are two-aircraft conflicts. It was assumed that the aircraft N°1 should make manoeuvres to avoid a conflict. The processes were discretized in time on 5 stages. The discretization step for the first 4 stages is equal to Δt=60 s. Being in a certain state at stages $j = \overline{1,4}$ aircraft N°1 is able:

- To make a left/right turn with bank angle β=20° (turning time is limited to 15 s) or to do not change the heading.
- And to increase/decrease the airspeed on ΔV=5 m/s or to do not change it.
- And to increase/decrease the vertical speed on ΔV_h=3 m/s or to do not change it.

The minimal value of weighting coefficients in the optimization problem (38) is equal to w_0=0.1.

The initial parameters of the aircraft flight and characteristics of predicted conflict situations are presented in Table 2. Results of computer simulation are represented in Fig. 10-11 and Table 3.

Conflict situation N°3 is the conflict between five aircraft that perform a horizontal flight on intersecting routes (star-type conflict). It was assumed that all involved aircraft perform manoeuvres to avoid the conflict. Diecretization and possible manoeuvres are the same as for described two-aircraft conflicts except of possibility to change vertical speed, i.e. vertical speed is equal to zero.

The initial parameters of the aircraft flight are presented in Table 4. Results of computer simulation are represented in Fig. 12-13 and Tables 5-6.

The results of computer simulation show that optimal conflict-free trajectories ensure the conflict avoidance and comply with defined optimality criteria, and prove the correctness, efficiency and adequacy of proposed methods.

Table 2. Parameters of the aircraft flight and characteristics of predicted conflict situations

Parameter	Conflict Situation N°1		Conflict Situation N°2	
	Aircraft N°1	Aircraft N°2	Aircraft N°1	Aircraft N°2
Heading, degrees	0	270	0	90
Airspeed, m/s	195	220	220	240
Vertical speed, m/s	9	0	-12	-8
Initial altitude, m	7200	9150	11000	10000
Assigned flight level (altitude, m)	350 (10650)	300 (9150)	230 (7000)	
Time interval of separation violation, s	[183, 250]		[184, 295]	

Figure 10. The aircraft trajectories: a), b) – conflict situation Nº1 in three-dimensional space and in horizontal plane respectively; c), d) – conflict situation Nº2 in three-dimensional space and in horizontal plane respectively; 1 – planed trajectory for the aircraft Nº1; 2 – planed trajectory for the aircraft Nº2; 3 – control point on the route; 4 – optimal conflict-free trajectory for the aircraft Nº1; 5 – states at the stages; 6 – assigned airspeed for aircraft Nº1 at the stages.

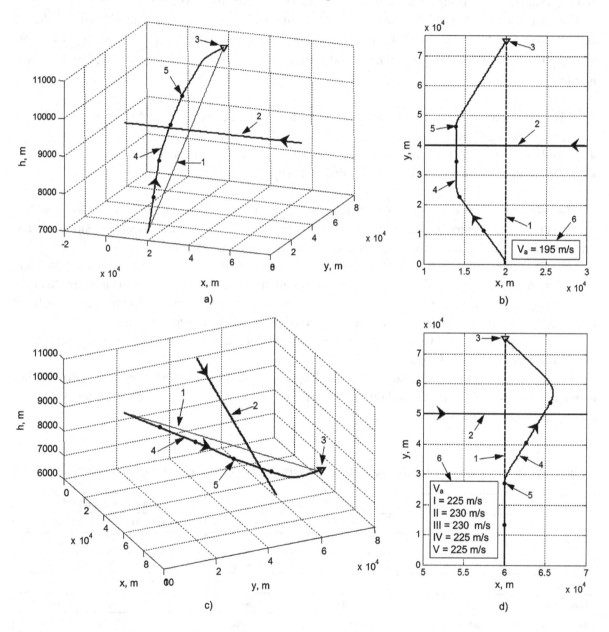

Figure 11. The dependence of distances between aircraft from time: a), b) – horizontal distance d and vertical distance Δh for conflict situation N°1; c), d) – horizontal distance d and vertical distance Δh for conflict situation N°2; 1 – during flight by planed trajectories; 2 – during conflict resolution; 3 – separation minimum; 4 – assigned vertical speed for aircraft N°1 at the stages.

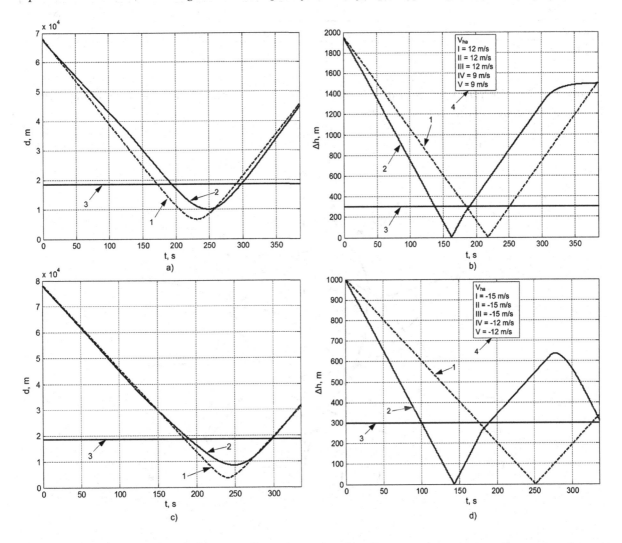

Table 3. The efficiency parameters of optimal conflict-free trajectory

Parameter	Conflict Situation N°1	Conflict Situation N°2
Deviation from the planned flight time, s	5,4	-2,9
Deviation from the assigned flight level at control point, m	0	0
Additional fuel consumption on the leg compared to planned trajectory, %	0,8	31
Number of flight profile changes	6	7

Table 4. Parameters of the aircraft flight

Parameter	Aircraft				
	N°1	N°2	N°3	N°4	N°5
Heading, degrees	0	72	144	216	288
Airspeed, m/s	220				
Flight level (altitude, m)	350 (10650)				

Figure 12. The trajectories for multi-aircraft conflict: 1 – planed trajectories; 2 – control points on the routes; 3 – optimal conflict-free trajectories

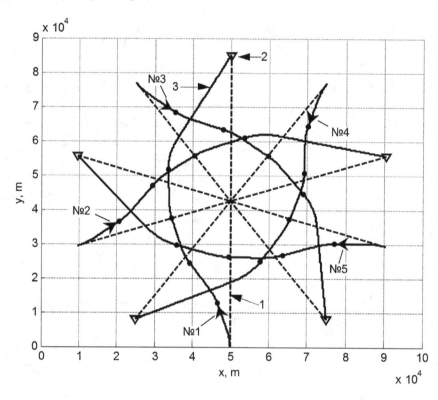

Cooperative Procedure for Air Traffic Conflicts Resolution in ATM System with Artificial Intelligence Tool

The cooperative procedure for conflicts detection and resolution in the medium-term period is proposed (Fig. 14). This procedure establishes the interactions between standard subsystems of the ATM system, multi-criteria conflict resolution subsystem (MTCR) and ATCO via controller working position (CWP).

The MTCR is a new artificial intelligence tool of the ATM system which uses the developed methods of air traffic conflicts resolution.

Sources of aircraft trajectory data are surveillance systems (primary and secondary surveillance radars, multilateration and automatic dependent surveillance -broadcast systems) that send plots/reports and local tracks to the surveillance data processing subsystem (SDPS). The SDPS performs multiradar/

Figure 13. The dependence of horizontal distances between aircraft from time: 1 – during flight by planed trajectories ; 2 – during conflict resolution; 3 – lateral separation minimum

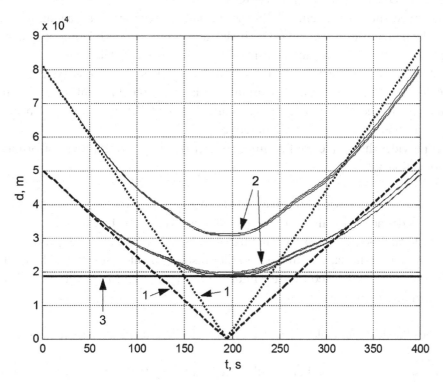

Table 5. The assigned airspeeds for conflict resolution

Assigned Airspeed	Stage				
	1	**2**	**3**	**4**	**5**
Aircraft N°1 and N°2, m/s	220	225	230	235	235
Aircraft N°3, N°4 and N°5, m/s	225	230	235	240	240

Table 6. The efficiency parameters of optimal conflict-free trajectories

Parameter	Aircraft N°1 and N°2	Aircraft N°3, N°4 and N°5
Deviation from the planned flight time, s	13,6	8,6
Additional fuel consumption on the leg compared to planned trajectory, %	14,7	17,4
Number of flight profile changes	8	9

multisensor tracking which results in system tracks that are transmitted to the medium-term conflict detection tool (MTCD) and monitoring aids (MONA).

Flight Data Processing Subsystem (FDPS) performs the processing and handling of flight plans, calculation of planned four-dimensional (4D) flight trajectories, and correlation of system flight plans with trajectory data. These data are also transmitted to MTCD and MONA.

The MTCD tool performs the prediction of 4D trajectories based on actual system tracks and flight plan data for detection of possible separation minima violations (potential conflicts) in the medium-term period. The data about detected conflicts is transmitted to the CWP for display. Conflict parameters, system tracks and flight plan data are transmitted to the MTCR.

The MTCR provides the synthesis of optimal conflict-free flight trajectories performing the following sequence of actions:

- Selection of aircraft that will perform avoidance maneuvers according to conflict resolution rules and priorities defined using flight plan data and inputted directly by ATCO.
- Synthesis of Pareto-optimal conflict-free trajectories (combinations of trajectories) according to defined optimality criteria, constraints, conflict resolution rules, trajectory data, aircraft performance, aeronautical and meteorological information. Synthesized trajectories are displayed on the CWP in formalized representation.

ATCO defines the decision-making method for conflict resolution from three possible:

- Automatic selection of the optimal conflict-free trajectory (combination of trajectories) – trajectory is selected from Pareto set by MTCR using defined multi-criteria optimization model.
- Manual selection of the optimal conflict-free trajectory (combination of trajectories) – trajectory is selected from Pareto set by ACTO.
- ATCO's own decision – ATCO independently determines the necessary maneuvers to resolve the conflict, forms the corresponding commands and inputs them on the CWP.

After the final approval the optimal conflict-free trajectories or ATCO's commands are formalized and transmitted to aircraft. For example, conflict-free trajectory can be converted into area navigation (RNAV) procedure.

Formalized conflict-free trajectory is transmitted directly to the on-board FMS using the controller-pilot data link communications system (CPDLC). These data are also transmitted to the FDPS subsystem, which updates system flight plans and recalculates planned 4D trajectories.

Monitoring of the execution of conflict-free trajectories or ATCO's commands is performed by MONA tools. When detecting deviations from selected conflict-free trajectory that do not lead to new conflicts, an appropriate signaling is created on the CWP. In such case ATCO defines the necessary maneuvers to eliminate deviations, forms the corresponding commands and inputs them on the CWP. If deviations from selected conflict-free trajectory lead to new conflicts all the procedure is repeated.

Figure 14. Operational procedure for air traffic conflicts detection and resolution

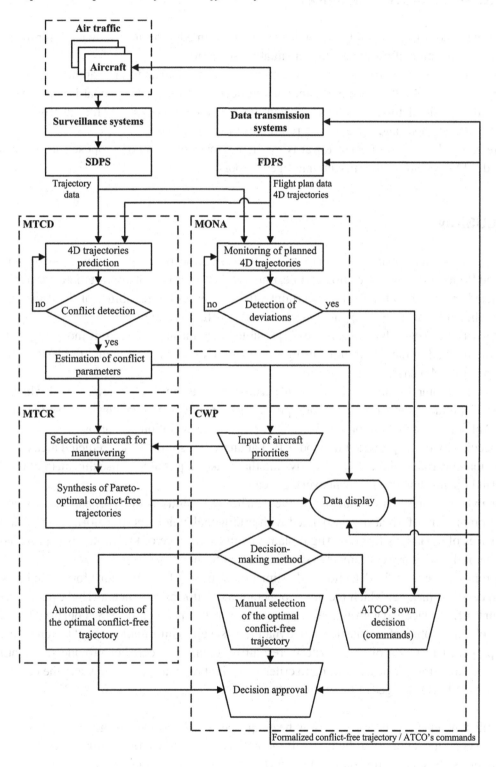

FUTURE RESEARCH DIRECTIONS

Development of conflict probability evaluation methods is aimed on a study of their use for maneuvering aircraft with simulation of different conflict situation scenarios.

Proposed methods for air traffic conflicts resolution are universal and can be applied for development of intelligent decision-making support functions in new ATM systems, on-board collision avoidance systems and ATC simulators, as well as for scientific researches, in particular for the efficiency assessment of various methods and systems of aircraft conflict resolution.

Future research may be directed towards the deployment of methods proposed in this chapter in automated ATM systems with artificial intelligence tools.

CONCLUSION

The method of conflict probability estimation, which takes into account the dynamic properties of aircraft motion and flight trajectory correlation in time, improves the adequacy of aircraft motion model, reduces the predicted area of aircraft position uncertainty, thus increases the reliability of the decision-making in solving the problem of detection and prevention of conflict situations.

The presence of flight dynamics makes it possible to synthesize the aircraft motion models and to develop the method in the direction of using information about navigation and the laws of trajectory control, including 4D navigation.

The proposed compositional method for conflict probability estimation is a new method, which removes a number of limitations inherent in existing probabilistic methods and enables to solve the problem of conflict probability estimation for a wider range of air traffic evolution scenarios.

The peculiarity of the proposed method is that at analytical solution the predicted areas of each aircraft position uncertainty are used separately, and that makes it possible to use this method not only for straight flight paths, but also for maneuvering areas.

An analytical expression for calculating the conflict probability is derived, and there is no need to perform complex transformations associated with orthogonalization and transformation of the covariance matrices of deviations from the flight plan. The independence of the variables is provided by the orientation of the auxiliary coordinate system in parallel to the track line.

The proposed methods for detection and estimation of the conflict situations forms the basis for cooperative decision-making, which includes cooperation of information such as flight plan agreed with ground air traffic service and with aircraft crew, information from on-board FMS about used law of trajectory control, information of stochastic flight dynamics, as well as information about the crew intentions.

The proposed methods of multi-criteria resolution of aircraft conflict provide the synthesis of conflict-free flight trajectories according to criteria of flight regularity, economy and the complexity of maneuvering. The advantages of these methods are:

- Conflict avoidance is performed using heading, speed and altitude change maneuvers.
- The special model of controlled aircraft motion is used for trajectories simulation.
- The multi-objective optimization of conflict-free trajectories is applied.
- Application of the dynamic programming for sequential synthesis of trajectories enhances the computational efficiency.

Cooperative procedure for air traffic conflicts resolution in ATM system with artificial intelligence tool represents a possible deployment of developed methods.

All proposed methods are the basis for development of artificial intelligence tools for automated ATM systems.

REFERENCES

Babak, V., Kharchenko, V., & Vasylyev, V. (2006). Methods of conflict probability estimation and decision making for air traffic management. *Aviation*, *10*(1), 3–9. doi:10.3846/16487788.2006.9635920

Bakker, G. J., Kremer, H. J., & Blom, H. A. P. (2001). Probabilistic approaches toward conflict prediction. In G. L. Donohue & A. G. Zellweger (Eds.), *Air Transportation Systems Engineering* (pp. 677–694). American Institute of Aeronautics and Astronautics, Inc.

Bellman, R. (1957). *Dynamic programming*. Princeton, NJ: Princeton University Press.

Bicchi, A., & Pallottino, L. (2000). On optimal cooperative conflict resolution for air traffic management systems. *IEEE Transactions on Intelligent Transportation Systems*, *1*(4), 221–232. doi:10.1109/6979.898228

Blin, K., Akian, M., Bonnans, F., Hoffman, E., Martini, C., & Zenghal, K. A. (2000). *A stochastic conflict detection model revisited*. AIAA Guidance, Navigation, and Control Conference and Exhibit. doi:10.2514/6.2000-4270

Blom, H. A. P., Bakker, G. J., Blanker, P. J. G., Daams, J., Everdij, M. H. C., & Klompstra, M. B. (2001). *Accident risk assessment for advanced air traffic management. Air Transportation Systems Engineering* (pp. 463–480). American Institute of Aeronautics and Astronautics, Inc.

Cafieri, S., & Durand, N. (2014). Aircraft deconfliction with speed regulation: New models from mixed-integer optimization. *Journal of Global Optimization*, *58*(4), 613–629. doi:10.100710898-013-0070-1

Cetek, C. (2009). Realistic speed change maneuvers for air traffic conflict avoidance and their impact on aircraft economics. *International Journal of Civil Aviation*, *1*(1), 62–73.

Eby, M. S. (1994). A self-organizational approach for resolving air traffic conflicts. *The Lincoln Laboratory Journal*, *7*(2), 239–254.

Eby, M. S., & Kelly, W. E. (1999). Free flight separation assurance using distributed algorithms. *Proceedings of 1999 IEEE Aerospace Conference, 2*, 429-441. 10.1109/AERO.1999.793186

Eurocontrol Experimental Centre. (2011). *User manual for the Base of Aircraft Data (BADA) (Revision 3.9). Brétigny-sur-Orge*. Author.

Hu, J., Prandini, M., & Sastry, S. (2000). Optimal maneuver for multiple aircraft conflict resolution: a braid point of view. *Proceedings of the 39th IEEE Conference on Decision and Control, 4*, 4164-4169. 10.1109/CDC.2000.912369

Hu, J., Prandini, M., & Sastry, S. (2002). Optimal coordinated maneuvers for three dimensional aircraft conflict resolution. *Journal of Guidance, Control, and Dynamics*, *25*(5), 888–900. doi:10.2514/2.4982

Kosecka, J., Tomlin, C., Pappas, G., & Sastry, S. (1997). Generation of conflict resolution maneuvers for air traffic management. *Proceedings of the 1997 IEEE/RSJ International Conference on Intelligent Robot and Systems, 3*, 1598-1603.

Kuchar, J. K., & Yang, L. C. (2000). A review of conflict detection and resolution modeling methods. *IEEE Transactions on Intelligent Transportation Systems, 1*(4), 179–189. doi:10.1109/6979.898217

Paielli, R. A., & Erzberger, H. (1997). Conflict probability estimation for free flight. *Journal of Guidance, Control, and Dynamics, 20*(3), 588–596. doi:10.2514/2.4081

Prandini, M., Lygeros, J., & Sastry, S. S. (2000). A probabilistic approach to aircraft conflict detection. *IEEE Transactions on Intelligent Transportation Systems, 1*(4), 199–220. doi:10.1109/6979.898224

Vasyliev, D. (2016). Method of multi-objective resolution of two-aircraft conflict in three-dimensional space based on dynamic programming. *Proceedings of the National Aviation University, 68*(3), 35–45. doi:10.18372/2306-1472.68.10907

Vasyliev, D.V. (2013). Matematychna model kerovanoho rukhu litaka dlia doslidzhennia protsesiv aeronavihatsiinoho obsluhovuvannia polotiv [Mathematical model of controlled aircraft motion for analysis of air navigation service processes]. *Systemy ozbroiennia i viiskova tekhnika, 23*, 63-67.

Zeghal, K. (1998). A review of different approaches based on force fields for airborne conflict resolution. *Proceedings of AIAA Guidance, Navigation, and Control Conference and Exhibit*, 818-827. 10.2514/6.1998-4240

ADDITIONAL READING

Eurocontrol (2003). *CORA 2 resolver final report*. Brussels: Author.

Eurocontrol (2017). *Eurocontrol specification for medium-term conflict detection* (Ed. 2.0). Brussels: Author.

Eurocontrol (2017). *Eurocontrol specification for monitoring aids* (Ed. 2.0). Brussels: Author.

Eurocontrol (2017). *Eurocontrol specification for trajectory prediction* (Ed. 2.0). Brussels: Author.

Prevot, T., Callantine, T., Lee, P., Mercer, J., Battiste, V., Johnson, W., . . . Smith, N. (2005). Co-operative air traffic management: a technology enabled concept for the next generation air transportation system. Retrieved from https://hsi.arc.nasa.gov/publications/Prevot-05-CO-ATM-ATM2005-115.pdf

KEY TERMS AND DEFINITIONS

Air Navigation Services (ANS): Services provided to air traffic during all phases of operations including air traffic management, communication, navigation and surveillance, meteorological services for air navigation, search and rescue and aeronautical information services.

Air Traffic Conflict (Aircraft Conflict): Converging of aircraft in space and time which constitutes a violation of a given set of separation minima.

Air Traffic Control (ATC): A service operated by appropriate authority to promote the safe, orderly and expeditious flow of air traffic.

Air Traffic Controller (ATCO): A person authorized to provide air traffic control services.

Air Traffic Control Service: A service provided for the purpose of preventing collisions between aircraft, and on the maneuvering area between aircraft and obstructions; and expediting and maintaining an orderly flow of air traffic.

Air Traffic Management (ATM): The aggregation of the airborne and ground-based functions (air traffic services, airspace management and air traffic flow management) required to ensure the safe and efficient movement of aircraft during all phases of operations.

Air Traffic Management System: A system that provides ATM through the collaborative integration of humans, information, technology, facilities and services, supported by air, ground and/or space-based communications, navigation and surveillance.

Air Traffic Service (ATS): A generic term meaning variously flight information service, alerting service, air traffic advisory service, air traffic control service (area control service, approach control service or aerodrome control service) and aerodrome flight information service.

Conflict Detection: The discovery of a conflict as a result of computation and comparison of the predicted flight paths of two or more aircraft.

Conflict Resolution: The determination of alternative flight paths which would be free from conflicts and the selection of one of these flight paths for use.

Multi-Criteria Decision-Making: Selection of a best alternative from several potential candidates in a decision subject to several criteria or attribute.

Chapter 5
Application of Deep Learning in the Processing of the Aerospace System's Multispectral Images

Heorhii Kuchuk

https://orcid.org/0000-0002-2862-438X

Kharkiv Polytechnic Institute, National Technical University, Ukraine

Andrii Podorozhniak

https://orcid.org/0000-0002-6688-8407

Kharkiv Polytechnic Institute, National Technical University, Ukraine

Daria Hlavcheva

Kharkiv Polytechnic Institute, National Technical University, Ukraine

Vladyslav Yaloveha

https://orcid.org/0000-0001-7109-9405

Kharkiv Polytechnic Institute, National Technical University, Ukraine

ABSTRACT

This chapter uses deep learning neural networks for processing of aerospace system multispectral images. Convolutional and Capsule Neural Network were used for processing multispectral images from satellite Landsat 8, previously processed using spectral indices NDVI, NDWI, PSRI. The authors' approach was applied to wildfire Camp Fire (California, USA). The deep learning neural networks are used to solve the problem of detecting fire hazardous forest areas. Comparison of Convolutional and Capsule Neural Network results was done. The theory of neural networks of deep learning, the theory of recognition of multispectral images, methods of mathematical statistics were used.

DOI: 10.4018/978-1-7998-1415-3.ch005

INTRODUCTION

There are many aerospace system's remote sensing tasks that use remote sensing multispectral images, including automated target detection, land cover and land-use classification, time-series analysis, and change detection (Ball, Anderson, & Chan, 2017). Many of these tasks use shape analysis, object recognition, dimensionality reduction, image enhancement, and other techniques, which are all amenable to deep learning neural networks approach. Multispectral images data classification is of major importance to aerospace system's remote sensing applications, so many of the deep learning results were on multispectral images classification. Some of the multispectral images deep learning approaches use both spectral and spatial information.

In the chapter data from remote sensing multispectral images are used for detecting fire hazardous forest areas applied to wildfire Camp Fire (California, USA).

BACKGROUND

Vegetation fires are widespread worldwide. They have been documented since prehistoric times, defining the composition and dynamics of ecosystems, including forests and open landscapes. The effects of fires have a significant impact on the environment and society. Fire emissions affect the composition of the atmosphere and the global climate, as well as human health and safety.

One of the largest wildfires happened in California (USA) on July 2018. It was called "Mendocino Complex". The total area of the fire was almost 460 thousand acres. In less than six months in this state, a wildfire called Camp Fire over 150 thousand acres took place again (The Department of Forestry and Fire Protection of California, 2018). Analyzing US statistics, (Hoover, 2018) compared to the 1980s, now, the number of acres on which fires have increased almost twice, while the number of fires has become smaller.

The primary losses from a large number of wildfires are human life and natural resources. On the other hand, it's worth remembering that wildfires cause considerable economic costs to overcome and eliminate them. For example, in 2017 in the USA, fires injured nearly $20 billion, about $ 2.5 billion was spent in 2008 and 2015, and in 2014, 2010 and 2009, it is less than a billion (Insurance Information Institute, 2017). It is important save human life and health, natural resources, ecology, and economic resources by preventing wildfires.

For monitoring and detection of wildfires terrestrial monitoring, aviation monitoring, space monitoring is being used. Nowadays U.S. Forest Service uses satellite data, laboratories, and research stations to fight wildfires (U. S. Forest Service. Managing Fire, 2019). But such an approach requires people to stay in potentially dangerous forest areas.

To solve the problem neural networks of deep learning will be used. They will be able to recognize fire hazardous forest areas. The multispectral images from the Landsat 8 will be used (U.S. Geological Survey, 2019). The required spectral indices that are able to allocate arid vegetation, moisture content and carbon will be calculated. Such an approach can lead to solve the problem of preventing wildfires.

LITERATURE REVIEW

Getting Earth Remote Sensing Data

Earth remote sensing is defined as a method of measuring the properties of objects on the earth's surface, which uses data obtained by airborne aircraft and artificial earth satellites (Shovengerdt, 2010).

Because of different categories of specialists' needs, a large number of remote sensing systems were developed, through which they study the spatial, spectral and temporal parameters of an object variety. Each of these systems is focused on a certain range of tasks and has its own peculiarities. For example, in meteorology, frequent surveys of areas with relatively low resolution are used. In cartography, on the contrary, the main requirement is the maximum spatial resolution, and the shooting frequency may be relatively low.

A multispectral image is an image that captures data at certain frequencies throughout the electromagnetic spectrum (Liubchenko, Podorozhniak, & Bondarchuk, 2017). The wavelengths can be separated by filters or by devices that are sensitive to certain wavelengths of light, including frequencies that are outside the visible light range. Spectral images may allow additional information that the human eye cannot detect. Multispectral images are the main type of image obtained by Earth remote sensing.

The Landsat program is the longest project, that provides satellite imagery of the planet Earth. Landsat data is used to solve a large number of thematic tasks, such as measurements of length and classification of vegetation, the determination of the crops state, geological mapping, soil erosion control in the coastal zone, etc. The most relevant data are from the satellite Landsat 8, an American satellite Earth sensing satellite. Data from this satellite is used to:

- Operational satellite control of natural resources.
- Research of the natural processes and phenomena course dynamics.
- Analysis of causes, forecasting of possible consequences and choosing ways to prevent emergencies.

The relevant data of the Earth remote sensing in the form of multispectral images are obtained from the satellite Landsat 8. Images have a resolution of 15 to 100 meters per pixel. Satellite images are publicly available (U.S. Geological Survey, 2019). Landsat 8 provides data in spectral bands, which presented in Table 1.

As input data the Camp Fire area of October 7, 2018 was taken. The fire began on 8th of November, 2018 (The Department of Forestry and Fire Protection of California, 2018). Coordinates of the fire - $39°50'51''N$, $121°23'42''W$.

CALCULATION OF SPECTRAL INDICES

To work with spectral data, index images are used. An image is constructed corresponding to the index value in each pixel based on a combination of brightness values in certain channels. This can allocate the object under research. Three spectral indices (NDVI, NDWI, PSRI) were used to do the analysis.

Three spectral indices (NDVI, NDWI, PSRI) were used to do the analysis.

Table 1. Landsat 8 Spectral Bands

No	Band	Wavelength, nm	Resolution, m
1	Coastal/Aerosol	435-451	30
2	Blue	452-512	30
3	Green	533-590	30
4	Red	636-673	30
5	NIR	851-879	30
6	SWIR1	1566-1651	30
7	SWIR2	2107-2294	30
8	Panchromatic	503-676	15
9	Cirrus	1363-1384	30
10	Thermal infrared 1	1060-1119	100
11	Thermal infrared 2	1150-1251	100

NDVI (Normalized Difference Vegetation Index) is the most well-known and widely used vegetation index. It is used in many tasks, including the discovery of green vegetation in multispectral images. The red spectral region has the maximum absorption of solar radiation by chlorophyll, and the near infrared zone is the maximum reflection of energy. High photosynthetic activity leads to lower values of reflection coefficients in the red spectral region and large values in the near infrared. Comparison of indicators in these spectral bands allows to allocate vegetation. The index is calculated according to the following formula:

$$NDVI = \frac{\rho_{NIR} - \rho_{RED}}{\rho_{NIR} + \rho_{RED}}, \tag{1}$$

where ρ_{NIR} – reflectance in the near infrared spectral band, ρ_{RED} – reflectance in the red spectral band (Bardish, & Burshtinska, 2014) .

NDWI (Normalized Difference Water Index) was based on the NDVI principle. This index characterizes the water content - a very important indicator, because the high moisture content is characteristic for healthy vegetation, which is grows faster and more resistant to fires. NDWI can be calculated by next equation:

$$NDWI = \frac{\rho_{NIR} - \rho_{SWIR}}{\rho_{NIR} + \rho_{SWIR}}, \tag{2}$$

where ρ_{NIR} – reflectance in the near infrared spectral band, ρ_{SWIR} – reflectance in the short wave infrared band.

When calculating NDWI for a multispectral satellite image processing, water objects have positive values, and objects of soil and terrestrial vegetation have zero or negative values due to their higher reflection coefficient in the infrared spectrum than in the region of green light. In this case, it is necessary to normalize and exclude negative values for saving information about water objects (McFeeters, 1996).

The PSRI (Plant Senescence Reflectance Index) belongs to Dry or Senescent Carbon group. These indices are designed to find the total amount of "dry" carbon in the form of lignin and cellulose. In large quantities this type of carbon is present in wood, dead or dry plant tissues. An increase in this indicator may reflect the process of "aging" and dying plants. To calculate the relative nitrogen content in the plant cover, the middle infrared range is used. The indices belonging to the Dry or Senescent Carbon group are widely used in the assessment of the fire hazardous areas. PSRI is calculated by the formula

$$PSRI = \frac{\rho_{RED} - \rho_{GREEN}}{\rho_{NIR}},$$

(3)

where ρ_{NIR} – reflectance in the near infrared spectral band, ρ_{RED} – reflectance in the red spectral band, ρ_{GREEN} – reflectance in the green band (Merzlyak, Gitelson, Chivkunova, & Rakitin, 1999).

After calculation (1-3) and normalization the index images are shown in Fig. 1.

DEEP LEARNING

Deep learning is a group of methods that allow multilayer computing models to work with data that has an abstraction hierarchy. The paper (LeCun, Bengio, & Hinton, 2015) provides an overview of the main trends of deep learning. This group of methods was used, for instance, in convolutional neural networks. Deep learning methods present their methodology, which consists of several simple but nonlinear modules - layers. For example, the image is displayed as an array of pixel values and the properties studied in the first layer are transmitted to the next. The second layer specifies the existing characteristics of the object being studied. The third layer can collect data already in the form of parts of objects, and the subsequent layers will be manifest as a combination of these parts. The deep learning has big success in solving many problems. It is used to recognize the language (Mikolov, Deoras, Povey, Burget, & Černocký, 2011; Hinton, Deng, Yu, Dahl, Mohamed, Jaitly, 2012; Sainath, Mohamed, Kingsbury, & Ramabhadran,

Figure 1. Distribution of Index Values: a) NDVI in research area. Darker shade matches green vegetation; b) NDWI in research area. Darker shade matches water objects; c) PSRI in research area. Darker shade matches higher carbon content.

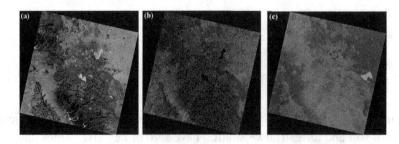

2013), to recognize faces (Sivaram, Porkodi, Amin Salih Mohammed & Manikandan, 2019) to predict the activity of potential molecules of drugs (Ma, Sheridan, Liaw, Dahl, & Svetnik, 2015), analysis of accelerator particles data (Ciodaro, Deva, De Seixas, & Damazio, 2012).

In the research two types of deep learning models had been chosen: Convolutional and Capsule Neural Networks. The first type demonstrates good results in image classification and is widely used nowadays (LeCun et al., 2015). The second type has been developed as an improvement of ConvNet and corrects its disadvantages (Sabour et al., 2017).

CONVOLUTIONAL NEURAL NETWORKS

Convolutional neural network (CNN, ConvNet) was developed by Yann Le Cun in 1988 (LeCun et al., 2015). Convolutional neural networks were created on the basis of "simple cells" in human brain. Such cells were discovered in 1962 by Torsten Nils Wiesel and David Habel (Hubel, & Wiesel, 1962). These cells react to straight lines at different angles. Then "complex cells" react to the activation of simple cells certain group.

The productivity of a network-based recognition system became a cause of using them by large technology companies, including Google, Facebook, Microsoft, IBM, Yahoo, Twitter and Adobe, as well as quickly increases the number of startups and projects that implements ConvNet-based image understanding products (LeCun et al., 2015).

Convolution is a mathematical operation that applies to two functions, and generates a third function (Podorozhniak, Lubchenko, Balenko, & Zhuikov, 2018). Having I, convolute image $I*K$ is calculated by putting the kernel on image. The sum of the multiplied elements of output image and kernel is written:

$$\left(I*K\right)_{xy} = \sum_{i=1}^{h}\sum_{j=1}^{w}K_{ij} \times I_{x+i-1, y+j-1}. \tag{4}$$

The simple architecture of ConvNet for image recognizing by k classes can be divided into two parts: a sequence of convolutional or alternating pooling layers and several fully-connected layers.

The main task of the pooling layer is to receive at the input of individual fragments of the image and their association in one value. Usually the maximization function is used (this layer is called MaxPooling) and the highest value is selected. The pooling layer does not take into account the details of the processed image, but only focuses on the data that gives the largest contribution.

One pair of the layers of convolution and puling affects this way: the length and width of a particular channel decreases, but its depth increases. On the output layer of the neural network the activation function of the neurons was used. In the research, this is a sigmoid activation function:

$$f\left(x\right) = \frac{1}{1+e^{-x}}. \tag{5}$$

In the process of training the neural network, a problem of overfitting could appear. To solve this problem the Dropout layer could be added in convolutional neural network. With probability of p p the neuron is excluded from the network at the time of current iteration (Yaloveha, Hlavcheva, & Podorozhniak, 2019).

In the research, for convolutional neural network the next architecture were used: three convolutional layers, three MaxPooling layers, Flatten layer, Dropout layer and two Fully Connected layers. The total number of parameters were 45,153. Architecture of CNN in the research is presented at Table 2:

Spyder IDE as a programming environment and Python 3.6 as a programming language were used. The frameworks were Tensorflow 1.6 and Keras 2.2.4. For training, an ASUS laptop was used. The laptop has 16Gb of RAM, one Nvidia GeForce 940MX GPU and an Intel Core i7 extreme 7th generation CPU (up to 3.5 GHz).

CAPSULE NEURAL NETWORKS

In recent years, Convolutional Neural Networks has achieved high results, but their architecture has disadvantages, which are highlighted by leading researchers in the field of artificial neural networks Jeffrey Hinton (Sabour et al., 2017).

As noted in CNN, there is a MaxPooling layer. When using such a layer, some information is lost. This information can be very useful for the neural network. For example, ConvNet is not able to take into account the spatial and hierarchical relationships between components that have the object being studied. The network may erroneously identify the object in the image if its structural elements are located in the wrong position relative to each other. Thus, the convolutional neural networks need to be trained in a large amount of training data so that it can identify the same object, transmitted to the corresponding value or rotated in space. Capsule neural networks are aimed at eliminating disadvantages of modern machine learning systems (Hlavcheva, & Yaloveha, 2018).

Table 2. Structure of convolutional neural network

Type	Input	Output	Parameters
1. Convolutional	(32x32x3)	(30x30x32)	896
2. MaxPooling	(30x30x32)	(15x15x32)	0
3. Convolutional	(15x15x32)	(13x13x32)	9248
4. MaxPooling	(13x13x32)	(6x6x32)	0
5. Convolutional	(6x6x32)	(4x4x64)	18496
6. MaxPooling	(4x4x64)	(2x2x64)	0
7. Flattening	(2x2x64)	256	0
8. Fully Connected	256	64	16448
9. Dropout	64	64	0
10. Fully Connected	64	1	65

Capsule neural network (CapsNet) is an artificial neural network, which was designed to improve the modeling of spatial relationships between features in different levels. Capsule is a function that tries to predict presence and specific features of object in a certain area. CapsNet does not work with the specific value of the neuron, but with entire sets of neurons (vectors). Capsules encapsulate all important information about the state of the feature they are detecting in vector form (LeCun et al., 2015).

To determine the probability of finding an object in the capsule image, the normalized length of the source vector is used. The direction of the resulting vector determines the spatial placement of the object under study relative to other objects. This allows you to correctly recognize the object, even if it was shifted in the image, since the probability of the resulting vector will remain constant, and only its direction will change (Sabour et al., 2017).

Capsule result vector:

$$v_j = \frac{s_j^2}{1 + s_j^2} \frac{s_j}{s_j},$$

(6)

where s_j – capsule j input vector.

For neurons in such a network the equivariance is used: the probabilities of determine an object remain unchanged while its position changes. Routing between capsules occurs using the dynamic routing algorithm. After the dynamic routing algorithm, the result is transferred to the upper level capsule, which is responsible for the specific characteristics of the object. In the training of the capsule network, the margin-loss function is applied:

$$L_k = T_k \max\left(0, m^+ - v_k\right)^2 + \lambda\left(1 - T_k\right)\max\left(0, v_k - m^-\right)^2,$$

(7)

where k – a class of object, $T_k = 1$, if object is present, $T_k = 0$, in another case; v_k – output vector of the upper level capsule; $\lambda = 0.5$ – down-weighting of the loss; $m^+ = 0.9$ – expected value of vector length if object is present; $m^- = 0.1$ – in another case.

In the research the architecture of the capsule network consists of one convolutional layer, PrimaryCaps layer, RouteCaps layer, Mask layer, three Fully Connected layers and output layer. The total number of parameters were 9,582,336.

Detailed architecture is presented at the Table 3.

The same programming environment and equipment as in previous item for CapsNet were used.

ILLUSTRATIVE EXAMPLE

Data Processing

The index images were calculated. These images were merged in one image with three channels. First (red) channel is responsible for PSRI, second (green) – for NDVI, third (blue) – for NDWI.

Table 3. Structure of capsule neural network

Type	Input	Output	Parameters
1. Convolutional	(32x32x3)	(24x24x256)	62464
2. PrimaryCaps	(24x24x256)	(2048x8)	5308672
3. RouteCaps	(2048x8)	(2x16)	528384
4. Mask	(2x16)	16	0
5. Fully Connected	16	512	8704
6. Fully Connected	512	1024	525312
7. Fully Connected	1024	3072	3148800
8. Reshape	3072	32x32x3	0

Knowing the exact coordinates of the fire hazardous area, it was prepared the final image with the territory on it where the Camp Fire occurred.

The total number of prepared images were 40,000. The 20,000 images form the fire hazardous area, others is the area where there was not the fire.

All images were divided into three parts:

- Training Data (28,000 images – 70% of total amount).
- Validation Data (6,000 images – 15%).
- Test Data (6,000 images – 15%).

The size of each image is (32×32) pixels in three channels.

The batch size is 16. On the figure 2 examples of input images for neural networks are presented.

Figure 2. Examples of images for neural network: a – fire hazardous area, b – area without fire hazardous

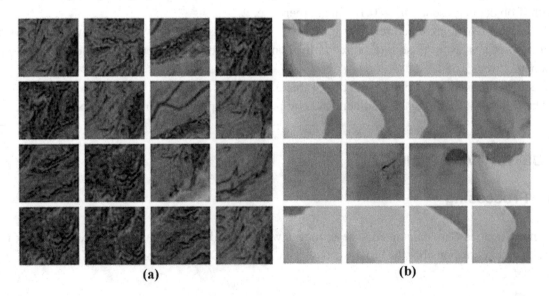

(a)　　　　　　　　　　　　　(b)

Results and Discussions

At the input to the ConvNet and CapsNet, images of the size (32×32) with three spectral channels were submitted, and therefore, a size of input matrix is $(32, 32, 3)$.

Both neural networks had been trained during 50 epochs. The average time of training for convolutional neural network was 60 seconds, for capsule neural network – 1020 seconds. The total training time was 60 minutes for ConvNet and 14 hours for CapsNet. A big difference in training time easy to explain because of bigger number parameters in capsule neural network. The graphs of the accuracy classification and loss on the number of periods for the training and validation data are shown on Figures 3 – 6.

It is obtained that convolutional and capsule neural networks are capable of effectively classifying objects belonging to two classes. ConvNet and CapsNet showed almost the same results on testing data. The classification accuracy for convolutional and capsule neural networks on this data is presented in the Table 4.

Figure 3. The graph of the classification accuracy dependence on the number of epochs on training data

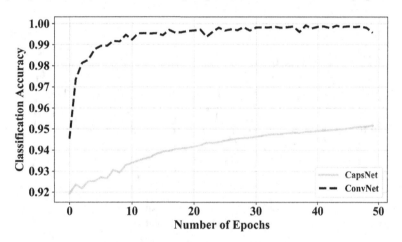

Figure 4. The graph of the classification accuracy dependence on the number of training epochs on validation data

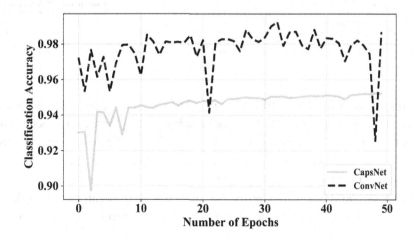

Figure 5. The graph of the loss dependence on the number of training epochs on training data

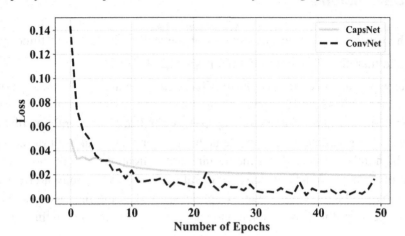

Figure 6. The graph of the loss dependence on the number of training epochs on validation data

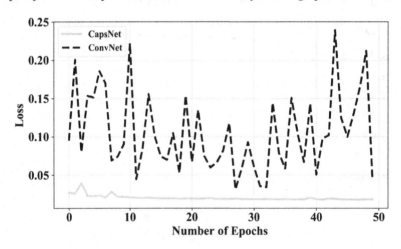

Table 4. The classification accuracy for convolutional and capsule neural networks on testing data

Neural Network	Classification Accuracy
Convolutional	0.943
Capsule	0.949

The accuracy for capsule neural network on training data grows fast after the first epochs. For capsule neural network, the growth is smooth and does not have big jumps during all epochs. The accuracy on validation data for ConvNet is an increase of 0.04, but there are big jumps on several epochs. For CapsNet it is not typical. These results may indicate that capsule neural networks study to summarize received information, more resistant to the overfitting problem and have tend to smooth increase the classification accuracy. Such behavior could be due to CapsNet architecture. In general, the loss value in ConvNet is lower in comparative with CapsNet. On the other hand, CapsNet demonstrates better result of loss on validation data (the difference between average values for all epochs 0.081 lower) than ConvNet and do not have a large range of jumps as ConvNet. An increase the number of epochs will lead to lower value of loss for CapsNet.

In (Mukhometzianov, & Carrillo, 2018) the capsule neural network was tested on CIFAR10. The achieved loss value was 0.106. The size of each image was 24x24 pixels. There were 64 different types of primary capsules. In the research, the loss value is less than in the paper because of using a smaller number of primary capsules and the networks classified images belonging only to two classes.

FUTURE RESEARCH DIRECTIONS

The main research directions are increasing the dimension of images and increasing number of images, by increasing number of potentially fire hazardous forest areas under research; using of Planet service, which provides everyday data from any territory of the Earth with resolution up to 72 cm, using of Arc-GIS. The purpose of future research is to improve neural networks for working with different Earth areas.

The main practical purpose of future research is developing an online service for real-time forest monitoring. It can be released in cooperation with volunteer organizations.

CONCLUSION

The scientific and practical problem of detecting fire hazardous areas of forests by using deep learning networks applied to Camp Fire (California, USA) is solved in the research.

The areas where the largest-scale wildfires took place recently are considered. The economic and ecological losses from wildfires were analyzed.

The spectral indices were calculated using multispectral images that were received from Landsat 8.

Convolutional and capsule neural networks for solving problem of detecting fire hazardous areas of forests were developed. The neural networks could be used as an integrated approach to solving the problem.

A comparative analysis of convolutional and capsule neural networks was conducted. The accuracy of classification on training data for convolutional neural network was 0.943, for capsule neural network – 0.949.

It has been established that capsule neural networks show less loss value on the validation data than the convolutional neural networks.

REFERENCES

Ball, J. E., Anderson, D. T., & Chan, C. S. (2017). Comprehensive survey of deep learning in remote sensing: Theories, tools, and challenges for the community. *Journal of Applied Remote Sensing*, *11*(4), 042609. doi:10.1117/1.JRS.11.042609

Bardish, B., & Burshtinska, H. (2014). Use of vegetation indexes for the identification of terrestrial objects. Modern achievements in geodetic science and production (2), 82-88.

Ciodaro, T., Deva, D., De Seixas, J., & Damazio, D. (2012). Online particle detection with neural networks based on topological calorimetry information. In *Proceedings of the Journal of Physics: Conference Series*. 10.1088/1742-6596/368/1/012030

Hinton, G., Deng, L., Yu, D., Dahl, G., Mohamed, A.-r., Jaitly, N., & Kingsbury, B. (2012). Deep neural networks for acoustic modeling in speech recognition. *IEEE Signal Processing Magazine*, 29.

Hlavcheva, D., & Yaloveha, V. (2018). Capsule neural networks. Control. *Navigation and Communication Systems*, 5(51), 132–135.

Hoover, K. Wildfire Statistics Congressional Research Service. Retrieved from https://fas.org/sgp/crs/misc/IF10244.pdf

Hubel, D. H., & Wiesel, T. N. (1962). Receptive fields, binocular interaction and functional architecture in the cat's visual cortex. *The Journal of Physiology*, 160(1), 106–154. doi:10.1113/jphysiol.1962.sp006837 PMID:14449617

Insurance Information Institute. Facts & Statistics: Wildfires: Wildfire losses in the United States. (2017). Retrieved from https://www.iii.org/fact-statistic/facts-statistics-wildfires

LeCun, Y., Bengio, Y., & Hinton, G. (2015). Deep learning. *Nature, 521*(7553), 436.

Liubchenko, N., Podorozhniak, A., & Bondarchuk, V. (2017). Neural network method of intellectual processing of multispectral images. *Advanced Information Systems*, 1(2), 39–44. doi:10.20998/2522-9052.2017.2.07

Ma, J., Sheridan, R. P., Liaw, A., Dahl, G. E., & Svetnik, V. (2015). Deep neural nets as a method for quantitative structure–activity relationships. *Journal of Chemical Information and Modeling*, 55(2), 263–274. doi:10.1021/ci500747n PMID:25635324

McFeeters, S. K. (1996). The use of the Normalized Difference Water Index (NDWI) in the delineation of open water features. *International Journal of Remote Sensing*, 17(7), 1425–1432. doi:10.1080/01431169608948714

Merzlyak, M. N., Gitelson, A. A., Chivkunova, O. B., & Rakitin, V. Y. (1999). Non-destructive optical detection of pigment changes during leaf senescence and fruit ripening. *Physiologia Plantarum*, 106(1), 135–141. doi:10.1034/j.1399-3054.1999.106119.x

Mikolov, T., Deoras, A., Povey, D., Burget, L., & Černocký, J. (2011). Strategies for training large scale neural network language models. Paper presented at the 2011 IEEE Workshop on Automatic Speech Recognition & Understanding. 10.1109/ASRU.2011.6163930

Mukhometzianov, R., & Carrillo, J. (2018). CapsNet comparative performance evaluation for image classification. *arXiv preprint arXiv:1805.11195*.

Podorozhniak, A., Lubchenko, N., Balenko, O., & Zhuikov, D. (2018). Neural network approach for multispectral image processing. In *Proceedings 2018 14th International Conference on Advanced Trends in Radioelectronics, Telecommunications, and Computer Engineering (TCSET)*. 10.1109/TCSET.2018.8336357

Sabour, S., Frosst, N., & Hinton, G. E. (2017). Dynamic routing between capsules. Paper presented at the Advances in Neural Information Processing Systems.

Sainath, T. N., Mohamed, A.-R., Kingsbury, B., & Ramabhadran, B. (2013). Deep convolutional neural networks for LVCSR. In *Proceedings 2013 IEEE International Conference on Acoustics, Speech, and Signal Processing*. 10.1109/ICASSP.2013.6639347

Shovengerdt, R. A. (2010). Remote sensing. Models and methods of image processing. Moscow, Russia: Technosphere, 560.

Sivaram, M., Porkodi, V., Mohammed, A. S., & Manikandan, V. (2019). Detection of accurate facial detection using hybrid deep convolutional recurrent neural network. *ICTACT Journal on Soft Computing*, 09(02), 1844–1850. doi:10.21917/ijsc.2019.0256

The Departmant of Forestry and Fire Protection of California. Public Information Map. Retrieved from http://cdfdata.fire.ca.gov/pub/cdf/images/incidentfile2277_4287.pdf

The Department of Forestry and Fire Protection of California. Top 20 Largest California Wildfires. Retrieved from https://www.fire.ca.gov/communications/downloads/fact_sheets/Top20_Acres.pdf

U. S. Forest Service. Managing fire (2019). Retrieved from https://www.fs.fed.us/science-technology/fire

U.S. Geological Survey. (2019). Retrieved from https://earthexplorer.usgs.gov/

Yaloveha, V., Hlavcheva, D., & Podorozhniak, A. (2019). Neural network method of intellectual processing of multispectral images. *Advanced Information Systems*, 3, 116–120. doi:10.20998/2522-9052.2019.1.19

KEY TERMS AND DEFINITIONS

An Artificial Neuron Network: Is a computational model based on the structure and functions of biological neural networks.

Deep Learning: Is a group of methods that allow multilayer computing models to work with data that has an abstraction hierarchy.

Earth Remote Sensing: Is the method of measuring the properties of objects on the earth's surface, which uses data obtained from aircraft and artificial satellites of the Earth.

Index Image: An image is constructed corresponding to the index value in each pixel based on a combination of brightness values in certain channels.

Multispectral Image: Is an image which captures data at certain frequencies throughout the electromagnetic spectrum.

Chapter 6
Intelligence–Based Operation of Aviation Radioelectronic Equipment

Oleksandr Solomentsev
National Aviation University, Ukraine

Maksym Zaliskyi
(iD) https://orcid.org/0000-0002-1535-4384
National Aviation University, Ukraine

Oleksii Zuiev
National Aviation University, Ukraine

ABSTRACT

This chapter presents the questions of aviation radioelectronic equipment operation. Operation is the main stage in the life cycle of equipment. This stage is the longest in time and very significant in terms of the costs for equipment reliability providing. Control influences are realized in operation system to provide the efficient functioning of the equipment. The operation system is complex in structure intelligence system and is the object of design and modernization. The chapter deals with the structure of the operation system and its elements interconnection and discusses the models of diagnostic variables, reliability parameters for the case of changepoint presence. An important issue during design and improvement of operation systems is the selection, evaluation, and support of the efficiency indicator. This chapter concentrates on the efficiency indicator substantiation taking into account the effectiveness and operational resources costs. Three data processing strategies analysis allowed choosing the most rational option from maximum efficiency point of view.

DOI: 10.4018/978-1-7998-1415-3.ch006

INTRODUCTION

Scientific and technical progress is developing towards the creation of intelligent automated systems that are aimed at increasing productivity, reducing production costs, etc. Such systems are based on the principles of adaptation and flexibility, system and process approaches, which are elements of artificial intelligence (Solomentsev, Zaliskyi & Zuiev, 2016).

It is known that civil aviation is a system of systems. One of them is air navigation service (ANS) system. The main component of the ANS is radioelectronic equipment (REE). This equipment contains communication, navigation and surveillance systems.

Air navigation service system is used for air traffic management, radio engineering support of flights, providing aeronautical and meteorological information in accordance with the standards and recommended practices of ICAO and Eurocontrol requirements. So REE is intended for the formation and transmission of flight information to aircraft board and air traffic controllers of flight control centers.

According to Doc 9859, one of the ways to reduce the risks of ANS is the technical condition monitoring for early recognition of its deterioration. To do this, it is necessary to perform procedures for collecting and processing statistics on reliability parameters and diagnostic variables of the REE.

To provide efficient use of REE for its designated purpose, the operation system (OS) is utilized. The purpose of the operation system (OS) is to ensure the stable functioning of ground-based radioelectronic equipment to provide the data and information to the consumers.

The OS includes:

1. Radioelectronic Equipment.
2. Processes.
3. Personnel.
4. Documentation.
5. Operational Resources.
6. Data Processing Algorithms (Solomentsev, Melkumyan, Zaliskyi & Asanov, 2015).

BACKGROUND

The main element of the operation system is radioelectronic equipment (Kuzmenko, Ostroumov & Marais, 2018). Therefore, all operational processes are directly related to REE (Nakagawa, 2005). The main process is the REE intended use. Other processes are a maintenance, repair, life extension, ground and flight tests, etc. (Dhillon, 2006).

According to ICAO and Eurocontrol documents requirements, all personnel must be certified. To do this, the initial training courses and the advanced training are implemented, the traineeship is organized and so on.

Regulatory documents play an important role during the safety and regularity of flights providing. The entire set of documents can be conditionally divided into international documents (ICAO, Eurocontrol), national domestic documents and documents developed by airlines taking into account the specifics of production activities.

Expendable resources are necessary to provide the REE serviceability in case of the influence of changing external conditions, degradation processes that lead to a deterioration in the technical condition of the equipment.

Data processing algorithms are very important in the operation systems since they implement the processing principle based on artificial intelligence (Sushchenko & Golitsyn, 2016). The timeliness and correctness of control and preventive actions aimed at the effective equipment operation for air navigation services largely depend on these algorithms.

The operation systems based on artificial intelligence principles must provide the given efficiency level of REE functioning in conditions that were unforeseen at the design stage (Jones, 2009).

OPERATION SYSTEM CONTENT

The generalized structural diagram of operation system based on the adaptability principle of system approach is shown in Figure 1.

The structure includes radioelectronic equipment designed to perform the requests flow from the air navigation services consumers (aircraft and air traffic controllers). The operation system contains a subsystem for the efficiency estimation and providing (for maintenance and repair processes, personnel, documents, etc.).

Figure 1. The generalized structural diagram of operation system

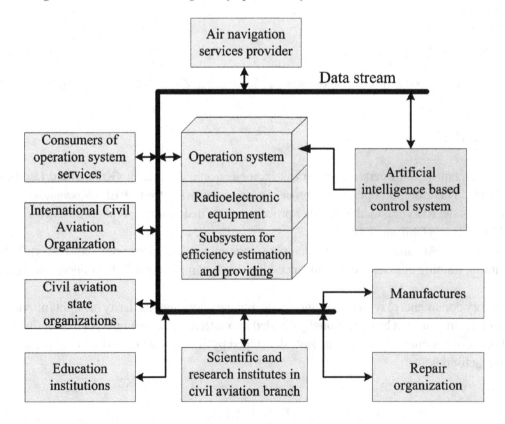

The operational data processing is carried out in the operation control system, which operates on the basis of the principles of artificial intelligence. From the system approach point of view, information is collected from the components of the subject area under consideration:

1. Air navigation service providers.
2. Consumers of operation system services.
3. International civil aviation organization (ICAO).
4. Civil aviation (CA) state organizations.
5. Educational institutions.
6. Repair organization.
7. Manufactures.
8. Scientific and research institutes in CA branch.

The control system performs data processing, and in the case of OS efficiency decrease, it generates and implements preventive and corrective actions. These actions can be formed and implemented depending on the change of requirements and tasks of the subject area.

OPERATION SYSTEM ELEMENTS MODELS

During operation systems analysis there are problems of these systems design and improvement. The main for this problem solvation is a methodological basis. The basis contains the set of principles, approaches, methods, assumptions, models, procedures for statistical operational data processing, etc.

During OS design and improvement it is necessary to use four main principles: adaptability, aggregation, system approach and process approach.

The adaptability is the main principle form intellectualization point of view. This principle adaptability involves the subject area elements condition monitoring with subsequent statistical data processing. Based on the results of the processing, decisions are made about corrective and preventive actions aimed at providing the required efficiency of OS and REE.

To implement the adaptability principle and catchall monitoring of subject area elements (Figure 1), it is necessary to know the models that describe these elements conditions. The models for REE condition description can be classified into two types:

1. Diagnostic Variables Models.
2. Reliability Parameters Models.

Let's consider the diagnostic variables model. This model takes into account possible degradation processes of REE aging, sudden failures and damages.

The mathematical model of diagnostic variable

$$y(t) = x(t) + z(t) + \vartheta(t),$$
(1)

where $y(t)$ is an observable random process of the diagnostic variable change; $z(t)$ is a deterministic signal component of the diagnostic variable, which describes the trend of its change (unobservable process); $x(t)$ is a random signal component of the diagnostic variable, which takes into account description inaccuracy; $\vartheta(t)$ is a noise component, which is due to measuring equipment errors.

Components $z(t)$, $x(t)$ and $\vartheta(t)$ have different fluctuation spectra. The component $\vartheta(t)$ has the widest spectra range.

In general case, the component $z(t)$ has two quasistationary intervals:

1. Interval, where the component $z(t)$ is equal to nominal value.
2. Interval, where the component $z(t)$ changes.

The transition point from one interval to another is the moment of random process $y(t)$ changepoint occurrence (Zaliskyi, Solomentsev, Kozhokhina & Herasymenko, 2017).

To build the mathematical model for deterministic signal component and to describe $z(t)$ in two intervals, it is necessary to use Heaviside step function

$$h(t) = \begin{cases} 1, & \text{if } t \geq 0, \\ 0, & \text{if } t < 0. \end{cases}$$

Then model for $z(t)$ will be:

1. In case of the linear trend:

$$z(t) = Z_0 h(t) + k(t - t_{sw})h(t - t_{sw}),$$

(2)

where Z_0 is a nominal value of the diagnostic variable; t_{sw} is a moment of changepoint occurrence; k is a diagnostic variable change rate in case of changepoint occurrence.

2. In case of the quadratic trend:

$$z(t) = Z_0 h(t) + k(t - t_{sw})^2 h(t - t_{sw}).$$

(3)

To take into account sudden failures, the initial model (1) of the diagnostic variable includes an additional component in the form of another Heaviside step function with given amplitude.

Then the mathematical model of the diagnostic variable will take the form:

1. In case of the linear trend:

$$y(t) = x(t) + Z_0 h(t) + k(t - t_{sw})h(t - t_{sw}) + \vartheta(t) + \sum_{i=1}^{n} a_i h(t - t_{fi}),$$

(4)

where t_{f_i} and a_i are random time moments of sudden failures or damages, and a drift in the diagnostic variable change associated with these failures or damages.

2.   In case of the quadratic trend:

$$y(t) = x(t) + Z_0 h(t) + k(t - t_{sw})^2 h(t - t_{sw}) + \vartheta(t).$$ (5)

Examples of the diagnostic variables realizations for the cases of linear and quadratic changepoints are shown in Figures 2 and 3, respectively.

Figure 2 shows the diagnostic variable realization that contains sudden failure at the time moment $t = 76$. In Figure 3 sudden failure doesn't occur during the observation interval.

Models of reliability parameters characterize the REE stability during a certain time interval (Hryshchenko, 2016). There are the following reliability parameters:

1.   Mean time to failure.
2.   Failure rate.
3.   Reliability function.
4.   Steady-state availability.
5.   Availability function.

The literature analysis shows that sufficient attention is paid to the failure rate trends. According to the considered approach about the changepoint presence, there are step-function model, linear, quadratic models, etc.

The mathematical model of failure rate:

1.   In case of step-function model:

$$\lambda(t) = \lambda_0 h(t) + d_0 h(t - t_{sw}),$$ (6)

Figure 2. The diagnostic variable realization in case of linear changepoint

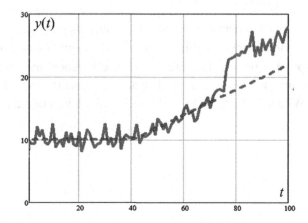

Figure 3. The diagnostic variable realization in case of quadratic changepoint

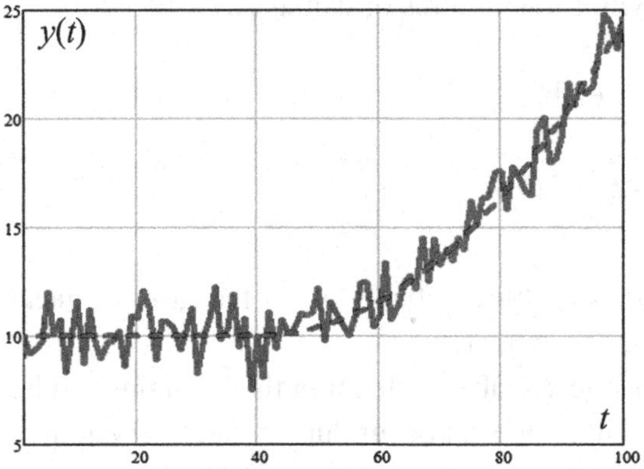

where λ_0 is an initial value of failure rate, d_0 is a parameter of technical condition deterioration for step-function model.

2.   In case of the linear model:

$$\lambda(t) = \lambda_0 h(t) + d_1(t - t_{sw})h(t - t_{sw}),\tag{7}$$

where d_1 is a parameter of technical condition deterioration for linear model.

3.   In case of the quadratic model:

$$\lambda(t) = \lambda_0 h(t) + d_2(t - t_{sw})^2 h(t - t_{sw}),\tag{8}$$

where d_2 is a parameter of technical condition deterioration for quadratic model.

USAGE OF INTELLECTUALIZATION PRINCIPLES IN OPERATION SYSTEMS

In operation system in Figure 1 important element are employees who make management decisions. The operation system based on artificial intelligence ideas allows processing large streams of data and information, to build adaptive structures that are aimed at efficiency providing. This allows making timely and correct control and preventive actions. The adaptation is the main feature of systems based on artificial intelligence (Wilamowski & Irwin, 2017). Different levels of adaptation are possible for:

1.   The model parameters.
2.   The models.
3.   The REE operation conditions.

4.   The new requirements of regulatory and normative documents, air navigation services consumers, etc.
5.   Structural changes in OS.

The artificial intelligence-based control system for REE operation can function in one of two modes: normal mode and optimal efficiency search mode. In normal mode, the database and knowledge base is used. This mode uses standard commands for all planned operation situations. The second mode is activated in case of efficiency decrease when the first mode doesn't provide its specified level. The optimal efficiency search mode is based on the use of neural networks, fuzzy sets procedures and heuristic procedures for operator-manager description (Padhy, 2005).

The block diagram of the control system in normal mode is presented in Figure 4. The generalized block diagram of the intelligent system can be represented in Figure 5.

Control system for REE operation has a complex structure. In general, it contains such subsystems:

1.   Data processing subsystem.
2.   Monitoring subsystem.
3.   The subsystem of REE technical condition inspection.
4.   Other subsystems included in the structural block diagrams for normal and intellectual modes.

The presented subsystems interact with the subsystem of efficiency estimation and providing, which is part of the operation system.

Figure 4. The block diagram of control system in normal mode

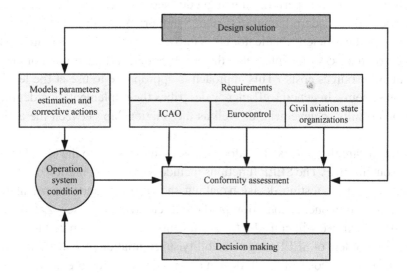

Figure 5. Block diagram of intelligent system

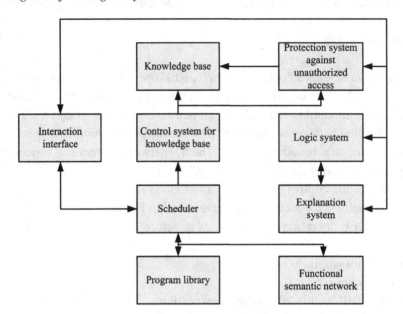

SUBSYSTEM OF EFFICIENCY ESTIMATION AND PROVIDING

The practical implementation of the introduction of new control methods can be achieved through the subsystem of efficiency estimation and providing (SEEP) for REE OS. This subsystem operates on the basis of international standards principles in the field of service quality.

In general, OS development is performed using a defined efficiency indicator. Different efficiency indicators are considered in modern literature (Galar, Sandborn & Kumar, 2017).

We will assume that the efficiency indicator takes into account the effect of reducing the average cost of resources for a certain observation interval in the presence of a certain number of unconformities and associated with them resources costs . This approach is appropriate to use at the level of the external design of operation systems. In general, efficiency describes the implementation degree of the planned activity and planned results achievement, as well as the relationship between the achieved result and resources costs.

The OS components are the objects of conformity assessment (CA) to the established requirements. This CA is carried out in SEEP. The SEEP functions include the implementation of the procedures: conformity assessment, hypothesis testing, decision making, the formation and implementation of corrective and preventive actions, components condition prediction, etc. (Solomentsev, Zaliskyi, Kozhokhina & Herasymenko, 2017). Next, we will consider the conformity assessment procedure.

Let's consider the problem of SEEP usage feasibility substantiation by calculating the OS efficiency for several CA strategies (the costs of SEEP design and implementation are neglected). The efficiency indicator takes into account the probability of correct detection of unconformities and CA costs. Assume that OS consists of N elements, each of which can be in one of two conditions: serviceable and failure (or with established requirements compliance and without compliance). Let at the current inspection time moment (the duration of OS functioning is T_{OS}) the system objectively contains n elements with failures $(0 < n \le N)$. The failures occur with a certain speed (rate) $v(t)$ (unconformity per time). Suppose that during the monitoring, new failures can't occur. The CA costs of each OS component will be denoted as C, they are equal to each other for all components.

In the general case, the numerical values of the efficiency indicator are random variables, so it is advisable to use its mean value.

The mathematical expectation of the probability of correct detection is defined as:

$$m_1(D) = \frac{m_1(n)}{v(t)T_{obs}} = \frac{m_1(n)}{n},$$

(9)

where $m_1(n)$ is a mathematical expectation of detected unconformities number during CA, T_{obs} is an observation interval (total time of OS functioning and CA procedure).

The mathematical expectation of the efficiency indicator for the whole OS will be determined as:

$$m_1(\text{Ef}) = \frac{m_1(D)}{C_{max} / (C_{max} - C_{\Sigma})},$$

(10)

where C_{Σ} is a total cost for OS components CA, C_{max} is a maximum allowable costs, which are determined at the design stage.

Variables $m_1(D)$ and $m_1(\text{Ef})$ are in the range $[0; 1]$.

Let's consider the CA case with SEEP usage (strategy S1). We assume that all N elements of the system are simultaneously controlled at the same CA time Δt. Then the observation time $T_{obs} = T_{OS} + \Delta t$, total costs $C_{\Sigma} = CN$. If there are no errors during the CA process, then all unconformities in the OS will be detected (i.e. $m_1(n) = n$), and the expressions (9) and (10) will take the form:

$$m_1(D(n, N, T_{obs} / \text{S1})) = \frac{m_1(n)}{v(t)T_{obs}} = \frac{n}{v(t)\left(T_{OS} + \Delta t\right)},$$

(11)

$$m_1(\text{Ef}(n, N, T_{obs}, C_{\Sigma} / \text{S1})) = \frac{m_1(D(n, N, T_{obs} / \text{S1}))}{C_{max} / (C_{max} - C_{\Sigma})} =$$
$$= \frac{n(C_{max} - C_{\Sigma})}{v(t)\left(T_{OS} + \Delta t\right)C_{max}} = \frac{n(C_{max} - NC)}{v(t)\left(T_{OS} + \Delta t\right)C_{max}}.$$

(12)

Let the second strategy (strategy S2): the OS components are checked in such a way that all N components are verified sequentially for m procedures of CA. The strategy S2 is the CA case without SEEP. We assume that during one CA procedure with duration Δt, which is carried out with periodicity T_{OS}, M components are verified, and $N / M = m$, where m is a positive integer $(m > 1)$. The quantity of CA procedures is m. The observation time $T_{obs} = m\left(T_{OS} + \Delta t\right)$.

If there are no errors in CA process, then all unconformities will be detected in the OS during m procedures, i.e. $m_1(n) = n$. The total costs $C_\Sigma = CN$.

The mathematical expectations of the probability of correct detection and OS efficiency for the second strategy:

$$m_1(D(n, N, T_{obs} / S2)) = \frac{m_1(n)}{v(t)T_{obs}} = \frac{n}{v(t)m\left(T_{OS} + \Delta t\right)}, \tag{13}$$

$$m_1(\mathrm{Ef}(n, N, T_{obs}, C_\Sigma / S2)) = \frac{m_1(D(n, N, T_{obs} / S2))}{C_{max} / (C_{max} - C_\Sigma)} =$$
$$= \frac{n(C_{max} - C_\Sigma)}{v(t)m\left(T_{OS} + \Delta t\right)C_{max}} = \frac{n(C_{max} - NC)}{v(t)m\left(T_{OS} + \Delta t\right)C_{max}}. \tag{14}$$

Let the third strategy (strategy S3): at each of m CA procedures, M components are verified randomly so that the total number of verified components is a random variable in the interval $[M; N]$. So during the observation time, any OS component can be checked for any random number of times: from 0 to m.

The system condition for the third strategy can be described using the graph (Figure 6), and the definition of the calculation formulas is carried out using the Markov model, for which the final probabilities are determined on the basis of the transition matrix of conditional probabilities.

The vector of initial probabilities $p^{(i)}$, where $i = \overline{0, m}$, can be arbitrary, but in this case, we assume that with the probability equal to one, the system is in a state when no unconformities are detected. Therefore, the vector has the form:

$$p^{(0)} = \begin{pmatrix} 1 & 0 & 0 & ... & 0 & 0 \end{pmatrix}^{\mathrm{T}}.$$

Figure 6. The system condition graph: condition 0 – no unconformities were detected, condition 1 – only one unconformity was detected, ..., condition n – all n unconformities were detected

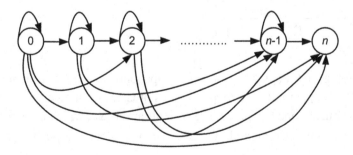

The transition of the system to another condition will be described by the transition matrix A of conditional probabilities:

$$
A = \begin{pmatrix}
P_{00} & P_{01} & P_{02} & \cdots & P_{0\,n-1} & P_{0\,n} \\
0 & P_{11} & P_{12} & \cdots & P_{1\,n-1} & P_{1\,n} \\
0 & 0 & P_{22} & \cdots & P_{2\,n-1} & P_{2\,n} \\
\cdots & \cdots & \cdots & \cdots & \cdots & \cdots \\
0 & 0 & 0 & \cdots & P_{n-1\,n-1} & P_{n-1\,n} \\
0 & 0 & 0 & \cdots & 0 & 1
\end{pmatrix},
$$

where $P_{i\,j}$ is a probability that $j - i$ unconformities will be detected during one stage of inspections in condition that i unconformities have been detected before the inspection.

Then

$$
p^{(i)} = A p^{(i-1)}.
$$

To determine conditional probabilities in the transition matrix A, let's consider the problem of a random selection of M elements from N among which n elements do not confirm the requirements.

In case when $M > n$ the probability mass function of random variable X (number of elements with unconformities among selected items) has the form:

$$
P_N(0) = \frac{(N - n)!(N - M)!}{N!(N - n - M)!};
$$

$$
P_N(1) = nM \frac{(N - n)!(N - M)!}{N!(N - n - M + 1)!};
$$

$\cdots$

$$
P_N(i) = \frac{n!(N - n)!\,M!(N - M)!}{i!(n - i)!(M - i)!\,N!(N - n - M + i)!};
$$

$\cdots$

$$
P_N(n) = \frac{(N - n)!\,M!}{(M - n)!\,N!}.
$$

Then conditional probabilities $P_{i\,j}$ (for $i \le j$):

$$P_{ij} = \frac{(n-i)!(N-n+i)!M!(N-M)!}{(j-i)!(n-j)!(M-j+i)!N!(N-n-M+j)!}. \tag{15}$$

The matrix of final probabilities depends on the inspections number m and has the following form:

$$p_j^{(m)} = \sum_{r_{m-1}=0}^{j}\left[P_{0\,r_{m-1}}\sum_{r_{m-2}=r_{m-1}}^{j}\left[P_{r_{m-1}\,r_{m-2}}\sum_{r_{m-3}=r_{m-2}}^{j}\left[P_{r_{m-2}\,r_{m-3}}\cdot\ldots\cdot\sum_{r_1=r_2}^{j}\left(P_{r_2\,r_1}P_{r_1\,j}\right)\right]\right]\right], \tag{16}$$

where $m \in \left[1, \dfrac{N}{M}-1\right]$.

The analysis showed that the expressions (15) and (16) satisfy the normalization condition, that is, the corresponding sums are equal to one.

The mathematical expectation of detected unconformities number is

$$m_1(n) = \sum_{j=0}^{n} j p_j^{(m)}.$$

Let's consider an example of calculation of the mathematical expectation of detected unconformities number for the case of three conformity assessment procedures ($m = 3$). Assume that OS consists of $N = 90$ components and there are $n = 20$ unconformities among them.

Using the equation for final probabilities, we can write the mathematical formula for the probability mass function of random variables X:

$$p_j^{(3)} = \sum_{r_2=0}^{j}\left[P_{0\,r_2}\sum_{r_1=r_2}^{j}\left(P_{r_2\,r_1}P_{r_1\,j}\right)\right].$$

After calculating probabilities P_{ij} for the transition matrix and substituting them in the formula for $p_j^{(3)}$, the probability mass function can be presented in Table 1.

The probability mass function of detected unconformities number is shown in Figure 7.

The mathematical expectation:

$$m_1(n) = \sum_{j=0}^{n} j p_j^{(3)} = 14.074.$$

In this case, fewer unconformities are detected than for the second strategy, and therefore, the second strategy efficiency will be higher than in case of the third strategy.

The mathematical expectations of the probability of correct detection and OS efficiency for the third strategy will be (taking into account that total costs are equal to $C_\Sigma = CN$):

Table 1. The probability mass function of detected unconformities number

j	0	1	2	3	4	5	6	7	8	9	10
$p_j^{(3)}$	0	0	0	0	0	0	0	0.0004	0.0018	0.0074	0.023

j	11	12	13	14	15	16	17	18	19	20
$p_j^{(3)}$	0.057	0.111	0.171	0.207	0.192	0.134	0.068	0.023	0.004	0.0004

Figure 7. The probability mass function of detected unconformities number

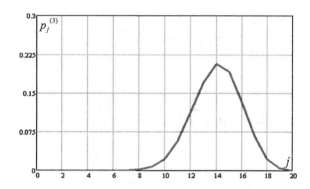

$$m_1(D(n, m, N, T_{\text{obs}} / \text{S3})) = \frac{m_1(n)}{T_{\text{obs}}} = \frac{\sum_{j=0}^{n} jp_j^{(m)}}{mv(t)\left(T_{\text{OS}} + \Delta t\right)}, \tag{17}$$

$$m_1(\text{Ef}(n, m, N, T_{\text{obs}}, C_\Sigma / \text{S3})) = \frac{m_1(D(n, m, N, T_{\text{obs}} / \text{S3}))}{C_{\text{max}} / (C_{\text{max}} - NC)} =$$
$$= \frac{(C_{\text{max}} - NC)\sum_{j=0}^{n} jp_j^{(m)}}{mv(t)\left(T_{\text{OS}} + \Delta t\right)C_{\text{max}}} = \frac{(C_{\text{max}} - NC)\sum_{j=0}^{n} jp_j^{(m)}}{mv(t)\left(T_{\text{OS}} + \Delta t\right)C_{\text{max}}}. \tag{18}$$

Let's consider the example of calculation of mathematical expectations of the probability of correct detection and OS efficiency for different strategies. According to previous calculations $N = 90$, $n = 20$, $m = 3$, $m_1(n) = \sum_{j=0}^{n} jp_j^{(3)} = 14.074$. Let the time of OS functioning $T_{\text{OS}} = 1000$ hours, CA duration $\Delta t = 1$ hour, total costs $C_{\text{max}} = 5000$ usd, CA costs $C = 10$ usd, unconformity rate $v(t) = 0.02$

unconformity per hour. According to the formulas (11) and (12) the mathematical expectations of the probability of correct detection and OS efficiency for the first strategy:

$$m_1(D(n, N, T_{obs} / S1)) = \frac{20}{0.02(1000 + 1)} = 0.999,$$

$$m_1(Ef(n, N, T_{obs}, C_\Sigma / S1)) = \frac{20(5000 - 90 \cdot 10)}{0.02(1000 + 1)5000} = 0.819.$$

According to the formulas (13), (14), (17) and (18) the mathematical expectations of the probability of correct detection and OS efficiency for the second and third strategies:

$$m_1(D(n, N, T_{obs} / S2)) = \frac{20}{0.02 \cdot 3(1000 + 1)} = 0.333,$$

$$m_1(Ef(n, N, T_{obs}, C_\Sigma / S2)) = \frac{20(5000 - 90 \cdot 10)}{0.02 \cdot 3(1000 + 1)5000} = 0.273,$$

$$m_1(D(n, m, N, T_{obs} / S3)) = \frac{14.074}{0.02 \cdot 3(1000 + 1)} = 0.234,$$

$$m_1(Ef(n, m, N, T_{obs}, C_\Sigma / S3)) = \frac{14.074(5000 - 90 \cdot 10)}{0.02 \cdot 3(1000 + 1)5000} = 0.192.$$

According to these calculations, the mathematical expectations of the probability of correct detection and OS efficiency in case of SEEP usage (the first strategy) more than corresponding values of other options without SEEP (the second and third strategies).

Comparing (11) – (14) and (17), (18), we can conclude that REE OS efficiency in case of SEEP usage more than efficiency for both considered second and third strategies, i.e.

$$\frac{m_1(Ef(n, N, T_{obs}, C_\Sigma / S1))}{m_1(Ef(n, m, N, T_{obs}, C_\Sigma / S2))} = \frac{\dfrac{n(C_{max} - NC)}{v(t)(T_{OS} + \Delta t)C_{max}}}{\dfrac{n(C_{max} - NC)}{v(t)m(T_{OS} + \Delta t)C_{max}}} = m,$$

$$\frac{m_1(\mathrm{Ef}(n, N, T_{\mathrm{obs}}, \mathrm{C}_\Sigma / \mathrm{S1}))}{m_1(\mathrm{Ef}(n, m, N, T_{\mathrm{obs}}, \mathrm{C}_\Sigma / \mathrm{S3}))} = \frac{\dfrac{n(\mathrm{C}_{\max} - NC)}{v(t)\left(T_{\mathrm{OS}} + \Delta t\right)\mathrm{C}_{\max}}}{\dfrac{(\mathrm{C}_{\max} - NC)\displaystyle\sum_{j=0}^{n} jp_j^{(m)}}{mv(t)\left(T_{\mathrm{OS}} + \Delta t\right)\mathrm{C}_{\max}}} = \frac{nm}{\displaystyle\sum_{j=0}^{n} jp_j^{(m)}}.$$

So the SEEP increases the REE OS efficiency exactly in m times compared with the second strategy and at least in m times compared with the third strategy (as $\displaystyle\sum_{j=0}^{n} jp_j^{(m)} \leq n$). The discussed analytical formulas for OS efficiency calculation for radioelectronic equipment can be considered as one of elements of the methodical basis for solving OS design and improvement problems.

Suppose that the OS at the moment of SEEP implementation is characterized by the basic levels of the probability of correct detection of unconformities and efficiency $\{\overrightarrow{D}_{\mathrm{bas}.N}, \overrightarrow{\mathrm{Ef}}_{\mathrm{bas}.N}\}$, where $\overrightarrow{D}_{\mathrm{bas}.N} = \{D_{\mathrm{bas}.1}, D_{\mathrm{bas}.2}, ..., D_{\mathrm{bas}.N}\}$, $\overrightarrow{\mathrm{Ef}}_{\mathrm{bas}.N} = \{\mathrm{Ef}_{\mathrm{bas}.1}, \mathrm{Ef}_{\mathrm{bas}.2}, ..., \mathrm{Ef}_{\mathrm{bas}.N}\}$ contains N component of different branches of activity.

The integral probability of correct detection and efficiency:

$$D_{\mathrm{bas}.\Sigma} = \sum_{i=1}^{N} B_i D_{\mathrm{bas}.i},$$

$$\mathrm{Ef}_{\mathrm{bas}.\Sigma} = \sum_{i=1}^{N} B_i \mathrm{Ef}_{\mathrm{bas}.i},$$

where B_i are weighting coefficients that characterize the contribution of each basic probability of correct detection and efficiency indicators by the direction of activity to the overall (integral) probability of correct detection and efficiency (in this case $\displaystyle\sum_{i=1}^{N} B_i = 1$).

According to formulas (9) and (10), in case of known costs for all OS C_Σ and $\mathrm{C}_{\max}$, and OS components $\mathrm{C}_{\Sigma i}$ and $\mathrm{C}_{\max i}$, the correlation between integral probability of correct detection and efficiency is:

$$\mathrm{Ef}_{\mathrm{bas}.\Sigma} = \frac{D_{\mathrm{bas}.\Sigma}}{\mathrm{C}_{\max} / (\mathrm{C}_{\max} - \mathrm{C}_\Sigma)},$$

$$\mathrm{Ef}_{\mathrm{bas}.i} = \frac{D_{\mathrm{bas}.i}}{\mathrm{C}_{\max i} / (\mathrm{C}_{\max i} - \mathrm{C}_{\Sigma i})},$$

$$C_{\Sigma} = \sum_{i=1}^{N} C_{\Sigma i},$$

$$C_{max} = \sum_{i=1}^{N} C_{max\,i}.$$

After SEEP implementation, new levels of probability of correct detection and efficiency will be:

$$D_{new\,i}(P) = D_{bas.\,i} + \Delta D_{\Sigma i}(P),$$

$$Ef_{new\,i}(P) = Ef_{bas.\,i} + \Delta Ef_{\Sigma i}(P),$$

where $\Delta D_{\Sigma i}(P)$ and $\Delta Ef_{\Sigma i}(P)$ are increments of probability of correct detection and efficiency due to the introduction of new monitoring methods P.

In this case

$$\Delta D_{\Sigma i}(\vec{P}) = \sum_{j=1}^{P} \Delta D_{i,j},$$

$$\Delta Ef_{\Sigma i}(\vec{P}) = \sum_{j=1}^{P} \Delta Ef_{i,j},$$

where $\Delta D_{i,j}$ and $\Delta Ef_{i,j}$ are increments of probability of correct detection and efficiency due to the j-th component action (can take both negative and positive values).

Therefore, the main task of creating SEEP is to find such a design decisions $DD_i^{(0)}$ from a set of all possible decisions Ω_{DD} ($DD_i^{(0)} \in \Omega_{DD}$), which will provide, in the framework of a defined criterion, a new probability of correct detection in the OS for the case when the efficiency is not worse than the basic level, i.e.:

$$D_{new\,i}(DD_i^{(0)}) > D_{bas.\,i} \text{ in condition } Ef_{new\,i}(DD_i^{(0)}) \geq Ef_{bas.\,i} \text{ for } i \in [1;\, N].$$

This approach sets strict requirements for design decisions. Another approach is to use the integral values of the probability of correct detection and efficiency for which

$$D_{new\,\Sigma}(DD^{(0)}) > D_{bas.\,\Sigma} \text{ in condition } Ef_{new\,\Sigma}(DD^{(0)}) > Ef_{bas.\,\Sigma}.$$

The design decisions $DD_i^{(0)}$ can include the following components: the selection of processes, organizational structures, indicators for monitoring, resources, development and modernization of documentation, recruitment and advanced training of personnel, etc.

The formation of design decisions can be carried out within the framework of certain strategies that specify directions for the OS improvement, for example, the application or absence of forecasting procedures for the numerical values of the organization's efficiency indicators.

DATA PROCESSING SUBSYSTEM

The initial information in the intelligent control system for OS is the data on the OS components condition. In most cases, these data are random variables (Levin, 1978).

The utilization of data processing algorithms is associated with removing indeterminacy condition. In the general case indeterminacy removing influences the risks of unconformities occurrence. Decreasing priori uncertainty increases OS efficiency (Goncharenko, 2017).

The main tasks of operational data processing are:

1. Detection.
2. Estimation.
3. Filtration.
4. Extrapolation.
5. Approximation taking into account heteroscedasticity, etc.

In the REE operation practice, the problems of changepoint detection have recently become topical. The changepoint is a process that is characterized by the presence of several quasistationary intervals. Changepoint detection is an important task since it is associated with the dynamics of OS components condition changes.

In the OS, changepoint problems can be solved by analyzing the diagnostic variables and reliability parameters (Solomentsev, Zaliskyi, Nemyrovets, & Asanov, 2015).

ILLUSTRATIVE EXAMPLES OF DATA PROCESSING

Let's consider two examples of solving changepoint detection problems.

Example 1. Diagnostic variables analysis.

Algorithm for changepoint detection in the trend of diagnostic variables corresponds to the condition-based maintenance with diagnostic variables inspection (Gertsbakh, 2005).

Let us suppose that technical condition is characterized by one diagnostic variable $y(t)$. Let us assume that normative documentation sets the allowance for possible changes of this variable in the form of lower and upper operating thresholds D_{o-} and D_{o+}. Preventive thresholds we will mark as D_{p-} and D_{p+}.

Let the model of diagnostic variable change will be described by the formula (4).

If $D_{o-} \leq y(t) \leq D_{o+}$, then it is assumed that REE is in operating condition. Otherwise, the failure appears. Let's assume that after the failure, REE recovery begins, and the value of the diagnostic variable will be equal to the initial value Z_0. The event when $D_{p+} \leq y(t) \leq D_{o+}$ or $D_{o-} \leq y(t) \leq D_{p-}$ is connected with REE technical condition deterioration. In this case, preventive maintenance is planned and fulfilled, after which the value of the diagnostic variable will be equal to initial value Z_0.

During the monitoring process, moments τ_1 and τ_2 of preventive and operating thresholds intersection are fixed. The difference between these moments $\Delta t = \tau_2 - \tau_1$.

We assume that preventive maintenance is carried out during the time ΔT. If $\Delta t > \Delta T$, then the failure doesn't occur, and the diagnostic variable is equal to initial value Z_0. Otherwise, the failure occurs. After the failure, the current repair is fulfilled. Duration ξ of current repair is a random variable.

To analyze the efficiency of the data processing algorithm, a computer simulation was performed. Initial parameters for simulation are $Z_0 = 0$, $k = 1$, $|D_o| = 500$, $C_r = 100$ usd, $C_m = 50$ usd, $\Delta T = 100$, t_{sw} is normally distributed random variable with parameters $m(t_{sw}) = 200$ and $\sigma(t_{sw}) = 60$, $a_i = 0$, $x(t)$ and $\vartheta(t)$ are uniformly distributed random variables in the range $[-20;20]$, ξ is exponentially distributed random variable with parameter $\lambda = 0.005$. During the simulation process, the number of failures, the number of successful preventive procedures, mean time between failures, mean time of repair, steady-state availability and average costs are calculated. Results of simulation are shown in Figures 8 and 9.

According to Figures 8 and 9, we can make such conclusions. Use of data processing algorithm, which realizes condition-based maintenance strategy with control of diagnostic variables, leads to increase of mean time between failures, and thus to steady-state availability increasing in case of preventive threshold value decreasing. Analysis of average costs dependence on preventive threshold shows that there is a minimum of function $m_1(C_\Sigma / T_{obs})$ for given initial parameters of condition-based maintenance strategy. Nature of this minimum can be explained by the existence of a balance between costs for repair and maintenance.

Figure 8. Availability estimation for different values of preventive threshold

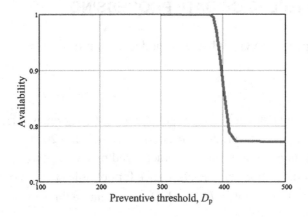

Figure 9. Average costs estimation for different values of preventive threshold

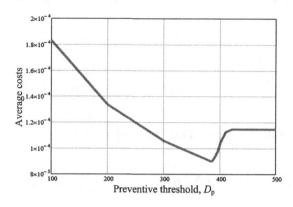

Example 2. Reliability parameters analysis.

Let's consider the case when the technical condition of the REE is characterized by the failure rate. Assume that the model failure rate description corresponds to formula (6). Initial parameters are times between failures. The observation interval consists of two intervals. At the first interval $[0; t_{sw}]$ times between failures are characterized by the probability density function (PDF) $f_1(t)$. At the second interval $[t_{sw}; T_{obs}]$ times between failures are characterized by PDF $f_2(t)$. Let PDFs are exponential, then

$$f_1(t_i) = \lambda_0 e^{-\lambda_0 t_i} \text{ for } i \in [1; m-1].$$

$$f_2(t_i) = (d_0 + \lambda_0)e^{-(d_0 + \lambda_0)t_i} \text{ for } i \in [m; n].$$

where t_i are times between failures; λ_0 is a failure rate before deterioration; d_0 is a coefficient of variation of the failure rate that should be detected with D probability; m is a number of failures after which the deterioration begins; n is a total quantity of observed failures.

The synthesis of changepoint detection procedure is based on the Neyman-Pearson criterion with fixed sample size n. In publications, there are changepoint detection procedures with a focus on a posteriori analysis of output data (with fixed sample size) and a sequential analysis of data (with infinite sample size). In this case, we consider changepoint detection procedure according to posteriori analysis. For this purpose, the decision about the presence or absence of changepoint is taken after the processing of the data of the entire sample of the volume n.

To check the simple hypotheses we make the following assumption. The hypothesis H_0 is deterioration absence; the alternative H_1 is technical condition deterioration at the observation interval.

Let's calculate the likelihood ratio for Neyman-Pearson criterion

$$\Lambda(\vec{t_n}, m, d_0) = \frac{\Phi(\vec{t_n} / H_1)}{\Phi(\vec{t_n} / H_0)},$$

where $\Phi(\vec{t}_n \, / \, H_1)$ is a likelihood function for H_1; $\Phi(\vec{t}_n \, / \, H_0)$ is a likelihood function for H_0.

After calculation we will get

$$\Lambda(\vec{t}_n, m, d_0) = \left(\frac{d_0 + \lambda_0}{\lambda_0}\right)^{n-m} e^{-d_0 \sum_{i=m}^{n} t_i}.$$

Typically the logarithm of the likelihood ratio function is used:

$$\ln \Lambda(\vec{t}_n, m, d_0) = (n - m)\ln\left(\frac{d_0 + \lambda_0}{\lambda_0}\right) - d_0 \sum_{i=m}^{n} t_i.$$

Let $\theta(d_0, \, \vec{t}_n, \lambda_0)$ is a decisive statistic that depends on the sample size n, parameters d_0, λ_0 and m. This means

$$\theta(d_0, \, \vec{t}_n, \lambda_0) = (n - m)\ln\left(\frac{d_0 + \lambda_0}{\lambda_0}\right) - d_0 \sum_{i=m}^{n} t_i.$$

In the formula for decisive statistics we need to know failure rate and coefficient of variation. For these parameters, the detection procedure should provide the necessary level of the probability of correct detection and other measures of the detection algorithm efficiency (Solomentsev, Zaliskyi, Herasymenko, Kozhokhina & Petrova, 2018).

In the decision-making scheme after calculating of decisive statistics for the complete set, the maximum value of decisive statistics is compared with the threshold decision level V_1. If the maximum value exceeds the threshold, then a decision is made about deterioration presence. Analysis problem consists of the calculation of numerical values of efficiency measure: the probability of type I errors α and the probability of type II errors β. Usually, for detection algorithms, they calculate the probability D of correct detection, that is $D = 1 - \beta$. In Figure 10 the dependency of the probability of correct detection of changepoint is shown. This probability depends on the level of the parameter d_0 for the given values of general set parameters of decisive statistics.

Figure 10. The dependence of probability of correct detection of changepoint D(d₀) on measure d₀.

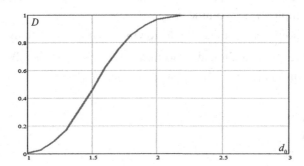

TECHNICAL CONDITION OF RADIO ELECTRONIC EQUIPMENT INSPECTION

Inspection is the process of establishing a match between the impartial assessment of the object state and a predetermined norm for the possible object states. This process is carried out by perception of a set of controlled parameters, processing received information, the formation and issuance of decisions on the results of the conformity establishment (Barlow & Proschan, 1965).

Technological inspection operation is one of the main among all the radio electronic equipment operation technological processes. After establishing the correspondence between the objective technical condition of the radio electronic equipment and the need for one or the other management influences. For example, regulation, adjustment, current repairs, planned repair, unscheduled flight inspection of the REE, etc. Figure 11 shows the operator circuit for monitoring the technical condition of radioelectronic equipment.

Suppose that the technical state of the OI objectively characterizes one diagnostic variable (DV), which is a random function of $\xi(t)$. The statistical model of DV was studied previously. Next, this model will also have the basic features of the DV trends.

The additive interaction $\xi(t)$ and noise $x_n(t)$ model can be written as

$$u(t) = \xi(t) + x_n(t).$$

The MDO operator performs the transformation of the form

$$\rho(t) = \text{MDO}(u(t)).$$

If the DV perceives the measuring device then possible measurement errors and random functions $\rho(t)$ and $u(t)$ differ.

Let the random function $u(t)$ be transformed by the MDO operator as an "unobservable process". In this case, the random function in $y(t)$ can be played back using an indicator device, for example, oscilloscope. Let us assume that the measurement errors $x_{m.e}(t)$ interact additively with the signal $u(t)$, that is,

$$\rho(t) = u(t) + x_{m.e}(t) = \xi(t) + x_n(t) + x_{m.e}(t) = \xi(t) + x_{\Sigma}(t),$$

Figure 11. Operator's diagram of inspection of the technical condition of the REE: OI – object under inspection; MDO – operator, which determines measuring device operation; DP – data processing operator; DM – decision making operator; ID – operator, which determines the indicator device operation; N – operator of measurement noise

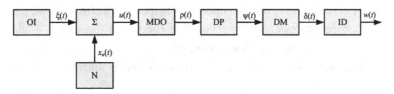

where $x_\Sigma(t)$ is a total random process that takes into account the presence of noises $x_n(t)$ and measurement errors $x_{m.e}(t)$.

Further, such transformations of signals are carried out with the help of the introduced operators

$$\psi(t) = \text{DP}(\rho(t)); \; \delta(t) = \text{DM}(\psi(t)); \; w(t) = \text{ID}(\delta(t)).$$

For solving the problems of synthesis and analysis of the operator circuit (Figure 11), it is necessary to detail the generalized form of the introduced operators, to assign indicators of efficiency of operations inspection of the technical condition of the REE.

We will assume that the objects of control are characterized by the fact that in the process of the REE technical condition assessing many parameters with a large dynamic range are used. Some parameters have been converted to standard view. To measure other parameters, special control-measuring equipment (CME) is used.

Thus, a significant number of DVs are characterized by a large dynamic range and DVs have a diverse physical nature.

Let us specify the certain components of the operator circuit (Figure 11). In general, the model DV $\rho(t)$ at the output of the operator $\text{MDO}(\cdot)$ has the form

$$\rho(t) = \xi(t) + x_\Sigma(t), \quad \xi(t) = z(t) + q(t),$$

where $\xi(t)$ is a signal component that objectively characterizes the change of the technical condition of the OC; $z(t)$ is an unobservable process that determines the trend of change $\xi(t)$; $q(t)$ is stochastic component $\xi(t)$, which takes into account the inaccuracy of the definition $\xi(t)$ only with the help of a component $z(t)$; $x_\Sigma(t)$ is interference component caused by interruptions in measuring channels and errors in inspection operations.

Suppose that the components $x_\Sigma(t)$ and $q(t)$ are characterized by Gaussian PDFs with $m_1(x)$ and $m_1(q)$ equal to zero, and variables $\mu_2(x)$ and $\mu_2(q)$ that are nonzero, ie, $\mu_2(x) \neq 0$ and $\mu_2(q) \neq 0$. The random function $z(t)$ can be described according to equation (2) or (3).

We will also assume that there is no component $q(t)$, and only the parameter a_0, having PDF $f(a_0)$ is used to determine $z(t)$. Then

$$\rho(t) = \xi(t) + x_\Sigma(t) = a_0 + x_\Sigma(t).$$

Figure 12 shows an example of the function $\rho(t)$ for the three observation intervals: I, II, III. On two intervals the parameter a_0 is within the operating thresholds D_{o-} and D_{o+}, and on one interval it is outside of these tolerances.

On the basis of the REE technical condition operation inspection definition, we assume that the norm of the possible control object states can be: serviceable (denote it through χ) and failure (denote it through $\bar{\chi}$). Taking into account the operator circuit, the function $\delta(t)$ can acquire two values, for example, $\delta(t) = 1$ in case of a decision on the technical condition χ and $\delta(t) = 2$ in case of a decision

Figure 12. An example of a function change $\rho(t)$

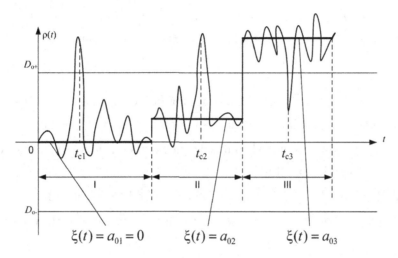

on the technical condition $\overline{\chi}$. Objective output technical condition for OI χ_0 and $\overline{\chi}_0$ meet the following inequalities:

$$\chi_0 \rightarrow \delta(t) = 1 \text{, if } D_{o-} < \xi(t) < D_{o+};$$

$$\overline{\chi}_0 \rightarrow \delta(t) = 2 \text{, if } \xi(t) \geq D_{o+} \text{ or } \xi(t) \leq D_{o-}.$$

Let's assume that the information processing operator $DP(\cdot)$ at clock speeds t_{ci} only performs the removal of information, without requiring any other data processing, i.e. $\psi(t) = \rho(t)$. Given this, we have

$$\psi(t_{ci}) = \rho(t_{ci}) = \xi(t_{ci}) + x_{\Sigma}(t_{ci}).$$

The constituent $\xi(t)$ in Figure 12 acquires the following values a_0 when the objective technical condition for OI at intervals I and II is χ_0. $\overline{\chi}_0$ is on the interval III. Taking into account the fact that there is an ingredient $x_{\Sigma}(t)$, the decisions on the technical state of the OI will not always coincide with the objective technical conditions χ_0 and $\overline{\chi}_0$. The examples of such situations are shown in Figure 12 for time points t_{c1}, t_{c2}, t_{c3}. An analysis of the equipment technical state controlling process should be considered using the graph of the OI technical condition inspection, which is shown in Figure 13.

The graph in Figure 13 characterizes the process of inspecting the technical condition of the OI based on the analysis of information about one DV. According to the results of the inspection of one objective technical condition, two decisions are taken: OI is serviceable or OI is with failure. But because there are two possible objective technical conditions of the OI, then in the end, when analyzing the efficiency of the inspection process we will already distinguish four technical conditions: χ_1, $\overline{\chi}_2$, χ_3, $\overline{\chi}_4$, namely:

1. "Serviceable" – $\chi_1 \to \delta(t) = 1$, if $D_{o-} < \rho(t_c) < D_{o+}$, if objectively $D_{o-} < \xi(t_c) < D_{o+}$.
2. "Failure" – $\overline{\chi}_2 \to \delta(t) = 2$, if $\rho(t_c) \geq D_{o+}$ or $\rho(t_c) \leq D_{o-}$, if objectively $D_{o-} < \xi(t_c) < D_{o+}$.
3. "Serviceable" – $\chi_3 \to \delta(t) = 1$, if $D_{o-} < \rho(t_c) < D_{o+}$, if objectively $\xi(t_c) \geq D_{o+}$ or $\xi(t_c) \leq D_{o-}$.
4. "Failure" – $\overline{\chi}_4 \to \delta(t) = 2$, if $\rho(t_c) \geq D_{o+}$ or $\rho(t_c) \leq D_{o-}$, if objectively $\xi(t_c) \geq D_{o+}$ or $\xi(t_c) \leq D_{o-}$.

When making decisions about the technical condition of the OI errors of the first kind occur when it is decided that the OI is in a condition "failure" – $\overline{\chi}_2$ if the objective technical condition of OI is χ_0. And there are errors of the second kind, when it is decided that OI is in a serviceable condition – $\overline{\chi}_2$ if the objective technical condition of OI is $\overline{\chi}_0$.

The possible probability $P(\overline{\chi}_2 / \chi_0) = \alpha$ determines the probability of making the error of the first kind. The possible probability $P(\chi_3 / \overline{\chi}_0) = \beta$ determines the probability of making the error of the second kind.

Probability states χ_1, $\overline{\chi}_2$, χ_3, $\overline{\chi}_4$ counted with the help of unconditional probabilities $P(\chi_1)$, $P(\overline{\chi}_2)$, $P(\chi_3)$, $P(\overline{\chi}_4)$. Each of these unconditional probabilities evaluates the probability that there are two events. For example, for $P(\chi_1)$ we have an objective technical condition of OI as χ_0 and a decision was made on the result of the inspection, so that OI has a technical condition as χ_1 and so on.

Figure 13. A graph of REE technical condition inspection

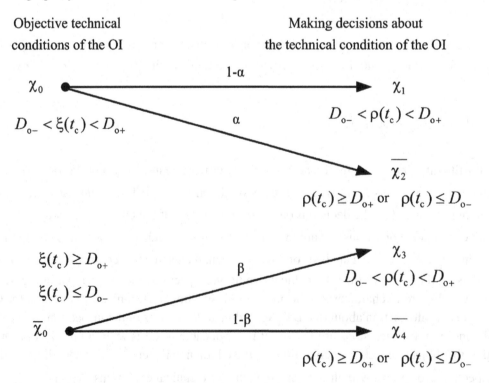

We will assume that for any moment of inspection of the OI technical condition t_c we know the PDF of the random variable $\xi - f(\xi)$. Also, for any moment of time t_c PDF of the random variable (RV) $x - f(x)$ is known. We assume that RVs ξ and x are independent.

Then, the expression for $P(\chi_1)$ will take the form

$$P(\chi_1) = \int\limits_{D_{o-}}^{D_{o+}} f(\xi) \left[\int\limits_{D_{o-}-\xi}^{D_{o+}-\xi} f(x)dx \right] d\xi . \tag{19}$$

The expression for $P(\overline{\chi_2})$ will look like

$$P(\overline{\chi_2}) = \int\limits_{D_{o-}}^{D_{o+}} f(\xi) \left[\int\limits_{-\infty}^{D_{o-}-\xi} f(x)dx + \int\limits_{D_{o+}-\xi}^{\infty} f(x)dx \right] d\xi . \tag{20}$$

The expression for $P(\chi_3)$ will look like

$$P(\chi_3) = \int\limits_{D_{o+}}^{\infty} f(\xi) \left[\int\limits_{D_{o-}-\xi}^{D_{o+}-\xi} f(x)dx \right] d\xi + \int\limits_{-\infty}^{D_{o-}} f(\xi) \left[\int\limits_{D_{o-}-\xi}^{D_{o+}-\xi} f(x)dx \right] d\xi . \tag{21}$$

The expression for $P(\overline{\chi_4})$ will look like

$$P(\overline{\chi_4}) = \int\limits_{D_{o+}}^{\infty} f(\xi) \left[\int\limits_{D_{o+}-\xi}^{\infty} f(x)dx + \int\limits_{-\infty}^{D_{o-}-\xi} f(x)dx \right] d\xi +$$
$$+ \int\limits_{-\infty}^{D_{o-}} f(\xi) \left[\int\limits_{D_{o+}-\xi}^{\infty} f(x)dx + \int\limits_{-\infty}^{D_{o-}-\xi} f(x)dx \right] d\xi . \tag{22}$$

In the scientific and technical literature in case of the Gaussian probability distribution $f(\xi)$ and $f(x)$ there are some tables defining parameters $P(\chi_1),..., P(\overline{\chi_4})$ by formulas (19), (20), (21), (22).

Let us consider the determining conditional probabilities

$$P(\overline{\chi_2} / \chi_0) = \alpha \text{ and } P(\chi_3 / \overline{\chi_0}) = \beta .$$

To do this, let us have a look at the REE graph of the inspection of the technical condition in Fig. 13. It follows from the graph (Figure 13) that

$$\begin{cases} P(\chi_1) = P(\chi_0)P(\chi_1/\chi_0) = P(\chi_0)(1-\alpha); \\ P(\chi_2) = P(\chi_0)P(\chi_2/\chi_0) = P(\chi_0)\alpha; \\ P(\chi_3) = P(\overline{\chi_0})P(\chi_3/\overline{\chi_0}) = P(\overline{\chi_0})\beta; \\ P(\chi_4) = P(\overline{\chi_0})P(\chi_4/\overline{\chi_0}) = P(\overline{\chi_0})(1-\beta). \end{cases} \qquad (23)$$

Probabilities $P(\chi_0)$ and $P(\overline{\chi_0})$ can be defined as follows:

$$P(\chi_0) = \int\limits_{D_{o-}}^{D_{o+}} f(\xi)d\xi;$$

$$P(\overline{\chi_0}) = 1 - P(\chi_0) = \int\limits_{-\infty}^{D_{o-}} f(\xi)d\xi + \int\limits_{D_{o+}}^{\infty} f(\xi)\,d\xi. \qquad (24)$$

It follows from the formulas (23) that

$$\alpha = \frac{P(\overline{\chi_2})}{P(\chi_0)} \text{ and } \beta = \frac{P(\chi_3)}{P(\overline{\chi_0})}. \qquad (25)$$

Probabilities $P(\overline{\chi_2})$ and $P(\chi_3)$ are determined by the formulas (20) and (21) accordingly.

The conditional probabilities α and β, calculated by the formulas (25), can be determined in such way. The current value of the parameter α depends on what value it will take RV ξ within tolerances $[D_{o-}; D_{o+}]$. Then to calculate α we will have a formula

$$\alpha(\xi) = \int\limits_{-\infty}^{D_{o-}-\xi} f(x)dx + \int\limits_{D_{o+}-\xi}^{\infty} f(x)dx.$$

Find the mean value of the parameter α considering PDF of RV ξ, which varies within tolerances $[D_{o-}; D_{o+}]$:

$$m_1(\alpha) = \alpha = \frac{\int\limits_{D_{o-}}^{D_{o+}} f(\xi)\left[\int\limits_{-\infty}^{D_{o-}-\xi} f(x)dx + \int\limits_{D_{o+}-\xi}^{\infty} f(x)dx\right]d\xi}{p(D_{o-} < \xi < D_{o+})}. \qquad (26)$$

The parameter $p(D_{o-} < \xi < D_{o+})$ must be set to comply with the valuation conditions for PDF of RV ξ, which describes the probabilistic properties of RV ξ in case of changing it in the range $[D_{o-}; \acute{D}_{o+}]$. It is easy to see that formula (26) coincides with expression (24) for a parameter α taking into account formulas (20) and (24), namely:

$$\alpha = \frac{P(\bar{\chi}_2)}{P(\chi_0)}.$$

Thus, we can assume that the parameter α in equation (23) is the average value of the possible probability of error of the first kind. Similarly, the parameter β in equation (23) is the average value of the possible probability of error of the second kind.

We will also consider other efficiency indicators that characterize the process of monitoring the technical condition of the OI. These include absolute veracity of inspection Z or total probability of correct decisions taken during the inspection of the technical condition of the OI; the veracity Z_χ of serviceable decision result; the veracity $Z_{\bar{\chi}}$ of failure decision result; the average risk R. The veracity quantitatively characterizes the degree of confidence in one or another decision. Parameter Z change range is $[0;1]$.

On the REE technical condition inspection graph (Figure 13) wrong decisions are connected with the states $\overline{\chi_2}$ and χ_3. Probabilities of these states are $P(\chi_2)$ and $P(\chi_3)$. The right decisions are related to the states χ_1 and $\overline{\chi_4}$. Probability of these states are $P(\chi_1)$ and $P(\overline{\chi_4})$, then the absolute veracity is determined as

$$Z = 1 - \beta - P(\chi_0)(\alpha - \beta).$$

When calculating the veracity of Z_χ or veracity of $Z_{\bar{\chi}}$ the proportion of correct decisions about one or another result in the overall set of decisions about the result is determined. Taking into account the graph presented in Figure 13, we have the following

$$Z_\chi = \frac{P(\chi_1)}{P(\chi_1) + P(\chi_3)} = \frac{P(\chi_0)(1 - \alpha)}{P(\chi_0)(1 - \alpha - \beta) + \beta}. \tag{27}$$

From the equation (27) it is clear that if there are no errors of the first and second kind, that is $\alpha = \beta = 0$, then the parameter $Z_\chi = 1$. The parameter $Z_{\bar{\chi}}$ is defined as follows:

$$Z_{\bar{\chi}} = \frac{P(\overline{\chi_4})}{P(\overline{\chi_4}) + P(\overline{\chi_2})} = \frac{P(\overline{\chi_0})(1 - \beta)}{P(\overline{\chi_0})(1 - \beta - \alpha) + \alpha}. \tag{28}$$

From equation (28) it is clear, that at the value $\alpha = \beta = 0$, parameter $Z_{\bar{\chi}} = 1$. Thus, in order to improve the quality of inspection, it is necessary to increase the numerical values of veracities Z, Z_χ, $Z_{\bar{\chi}}$.

In the scientific and technical literature (Rausand, 2004) two main ways to increase the following veracities are considered: improving the methodological component of inspection and the instrumental component of inspection. The methodological component of inspection is mainly related to the organization of inspection and technology (algorithms) of REE technical condition inspection. The instrumental component of inspection often involves the improvement of the hardware support for the REE inspection of its technical condition.

Taking into account these methods, it can be conventionally considered that, for example, absolute veracity Z contains a part of methodical probability and part of instrumental probability.

The main methods for increasing the instrumental component of veracity include:

- Increase of faultlessness and improvement of maintenance of measuring equipment.
- Reduction of errors of inspection and measuring operations.
- Reduction of the level of noise during measurement, etc.

To determine the average risk R it is necessary to know the parameters $P(\overline{\chi_2})$, $P(\chi_3)$, as well as the losses due to control errors (Smith, 2005). Then R will look like

$$R = C_1 P(\overline{\chi_2}) + C_2 P(\chi_3),\qquad(29)$$

where C_1 are costs caused by the fact that the serviceable product (by the results of inspection) is considered as with failure; C_2 are costs due to the fact that the product with failure is considered to be serviceable by the results of the inspection.

In a separate case, if $C_1 = C_2 = 1$, we have

$$R = P(\overline{\chi_2}) + P(\chi_3) = 1 - Z.$$

In the absence of control errors, that is if $\alpha = \beta = 0$, the average risk is zero.

For cases related to the operation of REE in civil aviation, costs $C_2 > C_1$. From this it follows that in order to reduce the average risk of R we must strive to reduce the parameter $P(\overline{\chi_2})$, that is, seek to reduce the probability β. Thus, to improve the inspection of the REE in terms of reducing the parameters α and β and it is advisable to pay more attention to the reduction of the parameter β.

Considering the operator scheme (Figure 11), we note some areas related to the improvement of inspection of the REE:

1. Perform a multi-parametric inspection of the technical condition of the REE for cases where inspected parameters $\rho(t)$ are independent or interdependent; choice of a rational list of parameters; indirect or direct methods of REE technical condition inspection.
2. Construction of models of measured parameters $\rho(t)$ for different models of inspection noise and models of measurement errors; multiplicative interaction $\xi(t)$ and $x_n(t)$.
3. The synthesis and analysis of the efficiency of optimal operators $DP(\cdot)$, allowing to perform a multi-alternative classification of the REE technical condition, which, in particular, involves determining the nature of changes in the REE technical condition, for example, the linear change of $\xi(t)$ without or without acceleration, etc. other application of Wald's sequential analysis with sampling truncation, as well as fixed-sample information processing algorithms.

FUTURE RESEARCH DIRECTIONS

Further researches may be directed to the development of data processing methodology in operation systems. The main attention should be paid to the methods of algorithms synthesis and analysis for the operational data processing in the cases of fixed and unknown sample sizes. Another aspect of the algorithms synthesis and analysis can be the wide use of the principles of intellectualization, depending on the different levels of a priori uncertainty.

Improving monitoring procedures may involve the use of the principle of adaptability. In this case, monitoring will involve the use of algorithms for operational data processing on the states of the OS elements to assess the degree of priori uncertainty.

CONCLUSION

In the chapter, the intellectualization questions for the operation systems of radioelectronic equipment are considered. The structure of the operation system is proposed taking into account the systems approach and the principles of artificial intelligence. The structure is built in such a way that data about the state of all elements are collected and processed and timely and necessary control actions are made. The intellectualization aspect allows the operation system to function in conditions that were not considered at the design stage.

One of the important tasks during the design of operation systems is to substantiate the efficiency indicator. In the chapter a new efficiency indicator was considered, this indicator takes into account the OS effectiveness, time and material resources costs. Analysis of the three inspection strategies of the OS elements state showed that inspection of all elements state is the most efficient.

In the chapter models for describing trends of diagnostic variables and reliability parameters are considered. This description takes into account the nonstationarity of their changes. For two options of the models, data processing algorithms were synthesized to detect changepoints in the REE technical condition.

The subsystem of REE technical condition inspection is considered, this subsystem measures and classifies the technical condition of REE taking into account errors of the first and second kind. Analytical expressions for assessing the inspection veracity for the case of two-alternative classification are obtained.

The results of the research can be used during the design and improvement of the REE OS.

REFERENCES

AdditionalREADING

Barlow, R. E., & Proschan, F. (1965). *Mathematical theory of reliability*. New York: John Wiley & Sons.

British Standards Institution. (2001). BS EN 13306: Maintenance Terminology, 31 p.

Dhillon, B. S. (2006). *Maintainability, maintenance, and reliability for engineers*. New York: Taylor & Francis Group. doi:10.1201/9781420006780

Galar, D., Sandborn, P., & Kumar, U. (2017). *Maintenance costs and life cycle cost analysis.* Boca Raton, FL: CRC Press. doi:10.1201/9781315154183

Gertsbakh, I. (2005). *Reliability theory: with applications to preventive maintenance.* New York: Springer. doi:10.1007/978-3-662-04236-6

Gnedenko, B. V., Belyayev, Y. K., & Solovyev, A. D. (1969). *Mathematical methods of reliability theory.* New York: Academic.

Goncharenko, A. (2017). Aircraft operation depending upon the uncertainty of maintenance alternatives. *Aviation, 21*(4), 126–131. doi:10.3846/16487788.2017.1415227

Hryshchenko, Y. V. (2016). Reliability problem of ergatic control systems in aviation. In *Proceedings of IEEE 4th International Conference on Methods and Systems of Navigation and Motion Control (MS-NMC).* (pp. 126 – 129). Kiev, Ukraine. 10.1109/MSNMC.2016.7783123

Jones, M. T. (2009). *Artificial intelligence: A systems approach.* Hingham, MA: Jones & Bartlett Learning.

Kapur, K. C., & Lamberson, L. R. (1977). *Reliability in Engineering Design.* New York: Wiley.

Kuzmenko, N. S., Ostroumov, I. V., & Marais, K. (2018). An accuracy and availability estimation of aircraft positioning by navigational aids. In *Proceedings of IEEE International Conference on Methods and Systems of Navigation and Motion Control (MSNMC),* (pp. 36 – 40). Kiev, Ukraine. 10.1109/MS-NMC.2018.8576276

Levin, B. R. (1978). *Theory of reliability of radio engineering systems.* Moscow, Russia: Radio. (in Russian)

Nakagawa, T. (2005). *Maintenance theory of reliability.* Springer.

Padhy, N. P. (2005). *Artificial intelligence and intelligent systems.* Oxford, UK: Oxford University Press.

Rausand, M. (2004). *System reliability theory: models, statistical methods and applications.* New York: John Wiley & Sons.

Safety Management Manual. (2013), Doc 9859, AN/474, 251 p.

Smith, D. J. (2005). *Reliability, maintainability and risk. Practical methods for engineers.* London, UK: Elsevier.

Solomentsev, O., Zaliskyi, M., Herasymenko, T., Kozhokhina, O., & Petrova, Yu. (2018). Data processing in case of radio equipment reliability parameters monitoring. In *Proceedings of IEEE International Conference on Advances in Wireless and Optical Communications (RTUWO 2018).* (pp. 219–222). Riga, Latvia. 10.1109/RTUWO.2018.8587882

Solomentsev, O., Zaliskyi, M., Kozhokhina, O., & Herasymenko, T. (2017). Efficiency of data processing for UAV operation system. In *Proceedings of IEEE 4th International Conference on Actual Problems of UAV Developments (APUAVD).* (pp. 27–31). Kiev, Ukraine. 10.1109/APUAVD.2017.8308769

Solomentsev, O., Zaliskyi, M., Nemyrovets, Yu., & Asanov, M. (2015). Signal processing in case of radio equipment technical state deterioration. In *Proceedings of Signal Processing Symposium 2015 (SPS 2015)*. (pp. 1–5). Debe, Poland. 10.1109/SPS.2015.7168312

Solomentsev, O., Zaliskyi, M., & Zuiev, O. (2016). Estimation of quality parameters in the radio flight support operational system. *Aviation*, *20*(3), 123–128. doi:10.3846/16487788.2016.1227541

Solomentsev, O. V., Melkumyan, V. H., Zaliskyi, M. Yu., & Asanov, M. M. (2015). UAV operation system designing. In *Proceedings of IEEE 3rd International Conference on Actual Problems of Unmanned Air Vehicles Developments (APUAVD)*, (pp. 95–98). Kiev, Ukraine. 10.1109/APUAVD.2015.7346570

Sushchenko, O. A., & Golitsyn, V. O. (2016). Data processing system for altitude navigation sensor. In *Proceedings of IEEE 4th International Conference on Methods and Systems of Navigation and Motion Control (MSNMC)*. (pp. 84–87). Kiev, Ukraine. 10.1109/MSNMC.2016.7783112

Tartakovsky, A., Nikiforov, I., & Basseville, M. (2015). *Sequential analysis: hypothesis testing and changepoint detection*. New York: Taylor & Francis Group.

Wilamowski, B. M., & Irwin, J. D. (2017). *Intelligent systems. The industrial electronics handbook*. Boca Raton, FL: CRC Press.

Zaliskyi, M., Solomentsev, O., Kozhokhina, O., & Herasymenko, T. (2017). Reliability parameters estimation for radioelectronic equipment in case of change-point. *In Proceedings of Signal Processing Symposium 2017 (SPSympo 2017)*. (pp. 1–4). Jachranka Village, Poland. 10.1109/SPS.2017.8053676

KEY TERMS AND DEFINITIONS

Adaptive Approach: The ability of the system to adapt to changing conditions.

Changepoint: The fact of transition of the controlled random process from one quasi-stationary state to another.

Control System: Complex to providing the required level of the REE reliability, which generates and performs corrective and preventive actions based on the results of statistical data processing.

Efficiency: The ability of the system to achieve the desired result taking into account the costs of resources.

Intellectualization: The property of the system to function effectively in conditions that were not foreseen at the design stage.

Maintenance: The operation objects serviceable state providing through the formation and implementation of control actions.

Monitoring: Observation for the process in time with using measurement and processing methods in order to two and multi-alternative classification of the technical condition.

Operation System: A set of devices, means of operation, personnel and documentation setting the rules of their cooperation for performing operation tasks. Operation is the stage of the device life cycle at which its quality is realized, maintained and restored.

Technical Condition: Set of possible classes of REE functioning.

Veracity: Quantitative assessment of the degree of confidence in the made decisions.

Chapter 7
Applications of Artificial Intelligence in Flight Management Systems

Ivan Ostroumov

 https://orcid.org/0000-0003-2510-9312
National Aviation University, Ukraine

Nataliia Kuzmenko
National Aviation University, Ukraine

ABSTRACT

Flight management system (FMS) is one of the key elements of the modern airplane. It is a computer-based system that helps a pilot with different routine operations. FMS includes numerous algorithms of Artificial Intelligence to support navigation, guidance, and control of aircraft. FMS hosts algorithms of airplane positioning by data from navigational aids and data fusion from multiple sensors. The internal memory of FMS includes global air navigation databases such as runways, airports data, air navigation charts, navigational aids, SIDs, STARs, approaches, and routes for automatic support of airplane operation.

BACKGROUND

Operation of each system of a modern airplane is connected with usage of numerous digital computation equipment. An airplane body can be referred to the complicated dynamic system. Numerous different systems are used to control an airplane state in three-dimensional space. Airframe orientation is provided by the flight control system. Operation of flight control surfaces is supported by hydraulic and pneumatic systems. Engine control system uses measurements of various pressure, temperature and flow sensors to provide required thrust value within safe engine operation. Fuel system provides the required fuel supply for normal engine operation and supports constant fuel flow between fuel tanks in order to guarantee airplane mass balancing. The electrical system provides generation and power distribution between all avionics equipment of an airplane. Each of these systems uses some elements of artificial intelligence

DOI: 10.4018/978-1-7998-1415-3.ch007

at different levels of control and data processing (Villarroel & Rodrigues, 2016). For example, at a low level of data processing, adaptive filters are used to reduce random noise. Search for outliers, filling the gaps, interpolation, and extrapolation of data is another place for artificial intelligence application (Solomentsev et al., 2018; Nanduri & Sherry, 2016). All functions of automatic control can be referred to artificial intelligence. In general, system control is based on a well-defined math model of process and set of parameters, interrogation with each can change a system state.

Airplane navigation is another important task for safe aviation. Navigation equipment supports safe take-off, flight by predefined trajectory and landing of the airplane within predefined time-frame . Airplane navigation within defined airspace takes significant attention from pilots. Airplane trajectory in most cases is predefined by specific documents and air navigation charts that need from pilots to spend a lot of time for airplane guidance and control. Flight management system (FMS) helps pilots with navigation and airplane guidance. FMS performs all technical routine operations with different airplane systems that are used during flight, and allows the pilot to spend more time on the flight control, rather than tuning and setting up avionics systems. FMS is a computer-based system that includes numerous important information for safe air navigation of the airplane and uses different sensor data to optimize flight trajectory in order to achieve better performance. Also FMS utilize algorithms of trajectory extrapolation and smoothing (Duanzhang at al., 2016). FMS is digital computation equipment that in basic configuration does not include any sensors. FMS uses specific software to run multiple algorithms of data processing and picture generation. And, of course, FMS utilizes multiple artificial intelligence operations at different levels. Introduction of FMS in aviation helped to reduce the number of persons in the cockpit that is needed for airplane control and guidance. Algorithms of FMS have utilized all duties of navigating officer (also called "air navigator" or "flight navigator") and makes possible to reduce the number of crew members to two persons only (captain and first officer). All heavy historical airplanes that are not equipped with FMS might required a flight engineer, radio operator or navigation officer to assist pilots for normal airplane operation (had to be a member of flight crew in commercial aviation on older aircraft, particularly used before 1990s, for example: MD-80, B-314, An-22, An-124, An-225). Therefore, FMS could be referred to only one system that fully replayed person responsibility on board of an airplane. In this case, FMS reduces the number of human functions with a help of deep automatization and leads to minimize human factor influence in aviation safety that is the main cause of the big amount of incidents that took place (Rizun & Shmelova, 2016; Wiegmann & Shappell, 2017).

Multiple advantages of FMS make it widespread in aviation. Nowadays each modern airplane is equipped with one of the FMS models and the future of avionics will not be imagined without FMS.

FMS supports multiple functions for big data analysis, data fusion, decision making support, automatic control, and navigation. Algorithms of FMS includes multiple application of artificial intelligence at different levels of data processing. The motivation and contribution of this chapter is to demonstrate application of artificial intelligence at different functions of FMS.

PLACE OF FMS IN AVIONICS

There are multiple commercially available architectures of FMS in civil aviation. In general, FMS includes FMS computer and Multifunction Control and Display Unit (MCDU) (Figure 1). FMS computer that utilizes all computational facilities is located at avionics compartment. Control and access to FMS

Figure 1. Data exchange in FMS

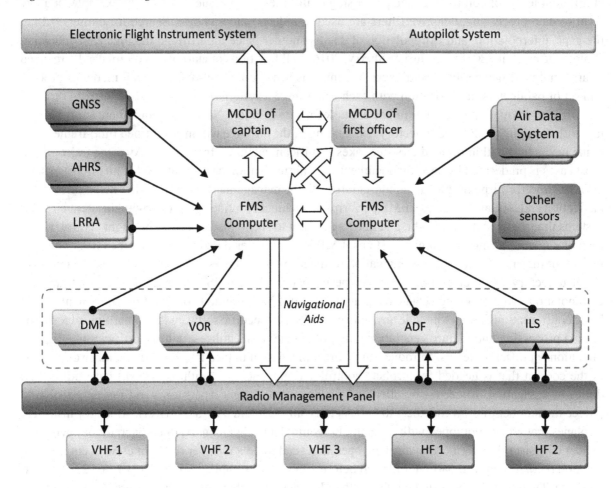

functionality are performed via MCDU placed in the cockpit. According to the minimum equipment list for avionics, airplane is usually equipped with two FMS computers and two MCDUs. Both sets are synchronized and work independently in parallel. Pilots use MCDU to get access to FMS functions. In common cockpit design, MCDU is located "near arms of pilot" at the central desk. Each of the pilots can use his MCDU to perform some actions independently from each other. It makes FMS flexible for parallel input and various operations.

The interaction of FMS with the pilot is provided by MCDU. This unit is used for data input and output in FMS. Common MCDU is shown on Figure 2. MCDU interface includes screen and a keyboard. The color screen displays only numerical and textual data. Keys on the MCDU allow to easily access to basic FMS menus and input required data for coding. Common FMS may include the following menus:

- Flight plan menu is used for operations with a flight plan.
- Navigation menu holds important data for the current part of the flight plan. It includes current and predicted navigational data for final crossed and next points of the flight plan.

Figure 2. MCDUs of FMS (Airbus 319)

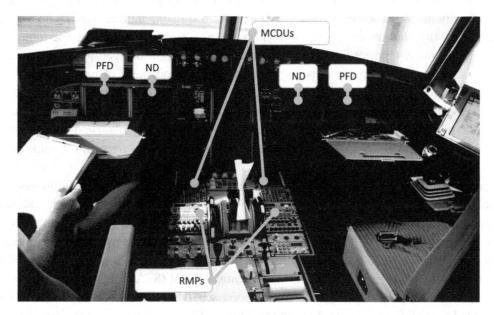

- The vertical navigation menu is used for setting and control trajectory of the airplane in a vertical plane.
- Radio control menu is used to control current and standby radio frequencies of communication equipment of airplane.

FMS uses data from multiple systems and sensors on board of an airplane. In particular, it supports data exchange with the following avionics (Figure 1):

- The receiver of Global Navigation Satellite System (GNSS) to obtain coordinates of airplane location.
- Attitude and heading reference system (AHRS) provides heading, orientation angles and airplane position in case of GNSS lock.
- On-board equipment of operation with navigational aids, such as:
  - Distance Measuring Equipment (DME) or Tactical air navigation system (TACAN) measures the distance to ground-based radio navigational beacons.
  - Very High Frequency (VHF) Omni-Directional Range (VOR) measures the radial angle to ground-based radio navigational beacons.
  - Automatic direction finder (ADF) measures angles to ground-based Nondirectional Beacons (NDB).
  - Instrument Landing System (ILS) is used for airplane landing and measures airplane deviation from predefined landing flight path "glideslope".
- Low Range Radio Altimeter (LRRA) measures airplane height Above Ground Level (AGL).
- Air Data System is a source of airflow data that includes airspeed, barometrical (or Mean Sea Level) altitude, vertical speed, and Mach number.
- Measurements of various sensors on board of an airplane.

Almost all of the heavy airplanes include two sets of each equipment in order to guarantee a high level of continuity in results of measurements. One of them is main, another one is a back-up system. FMS operation with radio navigational equipment required some operations with the tuning of radio frequencies of ground equipment operation. FMS includes global air navigational database that hosts all parameters for performing measurements including radio frequencies of navigation beacons, but setting-up of equipment is provided via Radio Management Panel (RMP). In common case, RMP is placed in the cockpit and is used as a universal panel for all radio and navigational onboard equipment. FMS generates guidance commands to RMP that changes settings of any radio and navigational equipment. According to settings from FMS, Navigational aids perform measurements of some parameters and send it back to FMS.

The internal memory of FMS holds the preplanned trajectory of the airplane. Before performing a flight, pilots should program FMS to the future flight plan. Mainly, pilots need to enter cleared flight plan to FMS database. Flight plan consists of a list of waypoints that define flight path in the horizontal plane and correspondent list of altitudes at which airplane should reach waypoints. The initial part of each flight plan includes a set of waypoints that define Standard Instrument Departure (SID) and the final part should end with waypoints of Standard instrumental Arrival (STAR) and approach. Each airport has specific predefined schemes of SID, STAR, and approach. Also, most waypoints of SID, STAR, and approach include specific requirements for altitude. Modern FMS includes global databases of runways, SID, STAR, and approach schemes. In this case, a pilot can choose only one of the cleared titles of SID, STAR, or approach scheme in FMS menu or can define crossing point of one of these schemes for code beginning and final parts of the flight plan. En-route part of flight plane is a flight path between SID and STAR defined by a list of waypoints. Air traffic uses a specific scheme of flight routes that is defined as a line curve between waypoints, except areas of free routes. Air navigation chart includes all available waypoints and defined routes between them. The internal memory of FMS also includes a global database of waypoints, therefore after setting SID, pilots need to enter each of waypoint according to the cleared flight plan. For example, FMS coding to fligh from UKBB (Boryspil international airport, Ukraine) to UKHH (Kharkiv international airport, Ukraine) can include next flight route:

RWY 18L – GOTAP – YHT – RS – SUTOR – KHR – RW 07,
Where:

- RWY 18L is a runway of UKBB for take-off.
- GOTAP is a waypoint of SID (50°15'18" N; 031°49'12" E; any altitude above FL 110).
- YHT ("Yahotyn") is en-route waypoint (DME placed at 50°15'54.4" N; 031°47'40.3" E).
- RS ("Rashivka") is en-route waypoint (NDB placed at 50°12'46" N; 033°53'07" E).
- SUTOR is a waypoint of STAR (49°55'02" N; 035°32'43" E; any altitude above 2150).
- KHR is a waypoint of STAR (VOR/DME placed at 49°55'44.1" N; 036°17'25.6" E; 1200).
- RW 07 is a runway of UKHH for landing.

Scheduled flights are usually performed by stable flight plans, therefore, FMS includes a database of plans approved by airline flight. Before the flight, pilots just need to choose required flight plan from a list of available flight plans of airline and make a setting for cleared altitudes. At some applications FMS can estimate the most efficient flight levels (Murrieta-Mendoza & Botez, 2015). According to flight plan, FMS automatically sets-up all radio navigational aids and tunes communication equipment. FMS automatically detects phases of flight and changes avionics settings.

The internal memory of FMS holds a mathematical model of airplane that takes into account engine actions and required options for flight simulation (Ghazi & Botez, 2019). Before each flight pilot enters into FMS data about airplane load and weight, cruise flight level, cost index, and some other coefficients. Based on the complexity of the math model, FMS estimates a set of important data for pilot. For example, FMS can estimate a set of critical speeds, orientation angles and altitudes for an airplane flight. Also, FMS solves an optimization task to estimate the most efficient flight parameters and engine modes. Based on current location and sensor data, FMS predicts the time of arrival, times of flight and time of arrival at each of waypoints. Taking into account distances between waypoints and altitude data, FMS can estimate fuel flow and predicts the fuel balance of the flight.

Results of FMS estimations are indicated on displays of Electronic Flight Instrument System (EFIS). Estimated values of critical parameters are indicated on the Primary Flight Display (PFD). For example, the estimated value of take-off airspeed is depicted as a red dot at the scale of airspeed, top of descent point can be marked at a barometrical altitude scale. Navigation data is indicated on Navigational Display (ND). Air navigational chart, flight path, SID, STAR, approach, measured data from navigational aids, and current location are presented in ND. The artificial intelligence of FMS automatically detects of current flight phase and represents only important data for airplane guidance.

FMS is connected with autopilot or flight director and engine. Estimated coordinates of the airplane are compared with the planned trajectory to identify unplanned deviations. The values of these deviations are transmitted to the autopilot system to correct the trajectory. Taking into account optimal speed, drift angle, and required time of arrival, FMS provides top-level data for engine control to maintain the optimal trajectory and time of arrivals.

NAVIGATION FUNCTIONALITY OF FMS

An exact airplane position is an important parameter to perform safe navigation. Multiple sensors are used on board of a modern airplane for positioning. FMS is a core of airplane coordinates estimation, fusion, and accuracy improving according to the required level of performance. The common civil airplane has three basic positioning techniques: GNSS, AHRS, and positioning by pair of navigational aids (DME/DME, VOR/DME, VOR/VOR, and ADF/ADF), sorted in the accuracy degrading side. Figure 3 represents the role of FMS in airplane position estimation.

Global Navigation Satellite Systems (GNSS) are considered today as the primary means of positioning and navigation in the airspace, due to its high accuracy, continuity, and availability on a global scale (ICAO, 2005). The most frequently used GNSS in civil aviation are GPS, GALILEO, and GLONASS. A uniformly distributed network of satellites at the Medium Earth orbit creates a field of navigational signals with approximately the same performance around the globe that gives equal positioning possibility for each airspace user. But, GNSS measurements depend on a large spatial variation of ionosphere delay, tropospheric delays (Kutsenko et al., 2018), interference of radio waves (Shvets et al., 2019) and unintentional jamming. All of these factors and multiple failures significantly decrease the accuracy of positioning and may result in GNSS receiver lock (Lubbers et al., 2015). In case of significant reduction of positioning performance, stand-by navigation equipment is used. During a short interruption in GNSS on-board equipment operation, an inertial navigation approach can be used. Attitude and Heading Reference System (AHRS) or Inertial Reference Unit (IRU) may utilize the main inertial navigation services. The inertial principle operation is limited due to the rapid increase in errors. Errors of the inertial navigation

Figure 3. Positioning function of FMS

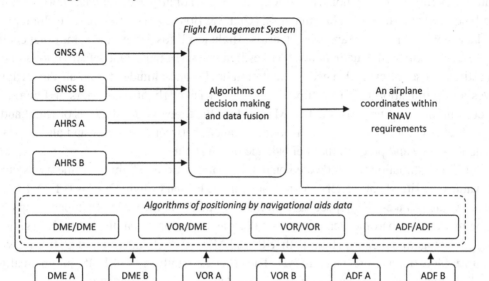

system (INS) become significant rapidly and FMS switches to positioning by data from navigational aids. Modern algorithms use a pair of navigational aids to simultaneously determine navigational parameters and estimate the coordinates of airplane location. An accuracy of alternative positioning methods depends on the geometry and distances in airspace (Ostroumov & Kuzmenko, 2018). However, under identical conditions of location, positioning accuracy using DME/DME is higher than accuracy using VOR/VOR or VOR/DME combinations (Kuzmenko et al., 2018; Kuzmenko & Ostroumov, 2018).

Algorithm of DME/DME positioning is grounded on simultaneous measurements of distances from airplane to two DME ground stations by time of arrival method. Obtained slant distances are transformed to horizontal with a help of altitude data. Navigation equation for positioning by a pair of DME A and B in the horizontal plane of the local reference frame is written as follows:

$$d^2_{DMEA} = (x_0 - x_{DMEA})^2 + (y_0 - y_{DMEA})^2$$

$$d^2_{DMEB} = (x_0 - x_{DMEB})^2 + (y_0 - y_{DMEB})^2 \qquad (1)$$

where d^2_{DMEA} and d^2_{DMEB} are horizontal distances to DME A and B correspondingly; x_{DME}, y_{DME} are coordinates of DME location; x_0, y_0 are aircraft coordinates.

Navigation equation for positioning by angular-based method for VOR/VOR pair can be represented as follows (Kuzmenko et al., 2018):

$$\begin{bmatrix} x_0 \\ y_0 \end{bmatrix} = \begin{bmatrix} tg(\alpha_{VORA}) & -1 \\ tg(\alpha_{VORB}) & -1 \end{bmatrix}^{-1} \begin{bmatrix} x_{VORA}tg(\alpha_{VORA}) - y_{VORA} \\ x_{VORB}tg(\alpha_{VORB}) - y_{VORB} \end{bmatrix}, \qquad (2)$$

where x_{VOR}, y_{VOR} are coordinates of VOR ground stations locations; α_{VOR} are VOR measurements.

In aviation, the VOR/DME positioning approach uses data from two navigational aids located at one point. The navigation equation for this method is the result of a relation in a right triangle:

$$x_0 = x_{VOR/DME} + d_{DME} \cos(\alpha_{VOR})$$

$$y_0 = y_{VOR/DME} + d_{DME} \sin(\alpha_{VOR}), \tag{3}$$

where d_{DME} is a horizontal distance from aircraft to VOR/DME ground station.

Current FMSs use different algorithms for operation (Ostroumov et al., 2018) and the selection of a pair of navigational aids. For example, scanning DME algorithm allows measuring the distances to all available navigational aids for a specified time frame (Herndon et al., 2011). Some navigational aids selection algorithms are grounded on a criterion of location proximity or on the choice of aids located ahead of the planned trajectory (Ruhnow & Goemaat, 1982). In addition, after DME ground station selection, the most commonly used approach of optimal pair selection will operate in the tracking mode of selected navigation pair and does not consider any possible change in navigational aids availability. Some algorithms are based on the estimation of navigational aids available set for navigation tasks comparing approximate coordinates of aircraft location with the characteristics of navigational aids network (Kuzmenko & Ostroumov, 2018). Then, the optimal pair is selected from the estimated set of navigational aids by means of computer simulation, which is used to set up the corresponding onboard equipment for navigation measurements, with a further estimation of aircraft location.

Maximum performance of positioning by navigational aids could be achieved by correct selection of their optimal pair in particular point of airspace (Ostroumov et al., 2018). In common case, the task of navigational aids optimal pair selection is a typical optimization problem which can be represented as an objective function and constraints in terms of integer linear programming (Schrijver, 1998). In the general case, the optimization problem can be written as follows:

$$WX^T \Rightarrow \min,$$

$$AX^T \geq 1$$

$$GX^T \geq 30° \tag{4}$$

$$-GX^T \geq -150°$$

$$IX^T = 1$$

$$X \geq 0$$

where $X \in \{0,1\}$.

The problem in the form of (4) is solved by one of the methods of linear programming, in case if a solution exists for a given set of pairs of navigational aids.

Results of optimal pair selection can be represented in form of graph. Vertices of graph include navigational aids identifications and edges show switch process from one pair to another. As an example

Figure 4 and Figure 5 show graphs of VOR/DME and VOR/VOR optimal pair selection during flight "AUI 58" Odessa (ODS) to Boryspil (KBP) operated by Boeing 737.

Requirements for positioning accuracy within area navigation are formulated in navigational specifications, such as RNP and RNAV. RNP specifications require onboard equipment to continuously monitor characteristics of the navigation system and the pilot's alerting if the system is not achieving the performance required for the operation at necessary safety level (ICAO, 2008). At that time, RNAV is a navigational specification that does not require monitoring and alerting (ICAO, 2008). Different types of RNP/RNAV specifications have been developed to meet navigation requirements for different parts of airspace. Type of navigational specification is determined by the number in nautical miles, which

Figure 4. Graph of optimal VOR/DME pair selection

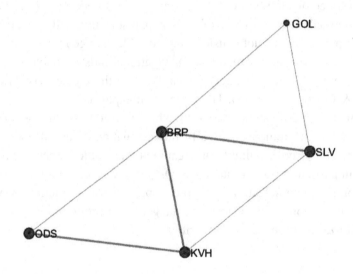

Figure 5. Graph of optimal VOR/VOR pair selection

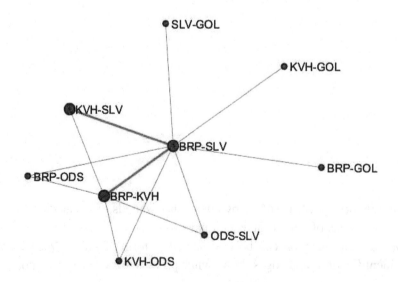

reflects the maximum value of the allowable deviation of aircraft from the preplanned trajectory during positioning. In particular, RNAV 10 or RNP 4 are used for oceanic and remote continental navigation applications; RNAV 5, RNAV 2, RNAV 1 are developed for en-route and terminal navigation applications. RNAV 1 and RNP 1 are used for various phases of flight (ICAO, 2008). Performance analysis of navigational aids ground network of Ukrainian airspace according to RNAV requirements is represented on Figure 6.

Performance of onboard navigation equipment is a subject for strict requirements. Performance-based Navigation (PBN) concept requires monitoring of RNP characteristics with an appropriate alerting in case of their deviation greater than established values (Ostroumov & Kuzmenko, 2018). There are requirements for accuracy, integrity, availability, and continuity of navigation data. Compliance of navigation system with the required safety level in accordance with the current RNP specifications at a certain point of airspace is performed by means of specific algorithms in each navigation system separately.

CONCLUSION

FMS takes an important place in avionics structure of modern airplane. FMS utilizes multiple operations of avionics control, guidance, decision support and data processing performed with a help of artificial intelligence. FMS helps a pilot with avionics and navigation. Multiple internal algorithms of FMS support

Figure 6. Performance of navigational aids network for Ukrainian airspace according to RNAV requirements

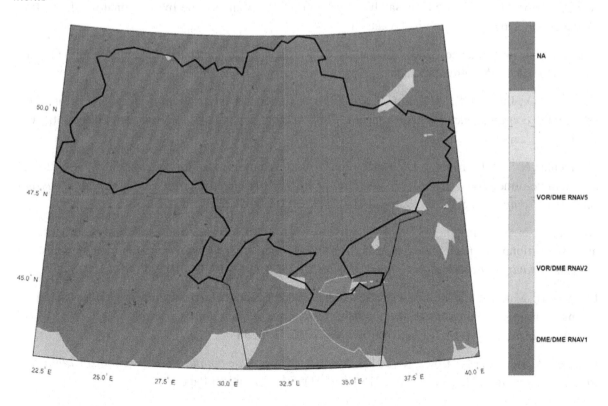

automatic avionics control and optimal navigation performance that minimize human factor influence in aviation safety. FMS makes navigation easier with a help of knowledge database located in internal memory. Tones of data inside of database are processed automatically and support efficient decision making for pilots with appropriate indication at EFIS.

This chapter also describes a problem of positioning data processing in FMS. The problem is more important in case of poor accuracy or GNSS lock. In this case positioning by navigational aids is only one possible solution. Thus, optimal DME or VOR pair selection with a help of artificial intelligence can guarantee an efficient airplane navigation.

REFERENCES

Duanzhang, L., Peng, C., & Nong, C. (2016). A generic trajectory prediction and smoothing algorithm for flight management system. In *Proceedings of the Chinese Guidance, Navigation and Control Conference* (CGNCC) (pp. 1969–1974). IEEE. 10.1109/CGNCC.2016.7829091

Ghazi, G., & Botez, R. M. (2019). Identification and validation of an engine performance database model for the flight management system. *Journal of Aerospace Information Systems*, 1-20.

Herndon, A. A., Cramer, M., Nicholson, T., & Miller, S. (2011). Analysis of advanced flight management systems (FMS), flight management computer (FMC) field observations trials: Area navigation (RNAV) holding patterns. In *Proceedings of the AIAA/IEEE Digital Avionics Systems Conference*, (pp. 4A11–4A117). 10.1109/DASC.2011.6096065

ICAO. (2005). Global Navigation Satellite System (GNSS) Manual. The International Civil Aviation Organization, Doc. 9849, Montreal, Canada.

ICAO. (2008) Performance-Based Navigation (PBN) Manual. The International Civil Aviation Organization, Doc. 9613, Montreal, Canada.

Kutsenko, O., Ilnytska, S., & Konin, V. (2018). Investigation of the residual tropospheric error influence on the coordinate determination accuracy in a satellite landing system. *Aviation*, *22*(4), 156–165. doi:10.3846/aviation.2018.7082

Kuzmenko, N. S., & Ostroumov, I. V. (2018). Performance analysis of positioning system by navigational aids in three-dimensional space. In *Proceedings of the International Conference on System Analysis and Intelligent Computing, SAIC 2018*, (pp. 101–104). 10.1109/SAIC.2018.8516790

Kuzmenko, N. S., Ostroumov, I. V., & Marais, K. (2018). An accuracy and availability estimation of aircraft positioning by navigational aids. In *Proceedings of the International Conference on Methods and Systems of Navigation and Motion Control, MSNMC 2018*, (pp. 36–40). 10.1109/MSNMC.2018.8576276

Lubbers, B., Mildner, S., Oonincx, P., & Scheele, A. (2015). A study on the accuracy of GPS positioning during jamming. In *Proceedings of the 2015 International Association of Institutes Navigation World Congress (IAIN)*, (pp. 1–6). IEEE. 10.1109/IAIN.2015.7352258

Murrieta-Mendoza, A., & Botez, R. (2015). *Aircraft vertical route optimization deterministic algorithm for a flight management system* (No. 2015-01-2541). SAE Technical Paper.

Nanduri, A., & Sherry, L. (2016). *Anomaly detection in aircraft data using Recurrent Neural Networks (RNN). In 2016 Integrated Communications Navigation and Surveillance (ICNS) (pp. 5C2-1)*. IEEE.

Ostroumov, I. V., & Kuzmenko, N. S. (2018). Accuracy assessment of aircraft positioning by multiple radio navigational AIDS. *Telecommunications and Radio Engineering, 77*(8), 705–715. doi:10.1615/TelecomRadEng.v77.i8.40

Ostroumov, I. V., & Kuzmenko, N. S. (2018). An area navigation (RNAV) system performance monitoring and alerting. In *Proceedings of the International Conference on System Analysis and Intelligent Computing, SAIC 2018*, (pp. 211–214). 10.1109/SAIC.2018.8516750

Ostroumov, I. V., Kuzmenko, N. S., & Marais, K. (2018). Optimal pair of navigational aids selection. In *Proceedings of the International Conference on Methods and Systems of Navigation and Motion Control, MSNMC 2018*, (pp. 90–93). doi:10.1109/MSNMC.2018.8576293

Rizun, N., & Shmelova, T. (2016). Decision-making models of the human-operator as an element of the socio-technical systems. In *Proceedings of the strategic imperatives and core competencies in the era of robotics and artificial intelligence* (pp. 167–204) doi:10.4018/978-1-5225-1656-9.ch009

Ruhnow, W. B., & Goemaat, M. L. (1982). VOR/DME automated station selection algorithm. *Navigation, 29*(4), 289–299. doi:10.1002/j.2161-4296.1982.tb00814.x

Schrijver, A. (1998). *Theory of linear and integer programming.* John Wiley & Sons.

Shvets, V., Ilnytska, S., & Kutsenko, O. (2019). Application of computer modelling in adaptive compensation of interferences on global navigation satellite systems. In T. Shmelova, Y. Sikirda, N. Rizun, & D. Kucherov (Eds.), *Cases on modern computer systems in aviation* (pp. 339–380). Hershey, PA: IGI Global; doi:10.4018/978-1-5225-7588-7.ch014

Solomentsev, O., Kuzmin, V., Zaliskyi, M., Zuiev, O., & Kaminskyi, Y. (2018). Statistical data processing in radio engineering devices operation system. In *Proceedings of the International Conference on Advanced Trends in Radioelectronics, Telecommunications and Computer Engineering*, TCSET 2018, (pp. 757–760). 10.1109/TCSET.2018.8336310

Villarroel, J., & Rodrigues, L. (2016). Optimal control framework for cruise economy mode of flight management systems. *Journal of Guidance, Control, and Dynamics, 39*(5), 1022–1033. doi:10.2514/1.G001373

Wiegmann, D. A., & Shappell, S. A. (2017). *A human error approach to aviation accident analysis: The human factors analysis and classification system.* Routledge. doi:10.4324/9781315263878

ADDITIONAL READING

Collinson, R. P. G., & Collinson, R. P. G. (2003). *Introduction to avionics systems*. Boston: Kluwer Academic. doi:10.1007/978-1-4419-7466-2

Fishbein, S. B. (1995). Flight management systems: The evolution of avionics and navigation technology. Westport.

Moir, I., Seabridge, A. G., & Jukes, M. (2013). Civil avionics systems.

Ostroumov, I. V., & Kuzmenko, N. S. (2018). Compatibility analysis of multi-signal processing in APNT with current navigation infrastructure. [English Translation of Elektrosvyaz and Radiotekhnika]. *Telecommunications and Radio Engineering*, 77(3), 211–223. doi:10.1615/TelecomRadEng.v77.i3.30

Robson, D. (2002). *Avionics and flight management systems for the professional pilot. Shrewsbury: Airlife Pub. Fishbein, S. B. (1990). Flight management systems*. Washington, D.C: Aeronautics Dept., National Air and Space Museum, Smithsonian Institution.

Tooley, M., & Wyatt, D. (2017). *Aircraft communications and navigation systems*. Routledge.

Chapter 8
Application SMART for Small Unmanned Aircraft System of Systems

Azad Agalar Bayramov

Armed Forces War College of the Azerbaijan Republic, Azerbaijan

Elshan Giyas Hashimov

(iD) https://orcid.org/0000-0001-8783-1277

Armed Forces War College of the Azerbaijan Republic, Azerbaijan

ABSTRACT

This chapter presents results of SMART for small Unmanned Aerial Vehicles System of Systems complex making, the development of complex control and Infrared communication algorithms, investigations of operation effectiveness of reconnaissance SMART UAVs System-of-System and of the distributed systems of wireless sensors. The development of UAV System-of-System complex and the military tactic tasks and reconnaissance application have been considered. The problems upon providing effective functioning of reconnaissance UAV during analysis and control of highly dynamic scenes are considered. It is shown that during the optimum regime of functioning and taking into account given limiting conditions on total radiation power of radio signals, the reconnaissance UAV should possess the current radiation power inversely proportional to the amount of signal-noise ratio. The technical realization of such optimum regimes is possible by way of development of adaptive control of transmitter power depending on the amount of useful signal detected during reconnaissance activity of UAV.

BACKGROUND

The SMART control System of Systems (SoS) for dynamic objects group can be based on the various logics. In recent years Systems of systems conception for dynamic objects group have been become a very important direction in various areas of science and industry, including military. SoS is consisted of components which solve one task and on jointing in the whole own properties and possibilities these

DOI: 10.4018/978-1-7998-1415-3.ch008

components create the complex system with new qualities. SoS is such system which during various operations combines and joints itself the synergetics of commands, control functions, network and feedback information change.

Since 1990s the possibilities and advantages of SoS ideology have very interesting and this ideology has been widely applied in the dynamic groups control and solution of various tasks, such as: conduct of policy, defense systems, transport, public health and other. In 1995 for the first time, the SoS ideology was applied in military area.

From the point of view of mathematics, moving in air, sea or space, the various automated robots including Unmanned Aerial Vehicles (UAVs) can be considered as dynamic systems (Bayramov, Hashimov & Nasibov, 2019). Such systems can execute both civil and military tasks. The control of these systems can be based on the various principles. The distinction of kind between by usual and SoS ideology control systems is that the criterion of usual system control is evaluation of UAV operation, but the criterion of SoS control is evaluation of effectiveness. Usual systems can not evaluate of changeable operation condition.

For example, UAVs group at 20 km distance from the distribution of troops and at a height of 1 km executes some task. The effectiveness of task solution is depended on a number of UAV, on space location, on environment conditions and events, etc. The task can be executed at own or enemy territory, on the mountainous or flat country. In these conditions, at each moment the rational decision making based on the UAV operation evaluation, UAV state and possibilities assessment are advantages of SoS ideology.

The combat task programs implemented by Unmanned Aerial Vehicles System-of-Systems have been considered and presented in this chapter. The software architecture of complex flight control is developed. The equations of motion of the given system have been obtained. The Infra Red (IR) wireless between UAVs, radio wireless between UAVs and land control post, and UAVs flight control conditions have been offered. It is shown that Unmanned Aerial Vehicle System-of-System complex can be applied for autonomy execution of various military combat tasks (Bayramov, Hashimov, Hasanov, Pashayev & Sabziev, 2018).

The developed and offered software architecture of UAVs SoS SMART complex include five main functions:

1. The control of combat task.
2. The control of sensors.
3. The control of safety.
4. The control of communication.
5. The control of UAV.

The most important distinctive property of developed UAVs SoS complex is intentionally turning off a radio wireless on the line of land control center after UAVs take off. In accordance with the written in memory software program, UAVs SoS is self-controlled by infra-red wireless and using "feedback" principle. They carry out a combat task independently. If some of UAVs loss IR wireless then it (or they) return land control center independently. If some of UAVs is destroyed under enemy fire then other UAVs continue to carry out a combat task.

Infra Red wireless communication can provide safety and security information transfer between UAVs in IR optic range: the frequency is ~ 300 THz and the wave length is ~ 980 nm. Usually, LED (Light Emitting Diodes) elements are used in Infra Red wireless communication for generation and transfer infrared waves. Infra Red wireless communication has next advantages in comparison with radio one:

- Is not sensible to electromagnetic noise.
- There is low energy necessary for waves generation.
- It is not necessary to order for additional communication channels.
- There are signals security transfer.
- There is information very much security transfer.
- İt is impossible outside intervention.

Infra Red optic channels can provide wireless communication upon the straight line at several kilometers away. Nowadays, the electronic modules transfer information in Infra Red range with 1 Gb/sec rate. This wireless communication can be formed between two (or more) movable UAVs or immovable land objects.

The weak sides of IR communication channel are:

- Dependence on signal transmission environment, climatic conditions (atmospheric precipitation, clouds, mist, etc.), atmospheric humidity and temperature.
- Signal transmission only on the line of sight.

The problems of security of the distribution system composed of wireless sensors (for example, one or more UAVs and Ground Control Station) used for gathering spatial-temporal information from controlled (investigated) territory have been considered. The task on providing minimum amount of integrated value of radiation power in the wireless sensors network system containing distributed on territory small size video sensors which connected to video signal processing center has been formulated. The presence of optimum functional dependence between the limiting value of signal - noise ratio of the relevant channel of shop and distance between the camera and procession center is shown. The presence of strong dependence between the minimum available integrated value of radiation power and level of difference between the limiting values of minimum and maximum amounts of the signal - noise ratio is shown. Increase of this difference leads to sharp increase of minimum achievable value of integrated amount of radiated power.

The problems upon providing of effective functioning of reconnaissance UAV during analysis and control of highly dynamic scenes are considered. It is shown that during the optimum mode of functioning and taking into account given limiting condition on total radiation power of radio signals, the reconnaissance UAV should possesses the current radiation power inversely proportional to amount of signal-noise ratio. The technical realization of such optimum regimes is possible by way of development of adaptive control of transmitter power depending on amount of useful signal detected during reconnaissance activity of UAV.

Thus, the radiation security of UAV carrying out the reconnaissance flights can be provided by way of organization of adaptive control of UAVs transmitter power for minimization of irradiated power.

The motivation and contribution of this chapter is to apply the results of SMART (Self-Monitoring Analysis and Reporting Technology) development for small Unmanned Aircraft (Unmanned Aerial Vehicles) System of Systems complex, to investigate the opportunity of military application, to develop of complex control and Infra Red communication algorithms. The development of UAV System-of-System complex (UAV SoS) and the tasks of the military tactic and reconnaissance application have been considered. IR communication has been considered and offered for information exchange between UAVs.

SOFTWARE ARCHITECTURE OF UAV SOS SMART COMPLEX FLIGHT CONTROL

There are many articles devoted UAVs and SoS (Hashimov, Bayramov & Abdullayev, 2017; Systems of systems engineering, 2009; Systems Engineering Guide for Systems of Systems, 2010; Pashayev & Sabziev, 2013). Let us consider some of the published works about UAVs SoS architecture conception. In order to obtain optimized flight vehicle concepts which meet SoS operation requirements, designers have to pay high attention to the impact of SoS at conceptual design stage. Perspectives and progresses of SoS oriented flight vehicle conceptual design are reviewed in paper (Liu Hu, Tian Yongliang, Gao Yuan, Bai Jinpeng & Zheng Jiangan, 2015). Such basic concepts of SoS as definition, characteristics, differences between systems engineering and SoS engineering, as well as SoS design process are introduced in this article. SoS engineering process model for research and development of flight vehicles and SoS design wheel model for conceptual design are proposed.

The use of UAVs in the military environments is predicted to grow much. The availability of more robust and capable vehicles that can perform multiple mission types will be needed. Militaries continue to demand more UAV capabilities for diverse operations around the world. Significant research has been performed and continues to progress in the areas of autonomous UAV control, for example results presented in article (Eaton, Chong & Maciejewski, 2016).

Interesting results about interpretation of development of the unmanned aerial vehicles using systems thinking are presented in (Ćosić, Ćurković, Kasać & Stepanić, 2013). Starting from general systems theory authors formulate classification of UAVs that properly groups diverse produced UAVs, along with their currently unproduced. Authors have structured the context of applications of UAVs using systems thinking. Then they divide UAVs according to their function in the environment: transfer of mass, energy and information. At last, they have analyzed the possible types of UAVs and have divided them based on the structure of their lift creating element, on their regulating programs, and on the type of their power plant.

In accordance of the offered program, on the base of UAVs the development of SMART SoS of dynamic flight objects, making and testing will been carried out. Usually, the Land Control Station manages UAVs by using of radio wireless. But it has shortcomings: 1) a radio wireless can be destroyed, and in the best case UAV does not execute the task and itself returns to base; 2) an enemy taps the radio wireless and carry out UAVs control. The creation of SoS on the basis of UAVs and SMART principles eliminates these flaws.

As it is mentioned above, a radio wireless intentionally turning off on the line of Land Control Station after UAVs take off. The electronic radio receiver block (detector) suspends work and is shaded in automatic mode. The purpose of this action is that the enemy can not intercept UAVs and lands intentionally unmanned vehicles.

After take-off UAVs control themselves in autonomous mode by using of inrfa-red electromagnetic waves. The control is carried out on the base of "feedback" principle (Astrom, Klein & Lennartsson, 2005): after the sent "Order" signal by infrared channel, the "Order executed" signal must be received. If in result of some reason infra-red signal is lost, then UAVs during some time (for example, 30 seconds) try to restore infrared communication. If infra-red communication does not restore, then this UAV comes back to Land Control Station. If some UAV is fire destroyed by enemy side, another UAVs continue to execute the programmed tasks.

One UAV can execute next military tasks: fighting, reconnaissance, control, communication facility, for example as booster converter in mountainous conditions. The tactical activities are offence, defence, encounter battle, march, posture of forces. The battle activities are offence, defence, encounter battle.

Let us assume that there are three reconnaissance UAVs in one UAVs SoS complex:

- UAVs flight on the prescribed altitude H ≥ 300 m, UAVs are painted blue colour, so at this height UAVs aren't seen by unaided eye (Yanoff, Myron; Duker & Jay, 2009) and the distance between each other is varied from 50 m to 200 m.
- UAVs are provided of radio wireless in "signal-output" mode for connection with Land Control Station and infra-red wireless in "signal-input-output" mode for interconnection between UAVs.

This UAVs SoS complex can execute combat and control tasks.

Combat Tasks

- Support of offence battle.
- Terrain monitoring during defence.
- İnterconnection between troops and units during battle activities, especially in mountains where there are many zones of radio silence.
- Reconnaissance tasks implementation during offence or defence activities.
- Electronic warfare, radio countermeasures, radio reconnaissance tasks.

Control Tasks

- İnterconnection between troops and units on the large area, especially in mountains where there are many zones of "radio silence".
- İnterconnection between troops and units during march especially in mountains where there are many zones of "radio silence".
- İnterconnection between troops and units during offence especially in mountains where there are many zones of "radio silence".

Enemy Electronic Warfare Means Destroy

Let us assume that UAVs SoS complex executes some combat tasks on the enemy territory. When collected reconnaissance data are sent to land control center and there is infra red inter communication between UAVs, the enemy's radar observations means reveal theirs and try to jam radio broadcasts that UAVs loss signal dispatch traffic. In this time, the electromagnetic sensors on UAVs board detect electromagnetic waves of radar observations means. In accordance with the written program in electronic memory of UAVs, there is an inter communication between UAVs by using of infra-red channel. The coordinates of enemy's radar observations means are determined by "triangle" mathematical method, this information is sent to land control center by radio communication. At last, in accordance with decision making, there are three possible variants: 1) by using of known coordinates and fire destruction, the enemy's radar observations means are destroyed or 2) advanced UAV inflicts a defeat on enemy's means.

The general UAVs SoS control (management) architecture has been defined for proposed autonomous complex system and has five primary functions (see fig. 1):

1. Combat task control (CTC) electronic block.
2. Sensors control (SC) electronic block.
3. Safety control (SfC) electronic block.
4. Communication control (CC) electronic block.
5. UAV control (UC) electronic block.

In accordance with defined combat tasks, missions and type of UAV, these five functions provide the sufficient operation of offered SMART UAVs SoS. The advantage of this SMART UAVs SoS complex is that it provides modular functionality of architecture that can be applied for any UAVs or any numerous vehicle types. The independent algorithms of SMART UAVs SoS will reside within the mission control functionality and will be dependent upon common interfaces and architecture (Figure 1).

1. **Combat Task Control**: The functions of combat tasks control are a basis of sufficient of the architecture of SMART UAVs SoS. It provides decision making at the highest level of execution of the tasks of UAVs SoS. There are planning, definition of flight traffic and execution of the tasks in combat task control (CTC).

Planning (Pl) Module: There are a tasks definition, an importance definition and a danger definition in Planning (Pl) module. The execution of these tasks is defined by the necessary elements of decision-making process. The data base of tasks importance and of safety definition can be included in planning. The data base of tasks planning can be formed before flight start, during execution of tasks or in the autonomous regime (in SMART UAVs SoS mode, in this case UAVs create tasks themselves). The defining tasks information provides execution of tasks. The importance of tasks defines a first priority of tasks queue. For instance, it is considered that the information about moving target is the most importance among of other common reconnaissance information. The definition of danger condition has sense that SMART system monitors environment, reveals a known type of danger and provides safety itself, or flights out off this zone, or flights round this zone.

Flight Traffic Determination: This module determines the movement direction and the flight regime of SMART UAVs SoS complex. During tasks execution this module selects the correct trajectory of flight in accordance with uninterruptedly obtained information. The trajectory determination algorithm is dynamic, that is, can permanently changed in accordance with in any time changed condition.

Tasks Execution: It is the most importance module of decision making for SMART UAVs SoS tasks execution. The function of this module is an execution of next task during movement or reply (movement change) to some event for the purpose with safety provide. Also, this module forms commands for in-flight maneuver by keeping communication between UAVs. This module is liked to man-operator.

Figure 1. The general architecture of SMART UAVs SoS control: FT - flight traffic; Pl – planning; Ts – task; FlC – flight control; ExC –execution control; El – electronic; Op – optic; Mc – mechanic, As – assessment; IRC – infra red communication; RdC – radio communication; Ts – tasks; DP - data processing; Cnt – control; Ex – execution; CSf – collision safety; FlCn – flight cancel.

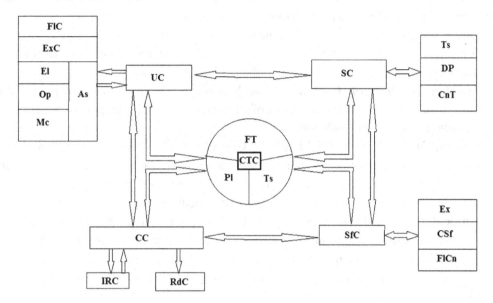

2. **Sensors Control**: The sensors control is a function of the module which controls all detectors with sensors in SMART UAVs SoS complex. The operation of this module is depended on all detectors' operations. The sensors transfer to this module information about current position of UAVs in space, the condition of the environment, each other communication, independence management under the control keeping and dangerous factor against UAV. These sensors are various kinds and purposes. They operate independently, do not dependence each other. By using of they, obtained data provide decision making process on the various levels.

There are three importance functions in the sensors control.

1) **Tasks Execution Sensors:** For purpose with support of combat tasks, the sensors gather data and transfer both to processing module and to tasks execution control module by using of interface. The function of sensors is depended on types of UAVs and tasks. Used in UAVs the types of sensors are electro optic, piezo electric, infra-red and sensitive to high and radio frequency electromagnetic waves, to pressure and humidity.

2) The module with sensors processing of obtained data. This module analyses obtained data and decision makes in accordance with the algorithm written in electronic memory. These data can be used for various tasks: in determination of a type and a movement of the target, in determination of state in space, in control of UAV flight, in each other communication.

3) The module with sensors kept control. This module keeps control over all sensors and connected with them tasks, provides connections with sensors, possibilities of sensors, assessment processing of data from sensors.

3. **Safety control (keep management)**: This module carries out a monitoring of UAV safety. This monitoring is depended on the types and the possibilities of sensors. The control of safety is one of the most importance tasks. The functions of this module play much part in execution of autonomic operation of SMART UAVs SoS complex. This module includes submodules which provide safety execution, avoidance of probably collisions risk and discontinuance of flight.

    ◦   **Safety Execution**: This submodule processes information obtained from danger detectors and generates commands for task execution. This submodule classifies dangerous events by importance and determines which first danger has to be anti actions.

    ◦   **Avoidance of Probably Collisions Module (Collision Safety):** The algorithm of this module assesses the UAV collisions risks of mountains, hills, buildings or some aerial objects, provides avoidance the collision events.

    ◦   **Discontinuance of UAV Flight (Flight Cancel):** This function of UAV landing or return back to ground control center can be executed due to various reasons: intensive gunfire in some flight zone, weather conditions worsening, change of flight plan, decreasing of electric power sources, some UAV's electronic or mechanical module breakage, etc.

4. **Communication Control**: This module provides bidirectional infra-red communication between UAVs and simplex radio communication between UAVs and ground control center. The "UAV – UAV" duplex infra-red communication on the sight line only between UAVs is kept. The "UAVs – gound control center" only simplex radio communication between UAVs and ground control center is kept.

5. **UAV Control**: This module is responsible for the control (management) of the SMART UAVs SoS complex as a whole. The functions of this module are the keeping of a flight control, the assessment of operations of the electronic, optical and mechanical sub modules and sub modules, the control of the UAV state in general. There is the sub module of UAV control execution, the sub module of flight control keeping, the sub module of the assessment (monitoring) of electronic, optical and mechanical operations sub modules in UAV control module.

SoS UAVs Dynamic Motion Equation: Let us consider SMART UAVs SoS flight as a motion of material points system. Hamilton (canonical) equations are one of the main equations of Newtonian mechanics for the investigated dynamic system, which in overall view describe a temporal change of the points coordinates of system (Zwillinger, 1997). Let write these canonical equations of motion as:

$$\frac{dp_i}{dt} = \frac{\partial H}{\partial q_i}, \quad \frac{dq_i}{dt} = \frac{\partial H}{\partial p_i} \tag{1}$$

It seems, that system is consisted of $2N$ first-order differential equations ($j = 1, 2, ..., N$) for the dynamic system described by N generalized coordinates which are system motion. Here: $H \equiv H(q_1, q_2, ..., q_N, p_1, p_2, ..., p_N, t)$ is a Hamilton function (Hamiltonian), t is time, q_i are generalized coordinates, p_i are generalized momentums. These generalized coordinates and momentums determine a state of system in phase space.

Since, a rectilinear flight of UAVs SoS (a material points system) is planar, then for simplicity, we assume that Hamiltonian H can be presented as a sum of kinetic $T(p)$ and potential $V(q)$ energies of UAVs SoS system:

$$T\left(p\right)=\frac{p^2}{2m} \text{ . and } V(q) = V(x)$$

In present case, $x = h$ – is a flight high of SoS PUAs. Let all UAVs are of the same mass, that is,

$$m_1 = m_2 = \dots m_N = m$$

Then, Hamilton equation can be written as:

$$H = \frac{1}{2m}\sum_{i=1}^{N}p_i^2 + N{\bullet}m{\bullet}g{\bullet}h, . \tag{2}$$

Here: g is the gravitational acceleration, $g = 9,8$ m/s^2. Therefore, by using (1) and (2) equations it is possible to describe of motion of UAVs SoS system. If it is necessary to take into account of rotational motion of UAVs then the equations system described in previously work of authors (Hashimov, Bayramov & Abdullayev, 2017) can used.

INFRA RED COMMUNICATION SYSTEM BETWEEN UAVS

Infra-red radio communication system (IR wireless transmission) is an equipment which provides information transfer between UAVs in optic IR range (Luftner, Kropl, Hagelauer, Huemer, Weigel & Hausner, 2003; Carruthers, 2002). The IR band of the electromagnet corresponds to 300 GHz and a wavelength of 980 nm (State Standard Specification, 2015). Usually, during IR wireless communication the LEDs are used for IR waves transmission. As opposed to radio wireless, the IR wireless is mot sensible to electromagnetic interference, has low power consumption, does not demand a communication channels reservation and provides signal hiding and high intercept security of information transfer.

In line of sight conditions, the IR channel can provide communication in the distance up to several kilometers. In present, the information-transferred modules have a capable of the rate of 1 Gbit/sec. in IR range. The propagation of light waves in this band can be used for a communication system (for transmission and reception) of data. This communication can be between two portable devices or between a portable device and a fixed device.

The IR wireless system includes IR radiator, IR transmitter, IR source and IR receiver modules (fig. 2). IR transmitter is a device, which transforms information to electrical signal for IR source modulating. IR source is infrared source for signals transfer. IR radiator is a device for the transformation of input electrical signal to infra-red radiation. IR receiver is a device with infra-red detector and signal processing which transforms or recodes (decrypt) input signal (information) for remote control (Figure 2).

For purpose of providing of the signal hiding and the high intercept security of information transfer and control mission between UAVs, the especial IR wireless system is offered in the given program (see Figure 3).

Let us consider two tubes provided IR wireless communication between two UAVs, IR-transmitter and IR- receiver. IR LED is a source of IR emission. IR R is a lIR light receiver. IR control signal is

Figure 2. IR wireless system: IR T – infra-red transmitter, IR rad – infra-red radiator, IR S – infra-red source, IR W – infra-red wave, IR R – infra-red receiver

Figure 3. The chart of IR wireless between UAVs. 1- small tubes of IR input-output radiation flux in UAVs with screened waveguide; 2 - the sources of IR emission (IR LED); 3 - lIR light receivers (IR R); 4 - IR light

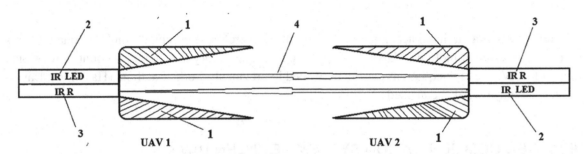

propagated directly in a line sight between UAVs. The tube form with screened waveguide provides a signal hiding and a high intercept security of information transfer. Moreover, this form of IR wireless system does not allow to execute a control from the enemy's side. Some radio or infra-red signal from ground cannot impact to this IR wireless system. Also, UAVs have radio wireless for sending output-signal to ground control post center (UAVs takeoff point). UAVs have screened electronic blocks and therefore cannot be controlled by ground enemy posts.

UAVs SoS MISSION PROGRAM

In this part of the Chapter let us consider UAVs SoS mission program. The SMART UAVs SoS mission program includes next stages (Figure 4):

1. Three UAVs are considered: UAV1, UAV2 and UAV3. These UAVs have a reconnaissance mission.
2. After takeoff all UAVs are controlled off - line by themselves, the ground station has not possibilities to control UAVs.
3. UAVs are provided by radio (only in output mode) and IR wireless (both input and output mode for information exchange only between UAVs) systems. UAVs can send information to gound control station by radio communication, but can not receive radio signal from outside. They keep only "input - output" IR communication channel.
4. After takeoff UAVs are controlled by themselves in accordance of SMART SoS principle.
5. All modules of UAV are covered by electromagnetic screen.

Figure 4. The test image of the land obtained by the high-precision digital electronic-optical photographic camera systems: flight height is 200 m

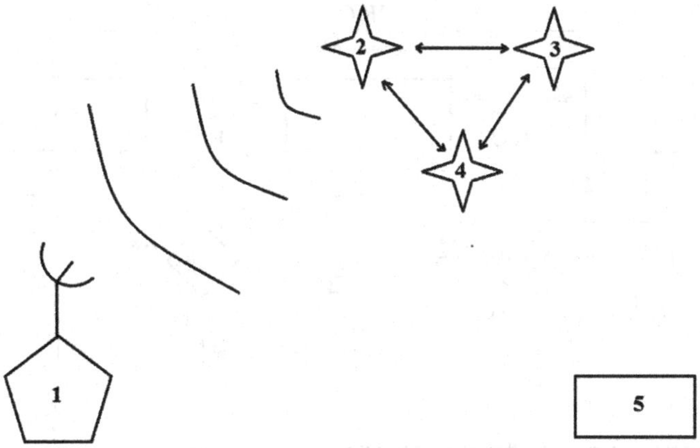

6. UAVs are provided by the high-precision digital electronic-optical photographic camera systems: Full Frame DSLRLike Camera a7R EMount Compact Camera I Sony US, 36 MP. Obtained by these camera photographies are sent to ground control station. As example, in result of the terrain investigation, obtained one test image of the land is shown in the photography 1; the flight height is 200 m.

7. After UAVs take off, UAVs regular exchange IR control signals in a "feedback" mode, it is necessary for handshaking. If for some reason the handshaking is broken, then information about it is sent to ground control station by radio wireless and this UAV comes back to ground control station.

8. After the take-off UAVs flight jointly in triangular form (see Figure 5). For testing of the reliable performance of IR communication, the size of this triangle is changed cyclically: from 50 m to 200 m and then again to 50 m (see Figure 6):

    ○ İn moment t_0, $l_0 = 50$ m.
    ○ İn moment t_1, $l_1 = 100$ m.
    ○ İn moment t_2, $l_2 = 200$ m.
    ○ İn moment t_3, $l_3 = 100$ m.
    ○ İn moment t_4, $l_4 = 50$ m.

Figure 5. The chart of UAVs SoS reconnaissance flight. 1 - ground control center; 2, 3 and 4 - UAVs; 5 - investigated enemy object

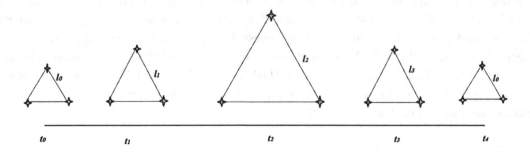

Figure 6. The chart of UAVs flight conditions: l_0, l_1, l_2, l_3 and l_4 are the distances between UAVs; t_0, t_1, t_2, t_3 and t_4 are the flight time slots

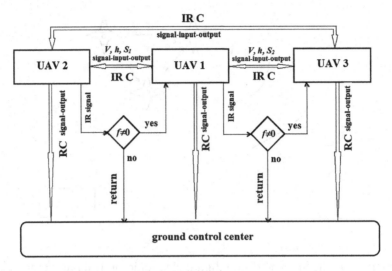

9. In 5 minutes after UAVs take off, IR transmitter of UAV1 is cut off; it continues by itself mission: flying, shooting of land and sends this information of images to ground control center. It will check the UAV autonomy.

The simplified chart of flight program of UAVs SoS is shown in fig. 6.

Let us take UAV1 as "commander" one. UAV1 gives flight speed (V), flight height (h) and distance (S_1 and S_2) between UAVs (in our case $S_1 = S_2 = S \in [50,200]$ m). UAV1 sends İR signal f to both UAV2 and UAV3. If there is IR handshaking ($f \neq 0$ - "yes"), then UAVs continue flight. If there is not IR handshaking ($f = 0$ – "no"), then this UAV returns to the ground control center.

It should be noted, that this only for test flight program of UAVs.

SUPPORT OF OPERATION EFFECTIVENESS OF RECONNAISSANCE SMART UAVs SoS

It is well known that the effectiveness of information system functioning is the most universal measure of this system of special-purpose. As it were noted in (Hyunkyung, Hayoung & Euiho, 2016; Bayramov, & Hashimov, 2018; Bayramov, & Hashimov, 2017; Abdulov, Bayramov, & Hashimov, 2019), there are many investigations concerning to the evaluation of functioning effectiveness of the reconnaissance UAVs and SMART UAVs SoS by modelling method (Figure 7). There are some UAV's characteristics used in these works for evaluation of UAV effectiveness during reconnaissance flights in the table 1.

It seems from data in table 1, the UAV flight hiding is pointed as one of the main properties necessary for UAV effectiveness functioning (Bayramov & Hashimov, 2017). In addition, the problem of hiding in reconnaissance information systems of the control of high dynamic scenes has a complex character and causes a necessity of next tasks solution:

Figure 7. The chart of flight program of UAVs SoS: 1- commander's UAV; 2- and 3 – UAVs; IR – infra-red; IR C – infra-red communication; RC – radio communication

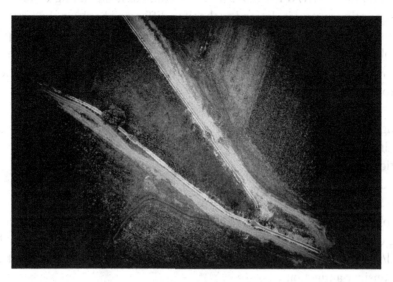

Table 1. UAV's characteristics for evaluation of effectiveness

Nº	Measure of Effectiveness	Characteristics
1	The ratio of a value of the given characteristic before outside impact to the value of the same characteristic after outside impact.	The distance of detection of the ground objects. The image quality of the ground objects. UAVs flight rate.
2	The number of detected sought for objects by UAV on the ground.	The number of UAV damages. The distance of revealing. The ability of achievement of the goal. The UAV flight hiding.
3	The level of UAV survival potential.	UAV flight rate. The UAV flight hiding. The UAV flight height. The distance of detection of the ground objects.

1. Minimization of hiding time of the high dynamic observable processes. As an example, it can be pointed the task of detection of launched enemy missiles. In this case, the minimization of detection time of launched enemy missiles and transfer of this information to the processing center has a very importance significance. In duration of this time, the fact of launch enemy missile is secretive and unknown.

2. Providing of UAVs SoS proper hiding which carry out reconnaissance activities on the battlefield.

In accordance with (WHITEPAPER, 2013), the hiding of high dynamic scenes and objects is determined by two factors:

1. The man response reaction T_r to light or audio (sound) impulse: it is 190 msec. for light impulse and 160 msec. for sound one.

2. The actuating time T_s of «sensing device (video camera, microphone, various sensors) – coder – modem - transmission line – modem - encoder - monitor» regenerating system.

At the same time, in duration of T_r and T_s the sought for object is remained unobservable.

In accordance with (WHITEPAPER, 2013), in common case, the hiding of observable objects on the monitor of regenerating system is defined as the time difference between the first and the last pixels in a frame. Such interpretation of the hiding of observable objects is true for the supervisory systems of high dynamic objects and processes. In common case, it should be noted, that the hiding problem of reconnaissance SMART UAVs SoS caused a necessity of the solution of next particular tasks of the hiding of UAVs SoS:

- Visual Hiding.
- Acoustic Hiding.
- Radio Location and Thermal Hiding.
- Radiation Hiding.

All technical solutions (adaptive protective coating, etc.) of the first three above pointed tasks often lead to increasing of the noisiness level of signals from UAVs SoS complex that is, decreasing of the ratio of the signal/noise. The solution of the fourth task provides for decreasing of radiation power. Therefore, let us consider a function

$$P = \varphi(\gamma) \tag{3}$$

Here: P is a radiation power of all devices and electronic modules set on SMART UAVs SoS; γ is a ratio of the signal/noise in the signal output module of UAVs SoS, $\gamma \in (\gamma_{min}, \gamma_{max})$.

Then, the next condition can be taken as limiting measures superposable on total radiation power

$$P_{lim} = \int_{\gamma min}^{\gamma max} \varphi(\gamma) d\gamma = C; \quad C = const. \tag{4}$$

On the other side, for the purpose of providing of the operation effectiveness, reconnaissance UAVs SoS can be included in the wireless reconnaissance sensors network (WSN) composed of static and dynamic components. In this case, the minimization of power consumption in WSN is strongly connected with the problem of power supply. It is provided by optimization of the hybrid supply of the network included of accumulators and solar energy transformers. In accordance with (Yu, Prasanna & Krishnamachari, 2006), if data gathering are hiding then should to find the optimal compromise between hiding level and power consumption.

Let us consider the optimization of energy consumption and a quality of transferred information. The optimization such networks demands to take into account an energy consumption and a quality of transferred image. Let us consider that video cameras are set on the separate UAVs and are directly connected by radio channel to the central module of development.

Each sensor has K pixels per string and H pixels per vertical. The digitization of one pixel demands N bit, and the frequency of frames is f. The transfer of video information from out the camera into the module of development can be calculated by below formula

$$C = f \cdot K \cdot H \cdot N \text{ (bit/sec)} \tag{5}$$

In the module of development, necessary threshold quantity of the signal/noise ratio γ are of the same for all cameras. In accordance with (Ibrahim & Alfa, 2017), for satisfaction of this demand, the camera power of transfer must be determined as

$$P = \frac{(4\pi)^2 \cdot N \cdot \gamma \cdot B}{\lambda^2 \cdot C_r \cdot C_t} \cdot d^2 . \tag{6}$$

Here: N is a spectral density of power of the white Gaussian noise; λ is a wave length of the carrier signal; C_r and C_t are the receiving and transmitting antennas power gains, respectively; B is a bandwidth of transferring frequency; d is a distance between the camera and the center of development.

It follows from (6) that

$$d = \sqrt{\frac{P \cdot \lambda^2 \cdot C_r \cdot C_t}{(4\pi)^2 \cdot N \cdot \gamma \cdot B}} \tag{7}$$

Therefore, it can be adopted, that there is some ordered set

$$\gamma = \{\gamma_i\}; i = \overline{1,m} . \tag{8}$$

Here:

$$\gamma_j = \gamma_{j-1} + \Delta\gamma; \Delta\gamma = const \tag{9}$$

In consideration of (8) and (9), let us formulate the next optimization task:
It should be calculated some optimal function of P and γ interconnection

$$P = \varphi(\gamma), \tag{10}$$

when, the new input performance criterion of reconnaissance UAVs would reach the extreme value. The offered criterion of effectiveness of reconnaissance UAV is based on the formula (7) and in the discrete mode it is expressed as below

$$D_d = \sum_{i=1}^{n} \left[\sqrt{\frac{P \cdot \lambda^2 \cdot C_r \cdot C_t}{(4\pi)^2 \cdot N \cdot \gamma_i \cdot B}} \right] \tag{11}$$

The physical sense of the criterion (11) is a sum of the distances between UAV and development center recommended for functioning of UAV at various values of γ_i, determined by set (8).

Let us conditionally to turn into streaming operation of recording and take into account (8), then the criterion of effectiveness can be written as below

$$D_s = \int\limits_{\gamma min}^{\gamma max} \frac{\lambda}{4\pi} \sqrt{\frac{\varphi(\gamma) \cdot C_r \cdot C_t}{N \cdot \gamma \cdot B}} d\gamma \tag{12}$$

In consideration of (4) and (11), let us write the criterion functional of unconditional variational optimization as below

$$D_{cf} = D_s + \chi \cdot P_{lim} \tag{13}$$

Here: χ is the Lagrange multiplier.

In consideration of (4), (12) and (13), there is below

$$D_s = \int\limits_{\gamma min}^{\gamma max} \frac{\lambda}{4\pi} \sqrt{\frac{\varphi(\gamma) \cdot C_r \cdot C_t}{N \cdot \gamma \cdot B}} d\gamma + \int\limits_{\gamma min}^{\gamma max} \varphi(\gamma) d\gamma \tag{14}$$

In accordance with Euler-Lagrange equation, leading functional (14) to its extreme value the optimal function $\varphi(\gamma)_{opt}$ provides execution next condition

$$F_1 = \frac{d\left\{ \dfrac{\lambda}{4\pi} \sqrt{\dfrac{\varphi(\gamma) \bullet C_r \bullet C_t}{N \bullet \gamma \bullet B}} + \chi\varphi(\gamma) \right\}}{d\varphi(\gamma)} = 0. \tag{15}$$

From (15) there has been obtained next expression

$$\frac{\lambda}{4\pi} \sqrt{\frac{C_r \bullet C_t}{N \bullet \gamma \bullet B}} \bullet \frac{1}{2\sqrt{\varphi(\gamma)}} + \chi = 0. \tag{16}$$

From (16) below formula follows

$$\varphi(y) = \frac{1}{\gamma} \bullet \frac{k^2}{2}. \tag{17}$$

Here:

$$k = \frac{\lambda}{4\pi} \sqrt{\frac{C_r \bullet C_t}{N \bullet B}}.$$

Then, the result of combination of (4) and (17) is:

$$P_{lim} = \int\limits_{\gamma min}^{\gamma max} \frac{1}{\gamma} \bullet \frac{k^2}{2} \, d\gamma = C; . \tag{18}$$

From (18):

$$P_{lim} = \frac{k^2}{\chi^2} \bullet ln\left(\frac{\gamma_{max}}{\gamma_{min}}\right) = C; \qquad C = const . \tag{19}$$

From (17) and (19):

$$\varphi(\gamma) = \frac{1}{\gamma} \bullet \frac{C}{ln\left(\dfrac{\gamma_{max}}{\gamma_{min}}\right)} \tag{20}$$

So, when (18) expression is satisfied, the criterion functional of (12) reaches the extreme value.

DISTRIBUTED SYSTEMS OF WIRELESS SENSORS

As it was mentioned in (Westhoff, Girao, & Sarma, 2017), the wireless sensor networks (WSN) use the tiny inexpensive measuring converters, possessing low radiation power, low energy consumption and specific functions of monitoring. Such networks can be used both in peaceful purposes, for example in agriculture, and in military areas, for example in safety warning systems of various zones, in SMART UAVs SoS etc. The main distinction of WSN is possibility to function without infrastructure or with minimum infrastructure. In typical case, WSN includes set of specific sensors, a wireless transmitter-receiver, a processor of common purpose and an energy supply (Manita, 2016). Sensors are installed on vast area and they form static self-organizing network. There are also portable versions of WSN. Typical WSN functions are data measuring and gathering; data processing and transferring, temporary data storage, transfer processed data as information to the upper level signal center. In the large WSN there have been application several data storages and several levels of sites of these data storages.

There has been pointed in (Westhoff, Girao, & Sarma, 2017) that, in accordance with the WSN safety model, in the typical configuration of WSN the next below safety providing tasks would be solved:

1. Providing of flexibility in choosing of the information transfer route and storage.
2. Providing of hiding of the stored data. The main problems formed in the framework of WSN are reaching of hiding of the stored data and the minimal energy consumption. The optimal solution is data processing in the measuring sites (modules) and data acquisition in the single storage device.
3. Providing of data acquisition on the base of changes query and many number of sensors. At the same time, it is reasonable to transfer of data of the several indicators (for example, temperature, humidity, radiation power, etc.) in the single transferring message.

4. Safety acquisition of data of the distributed sensors. At the same time, the possibility of a safety buffer storage of the temporarily not transferred data must be provided in database sites.

5. Preliminary exactly positioning of the measuring sites. The fact of the matter is that some part of information, it is possible, can be determined and stored only in the presence of information about exactly position of the measuring instrument in the topology of the network.

6. Providing of data adequacy. Providing of testing a possibility of authenticity of the receiving and stored data.

7. Testing of safety of the chosen data routing. It would be mathematical providing of such possibility.

8. Reliable data storage without leakage.

9. Testing of data authenticity by the establishment of pairwise or group interconnections.

10. WSN control possibility on stored information consumer's part.

The minimization of energy consumption in WSN is closely connected with the problem of energy supply. For the purpose of this, it is widely considering the problem of optimization of the hybrid supply of the network including accumulators and solar energy converters (Tala't, Yu, Ku & Feng, 2017). For this purpose, various mathematical models are applied, such as the dynamic programming (Ahmed, Ikhlef, & Schober, 2013), theory of Markoff processes (Ku, Chen, & Liu, 2015) etc. In (Yu, Prasanna, & Krishnamachari, 2016) the special process had been investigated, in the case when data gathering must be secretly carried out it would be found the optimal compromise between a hiding level and energy consumption. At the same time, it should be remembered that consumed energy for information transfer can be substantially reduced if reduce of power of a transmitter and, in the same time, it is longer duration of data transfer (Prabhakar, Uysal-Biyikoglu, & Gamak, 2000).

In accordance with (Feistel, Wiczanowski, & Stanczak, 2007), the minimization of transfer power does not mean the minimization of energy consumption, in particular, in WSN, which has a low rate of data transfer, the mentioned difference is expressed more clearly, since energy consumption of data processing sites is not negligible. In (Dousse, Mannersalo, & Thiran, 2004) article it is mentioned that energy saving of WSN sites is possible connected with the effectiveness of space coverage by sensors, reciprocal relationships. For this, there do appear to be sufficient reasons for organization of active and "latent" working modes of sensors. At that, the WSN hiding can be provided, also by removing of transfer of the strong correlated messages. It is reached by effective compression of transferred data (Sivagami, Pavai, & Sridharan, 2010).

CONCLUSION

In given chapter, the development of combat task program of the offered version of SMART Unmanned Aerial Vehicles System-of-Systems has been presented. The software architecture of complex flight control is developed. The Hamilton equations describing of the SMART UAVs SoS motion have been obtained. It is shown that SMART UAVs SoS complex can be applied for autonomy implementation of various military combat tasks. The IR wireless between UAVs in "signal-input-output" mode, radio wireless between UAVs and land control post in "signal-output" mode, and UAVs flight control conditions have been offered.

In accordance of this program SMART UAVs SoS complex can autonomy flight by control itself.

For providing of safety purpose UAVs are located in triangular form each other, the distance between them is changed and this system flights in such modes. The test program of SMART UAVs SoS complex is offered.

The questions about providing of effective functioning of surveying UAV in analysis and control of studying highly dynamic scenes are considered. It is shown that in optimum regime of functioning and taking into account the limiting condition imposed on total radiating power of radio signals the surveying UAV should possesses the current radiating power inversely proportional to amount of signal-noise ratio. The technical realization of such optimum regimes is possible by way of development of adaptive control of power transmitters depending on amount of useful signal detected during surveying activity of UAV. Thus, the radiating privacy of UAV carrying out surveying flights can be provided by a way of organization of adaptive control of power of UAVs transmitter for minimization of irradiated power.

There have been investigated a problem of a hiding of distributed system composed of wireless sensors utilized for gathering spatial-temporal information from controlled territory. The task on providing the minimum amount of integrated value of radiation power in the system containing distributed on area the small size video sensors connected to video signal processing center is formulated. The presence of optimum functional dependence between the limiting value of the signal-noise ratio of the relevant channel of procession center and distance between the camera and procession center is shown. The presence of a strong dependence between the minimum available integrated value of radiation power and a level of the difference between the limiting values of minimum and maximum amounts of the signal-noise ratio is shown. In line with above increase of indicated difference leads to strong increase of minimum achievable value of integrated amount of radiation power.

REFERENCES

Abdulov, R. N., Bayramov, A. A., & Hashimov, E. G. (2019). Providing of effective operation of reconnaissance UAVs during analysis and control high dynamic scene. *National Security and Military Sciences*, 5(3), 27–34.

Ahmed, I., Ikhlef, A., Ng, D. W. K., & Schober, R. (2013). Power allocation for an energy harvesting transmitter with hybrid energy sources. *IEEE Transactions on Vehicular Technology*, 12, 6255–6267.

Astrom, K. J., Klein, R. E., & Lennartsson, A. (2005). Bicycle dynamics and control. *IEEE Control Systems Magazine*, 25(4), 26–47. doi:10.1109/MCS.2005.1499389

Bayramov, A. A., & Hashimov, E. G. (2017). The numerical estimation method of a task success of UAV reconnaissance flight in mountainous battle condition. *Advanced Information Systems*, 1(2), 70–73. doi:10.20998/2522-9052.2017.2.12

Bayramov, A. A., & Hashimov, E. G. (2018). Assessment of invisible areas and military objects in mountainous terrain. *Defence Science Journal*, 68(4), 343–346. doi:10.14429/dsj.68.11623

Bayramov, A. A., Hashimov, E. G., Hasanov, A. H., Pashayev, A. B., & Sabziev, E. N. (2018). SMART control system of systems for dynamic objects group. In Proceedings of Military Academy after "Георги Стойков Раковски" on National Security (II part. pp. 121-123). Sofia, Military Academy Press, Bulgaria.

Bayramov, A. A., Hashimov, E. H., & Nasibov, Y. A. (2019). Unmanned aerial vehicles application for military GIS tasks solution. In S. Tetiana, Y. Sikirda, N. Rizun, D. Kucherov, & K. Dergachov (Eds.), *Automated systems in the aviation and aerospace industries* (pp. 273–296). Hershey, PA: IGI Global. doi:10.4018/978-1-5225-7709-6.ch010

Carruthers, J. B. (2002). *Wireless infrared communications. Wiley encyclopedia of telecommunications.* Boston, MA: Department of Electrical and Computer Engineering Boston University.

Ćosić, J., Ćurković, P., Kasać, J., & Stepanić, J. (2013). Interpreting development of unmanned aerial vehicles using systems thinking. *Interdisciplinary Description of Complex Systems, 11*(1), 143–152. doi:10.7906/indecs.11.1.12

Dousse, O., Mannersalo, P., & Thiran, P. (2004). Latency of wireless sensor networks with uncoordinated power saving mechanisms. In *Proceedings of the 5th ACM International Symposium on Mobile Ad Hoc Networking and Computing.* (May 24-26, pp. 109-120), Tokyo, Japan. 10.1145/989459.989474

Eaton, C. M., Chong, E. K. P., & Maciejewski, A. A. (2016). Multiple-scenario unmanned aerial system control: A systems engineering approach and review of existing control methods. *Aerospace, 3*(1), 2–26. doi:10.3390/aerospace3010001

Feistel, A., Wiczanowski, M., & Stanczak, S. (2007). *Optimization of energy consumption in wireless sensor networks.* Retrieved from https://pdfs.semanticscholar.org/59b3/e2a10ada19151f5730f-153661b46812471c5.pdf

Hashimov, E. G., Bayramov A. A., & Abdullayev, F. A. (2017). Development of the multirotor unmanned aerial vehicle. *National security and military sciences.* Baku, 3(4), 21-31.

Hu, L., Tian, Y., Yuan, G., Bai, J., & Zheng, J. (2015). System of systems-oriented flight vehicle conceptual design: Perspectives and progresses. *Chinese Journal of Aeronautics, 28*(3), 617–635. doi:10.1016/j.cja.2015.04.017

Hyunkyung, M., Hayoung, J., & Euiho, S. (2016). An analysis for measure of effectiveness of an unmanned aerial vehicle using simulation. Research & Reviews. *Journal of Engineering and Technology. RRJET, 5*(3), 69–82.

Ibrahim, A., & Alfa, A. (2017). *WSN optimization. Sensors.* doi:10.339017081761

Jamshidi, M. (Ed.). (2009). *Systems of systems engineering. Principles and applications.* Boca Raton, FL: CRC Press Taylor & Francis Group.

Ku, M. L., Chen, Y., & Liu, K. J. R. (2015). Data–driven stochastic models and policies for energy harvesting sensor communication. *IEEE J. Sel. Areas Commun.* (33), 1505-1520.

Luftner, T., Kropl, C., Hagelauer, R., Huemer, M., Weigel, R., & Hausner, J. (2003). Wireless infrared communications with edge position modulation for mobile devices. *IEEE Wireless Communications, 10*(2), 15–21. doi:10.1109/MWC.2003.1196398

Manita, L. (2016). Optimization problems for WSNs: Trade-off between synchronization errors and energy consumption. *Journal of Physics: Conference Series.* pp. 1-6.

Pashayev, A. B., & Sabziev, E. N. (2013). Description of fluid flow as a system of systems. In *Proceedings of 8th International Conference on System of Systems Engineering*, June 2-6, pp. 28-33, Maui, Hawaii.

Prabhakar, B., Uysal-Biyikoglu, E., & Gamak, A. E. (2000). Energy-efficient transmission over a wireless link via lazy packet scheduling. In Proceedings IEEE InfoCom, (Apr. vol. 1, pp. 386-394).

Sivagami, A., Pavai, K., & Sridharan, D. (2010). Latency optimized data aggregation timing model for wireless sensor networks. *IJCSI International Journal of Computer Science Issues, 7*(3), no. 6.

Systems engineering guide for systems of systems. (2010). *Essentials*. Washington, D.C.: Office of the Director, Defense Research and Engineering, Director of Systems Engineering.

Tala't, M., Yu, Ch.-M., Ku, M.-L., & Feng, K.-T. (2017). On hybrid energy utilization in wireless sensor networks. *Energies, 10*(12), 1940. doi:10.3390/en10121940

Transmission of audio and/or video and related signals using infra-red radiation. (2015). *State standard specification IEC 61603-1-2014*. Part 1. Moscow, Russia.

Westhoff, D., Girao, J., & Sarma, A. (2017). *Security solution for wireless sensor networks. General papers*. Retrieved from https://www.nec.com/en/global/techrep/journal/g06/n03/pdf/t060322.pdf

WHITEPAPER. (2013). *Low latency systems*. Ittiam. Systems Pvt. Ltd. February 2013. Retrieved from https://www.ittiam.com/wp-content/uploads/2017/12/WP005_low-latency-systems.pdf

Yanoff, M., & Duker, J. S. (2009). Ophthalmology. 3rd Edition. MOSBY Elsevier. p. 54.

Yu, Y., Prasanna, V. K., & Krishnamachari, B. (2006). Energy minimization for real–time data gathering in wireless sensor networks. *IEEE Transactions on Wireless Communications, 5*(11), 3088–3096. doi:10.1109/TWC.2006.04709

Zwillinger, D. (1997). *Handbook of differential equations* (3rd ed.). Boston, MA: Academic Press.

Chapter 9
The Method of Evaluation of the Resource of Complex Technical Objects, in Particular, Aviation

Lyudmyla Kuzmych

https://orcid.org/0000-0003-0727-0508

National Aviation University, Ukraine

Volodymyr Kvasnikov

National Aviation University, Ukraine

ABSTRACT

The method of estimating the resource of complex technical objects, in particular, aviation technique, is proposed, which includes the construction of a mathematical model of the object's functioning to determine its actual technical condition and residual resource for planning, design, production, operation, repairs, and modernization of aviation technique. To determine the actual state and estimate the residual life of structures, it is proposed to simultaneously evaluate several characteristics of the material of the object: the characteristic parameters of the structure of the material, integral parameters of the material, related to strength (for example, hardness), the presence and nature of macro defects, the degree of corrosion wear of the metal. Limit values of the selected diagnostic parameters are determined by available standards or technical conditions for objects. The dynamics of changes in diagnostic parameters can be monitored and modeled on the basis of the data of periodic inspections of the control object.

BACKGROUND

The modern level of scientific and technological progress allows us to create objects that are endowed with high reliability. The basis for this is the set of measures that are used at the stages of design, manufacturing, installation, and operation. The problem of forecasting and providing technical resources of complex technical objects remains the most actual problem.

DOI: 10.4018/978-1-7998-1415-3.ch009

The complex technical objects mean the systems in which, when securing or adding components, qualitatively new properties of the system arise. Each component of the system can be described by a set of characteristic features whose values determine the current state of the component and the system as a whole. Thus, the behavior of the system is described by a vector of values of characteristic features, taking into account the influence of external factors on the behavior of the system (Babich, Dovbenko, et al., 2017; Kuzmych & Kvasnikov, 2017; Kuzmych, 2018)

The most versatile characteristic of a complex technical object is its reliability, that is, the property of the object to store in time within the established limits the values of all parameters that characterize the ability to perform the required functions in the given modes and conditions of application, maintenance, storage and transportation (Ching, 2009; Makhutov & Reznikov, 2011).

An overwhelming majority, modern technical objects are very complex technical systems consisting of thousands or even tens of thousands of elements. The consequence of such complexity is their high cost, large material costs for their design, fabrication, and operation (Kuzmych, Kobylianskyi, et al., 2018; Kuzmych, 2018; Wu, Clements-Croome, et al., 2006).

The creation of such systems requires the creation and development of scientific and methodological foundations of the theory of their analysis and synthesis, which is based on the development of scenarios of the emergence and development of dangerous conditions and failures.

In the conditions of modern production, an important task is to evaluate the resource of complex potentially dangerous engineering structures at the design stage, to evaluate and forecast the residual resource during operation, to extend the term of the operation and service after the ending of the normative term of operation. Operating conditions of such structures are characterized by various influences, which lead to the degradation of structural materials and the exhaustion of the resource structures. The process of resource exhaustion is multistage, nonlinear, and interconnected (Kuzmych, 2018; Kuzmych, Kobylianskyi, et al., 2018; Makhutov & Reznikov, 2011).

For evaluating and substantiating the resource of complex technical constructions the following tasks should be solved:

- Elemental analysis of the design, operating conditions, manufacturing technology, and initial defects.
- Detection of dominant nodes, which lead to the exhaustion of the resource of the entire design.
- Modeling the exhaustion of the resource, taking into account the degradation of structural materials and operating conditions.
- Making calculations and experimental studies for the detection of hazardous zones in structures and determining the specific characteristics of the process of exhaustion of the resource in these zones.
- The development of appropriate databases to gather information on the processes of resource exhaustion in complex technical constructions and obtain the necessary information for operational evaluation and forecasting of the residual resource during operation.
- Creation of methods and systems of operative estimation and forecast of the residual resource during the operation of complex technical constructions.
- Obtaining the necessary information for deciding on the extension of the life of the structures, their repair, reconstruction or replacement.

The main issues for solving these tasks are:

- The choice of macro- variables that reflect the essential processes of exhaustion of the resource at the micro- and macro- levels.
- Substantiation of the terms of exhaustion of the resource, taking into account the degree of damage (failure).
- Simulation of processes of deformation and exhaustion of the resource.
- Identification of the main factors that influence the speed of processes of resource exhaustion and the establishment of specific dependencies for these impacts.
- Simulation of the velocity dependence of the resource drainage processes at all stages of its development from operating conditions.
- Establishing the principles of equivalence of processes of exhaustion of resources between themselves and experimental data of researches.
- Establishing correlation relations between the parameters of the mathematical models of exhaustion of the resource and the parameters of the diagnostics of the condition of the structural material.
- Development of methodology and algorithm for operational estimation of spent resource and forecast of residual resource of structures in the process of exploitation based on received preliminary information.
- Determination of real operational parameters of complex technical constructions throughout its entire operation.

METHODS AND TECHNIQUES

Reliability, including durability and survivability, is ensured by simultaneous fulfillment of the requirements for the selection of materials, constructive and bulk-planning solutions, to the methods of calculation, design and quality control of work in the manufacture of structures, their construction, as well as compliance with technical rules exploitation, supervision and care (Chybowski, Żółkiewski, et al., 2015; Kuznetsova & Barzilovicha, 1990).

Of all existing methodological approaches to the analysis of complex systems, the system is closest to the problem of the analysis of the laws of development and the structure of technical systems. The system approach is a natural scientific method based on the formal derivation and quantitative evaluation of the properties of the system. This method allows you to define the task, describe the object with the help of certain techniques and, finally, find a solution to the task.

The complexity of technical constructions is determined by a large number of its possible states. The cost includes costs for creation, production, and operation. The multipurpose character of complex technical constructions leads to the need to characterize their properties in many indicators, the requirements of which are often contradictory.

The plurality of states is characterized by a system, which is a reflection of its dynamism, plurality of variants of development. The more varied possible states of the system, the more difficult it behaves (Kuzmych & Kvasnikov, 2017; Makhutov & Reznikov, 2011).

A typical approach in the diagnosis of complex technical systems is the development of scenarios for the emergence and development of hazardous states and failures that characterize the transition in time t from the normal (normal) stages of the operation of complex systems to failures and accidents (Ching, 2009; Kuzmych, 2018; Wu, Clements-Croome, et al., 2006).

Based on the analysis of recent studies on the emergence and development of hazardous conditions and failures (Babich, Dovbenko, et al., 2017; Kuzmych, 201; Makhutov & Reznikov, 2011), we distinguish three main groups of scenarios:

1) Scenarios of monotone transitions, when the current parameters of external influences, reactions, and states of the systems change monotonously (as a rule, on the dangerous side) and the risks of R (t) increase continuously to the critical values of $R_c(t)$ (Figure 1, a, curve 1);

2) Scenarios with sharp transitions (with aggravation), when the parameters of perturbing influences or reactions change spin-free (almost instantaneously) at short intervals, causing a sharp increase in the risk R (t) to the critical values of $R_c(t)$ (Figure 1, a curve 2);

3) Scenarios with bifurcation (soft and hard) transitions, when in some unstable states of systems changes can be made with a complex trajectory of risk changes from R (t) to $R_c(t)$ (Figure 1, b, curve 3).

In terms of security and reliability, the most manageable and easy to manage are scenarios with monotone transitions (Figure 1, a, curve 1), and the most complex scenarios with bifurcation transitions (Figure 1, b, curve 3)

For example, the growth of engineering and technical complexity of systems, their energy saturation, and the branching of information and control flows leads to the scenario of the emergence and development of dangerous states and failures of the third type (Figure 1, b, curve 3).

With the help of similar diagrams, diagnostic parameters and their possible dynamics can be presented. The nature of the change in the diagnostic parameter determines the possibility of automatic protection systems to prevent accidents and possible subsequent man-made disasters in hazardous production facilities.

The complicated technical system when the failure of individual elements and even entire subsystems does not always lose their ability to work, often only reduced the characteristics of its efficiency. This is the property of large complex technical designs. Based on the previous, if an accident occurs under the "failure" or bifurcation scenario, diagnostic parameters should be introduced that would evaluate the factors that are manifested at the stage of the origin of the precondition for a transition to a new mode

Figure 1. Diagram of scenarios for the emergence and development of hazardous states and failures

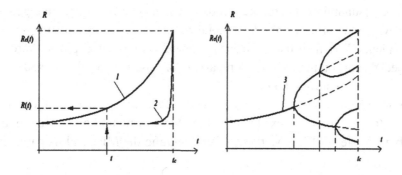

of operation or allow estimating the likelihood of such a transition. Thus, the prevention of dangerous processes can only be effective if the timely and objective estimation of systems and processes and the prediction of their future state are possible.

For qualitative research of the technical condition, complex systems should be divided into subsystems (complexes) and elements (nodes). As part of the subsystems, it can be considered constructively and functionally completed components of the system, the interaction of which ensures the achievement of the goal when performing the planned task. Each element of the system is associated with other elements in a certain way, and identical elements may have different characteristics in different systems. Therefore, first of all, it is necessary to identify the relationships and conduct their structural analysis. The information thus obtained can be provided in the form of various schemes, drawings, technical descriptions, logic schemes, etc.

Complex technical systems are characterized by complex nonlinear interactions between elements that are part of them, complex chains (scenarios) of causal relationships between dangerous events and processes occurring during operation and are probabilistic.

In the process of functioning, complex technical objects are subject to continuous impacts with the constant deterioration of their technical condition. This is due to the effect of technological and external operational loads and some environmental factors that are accidental (Kuzmych, 2018; Kuznetsova & Barzilovicha, 1990).

These influences lead to the emergence and accumulation in the elements of technical constructions of various types of damaged(τ) (fatigue, physical aging, the appearance of cracks, etc.). Upon reaching the critical level, the damage resulting from the operation of the operation results in disruption of the working condition of the equipment, the operation, and exhaustion of the resource, and as a result, in emergencies (refusals, accidents, and disasters).

Influence is any cause that changes internal stresses, deformations or other parameters of the structure in the design.

Mechanical influences that are included in the calculation directly, are considered as a combination of forces applied to the design (load), or as forced displacement and deformation of the structural elements. Other impacts of non-mechanical nature (for example, the effects of aggressive environments) are usually counted indirectly.

Depending on the cause of the impact, the effects are divided into the main and emergency.

Major influences are the consequences of natural phenomena or human activity. This may be the effect of water or temperature on the object, from the technological manipulation of the mode of operation of the object, etc.

Accident influences are undesirable results of human activity (the consequences of gross mistakes), or the results of unfavorable coincidence of circumstances (very rare effects of natural origin such as stress from tornadoes, tsunamis, catastrophic floods, etc. can be attributed to emergency ones).

Accidents, for which the project provides special means of active management and protection, are called Project Accidents (PA), their list and basic parameters (fire load, explosion force, flood, etc.) are determined by special rules on the basis of comparison of possible social losses and material damage with the necessary means to prevent them.

In addition to the parameters of the project accident - PA, for a particular object should set parameters as possible in the conditions of existence of the object of natural and/or man-made disaster (the Maximum Possible Disaster - MPS). Methods of determining the MPS and its parameters are also set by special rules.

It is allowed to accept the parameters of the maximum possible disaster - MPS, based on the probability of their occurrence 100 times less than the accepted probability of occurrence of PA.

$$P (MPS) \leq 0.01 P (PA), \tag{1}$$

Where:

P (MPS) - Parameter of the maximum possible disaster – MPS.
P (PA) - Parameter of the design accident - PA.

When designing special norms and defining the parameters of PA and MPS should consider the phenomena that may be caused by the following output events:

- The catastrophic excessive intensity of natural influences of the level established by the current norms for the construction area.
- Man-made disasters (vehicle accidents, explosions, fires, leakage of molten metal, etc.) occurring within the facility or in its immediate surroundings.
- Gross personnel mistakes in the stages of design, construction or operation of the facility.
- Serious shortage or sharp inconsistency of characteristics of building materials and products, elements of equipment requirements of normative and technical documentation.

When considering and classifying the causes of PA and MMK must also take into account the influence of secondary factors (explosions, the occurrence of fires, the destruction of protective obstacles, impacts from falling elements, etc.), the cause of which was the initial accident. It is recommended to develop and analyze scenarios for the development of accidents.

The calculation of damage characteristics and reliability of structural elements and nodes of complex technical systems is usually based on statistical and deterministic models that do not take into account changes in the stress-strain state and properties of the material, as well as the kinetics of damage to parts and elements of systems in time τ. Most processes of accumulation of damages d (τ) occur during long intervals of time with multiparameter influences.

The introduction of the time factor τ into the analysis of the damages d (τ) of structures is possible on the basis of kinetic representations about the process of accumulation in the material of the structure of irreversible damage of mechanical, physical and chemical origin, taking into account scenarios of multifactorial static, cyclic and dynamic effects of working processes and the environment. The basis for solving this problem is the development of methods for analyzing, calculating and predicting the boundary states of individual elements and the system as a whole, as well as the development of quality of damage - methods of technical diagnostics that allow numerical evaluation of the degree of damage to parts of different nature and character (Ching, 2009; Kuzmych, 2018; Kuzmych, Kobylianskyi, et al., 2018).

According to the results of the forecasting at the exploitation stage, an increase in the resource and timing is a possible operation of elements of technical systems. The solution to this problem involves the establishment of qualitative and quantitative patterns that determine the resource, the development of methods for assessing the impact of various factors on the average resource and loss of resources during the operation of the object [6-9].

Improving the methods of calculation requires, in particular, deduction of stochastic variability of properties and structure of the system, as well as the variability of other random factors. The necessary stage of development of a stochastic approach should be the development within the framework of existing deterministic schemes for calculating the methodology for assessing the reliability, taking into account the variability of the properties of the material and significant uncertainty of the original data.

Proceeding from the above, we can propose the following scheme for constructing a system for determining the final resource that will be split into two stages: design and operation:

1. For the object at the design stage and the object being in operation, mathematical models of resource loss are constructed. To do this, in cases, the recommended operating mode, periodicity of control, a list of technical parameters that characterize the state of the object, as well as a set of means of control and measurement is determined.

2. For the object, at the design stage, a mathematical model of functioning and wear in the form of a tree is constructed. When constructing the model, the strength characteristics of the materials are taken into account; predictable operating modes (magnitude and distribution of loads in time, etc.) and operating conditions (climatic conditions, corrosion factors, etc.).

3. For modeling an object during the period of operation, the required periodicity of control over the actual state of the object is estimated. Criteria may be: limit values of some physical parameters and components of the object; initial data about the project resource of the object, taking into account the term and conditions of operation of the object; operating conditions; statistical data obtained during the operation of the same type of technical systems and objects; technical condition at the current time; expert opinion on exploitation and non-destructive testing.

Then the mathematical model of the system for evaluating the actual state of the object is constructed and operates based on conditions and hypotheses (Kuzmych, 2018; Kuznetsova & Barzilovicha, 1990; Makhutov & Reznikov, 2011).

The current set of technical parameters of the object depends on what these parameters were at the beginning of the operation, from the mode of operation of the object and the history of conditions operation and operating modes. Under the conditions of operation, in this case, should be understood workloads, systematic and random corrosion factors, etc., which, in general, can also be called a set of destructive impacts. Under the mode of operation of the object is the scanning in time of a set of technical processes, each of which is characterized by a set of operating parameters. Changing the technical parameters of an object is described by the equation:

$$x\left(t\right) = F\left(x\left(t_0\right), u_{[t_0, t]}, K\right). \tag{2}$$

Where:

$x(t)$ - Vector of technical parameters.

$u_{[t_0, t]}$ − The condition of operation of the object at a specific time interval $[t_0, t]$.

K - Vector characterizing the mode of operation of the object.

The set of technical parameters of the object is judged by the results of direct or indirect measurements of physical parameters. The set of measurement results depends on the relevant technical parameters of the object at the time of measurements and conditions in which measurements were made. This stage is described by the equation:

$$y(t) = G\big(x(t), u(t)\big). \tag{3}$$

Where:

$y(t)$ - Measurements during control; it is a random value.
$u(t)$ - The condition of operation of the object at the moment.

On the received set of measurements, the estimation of true values of the technical parameters of the object is constructed. This process is described by the equation of estimation:

$$\hat{x}(t) = H\big(y(t)\big). \tag{4}$$

Where: $\hat{x}(t) - .$Estimation of the technical state vector.

Then the actual status of the object is estimated, which is judged by the set of estimates of the true values of the technical parameters of the object, obtained in the specific conditions (Kuzmych, 2018; Makhutov & Reznikov, 2011):

$$\vdots\,(t) = \ddots\,\big(\hat{x}(t), u(t)\big). \tag{5}$$

Where: $\Phi(t)$ - An estimate of the actual state of the object at the time t.

The final resource of the object is calculated on the constructed mathematical model and is determined by a set of estimates of the technical parameters of the object, the state equation, conditions of operation, the actual state of the object and a set of limiting technical parameters:

$$R(t) = W\big(t, \hat{x}(t), u(t), \bar{x}, \vdots\,(t)\big). \tag{6}$$

Where:

R(t) - An estimate of the final resource at time t.
$\bar{x} -$ Limit values of technical parameters.

Then the mathematical model of the system for evaluating the actual state of the final resource has the form:

$$\begin{cases} x(t) = F\Big(x(t_0), u_{[t_0,t]}, K\Big) \\ \quad y(t) = G\Big(x(t), u(t)\Big) \\ \qquad \hat{x}(t) = H\Big(y(t)\Big) \\ \quad \Phi(t) = \Psi\Big(\hat{x}(t), u(t)\Big) \end{cases} \qquad (7)$$

$$R(t) = W\Big(t, \hat{x}(t), u(t), \overline{x},\vdots\ (t), K\Big). \qquad (8)$$

Where (2) - State equation (maybe as an evolutionary or differential equation); (3) - measurement equation (stochastic); (4) - equation of estimates (deterministic); (5) - equation for estimating the actual state of an object (involves an algorithm for assigning an object to a certain class, constructed on the basis of probabilistic methods); (8) - equation for estimating the residual resource (prediction algorithm using equation (2).

In the equation (2) when constructing a mathematical model, (t_0).will be the moment of commencement of the operation of the object is taken, and when determining the residual resource, the moment of evaluation of the technical condition of the object.

In order to estimate the residual resource, it is necessary to indicate the estimation $\hat{x}(t)$.from equation (4) in the equation (2) as an initial condition, as the operating load - the load is scheduled, to calculate the trajectory of the change of the technical parameters and, accordingly, the period of time, during which neither one of the technical parameters will not reach its limit value or the physical state vector will not go beyond the permissible state.

This model is built for each component of the design. The residual life of the whole design is estimated by component and worst condition.

Since measurements during control are random values, the described model cannot be considered completely deterministic, therefore a mathematical description of statistical regularities of measurements and their relationships with indicators of the state of the test object is necessary to assess the actual state of this object. The assessment of the true values of technical parameters following equation (4) is the task of recognizing the state in which the object of control is located, for the solution of which the probabilistic approach can be applied (Kuznetsova & Barzilovicha, 1990).

It should be noted that, by choosing the function u (t) (operating conditions) and the vector K (operation parameters) in equation (2), one can control the change in technical parameters and, therefore, the residual resource.

When constructing a system for estimating the residual resource when designing a new (especially complex) facility, it is necessary to introduce means of monitoring the current state at the most critical points (in the calculated locations of stress concentration, known by the example of previously constructed and in operation objects, possible centers of corrosion, etc.).

If it is impossible to introduce monitoring tools, it is necessary to provide constructive access to critical points for the implementation of planned diagnostic activities.

Solving the tasks requires adequate information support at the expense of improving the reliability of the operation of structures, the use of systems of internal control and operational diagnostics of the state of technical means. However, all these measures do not fully solve the problems, especially in conditions of a slow evolution of the characteristics of the object of control, their large dimension, the complexity of the monitoring of the processes and the variability of the factors affecting them, information which can be categorized fuzzy sets. The very models of processes are empirical and semi-empirical (Kuzmych, Kobylianskyi, et al., 2018).

The evolutionary damage equations take into account the physical stages of the process of accumulation of damage, the influence of the parameters of the stress-strain state, temperature, the type of the deformation trajectory and the effect of deformation on the rate of accumulation of damage, its nonlinear character, the nonlinear summing of damage when changing the load regime and from various mechanisms of resource exhaustion (Chybowski, Żółkiewski, et al., 2015; Wu, Clements-Croome, et al., 2006).

The evaluation of the residual resource is a multilevel iterative process of sequential detailing and optimization of diagnostic solutions. Insufficient study of individual phenomena does not allow having a fully mathematically formalized description of the design. Since evaluation is performed based on the function of damage (failure), for its definition it is necessary to take into account the uncertainties that arise due to the variability of the technological situations in which the diagnoses are presented.

The effectiveness of the operation of a technical object primarily associated with its intended use, so the main task of managing the process of technical use is to increase the length of time used at acceptable costs for maintenance and repair (Kuzmych, 2018; Makhutov & Reznikov, 2011; Wu, Clements-Croome, et al., 2006).

The modern level of scientific and technological progress allows you to create objects that have high reliability. The basis for this is a set of measures that are used at the stages of designing, manufacturing, installation, and operation. The most actual problem is the forecasting and provision of the technical resource of the object. The general scheme for evaluating the residual resource is presented in (Figure 2).

Controlled parameters can be either directly measurable values of damage (depth of corrosion or wear of the part), or output parameters of equipment (performance, efficiency, degree of purification, etc.) and other quantitative quality indicators. Control of the change of these parameters as they approach their limits to the maximum allowable allows predicting the moment of failure (Biondini, Frangopo. 2008; Frangopo, Kawatani, et al., 2007; Adams, 2007). The assessment of reliability, in this case, is carried out by carrying out periodic inspections of the object, measuring the values of the defining parameter, statistical processing of the measurement results and further calculation of reliability indicators.

As it was noted, the main sources of damage development are areas of stress concentration. It is in them that the structural and mechanical properties of the material need to be investigated in the first place.

At the stage of operation of machines, equipment and constructions, taking into account the change in the state of structural elements and the accumulation of operational damage, tests both individual units and the entire object, determine residual strength, resource, and fracture resistance. To extend the life of safe operation is possible by applying all types of stocks - at rated voltage, local stresses and deformations, crack resistance, time and number of cycles, etc.

It should be noted that significant influence on the accumulation of damage, and, accordingly, and on the resource, have both constructive factors, including stress concentrators, and technological, determined by the mechanical properties of the material. In the absence of macro defects, the boundary state is determined by the critical values of local stresses or deformations, taking into account stress concentration zones and the allocation of characteristic points and voltage values. Introduction into the calculation according to the criteria of static, long and cyclic strength of the stock factors by local stresses and deformations allows establishing a safety permit from total load over time, number of cycles and temperature and estimates the resource of safe operation, including in emergency modes.

Figure 2. The scheme of the assessment of a residual resource

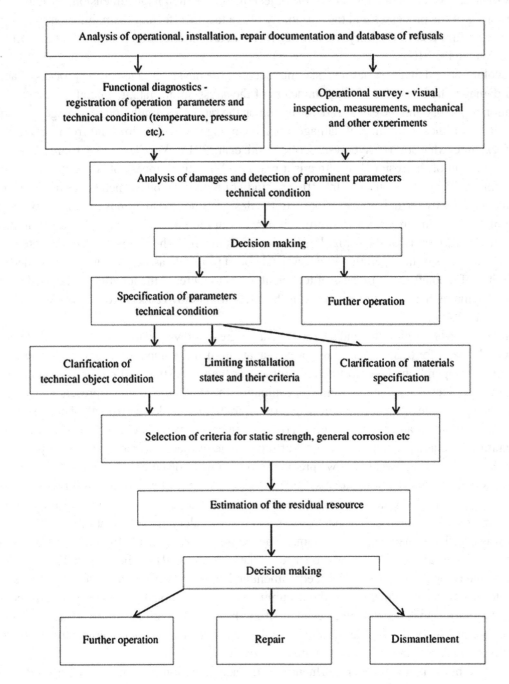

Illustrative Example

As an example, the calculation of reliability by taking into account random factors based on materials from an examination of the technical condition of the distant column, which has been in operation for 29 years.

Specifications are as follows: operating pressure – 0,6 kgf / mm^2 or vacuum; working space – latex; volume - 29 650 m^3; temperature - 30 - 65 °C, actual thickness of the metal is 12 $\pm$ 0,2 mm (Frolov, 1992; Avyrom, 2001).

The mean square deviation of the metal thickness is based on the assumption of normal distribution according to the "three-sigma" rule:

$$\sigma_{init} = \frac{S_{max} - S_{min}}{6} = \frac{12,2 - 11,8}{6} = 0,067\, mm\,., \tag{9}$$

Mathematical expectation μ_{init} the average value of $S_{av}=12mm$.

According to the results of measurements after $t=29$ years of operation, the maximum and minimum thickness, average value $\mu_{op}=S_{av}=6.37mm$ and standard deviation will be:

$$\sigma_{op} = \frac{S_{max} - S_{min}}{6} = \frac{6,84 - 5,83}{6} = 0,17\, mm\,., \tag{10}$$

Based on the found values of σ_{init}, σ_{op}, μ_{inig} and μ_{op} using the MathCAD random number generator, 1000 initial and actual values of the thickness S_{init} op were obtained.

When evaluating a resource by a determining parameter $X(t)$. (in the case under consideration, this is the thickness of the metal) uses as:

$$X(t) = X_0 - \gamma t^n. \tag{11}$$

Where:

X_0 - The initial value of the determining parameter.
γ - The rate of change of the parameter.

degree n take from the interval 1 to interval 2 (Yablonsky. 1988; Burau, Pavlovsky, et al., 2013).
In two limiting cases with $n=1$ or $n=2$ we will have:

$$X(t) = X_0 - \gamma t, \quad (t) = X_0 - \gamma t^2. \tag{12}$$

Then the estimate of the total resource T for a known value γ is produced by solving equations (12) for $X=X_{lim}$., the estimate of residual life at $X=X_{op}$:

$$T = \frac{X_0 - X}{\gamma} = \sqrt{\frac{X_0 - X}{\gamma}} \tag{13}$$

The value of limited exploitation X_{lim} is determined from strength conditions, requirements of normative and technical documentation or based on safety requirements.

Estimated allowable metal thickness from strength conditions is defined as:

$$X_{lim} = \frac{P_{av} D}{2\varphi\left(\sigma_{al} - P_{av}\right)} = \frac{0,6 \bullet 2400}{2 \bullet 1 \bullet \left(168 - 0,6\right)} = 4,29\, mm \,., \tag{14}$$

Where:

P_{av}=0.6kgf/mm^2 - Available working pressure.
D=2400mm - Inner diameter of the object.
φ=1 - Coefficient of weld strength.
σ_{al}=168MPa - Allowable stress for the design temperature.

To find the function of the determining parameter $X(t)$ it is necessary to know the joint distribution law $f(X)$ of the random variable X.

In the general case, the distribution law of random quantity X has the form:

$$F\left(X\right) = P\left[\left(X_0, \gamma\right) \subset D\right] = \int_{D\left(X \leq x\right)} f\left(X_0, \gamma\right) dX_0 d\gamma. \tag{15}$$

The mathematical problem of determining $F(X)$ is reduced to the solution of the double integral. For the case, when $=X_0 + \gamma t$, the integral is taken over the domain D where $X_0 + \gamma t < x$. Thus, we obtain specific limits integration:

$$F\left(X\right) = \int_D \int f\left(X_0, \gamma\right) dX_0 d\gamma = \int_{-\infty}^{\infty}\left[\int_{-\infty}^{x} f\left(X_0, \gamma\right) d\gamma\right] dX_0 ., \tag{16}$$

We obtain the distribution density $f(X)$ by differentiating concerning x which is included as a parameter to the upper limit of the integral. Since the random variables X_0 and γ are independent and equal, the density distribution is like:

$$f\left(X\right) = \int_{-\infty}^{\infty} f_1\left(X_0\right) f_2\left(X - X_0\right) dX_0 ., \tag{17}$$

or

$$f\left(X\right) = \int_{-\infty}^{\infty} f_1\left(X - \gamma\right) f_2\left(\gamma\right) d\gamma ., \tag{18}$$

Where f_1 and f_2 - The distribution densities of the arguments and possible argument values are non-negative.

Accepting the normal distribution law for them, we obtain:

$$f(X_0) = \frac{1}{\sigma_{Xo}\sqrt{2\pi}} exp\left\{-\frac{[X_0 - \mu_{Xo}]^2}{2\sigma_{Xo}^2}\right\}., \tag{19}$$

$$f(\gamma) = \frac{1}{\sigma_{\gamma}\sqrt{2\pi}} exp\left\{-\frac{[\gamma - \mu_{\gamma}]^2}{2\sigma_{\gamma}^2}\right\}., \tag{20}$$

The distribution law of a random variable X is defined as:

$$f(X) = \frac{1}{2\pi\sigma_{Xo}\sigma_{\gamma}} \int_{-\infty}^{\infty} exp\left\{-\frac{[X_0 - \mu_X]^2}{2\sigma_X^2}\right\} \times exp\left\{-\frac{[X_o - X - \mu_{\gamma}]^2}{2\sigma_{\gamma}^2}\right\} dX_0, \tag{21}$$

After the transformations we get:

$$f(X) = \frac{1}{\sigma_X\sqrt{2\pi}} exp\left\{-\frac{[X - \mu_X]^2}{2\sigma_X^2}\right\}., \tag{22}$$

where

$$\mu_X = \mu_{Xo} - t\mu_{\gamma} \quad \text{or} \quad \mu_X = \mu_{Xo} - t^2\mu_{\gamma} \tag{23}$$

$$\sigma_X = \sqrt{\sigma_{Xo}^2 - t^2 \bullet \sigma_{\gamma}^2} \quad \text{or} \quad \sigma_X = \sqrt{\sigma_{Xo}^2 - t^4 \bullet \sigma_{\gamma}^2} \tag{24}$$

$$f(X) = \frac{1}{\sqrt{\sigma_{Xo}^2 - t^2 \bullet \sigma_{\gamma}^2}\sqrt{2\pi}} exp\left\{-\frac{\left[X - \left(\mu_{Xo} - t\mu_{\gamma}\right)\right]^2}{2\left(\sigma_{Xo}^2 - t^2 \bullet \sigma_{\gamma}^2\right)}\right\}$$

or

$$f(X) = \frac{1}{\sqrt{\sigma_{Xo}^2 - t^4 \bullet \sigma_{\gamma}^2}\sqrt{2\pi}} \times exp\left\{-\frac{\left[X - \left(\mu_{Xo} - t^2\mu_{\gamma}\right)\right]^2}{2\left(\sigma_{Xo}^2 - t^4 \bullet \sigma_{\gamma}^2\right)}\right\}. \tag{25}$$

Next, it needs to determine the type and function of resource allocation. Since the conclusion of analytical dependencies presents certain difficulties it makes sense to solve this problem in numerical form. When t, μ_{Xo}, μ_X, σ_{Xo}, and σ_X are known using formulas (23), (24) we define μ_γ and σ_γ.

$$\mu_\gamma = \frac{\mu_{Xo} - \mu_X}{t} \quad \text{or} \quad \mu_\gamma = \frac{\mu_{Xo} - \mu_X}{t^2} \tag{26}$$

$$\sigma_\gamma = \pm \frac{\sqrt{\sigma_{Xo}^2 - \sigma_X^2}}{t} \quad \text{or} \quad \sigma_\gamma = \pm \frac{\sqrt{\sigma_{Xo}^2 - \sigma_X^2}}{t^2} \tag{27}$$

After determining the values of μ_γ and σ_γ using MathCAD random number generator 1,000 values of the rate of change of the determining parameter and 1,000 values of the resource were obtained. After testing the hypothesis of the distribution law with using the Kolmogorov criterion, a resource distribution function is constructed that is well approximated by the normal distribution function (Klyuyev, Parkhomenko, et al., 1989; Pontryagin, 1986):

$$f(T) = \frac{1}{\sigma_t \sqrt{\pi}} exp\left\{ -\frac{[T - \mu_t]^2}{2\sigma_t^2} \right\}., \tag{28}$$

Based on the data obtained, the parameters are determined distribution function of total and residual resources (Table 1).

FUTURE RESEARCH DIRECTIONS

With the help of similar diagrams, diagnostic parameters and their possible dynamics can be presented. The nature of the change in the diagnostic parameter determines the possibility of automatic protection systems to prevent accidents and possible subsequent man-made disasters in hazardous production facilities.

CONCLUSION

The basis of the theory of reliability and system analysis of complex technical constructions is the consideration of scenarios for the emergence and development of hazardous states and failures.

Table 1. Parameters of the distribution function of the complete and residual resources

Function				
$X(t) = X_o - \gamma t^2$	39,769	0,756	29,165	4,026
	33,959	0,322	29,013	2,006

With the increase in the level of engineering and technical complexity due to energy saturation, branching of information-controlling system flows, etc., the probability of developing a scenario of the emergence of dangerous states and failures of the third type increases.

Described scenarios for the emergence and development of hazardous states and failures include maintaining the history of the operation of the object, monitoring of diagnostic parameters, the transmission of received data, their processing and interpretation.

The described approach to constructing a system for determining the actual state and assessing the residual resource involves maintaining the history of the operation of the object, measuring and monitoring a sufficiently larger number of diagnostic parameters, the transmission of received data, their processing and interpretation.

To determine the actual state and estimate the residual life of structures, it is proposed to simultaneously evaluate several characteristics of the material of the object: the characteristic parameters of the structure of the material, integral parameters of the material, related to strength (for example, hardness), the presence and nature of macro defects, the degree of corrosion wear of the metal.

Limit values of the selected diagnostic parameters are determined by available standards or technical conditions.

The dynamics of changes in diagnostic parameters can be monitored and modeled based on the data of periodic inspections of the control object.

The space of the diagnostic parameters is broken down into the area, taking into account the rate of change of each of the parameters, while critical areas are determined. The state of the control object is connected to a point in the parameter space. The change in the state of an object is characterized by a trajectory that shows its movement in the parameter space. Getting a trajectory in a critical area indicates an emergency state of the control object.

REFERENCES

Adams, D. (2007). *Health monitoring of structural materials and components: methods with applications* . John Wiley & Sons. doi:10.1002/9780470511589

Avyrom, L. S. (2001). Design reliability of prefabricated buildings and structures. London, UK: Publishing House on Construction.

Babich, V., Dovbenko, V., Kuzmych, L., & Dovbenko, T. (2017). Estimation of flexures of the reinforced concrete elements according to the National Ukrainian & European standards. MATEC Web of Conferences, Vol. 116, July 10, 2017, Article number 02005.

Biondini, F., & Frangopol, D. (Eds.). (2008). *Life-cycle civil engineering* (p. 970). London, UK: Taylor & Francis Group. doi:10.1201/9780203885307

Burau, N. (2013). Structural and functional synthesis of systems of diagnostics of structures in operation. In N. Burau, O. Pavlovsky, & D. Shevchuk (Eds.), Bulletin of TNTU. 72, 4. pp. 77-86.

Ching, J. (2009). Equivalence between reliability and factor of safety. *Probabilistic Engineering Mechanics, 24*(2), 159–171. doi:10.1051/matecconf/201711602005

Chybowski, L., & Żółkiewski, S. (2015). *Basic reliability structures of complex technical systems.* Springer International Publishing Switzerland. doi:10.1007/978-3-319-16528-8_31

Frangopol, D. M., Kawatani, M., & Kim, C. W. (Eds.). (2007). Reliability and optimization of structural systems: assessment, design, and life-cycle performance. Taylor & Francis Group, London, UK, p. 269.

Frolov, K. V. (1992). Problems of safety of complex technical systems.

Frolov, K. V., & Makhutov, N. A. (n.d.). Problems of machine building and machine reliability, 5. pp. 3-11.

Klyuev, V. V., Parkhomenko, P. P., & Abramchuk, V. E. (1989). Technical means of diagnosis: a reference book. In V. V. Klyueva (Ed.), *Mashinostroenie.* Moscow, Russia.

Kuzmych, L., Kobylianskyi, O., & Duk, M. (2018, Oct. 1). Current state of tools and methods of control of deformations and mechanical stresses of complex technical systems. *Proc. SPIE 10808, Photonics Applications in Astronomy, Communications, Industry, and High-Energy Physics Experiments.* doi:10.1117/12.2501661

Kuzmych, L., & Kvasnikov, V. (2017). Study of the durability of reinforced concrete structures of engineering buildings. Advances in Intelligent Systems and Computing. Vol. 543, pp. 659-663. In *Proceedings International Conference on Systems, Control and Information Technologies, SCIT 2016;* Warsaw; Poland; May 20-21, 2016. 10.1007/978-3-319-48923-0_70

Kuzmych, L. V. (2018). Scripts of emergence and development of dangerous states of complex technical systems [Text] / L. V. Kuzmych / Zbirnyk naukovykh prats Viiskovoho instytutu Kyivskoho natsionalnoho universytetu imeni Tarasa Shevchenka. Vol. 62. pp. 35–40.

Kuzmych, L. V. (2018). Mechanical effects on the reliability of complex technical systems [Text] / L. V. Kuzmych // Technical sciences and technologies. Vol. 4(14). pp. 28-33. doi:10.251402411-5363-2018-4 (14) -28-33.

Kuzmych, L. V. (2018). Approaches for estimating the resource of technical objects and systems [Text] / L. V. Kuzmych / Bulletin of the Zhytomyr State Technological University. Series: Technical Sciences. Vol. 2(82). pp. 204-207. doi: (82) -204-207. doi:10.26642/tn-2018-2

Kuznetsova, V. I., & Barzilovicha, E. Y. (1990). Nadezhnost I effectivnost v tehnike. Moscow, Russia: Mashinostroenie, 8, 248.

Makhutov, N. A., & Reznikov, D. O. (2011). A comparative assessment of specification-based and risk-management- based approaches to the security assessment of complex technical systems. *Journal of Machinery Manufacture and Reliability, 40*(6), 579–584. doi:10.3103/S1052618811060124

Pontryagin, L. S. (1986). *Generalization of numbers. Moscow, Russia*: Science.

Wu, S., Clements-Croome, D., Fairey, V., Albany, B., Sidhu, J., Desmond, D., & Neale, K. (2006). Reliability in the whole life cycle of building systems. *Eng. Constr. Architectural Manage, 13*(2), 136–153. doi:10.1108/09699980610659607

Yablonsky, S. V. (1988). Some problems of reliability and control of control systems. *Mathematical Problems of Cybernetics, 1.* pp. 5-25.

Chapter 10
Computer–Aided Design of Intelligent Control System of Stabilizing Platforms With Airborne Instrumentation

Olha Sushchenko

National Aviation University, Ukraine

ABSTRACT

The chapter includes principles of the computer-aided design of the intelligent control system for stabilizing platform with airborne instrumentation. The emphasis is given to the design of the robust controller. The design procedures including mathematical models development, simulation, analysis, and making a decision using such artificial intelligent approaches as soft computing and expert assessments of obtained results are considered. Two basic problems such as the design of new systems and modernization of operating systems are researched. The first problem can be solved by means of the robust parametric optimization, and the second by means of the robust structural synthesis. The basics of algorithmic, software, and method supports are represented. The generalized structural schemes of the interactive design process are given. This approach allows decreasing engineering time and improving the quality of design systems.

INTRODUCTION

The motivation of the proposed chapter is to fill gaps in approaches to design of robust inertially stabilized platforms with aviation instrumentation. Design of robust systems is accompanied by complex calculations and transformations. It requires also artificial intelligence procedures such as expert assessments, decision-making, modern optimization methods.

The problem of the design of control systems of dynamic objects is characterized by transition from adaptive control paradigm to intellectual control paradigm. This is caused by a complication of plants, operating conditions, and increased requirements to reliability and efficiency of control. To take above-stated factors into consideration it is possible on the basis of transition from the determinate algorithms of parametrical and structural synthesis to control with artificial intelligence approaches.

DOI: 10.4018/978-1-7998-1415-3.ch010

Now the inertially stabilized platforms are widely used to stabilize and point sensors, cameras, antennas and weapon systems operated at the vehicles of the different type (Hilkert, 2008). The further progress of the above-listed information and measuring systems by accuracy and other operating performances is impossible without stabilization of a base, at which they are mounted. This proves the actuality of the inertially stabilized platforms technology development. The above-stated observation equipment is operated in difficult conditions of influence of aerodynamic disturbances. Moreover, in the case of platforms with sufficient mass and dimension, there are sufficient changes of platform parameters (especially of the inertia of moment). This can lead to losses of control not to mention about accuracy. It is expedient to solve the problem of the vehicle stabilized platforms design using principles of robust control.

There are two important problems of stabilized platforms creation such as modernization and design of new perspective systems. The first problem can be solved on the basis of robust parametrical optimization. The second problem can be solved using the robust structural synthesis. Choice of the method of the synthesis depends on features of the designed system and conditions of its operation. Designing the inertially stabilized platforms, it is necessary to take into consideration two factors. Firstly, some parameters of the vehicle stabilized platforms vary in the wide range. Secondly, the researched systems operate in the difficult conditions of disturbances caused by the external environment.

Creation of the modern inertially stabilized platforms, as a rule, is implemented by means of Mat-Lab software including special toolboxes for the automated design of the robust systems (Gu, Petkov & Konstantinov, 2005).

REVIEW OF PREVIOUS LITERATURE

The main directions of the design of inertially stabilized platforms are given in (Hilkert, 2008). The problems of robust control are not mentioned in this paper although robustness is very important for aviation inertially stabilized platforms, which function in difficult conditions of real operation. The features of the design of robust stabilization system depend significantly on the type of the researched vehicle. For example, papers (Wu & Yu 2018) and (Kim, Cho & Him et al., 2018) deal with such different moving vehicles as the hypersonic vehicle and multirotor.

Principles of the design of aviation automated control systems differ significantly for such objects as aircraft and moving platforms with a payload of the different type. Researches represented in (Sushchenko & Goncharenko, 2016), deal with the design of aviation robust platforms, which stabilize observation equipment. This chapter improves and represents in details approaches to computer-aided design of robust inertially stabilized platforms with instrumentation. Basic concepts of computer-aided design of robust stabilized platforms are given in (Sushchenko, 2017). Concepts of robust design of stabilization systems are represented in (Skogestad & Postlethwaite, 2001). Information about approaches to signal processing can be found in many papers including (Zaliskyi & Solomentsev, 2014).

BACKGROUND

The computer-aided design of any complex system includes the following tasks (Solnitsev, 1991):

1)   Forming the appearance of the system.

2) Choice of the functional scheme.
3) Choice of the structural scheme and basic technical facilities.
4) Choice of the computing device.
5) Development of functioning algorithms.
6) Design of the controller with definite properties, for example, the robust controller.

Usually, the problem of forming the appearance of the system is based on researcher experience and analog systems. The problem statement can be formulated as achieving a compromise between ergonomic indices and cost expenses

$$\max E(\mathbf{X}), \mathbf{X} \in D_x, \tag{1}$$

$$D_x = \{\mathbf{X} \mid C(\mathbf{X}) < C_{per}\},$$

where $E(X)$ is the objective function of the ergonomic indices; $\mathbf{X}$ is the vector of design parameters. D_{per} is the region of permissible values; $C(\mathbf{X})$ is the function of the cost constraints.

During the choice of the functional scheme, it is convenient to consider basic functions necessary for system operation and additional functions able to extend functional possibilities of a system. The basic compromise is between the quantity of carried out functions and mass and dimension losses necessary for their implementation. The formalized problem statement becomes

$$\min[C(\mathbf{f}_1) + C(\mathbf{f}_2)], \mathbf{f}_1 \in F_{x1}, \mathbf{f}_2 \in F_{x2}; \tag{2}$$

$$D_{x1} = \{\mathbf{f}_1 \mid V(\mathbf{f}_1) < V_{per1}\}, D_{x2} = \{\mathbf{f}_2 \mid V(\mathbf{f}_2) < V_{per2}\},$$

where $\mathbf{f}_1$ are basic functions; $\mathbf{f}_2$ are additional functions; $C(\mathbf{f}_1)$, $C(\mathbf{f}_2)$ are loss functions; $V(\mathbf{f}_1)$, $V(\mathbf{f}_2)$ are mass and dimension constraints.

Choice of the structure and hardware is implemented by search of possible variants based on characteristics stored in data ware. For example, functions of measurement of the angular rate can be realized by sensors, characteristics of which are given in data ware, as it is shown in Table 1

The formalized problem statement of the structural scheme choice must provide the maximum reliability under conditions of fulfilment of constraints imposed on design parameters including mass and dimensions losses. This problem can be represented in the form

$$\max P(\mathbf{X}) = \prod_{i=1}^{n} P(x_i), \mathbf{X} \in D_x; \tag{3}$$

$$D_x = \{\mathbf{X} \mid V(\mathbf{X}) < V_{per}, \Delta(\mathbf{X}) \leq \Delta_{per}\},$$

where $P(\mathbf{X})$ is the objective function of reliability; $V(\mathbf{X})$, $\Delta(\mathbf{X})$ are constraints on mass, dimensions, and accuracy, V_{per}, Δ_{per} are permissible values.

Table 1. Functions of measuring instruments

Functions	Sensors	Characteristics
Measurement of angular rate	GT (gyrotachometer).	• Measuring range. • Sensitivity threshold. • Resistance to shocks. • Mass and dimensions. • Cost.
	MEMS-gyroscope.	
	FOG (fiber-optic gyroscopes).	

The formalized problem statement of a computer choice can be stated as the problem of achieving minimum permissible operation of speed under conditions of constraints on memory capacity and cost expenses

$$\min \Delta t(\mathbf{X}), \mathbf{X} \in D_x; \tag{4}$$

$$D_x = \{\mathbf{X} \mid V_{\text{ROM}}(\mathbf{X}) \geq V_{per}, V_{\text{RAM}}(\mathbf{X}) \geq V_{per}, C \leq C_{per}\},$$

where $\Delta t(\mathbf{X})$ is the objective function on the operation of speed; $V_{\text{ROM}}(\mathbf{X})$, $V_{\text{RAM}}(\mathbf{X})$, C are constraints on memory capacity and cost.

The problem of computer-aided design of functional algorithms can be stated as the problem of decrease of operation of the speed of algorithms with the goal to free time for implementation of additional algorithms such as operating tests and checks:

$$\min \Delta t(\mathbf{X}), \mathbf{X} \in D_x; \tag{5}$$

$$D_x = \{\mathbf{X} \mid V_{\text{ROM}}(\mathbf{X}) \geq V_{per}; V_{\text{RAM}}(\mathbf{X}) \geq V_{per}, \Delta(\mathbf{X}) \leq \Delta_{per}\},$$

where $\Delta t(\mathbf{X})$ is the objective function on a speed of operation of basic algorithms of system functioning; $V_{\text{ROM}}(\mathbf{X})$, $V_{\text{RAM}}(\mathbf{X})$, Δ_{per} are constraints on memory capacity and accuracy.

The problems (1) – (5) represent the general tasks of computer-aided design.

The specific feature of designing robust systems is the necessity to carry out a great quantity of transformations and calculations. Solving this problem requires using computer-aided design procedures with elements of artificial intelligence including expert assessments, decision-making, and analysis of obtained results.

The general structural scheme of the inertial stabilized platform is shown in Figure 1 (Hilkert, 2008).

In the case of modernization of the stabilized platforms, the problem statement can be reduced to the determination of the controller $\mathbf{K}(s)$, which provides stability of the closed-loop system with a transfer function $\Phi(s)$ (Polyak & Shcherbakov, 2002)

$$\mathbf{K}(s) \in D, \, D : \text{Re} \mid \text{eig}(\mathbf{I} + \mathbf{L}(s)) \mid < 0, \tag{6}$$

where $\mathbf{L}(s) = \mathbf{K}(s)\mathbf{W}(s)$; $\mathbf{W}(s)$ is the transfer function of the plant.

Figure 1. The general structure of the stabilized platform: $\mathbf{W}(s)$ *is the transfer function of a plant with the actuator and measuring system;* $\mathbf{K}(s)$ *is the transfer function of a controller;* $\mathbf{\Phi}(s)$ *is the transfer function of the closed-loop stabilization system; w is a disturbance; z is the output vector assigned for assessment of stabilization system quality; u is the vector of feedback signals; y is the observed output vector*

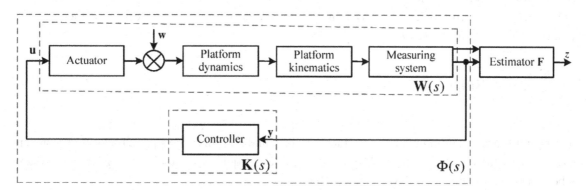

For practical applications, the problem of the design of the controller with the given structure (6) under conditions of some constraints imposed on the controller parameter is very important. For PID-controller $K(s) = q_1 + q_2 / (s) + q_3(s)$, it is possible to impose constraints $a_i \leq q_i \leq b_i$, $i=1,\dots,3$, a_i, b_i, $i=1,..,3$ on controller coefficients.

The problem of static stabilization deals with the representation of the stabilization system in the state space (Polyak & Shcherbakov, 2002)

$$\dot{\mathbf{x}} = \mathbf{A}(q)\mathbf{x} + \mathbf{B}(q)\mathbf{u} \; ; \tag{7}$$

$$\mathbf{y} = \mathbf{C}(q)\mathbf{x} \, ,$$

where $\mathbf{x}$ is the state vector; $\mathbf{A,B,C}$ are matrices of state, control, and observation; $\mathbf{u}$ is the vector of controls; $\mathbf{y}$ is the vector of observations. It is necessary to determine the controller $\mathbf{K}$, which provides stabilization by means of feedback $\mathbf{u}=\mathbf{Ky}$. From the mathematical point of view, the problem of static stabilization reduces to determination of such a matrix $\mathbf{K}$, which provides stability of the closed-loop system $\mathbf{A}_c=\mathbf{A}+\mathbf{BKC}$ for the given matrices $\mathbf{A,B,C}$. This is a hard problem. Therefore it is convenient to determine control in the form $\mathbf{u} = \mathbf{K}\hat{\mathbf{x}}$, where $\hat{\mathbf{x}}$ is an assessment of the system state by the observed outputs.

The static stabilization by means of feedback can provide not only stabilization and the given arrangement of poles of the closed-loop system. In this case, the controller $\mathbf{u}=\mathbf{Kx}$ is searched for the system (7). From the mathematical point of view, the problem of static stabilization reduces to determination of the such a matrix $\mathbf{K}$, which provides stability of the closed-loop system $\mathbf{A}_c=\mathbf{A}+\mathbf{BKC}$ for the given matrices $\mathbf{A,B}$.

It is convenient to extend this problem for practical applications and consider the matrix $\mathbf{K}$ as a supplement to the given structure of the controller with the goal to stabilize the system. The latter problem can be reduced to the problem of the matrix robust stabilization (Polyak & Shcherbakov, 2005). If to suppose that some constraints $q \in Q$ of the kind $\{a_i \leq q_i \leq b_i, i=1,\dots,n\}$ are imposed on controller parameters, the problem of the matrix robust stabilization reduces to the determination of the matrix $\mathbf{K}$, for which all the matrices $\mathbf{A}(q)+\mathbf{B}(q)\mathbf{K}$, $q \in Q$ are stable. For modernization of stabilized platforms is important the problem statement, when for constraints $q \in Q$ and adjustable controller parameters $\mathbf{h}$ are determined values $\mathbf{h}^*$, for which the system is stable.

The quality of deterministic control systems of the wide class can be characterized by the integrated quadratic index (Kwakernaak, 1993)

$$J_d = \int\limits_0^\infty (\mathbf{x}'\mathbf{Q}\mathbf{x} + \mathbf{u}'\mathbf{R}\mathbf{u})dt , \tag{8}$$

where $\mathbf{Q}$ and $\mathbf{R}$ are weighting matrices of state variables and controls. The square root of (8) represents the H_2-norm of the dynamic system. The quadratic index of quality of stochastic signals and systems can be represented in the form (Kwakernaak, 1993)

$$J_S = \mathrm{M}[\mathbf{x}'\mathbf{Q}\mathbf{x} + \mathbf{u}'\mathbf{R}\mathbf{u}],$$

where M is the mathematical expectation.

The problem of optimal design of the deterministic stabilization system can be formulated in the following way (Polyak & Shcherbakov, 2005). It is necessary to obtain the controller $\mathbf{K}(s)$, which stabilizes a system, and to determine parameters of a controller of the given structure $\mathbf{k}$. The index quality (3) achieves minimum under the condition (1)

$$\mathbf{k}^* = \arg \inf_{\mathbf{K}(s) \in D} J_d(\mathbf{x}, \mathbf{u}, \mathbf{k}).$$

If the mathematical description of the system has uncertainty

$$\dot{\mathbf{x}} = \mathbf{A}(q)\mathbf{x} + \mathbf{B}(q)\mathbf{u} \quad q \in Q,$$

where Q is uncertainty of some type, the problem of robust stabilization arises. After solving this problem, the controller of the given structure provides the assured index of quality for $q \in Q$. For the system represented in the state space by equations (7), it is necessary to determine a controller of the given structure $\mathbf{u} = \mathbf{K}(s, \mathbf{h})\mathbf{y}$, where $\mathbf{h}$ is the vector of adjustable parameters. The functional of this problem can be represented as $J_d(\mathbf{x}, \mathbf{u}, \mathbf{h})$. The problem of the modal parametric optimization is a problem of the search of the adjustable parameters of the optimal controller $\mathbf{u} = \mathbf{K}(s, \mathbf{h})\mathbf{y}$. It can be formulated in the following way: to find $\mathbf{h}^*$, $h \in \Omega_h$ that minimizes functional $J_d(\mathbf{x}, \mathbf{u}, \mathbf{h})$, that is

$$\mathbf{h}^* = \arg \inf_{\mathbf{h} \in \Omega_h} J_d(\mathbf{x}, \mathbf{u}, \mathbf{h}),$$

where Ω_h is a set of permissible values of adjustable parameters, for which roots of the characteristical equation of the closed-loop stabilization system are in the given place of the left semi-plane of the complex variable.

One of the modern approaches to the formalization of quality functional is using of matrix norms of transfer functions of the multi-dimensional closed-loop system. The generalized formalized problem statement of optimal controller design can be formulated in the following way (Veremey, 2010)

$$J_H(\mathbf{K}) = \| \Phi(\mathbf{K}) \|, \tag{9}$$

$$\mathbf{K}^* = \arg \inf_{\mathbf{K} \in D} J_H(\mathbf{K})$$

under condition of system stability.

Type of the concrete problem depends on the choice of the norm in the expression (9). So, the choice of H_2-norm is typical for the problem of LQR-synthesis, choice of H_∞-norm – for the problem of H_∞-synthesis. The above-mentioned norms with weighting coefficients can be used also for problems of mean-square optimal synthesis. LQR-problem can be equivalent to the problem of H_2-optimization under some conditions when disturbance represents the white time-invariant noise with noncorrelated components of unit intensity (Egupov, 2002).

Nowadays γ-optimal control becomes widespread (Skogestad & Postlethwaite, 2001). In this case, a controller provides stabilization of the plant and some given small number γ for H_∞-norm of the transfer function of the closed-loop system. A number γ is called the tolerance level. In these cases, the suboptimal solution is determined. It is possible to consider the family of suboptimal controllers, which satisfy the condition $H_\infty < \gamma$. The optimization criterion of the problem of robust controller design under condition (1) can be formulated in the following way

$$J_H = \| \Phi(\mathbf{K}) \|_\infty < \gamma .$$

The efficiency of modernization of stabilization systems can be achieved by means of the mixed H_2/H_∞-optimization. Such an approach provides the design of the controller of optimal quadratic quality, which is adaptable for functioning under conditions of the maximum disturbance. In this problem, it is convenient to use the combined criterion of quality (Tunik, Rye & Lee, 2001)

$$J_{H_2/H_\infty}(\mathbf{K}) = \varkappa_d J_d^2 + \varkappa_s J_s^2 + \lambda_\infty \| \Phi \|_\infty, \tag{10}$$

where λ_d, λ_s are weighting coefficients of quality indices; λ_∞ is the weighting coefficient of robustness. Problem statement of H_2/H_∞ optimal stabilization with optimization criterion (10) becomes

$$\mathbf{K}^* = \arg \inf_{\mathbf{K} \in D} J_{H_2/H_\infty}(\mathbf{K}) .$$

The robust structural synthesis is based on the solution of two Riccati equations, checking of some conditions and minimization of H_∞-norm of the function of the mixed sensitivity of the system that includes plant **G**, controller **K**, and also is represented by the vector of outputs **z**, which characterizes inputs system quality, vector of inputs **r**, and vectors of control **u** and observation **y** (Skogestad & Postlethwaite, 2001). The modern approach to solving the problem of the robust structural synthesis is based on forming the desired frequency characteristics of the system that is implemented by means of forming augmented plant due to introducing weighting transfer functions. H_∞-norm of the function of the mixed sensitivity is used as the optimization criterion

$$J_{H_\infty s} = \left\| \begin{bmatrix} \mathbf{W}_1 \mathbf{S} \\ \mathbf{W}_2 \mathbf{R} \\ \mathbf{W}_3 \mathbf{T} \end{bmatrix} \right\|_\infty . \tag{11}$$

In the expression (11) $\mathbf{W}_1, \mathbf{W}_2, \mathbf{W}_3$ are weighting functions; $\mathbf{S}, \mathbf{R}, \mathbf{T}$ are the function of sensitivity, the function of sensitivity by control, and the complementary function of sensitivity.

To design systems, which provide both tracking and stabilization, it is necessary to use the two-degrees-of-freedom controller. The generalized structural scheme of such a system is represented in Figure 2.

The problem statement of design of two-degrees-of-freedom controller can be formulated in the following way. The signal **y**., which is the output signal of the plant $\mathbf{W}(s)$, tracks the reference signal **r**. To provide stabilization of the plant and tracking of the signal **r**, the controller $\mathbf{K} = [\mathbf{K}_1(s)\ \mathbf{K}_2(s)]$ is used, where $\mathbf{K}_1(s)$ is the transfer function of the prefilter; $\mathbf{K}_2(s)$ is the transfer function of the feedback controller. Tracking error is defined by the expression **e**=**r**-**y**.

Figure 2. Structural scheme of the two-degrees-of-freedom robust controller: $\mathbf{W}(s)$ is the transfer function of the plant with actuator and measuring system; $\mathbf{K}_1(s)$ is the transfer function of the correction filter or prefilter; $\mathbf{K}_2(s)$ is the transfer function of the feedback controller; $\Phi(s)$ is the transfer function of the closed-loop system; w is disturbance; z is the output vector for assessment of the quality of stabilization process; u is the vector of feedback signals; y is the observed output vector; e is the error vector; $\mathbf{T}(s)$ is the transfer function of the reference system

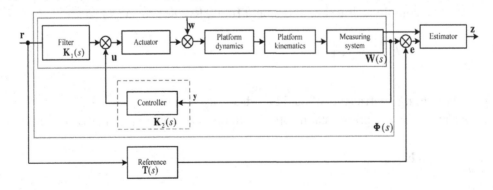

The problem of the two-degrees-of-freedom controller design can be formulated in the following way (Skogestad & Postlethwaite, 2001).

$$\left\| \begin{bmatrix} \mathbf{W}_z^w \\ [\mathbf{I} - \mathbf{W}\mathbf{K}_2]^{-1}\mathbf{W}\mathbf{K}_1 - \mathbf{T} \end{bmatrix} \right\|_\infty \leq \gamma,$$

where $\mathbf{T}(s)$ is the transfer function of the reference system.

AUTOMATED DESIGN OF ROBUST INERTIALLY STABILIZED PLATFORMS WITH INSTRUMENTATION

The design of the intelligent robust control system includes specific procedures such as the development of mathematical models, synthesis of the robust controller, analysis of stability, performance, and robustness. The most important procedures are systems of expert assessments and optimization procedures by means of genetic algorithms (Sushchenko, 2017). The process of the design of the robust inertially stabilized platform with instrumentation is illustrated by the block-scheme represented in Figure 3.

Figure 3. Process of designing robust inertially stabilized platform with aviation instrumentation

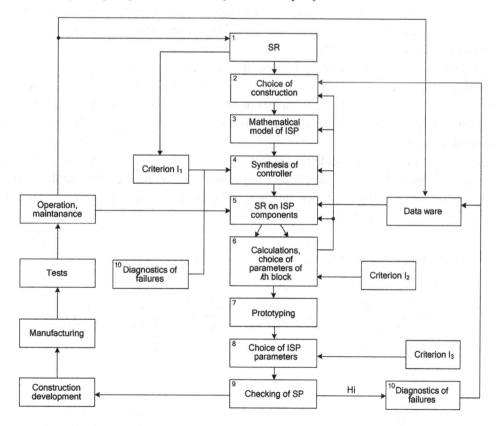

The stabilization plant is a platform with observation instrumentation such as direction finders, photo cameras, laser scanners, and so on. The stabilization and control of platform motion are implemented by means of servo systems, which consist of inertial sensors, controllers and actuators. The first steps of the design of the platform stabilized in the inertial space include the choice of the functional and kinematical schemes based on the specification of requirements and experience of the previous researches. The next step is the development of mathematical models including nominal and parametrically disturbed models, and also models of disturbances. The most important stage of the design of the intelligent automatic control system is the synthesis of the robust controller using expert assessments and optimization procedures based on the genetic algorithms. The further design procedures include calculations, prototyping, design and technological developments, manufacturing of the experimental samples and tests, which are carried out for separate components of the control system and a system as a whole.

The presented research deals with 2 – 4 blocks of the scheme represented in Figure 3. The main goal of the carried our researches is the creation of computer-aided procedures directed to the design of the intelligent robust inertially stabilized platform with aviation instrumentation.

Systems of the researched type have some features, which complicate procedures of analysis and synthesis. These features are complex multi-channel structure of a system and control laws respectively, the significant quantity of operating modes, the high order of a system (a set of differential equations representing a system's model), and rigid requirements to the quality of the transient processes. Structure of the robust platform software is represented in Figure 4.

Figure 4 uses following notations such as GD – gyroscopic devices including a sensitive element (SE), angle transmitter (AT), torque moment (TM); accelerometers (A); a gimballed platform (GP) including AM (actuating mechanism), EM (electromotor), reducer (R); control unit (CU); electronic devices (ED) including digital-analog converter (DAC), analog-digital converter (ADC), and pulse-width-modulator (PWM); sensor of relative motion (SRM).

A specific feature of a process of the optimal design of complex multi-loop systems is the existence of both global and local extremes. Such a situation is caused by limits imposed on the space of design parameters. In some cases, the necessity to carry out repeated procedures of the parametrical optimization or the structural synthesis can be arising. Solving this problem requires the usage of artificial intelligence design procedures.

In the general, creation of robust platforms with aviation instrumentation requires some artificial intelligence procedures such as the statement of the problem of optimal design; analysis of requirements given to a system; forming objective and penalty functions; choice of the optimization method;

Figure 4. Structure of the robust platform hardware

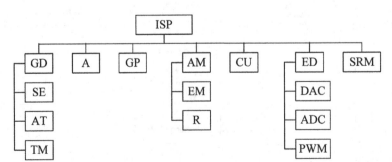

and analysis of the design results. The design process accompanied with creation of the mathematical description including the full mathematical model taking into consideration all nonlinearities inherent to real systems, linearized model in the state space, models of disturbances. Creation of computer-aided design procedures with elements of artificial intelligence requires the usage of the complex of models with the different properties.

Design of any complex dynamic object is based on an optimization criterion. Designing modern stabilized platforms with instrumentation it is necessary to take into consideration that these objects operate under the influence of structural parametric and external coordinate disturbances. In this case, it is convenient to use quality indices of robust systems such as H_∞-norm of the function of the sensitivity of the closed-loop system. It should be noted that design of the robust control systems in general and robust stabilized platforms in particular by means of H_2/H_∞-parametrical optimization and H_∞-structural synthesis is based on the frequency methods, state space methods and elements of artificial intelligence. These methods are characterized by complex transformations and calculations. Therefore, they require the usage of computer-aided design procedures based on software environment such as MATLAB. The greatest advantages of MATLAB are the simplicity of the user interface, usability, and a great number of specialized calculating procedures directed to the design of robust control systems.

Computer-aided design procedures of intelligent robust systems of control of platform motion include methodological, mathematical, algorithmic, and program supports. Improvement of the stabilized platforms can be implemented in two directions including modernization of operating systems and design of new perspective systems. It is convenient to implement modernization of systems of the researched type on the basis of the vector robust parametric optimization. Design of perspective systems requires the usage of the robust structural synthesis.

Methods of computer-aided design based on the vector robust parametric optimization provide determining of the controller sought parameters including the adjustable coefficients. Methods of computer-aided design based on the robust structural synthesis provide the determination of the structure and parameters of the robust controller. These methods foresee design of both one-degree-of-freedom and two-degree-of-freedom robust controllers.

The block-schemes, which explain the above-mentioned methods, are represented in Figure 5 and Figure 6.

Block-schemes represent basic artificial intelligence procedures and their interconnections. The procedures implement the creation of mathematical models, choice of the optimization criteria, bounds imposed on the optimization task, and interactive process of searching an optimal solution taking into consideration the technical requirements. The computer-aided design of the robust stabilized platforms includes the following stages.

1. The problem statement of the optimal design.
2. Creation of the full mathematical models taking into consideration nonlinearities inherent to real systems.
3. Creation of the linearized mathematical model in the state space.
4. Analysis of the technical requirements and forming objective and penalty functions.
5. Creation of mathematical models of disturbances.
6. Choice of the optimization method.
7. Creation of an algorithm of robust platform design, which is adapted to MATLAB.
8. Simulation and analysis of the obtained results.

Figure 5. The method of the computer-aided design based on the vector robust optimization

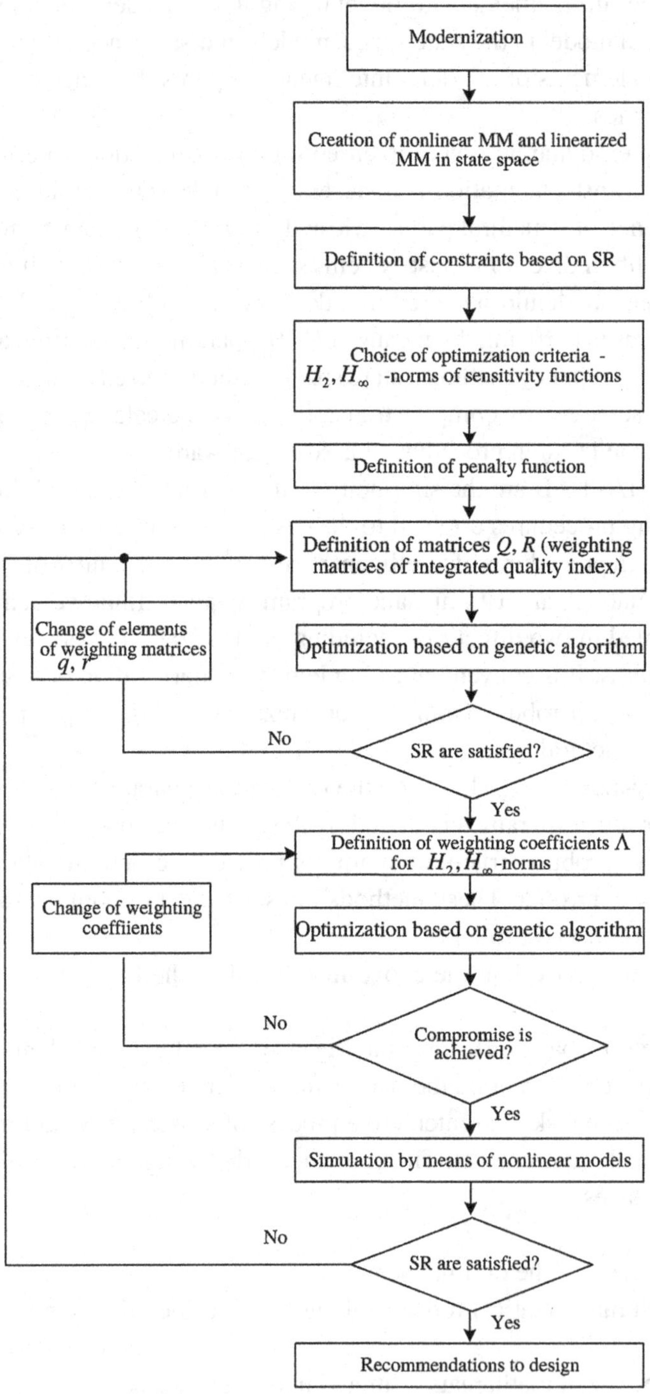

Figure 6. The method of the computer-aided design based on robust structural synthesis

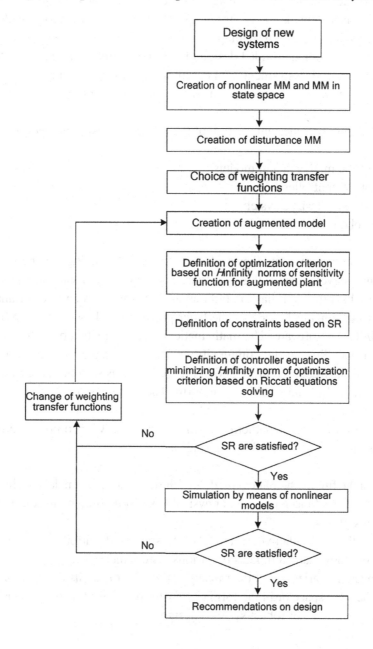

The proposed conception of computer-aided design of robust platforms foresees the development of the following algorithms.

1. The algorithm of the computer-aided design of the continuous system for the stabilization of a platform based on the vector robust parametrical H_2/H_∞-optimization.
2. The algorithm of the computer-aided design of the discrete system for the stabilization of a platform based on the vector robust parametrical H_2/H_∞-optimization.

3. The algorithm of the computer-aided design of the one-degree-of-freedom system for the stabilization of a platform based on the robust structural H_2-synthesis taking into consideration the method of the mixed sensitivity and loop shaping.
4. The algorithm of the computer-aided design of the two-degree-of-freedom system for the stabilization of a platform with instrumentation based on the robust structural H_2-synthesis taking into consideration the method of the mixed sensitivity and loop shaping.

The represented system provides automation of the following design procedures:

1) Development of the mathematical models.
2) Synthesis of robust controllers.
3) Simulation of the synthesized system.
4) Analysis of the obtained results.

The basic design procedure is synthesis of a controller. Depending on the type of the problem (modernization or new development), this procedure is based on the robust parametrical optimization or the robust structural synthesis. It has the tight connections with procedures of analysis, simulation, and development of mathematical models. The latter procedure is based on a combination of analytical methods and MATLAB software. The mathematical models create a basis for all above-mentioned design procedures. Depending on the complexity level, mathematical models can be represented by the two-level hierarchical structure represented in Figure 7. In this structure, the first level corresponds to the mathematical model of the stabilized platform. The second level corresponds to mathematical models of platform units.

The procedure of development of mathematical models necessary for computer-aided design of robust stabilized platforms must include the following stages:

1) Development of the full mathematical models of the designed system devices in the analytical form on the basis of classical mechanical and physical laws taking into consideration all nonlinearities inherent to real systems.
2) Development of the full mathematical model of the stabilized platform based on models of its unit.
3) Development of mathematical models of various disturbances.
4) Simplification and linearization of the models of the system and its devices for the representation of models in the state space or in the form of transfer functions for the further implementation of the automated procedures of robust systems design.

Figure 7. The hierarchical structure of mathematical models

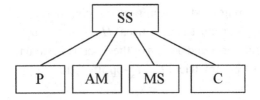

5) Creation of the models of the closed-loop and open-loop stabilization systems in the form suitable for the automated design taking into consideration structural connections of the system's units by means of MATLAB functions.

The mathematical support of the system of the computer-aided design of the stabilized platform with instrumentation must include the following models.

1. The mathematical model of the continuous stabilization system and its units taking into consideration all nonlinearities inherent to real systems.
2. The linearized mathematical model of the continuous stabilization system.
3. The mathematical model of the discrete stabilization system and its units taking into consideration all nonlinearities inherent to real systems.
4. The mathematical model of the discrete stabilization system and its units taking into consideration all nonlinearities inherent to real systems.
5. Mathematical models of disturbances.

All the above-stated models are implemented in MATLAB. The full mathematical models taking into consideration nonlinearities inherent to real systems are developed in Simulink.

The automated design of the robust stabilized platform with instrumentation requires tight connections between procedures of the development of mathematical models, synthesis, simulation, and analysis, as it is shown in Figure 8.

Accuracy of the system of the researched type can be estimated by the method of comparing amplitude-frequency characteristics.

The designed procedure of analysis must implement (Sushchenko, 2017):

1) Analysis of the functionality of the designed system and its correspondence to the given requirements.
2) Research of influence of the structured parametrical and external coordinate disturbances on characteristics of the designed system.
3) Assessment of the stability, dynamical and static errors of the synthesized control systems.
4) Making a decision about the acceptability of the designed system.

Figure 8. Interconnections of design procedures for development of the robust stabilization system

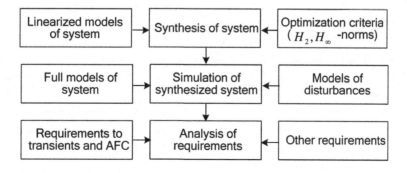

Design of systems of the researched type is a complex problem as it is solved in conditions of uncertainty. This causes the necessity to use heuristic approaches during the implementation of the proposed methods of the automated design. Usually, decision-making based on the experience and intuition of the researcher is combined with the possibility of quick checking its correctness by means of the computer technique. The new approach is using artificial intelligence functions, which can realize heuristic methods. The block-scheme of basic designed procedures for development of stabilized platforms is represented in Figure 9.

Taking into consideration complexity of determination of the optimal solution for tasks of the researched type and also rigid requirements to time and cost of designed works, it is the most efficient to use the interactive mode of the design. This mode combines automated design procedures and artificial intelligence functions developed by means of MATLAB software. The artificial intelligence functions implement procedures, which are difficult for formalization and automation such as decision-making, estimation of results, changing and introducing conditions and data for the further design process.

Usage of the proposed methods of automated design increases the efficiency of design process due to decreasing design time, improving quality of design works due to decrease of engineering errors and also a decrease of prototype and experimental samples due to the possibility to change prototyping and tests by means of simulation.

ANALYSIS OF MODERN COMPUTING FACILITIES AND OPTIMIZATION METHODS

Design of robust systems requires complex calculations and transformations of matrix transfer functions. It should be noted that the MATLAB system includes special toolboxes directed to the implementation of procedures for the design of robust systems. Control System Toolbox is assigned for modeling, analysis, and synthesis of control systems of the wide class. Advantages of this toolbox are the possibility to use both traditional frequency methods and methods of the modern control theory. Control System Toolbox includes a great number of program realizations of algorithms for control systems analysis and synthesis. Optimization Toolbox provides the possibility to choose an optimization method taking into consideration the features of the concrete optimization problem. As a rule, robust synthesis of control systems it is convenient to implement on the basis of the simplex Nelder-Mead method or the genetic algorithm. Namely, genetic algorithms can be used in artificial intelligence procedures of the automated design. The powerful instrument of robust systems development is Robust Control Toolbox, which provides complex calculations and transformations necessary for parametrical optimization and structural synthesis on the

Figure 9. Basic design procedures

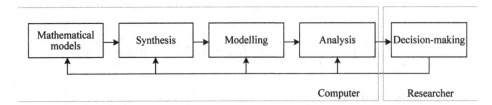

basis of H_2, H_∞-norms. For analysis of the synthesized system, it is necessary to use models taking into consideration nonlinearities inherent to real systems. MATLAB environment has wide possibilities for the development of such models using Simulink Toolbox.

One of the important stages of the robust system's design is the choice of the optimization method. The most widespread optimization methods are searching by the method of the gold section, method of the quadratic approximation, Nelder-Mead method, method of the quickest descent, Newton method, conjugate gradient method, model hardening method, and genetic algorithm (Yang, Cao & Chung et al., 2005). The method of the gold section is used for minimization in limits of some given interval under the condition of the objective unimodal function. The basic feature of the method of quadratic approximation is the approximation of an objective function by the quadratic function. The Nelder-Mead method is used for minimization of a multi-variable objective function when the method of the gold section and the method of quadratic approximation cannot be used.

In the general case, it is necessary to change initial conditions and to determine the global minimum among all local minima. Such a situation requires using artificial intelligence computing procedures, which can be developed on the basis of the genetic algorithm. It represents a method of the controlled search based on modeling of evolution-selection processes in direction of survival the best individual. Genetic operators deal with population individuals during some generations with the goal of the best improving. Individuals from the possible solutions can be considered as chromosomes and are represented as a string of binary codes. The genetic algorithm allows determining a global minimum even in the case when the objective function has some extremes including maxima and minima.

The problem of optimal design of any system is characterized by limits on design parameters. The most widespread approaches to determination of such limits are Lagrange multiplier method and method of penalty functions. The more practical importance has the method of penalty functions. This method can be applied for many optimization problems including bounds in the form of both equalities and inequalities. This method is effective for optimization problems with fuzzy and free constraints. The method of penalty functions is implemented in two stages. The first stage includes the determination of the new objective function with components, which take big values if the given constraints are not satisfied. At the same time, the objective function is not changed if the given constraints are true during the optimization process. The second stage minimizes a new objective function by means of the optimization method, which is used for solving optimization problems without bounds.

ELEMENTS OF ARTIFICIAL INTELLIGENCE ON PROCEDURES OF ROBUST PARAMETRIC OPTIMIZATION

As stated above, the robust parametric optimization can be successfully implemented on the basis of combined criterion "performance-robustness" (Tunik, Rye, & Lee, 2001). The combined optimization criterion includes indices of the robustness of the nominal system and the system disturbed by the parametric structured disturbances.

Then the combined criterion becomes (Tunik & Sushchenko, 2013)

$$J_{H_2/H_\infty} = \lambda_2^{\text{nom d}} \| \Phi_{S1}(\mathbf{K}, \mathbf{x}, \mathbf{u}, j\omega)^{\text{nom d}} \|_2 + \lambda_2^{\text{nom s}} \| \Phi_{S2}(\mathbf{K}, \mathbf{x}, \mathbf{u}, j\omega)^{\text{nom s}} \|_2 +$$

$$\lambda_\infty^{\text{nom}} \| \Phi_T(\mathbf{K}, \mathbf{x}, \mathbf{u}, j\omega)^{\text{nom}} \|_\infty + \sum_{i=1}^{n} \lambda_{2_i}^{\text{par d}} \| \Phi_{S1}(\mathbf{K}, \mathbf{x}, \mathbf{u}, j\omega)_i^{\text{par d}} |_2 + \tag{12}$$

$$\sum_{i=1}^{n} \lambda_{2_i}^{\text{par s}} \| \Phi_{S2}(\mathbf{K}, \mathbf{x}, \mathbf{u}, j\omega)_i^{\text{par s}} \|_2 + \sum_{i=1}^{n} \lambda_{\infty i}^{\text{par}} \| \Phi_T(\mathbf{K}, \mathbf{x}, \mathbf{u}, j\omega)_i^{\text{par}} \|_\infty + PF,$$

where $\| \Phi_{S1}^{\text{nom d}} \|_2, \| \Phi_{S2}^{\text{nom s}} \|_2, \| \Phi_{S1_i}^{\text{par d}} \|_2, \| \Phi_{S2_i}^{\text{par s}} \|_2$ are H_2-norms of matrix functions of sensitivity of the closed-loop system such as nominal and disturbed by the parametric structured disturbances for both deterministic and stochastic situations; $\| \Phi_T^{\text{nom}} \|_\infty, \| \Phi_{T_i}^{\text{par}} \|_\infty$ are H_∞-norms of matrix functions of the complementary sensitivity of the closed-loop systems such as nominal and disturbed by the parametric structured disturbances; $\lambda_2^{\text{nom d}}, \lambda_2^{\text{nom s}}, \lambda_\infty^{\text{nom}}, \lambda_{2_i}^{\text{par d}}, \lambda_{2_i}^{\text{par s}}, \lambda_{\infty_i}^{\text{par}}$ are weighting coefficients of respective norms; n is the quantity of models of the system disturbed by parametric structured disturbances; PF is the penalty function, which provides fulfilment of the condition of system stability during optimization process; $\mathbf{K}$ is the vector of controller parameters.

Weighting matrices for determination of optimization criterion (12) in the deterministic situations are chosen on the basis of Gramian of controllability of the balanced model.

Usage of H_∞-norm of parametrically disturbed models in the optimization criterion guarantees a definite non-sensitivity of the synthesized system to parameter changes in a range of permissible values. It is known that requirements to control performance and robustness are mutually conflicting (Kwakernaak, 1993). Therefore, the problem of H_2/H_∞-optimization of the stabilized platform is the search of a compromise between performance and robustness of the system. This compromise can be achieved by means of using the combined optimization criterion with changing weighting coefficients (1).

The goal of the robust parametrical optimization is the minimization of the optimization criterion (12) for various combinations of numerical values of system parameters. H_2/H_∞-optimization is based on the principle of the assured result (Balandin & Kogan, 2007), which does not depend on the influence of external disturbance. It is determined by H_∞-norms from disturbance to the system input

$$\frac{\| \Phi(s)\mathbf{w}(s) \|_\infty}{\| \mathbf{w}(s) \|_\infty} < \gamma. \tag{13}$$

In (13) $\Phi(s)$ is the matrix transfer function of the closed-loop system; $w(s)$ is the transfer function of disturbance; γ is a small number.

Consider features of the robust H_2/H_∞-optimization by the criterion (12). This problem belongs to "hard" tasks of control theory. Such problems are non-convex and *NP*-hard (Polyak & Shcherbakov, 2005). The latter property (*NP* – non-deterministic polynomial hard) means that problem solution including n output data cannot be obtained on Turing machine in for a time of the order $O(_n^k)$, where k is an arbitrary constant, which does not depend on the output data. Therefore it is impossible to obtain the accurate solution of the optimization problem. It should be noted that solution in some cases it is simply impossible to obtain. So, refusal from the search of the accurate solution leads to the necessity to search a solution, which satisfies input data of the previous design. Nevertheless, this solution is effective from the point of view of the researcher.

The optimization criterion (12) represents the transformation of the set of scalar (local) criteria in the single global criterion. It should be noted that the scalar criteria are conflicting. It is known (Kwakernaak, 1993) that minimization of the performance criterion (H_2-norm of the sensitivity function **S**) leads to maximization of the robustness criterion (H_∞-norm of the complementary sensitivity function **T**). Moreover, performance criteria based on H_2-norms of the sensitivity function **S** for stochastic and deterministic situations are also conflicting. Really, minimization of this norm for the deterministic case means the decrease of H_2-norm impulse transient characteristic of the closed-loop system. For the stochastic case, this means the decrease of the variance of the error at the output of the closed-loop system during the influence of the white noise. This leads to the decrease of the effective bandwidth of the closed-loop system and to increase of H_2-norm of the impulse transient characteristic.

The optimization criterion (12) can be represented in the scalar form by means of weighting coefficients λ_i, which in the theory of multi-criteria optimization are called coefficients of importance of local criteria. It is convenient to consider the optimization criterion H_2/H_∞ (17) from the point of view of multi-criteria optimization.

Taking into consideration contradictory local criteria of the optimization criterion (12), they can be represented in the following form (Tunik & Sushchenko, 2013)

$$J_{H_2}^d = \begin{bmatrix} \lambda_2^{\text{nom d}} & \lambda_{21}^{\text{par d}} & \cdots & \lambda_{2n}^{\text{par d}} \end{bmatrix} \begin{bmatrix} \| \Phi_{S1}(\mathbf{K}, \mathbf{x}, \mathbf{u}, j\omega)^{\text{nom d}} \|_2 \\ \| \Phi_{S1}(\mathbf{K}, \mathbf{x}, \mathbf{u}, j\omega)_1^{\text{par d}} \|_2 \\ \cdots \\ \| \Phi_{S1}(\mathbf{K}, \mathbf{x}, \mathbf{u}, j\omega)_n^{\text{par d}} \|_2 \end{bmatrix}; \tag{14}$$

$$J_{H_2}^s = \begin{bmatrix} \lambda_2^{\text{nom s}} & \lambda_{21}^{\text{par s}} & \cdots & \lambda_{2n}^{\text{par s}} \end{bmatrix} \begin{bmatrix} \| \Phi_{S2}(\mathbf{K}, \mathbf{x}, \mathbf{u}, j\omega)^{\text{nom s}} \|_2 \\ \| \Phi_{S2}(\mathbf{K}, \mathbf{x}, \mathbf{u}, j\omega)_1^{\text{par s}} \|_2 \\ \cdots \\ \| \Phi_{S2}(\mathbf{K}, \mathbf{x}, \mathbf{u}, j\omega)_n^{\text{par s}} \|_2 \end{bmatrix}; \tag{15}$$

$$J_\infty = \begin{bmatrix} \lambda_\infty^{\text{nom}} & \lambda_{\infty 1}^{\text{par}} & \cdots & \lambda_{\infty n}^{\text{par}} \end{bmatrix} \begin{bmatrix} \| \Phi_T(\mathbf{K}, \mathbf{x}, \mathbf{u}, j\omega)^{\text{nom}} \|_\infty \\ \| \Phi_T(\mathbf{K}, \mathbf{x}, \mathbf{u}, j\omega)_1^{\text{par}} \|_\infty \\ \cdots \\ \| \Phi_T(\mathbf{K}, \mathbf{x}, \mathbf{u}, j\omega)_n^{\text{par}} \|_\infty \end{bmatrix}. \tag{16}$$

Using expressions (18) – (20), it is possible to rewrite the criterion (17) in the form

$$J_{H_2/H_\infty} = J_{H_2}^d + J_{H_2}^s + J_\infty + PF. \tag{17}$$

Based on the Pareto principle, denote optimal solution of the problem of design controller with a vector of adjustable coefficients as $\mathbf{K}_p$. In accordance with the property of the *NP*-complexity, the strict solution of the above-mentioned problem can be absent. It is convenient to find the solution acceptable from the point of view of requirements to the designed controller. This solution is compromised from the point of view of the researcher of the system. Such a solution can be considered as some "engineering" analog of the Pareto optimal solution. The search of such a solution is implemented in the space of coefficients, which are determined by vectors λ_d, λ_s and λ_∞, and also elements of the weighting matrices $\mathbf{Q}$ and $\mathbf{R}$ in H_2-norms calculated for local criteria (14) and (15). Denote $\Lambda = [\ \lambda d \ \lambda s \ \lambda\infty\ ]^T$ Then the problem of design of H2/H∞ controller for stabilization system can be formulated in the following way (Tunik, Sushchenko, 2013)

$$\mathbf{K}_p = \arg \min_{\mathbf{K}} {}_p J_{H_2/H_\infty} (\mathbf{K}, \mathbf{Q}, \mathbf{R}, \rangle, \mathbf{x}, \mathbf{u}, j\acute{E}), \tag{18}$$

$$\mathbf{K} \in D, \ D: \mathrm{Re}[\mathrm{eig}_i (\mathbf{I} + \mathbf{L}(s))] < 0, \ i \in 1, ..., i_0, \tag{19}$$

$$x_i < x_{i0}, \ i = 1, ..., n_0; \ u_j < u_{j0}, \ j = 1, ..., m_0, \tag{20}$$

here $\mathbf{L}(s) = \mathbf{P}(s)\mathbf{W}(s)$; $\mathbf{P}(s)$ is the matrix of controllers; $\mathbf{W}(s)$ is the matrix of transfer functions of the plant taking into consideration actuator and measuring system; D is area of stability in the space of parameters; i_0 is an order of the set of differential equations in the Cauchy form; $n_0 \times 1$ is dimension of the state vector; $m_0 \times 1$ is dimension of control vector.

Constraints (19) are determined by conditions of stability of the closed-loop system. Constraints (20) are determined by the technical requirements to the controller. The symbol $\min_p$ denotes "non-strict" minimum, which is defined by the concrete requirements to the designed system.

The process of search of a solution of the problem (18) represents a procedure of the repeated execution of minimization of the criterion (17) by one of the known optimization methods. In this case, it is convenient to use the genetic algorithm. The advantage of this method is the possibility to find the global minimum.

The artificial intelligence procedure of the vector optimization is divided into two stages. At the first stage a solution acceptable from the point of view of the technical requirements to the controller. At this stage components of the vector Λ are believed to be unit values. For the fixed values of matrix elements Q and R the procedure of minimization by means of one of above-mentioned criteria is carried out. After every minimization procedure checking of constraints (20) is carried out. If the constraints are not satisfied, an increase of weighting coefficients qi (for the concrete component $xi_>xi_{0)}$ and rj (for $uj_>ui_{0)}$ is implemented. The procedure is repeated until non-equalities (20) will be satisfied. Fulfilment of condition (20) for all xi and uj means that the solution is permissible from for the technical requirements given to the system.

At the second stage of the vector optimization, variations of the coefficients of weighting coefficients Λ for the fixed values of elements of matrices Q and R obtained at the previous stage are determined. A compromise between the criteria is determined on the basis of some advantages connected with operation conditions and construction features. For example, the advantage is given to the criterion $J_{H_2}^s$, if the operation is carried out in conditions of intensive random disturbances. In other situations, it is desired to decrease J_∞ for significant parametrical disturbances. To find the desired compromise it is sufficient to implement some cycles of minimization procedure execution. This process is accompanied with variation of weighting coefficients (components of the vector Λ).

In accordance with (19) during parametric optimization and search of the optimal vector $\mathbf{K}^*$ it is necessary to keep stability of the closed-loop system in the process of variation of controller parameters. Therefore, the criterion (17) includes the penalty function, which provides arrangement of poles of the closed-loop system in the left semi-plane of the complex variable. To determine the penalty function, it is necessary to check arrangement of system poles in the section of the semi-plane of the complex variable, which satisfies the criterion of system stability.

Quality of stabilization depends on the mutual arrangement of the system poles. Arrangement of poles is defined by parameters η, ξ, and variability of the system $\mu = \text{tg}\psi$, a parameter η is called a level of stability (Pupkov, 2004). To decrease system error in for time t_p to m percents from initial values it is necessary to provide the arrangement of system poles in the left semi-plane of the complex variable on the distance $|\eta_0| = \dfrac{\ln 0,01m}{t_p}$ from the imaginary axis. The oscillatory of the transient process is determined by the presence of complex-conjugated roots $\alpha \pm j\beta$. The index of oscillatory is defined as the ratio of the imaginary part of the root, which characterizes the angular rate of oscillations, to the real part of a root. Attenuation of the oscillations is characterized by the level of amplitude decrease in for one period $f = 1 - e^{-\frac{2\pi}{\mu}}$ (Pupkov, 2004).

APPLICATION OF GENETIC ALGORITHM

The procedure of genetic algorithm consists of next steps (Booker, 1987).

1. It is necessary to create the initial population of individuals of size m ($m < N$), where N is a dimension, in search space E^N. The initial population is usually created in a random way in a symbol form.

2. Then each coordinate of the ith vector $X_i = [x_{1i} \, x_{21} \, \dots \, x_{li}]$, $i \in \overline{0, \mu}$ is transformed $X_i = [x_{1i} \, x_{21} \, \dots \, x_{li}]$; $i \in 0, \mu$ from the symbol into decimal form and the fitness function for each coordinate point $f_i(X_i)$, $i \in \overline{0, \mu}$ is calculated.

3. After that, it is necessary to estimate the population degeneracy. The population degeneracy is determined as a difference between fitness function maximum $f^{\text{max}+}$ and minimum $f^{\text{min}-}$. If the condition

$$\left| f^{\text{max}+} - f^{\text{min}-} \right| \le e \left| f^{\text{max}} - f^{\text{min}} \right|$$

is satisfied (e is a sufficiently small number) the population degenerates into the point corresponding to problem solving. Otherwise, the next step is carried out.

4. Here the least adapted individuals are deleted taking into account their fitness function value, where c is elimination coefficient (usually it equals 0.1). The rest $(1-\rho) \cdot \mu$ individuals compose the new parental group that is used for descendant generation (new coordinate points).

5. From the parental group, equiprobable individuals are selected for parental couples, to which correspondingly the genetic operators are applied. As a result of genetic operations, we obtain descendants (new coordinate points). Obtained descendants are set in the initial population and they are valued at fitness function.

6. The algorithm goes to 3rd step beginning a new evolution stage.

The genetic algorithms search the global extreme on the basis of the probabilistic approach. Therefore it is expedient not to talk about a global extremum but about the best-achieved solution in the accepted search range. The success in genetic algorithm procedure is provided first of all with the collective search idea, i.e. the search provided by means of population of searching points and genetic operators taken from nature. The genetic operators affecting with some probability on parental chromosomes provide from the one side the information transfer to descendants about population state and from the other side – support the sufficient level of changeability, this factor retains the algorithm's searching ability. The genetic algorithms searching ability to a considerable extent depends on the population size. It is obvious that the bigger population size, the higher approximation probability to the searched global extreme. However, in practice, the population size is bounded by computer technology opportunities and keeps in range 10 … 500 individuals (Kochenderfer & Wheeler).

One of genetic algorithms important peculiarities is that no one of genetic operators (crossover, mutation, inversion) during generation process relies on information about local relief of fitness function surface (Kochenderfer & Wheeler). The descendant formation happens in a random manner and there is no guarantee that the found solutions will be better than the parental ones. Therefore, during the evolution process, one can meet the "unsuccessful" descendants which extent the fitness function call number and thereby the global extreme search time. In fact, the genetic algorithms have mainly the particularized application in neural network technologies for multi-parametric problems solution. However, development of the simple superficial conception genetic algorithms requires considerable efforts in order to adapt them to a certain problem. First of all, adaptation is required in genetic operator application probability.

In control system problems intended to regulate deterministic disturbances an integral performance criterion is used as a fitness function, calculated at a transient process time interval and requiring a considerable calculation capacity. For such problems, one makes strict requirements to a genetic algorithm concerning numbers of the fitness function. Taking into account the above-mentioned supposition, one proposes the genetic algorithm modification for universal application to problems having comparatively small dimensions. The modified genetic algorithm retains genetic qualities of static searching points of population selection. In order to exclude unsuccessful descendants there realized the local extremes regular search procedure with the usage of deformable polyhedron operators.

Checking of efficiency of two methods can be done using the MATLAB environment. The program is composed of a few files. One of them is fitness function, which includes the model of the stabilization system. The model is created using MATLAB functions *append* and *connect*. During the optimization process the objective and penalty functions are formed, the weighting transfer functions are determined, and compute H_2- and H_∞-norms. We use these calculations in order to evaluate the complex "performance – robustness" criterion, that is the very fitness function, which is the sum of all these factors and norms. Next step is controller optimization using the Nelder-Mead method. In MATLAB the Nelder-

Mead method is used inside *fminsearch* function, which takes as arguments the fitness function and the start point, from which the search begins. So, here we take as a start point the vector with initial values. After running the code we obtain such a set of parameters of the controller such as $k1=0.2986$; $k2=0.0802$; $k3=0.3024$.

Now, we will consider the parametric optimization by means of a genetic algorithm. Unlike the Nelder-Mead method, where we should set the starting point, in genetic algorithm, one should set a number of variables and the initial and final values of the variable. The genetic algorithm also has other parameters, which are intended to modify it for a certain problem. Doing the optimization with the help of genetic algorithm it is expedient to mention that this algorithm is universal, as it does not impose constraints for fitness function type. In addition, it gives us an opportunity to perform the repeating procedures. Otherwise, there are such situations, when one should terminate the algorithm because of such reasons as the achievement of a certain number of populations; the evolution time expiration; the population convergence. First two criteria depend on the problem type, and sometimes there occurs a situation, when the algorithm cannot find the function extreme or when the obtained after some number of populations result satisfies the requirement. Under the population convergence, one means that neither crossover nor mutation operations make the change into algorithm result during a few populations creation. Results of two optimization methods comparing are given in Table 2.

ARTIFICIAL INTELLIGENCE PROCEDURES OF ROBUST STRUCTURAL SYNTHESIS

Basic principles of robust structural synthesis of inertially stabilized platforms with aviation instrumentation are represented in (Sushchenko, 2019). Basic stages of design of two-degrees-of-freedom robust stabilized platforms are represented in (Sushchenko, 2015) based on concepts represented on (Skogestad & Postletwaite, 2001). Taking into consideration the complexity of calculations and transformations of this process, it is necessary to develop computer-aided design procedures for its implementation. The scheme of design of the two-degrees-of-freedom stabilized platform by means of H_∞-synthesis with elements of artificial intelligence is represented in Figure 10.

Table 2. Results of comparison of two optimization methods

Characteristics	Optimization Methods	
	Genetic Algorithm	Nelder-Mead Method
H_2-norm	0.207	0.399
H_∞-norm	0.632	0.793
Settling time (s)	0.59	0.727
Oscillation factor	3.5	2.91
Number of oscillations	3	3
Delay time (s)	0,0542	0,0543
Rise time (s)	0,0315	0,0314

Figure 10. Block-scheme of the algorithm of computer-aided design of two-degree-of-freedom stabilized platform

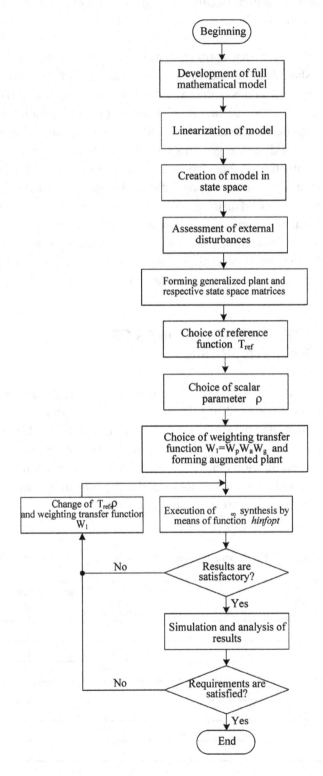

Nowadays, discrete controllers are the most widespread for practical applications. There are two the most known approaches to design of discrete controller (Balas, Chiang & Packard et al., 2008). Schemes of both approaches to designing discrete controllers are shown in Figure 11 and Fugure 12.

The first approach is based on the digitization of the plant including weighting transfer functions (augmented plant). Further, w-transform, and H_∞-synthesis for the continuous controller are implemented. z-representation of the controller can be obtained by means of the inverse w-transform. All the listed operations can be realized by means of the function *dhinfopt*, which belongs to Robust Control Toolbox.

The second approach to the computer-aided design of the discrete robust controller is carried out by means of H_∞-synthesis of continuous systems. The discrete controller can be obtained on the basis of z-transform. Automation of this transformation is implemented by means of function *c2d*. There are some requirements to this approach (Balas, Chiang & Packard et al., 2008). Firstly, Tustin bilinear transform must be used, as H_∞-norm stays invariant in the case of its usage. Secondly, a period of sampling must be in sometimes bigger, then a bandwidth of the stabilization system.

In accordance with simulation results, the design of the discrete controller on the basis of the second approach is more convenient and apparent for a researcher.

ILLUSTRATIVE EXAMPLE

The software of proposed algorithms is grounded on MATLAB possibilities. To create effective design procedures of robust inertially stabilized platforms is possible combining functions of Control System Toolbox and Robust Control Toolbox. The software is created by means of MATLAB programming language using the above-mentioned functions. The finite results include recommendations to design of inertially stabilized platforms. Obtaining control laws is one of the most important results. As a rule, these laws are represented as quadruples of discrete matrices in the state space. This information corresponds to difference equations, which can be easily programmed by means of programming language C. This programming language is widely used for microcontrollers programming. It is possible to transform the obtained results in code directly using the embedded facilities of MATLAB environment.

The hardware of the researched stabilization systems includes servo systems with measuring instruments, actuators, and controllers. The most responsible task is the choice of inertial sensors. It is known that laser gyroscopes are not used for platform stabilization. The fiber-optic inertial technologies are widely used in the systems of payload stabilization. Absence of moving parts, small readiness time, high sensitivity and accuracy are advantages of such gyroscopic sensors. But gyroscopic sensors built on the fiber-optic technologies have significant mass and dimensions due to usage of receiving-transmitting units. At the same time, they have high cost caused by hand operations in manufacturing. Usage of gyroscopic angle rate sensors built on MEMS-technologies is one of the basic trends of modern stabilized platforms design. Such gyroscopic sensors have a wide area of application including stabilization of platforms with information and measuring devices. Basic advantages of such gyroscopic sensors are the simplicity of operation and low cost. But MEMS-gyros have some disadvantages. Their characteristics have some statistical dispersion caused by deviation of manufacturing conditions from given in the technical documentation. Furthermore, processes of separate sensors aging have different rates. Therefore, the gyroscopic sensors of this type require compensation of bias. The above-listed disadvantages

Figure 11. Block-scheme of the first approach to the development of the computer-aided design of discrete stabilization system

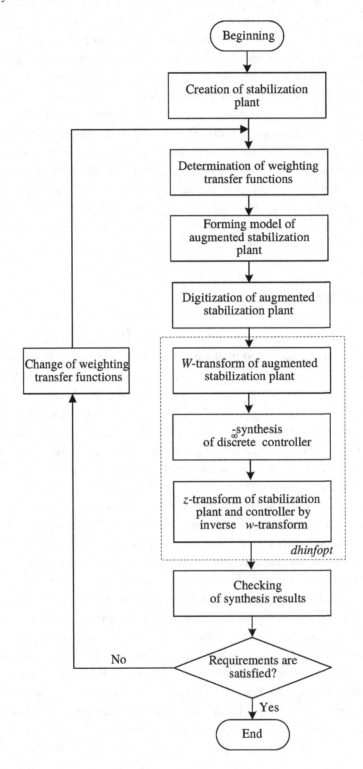

Figure 12. Block-scheme of the second approach to the development of the computer-aided design of discrete stabilization system

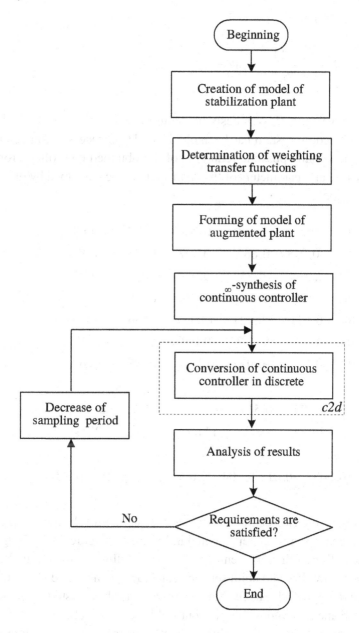

of MEMS-gyros have eliminated in the perspective digital Coriolis vibratory gyro with the metallic resonator. Development of this gyro is implemented by the Aerospace Control Systems Department of the National Aviation University together with Kyiv Automatic Plant (Kyiv, Ukraine).

Consider results of design based on the example of the two-degrees-of-freedom robust inertially stabilized platform. Choice of the weighting transfer functions is one of the most complex steps of the design procedure. It requires decision-making and respectively artificial intelligence approach. To form the augmented stabilization plant, the following transfer functions were used:

$Gs = W_2 G W_1,$

where $W_2=1$, $W_1=W_p W_a W_g$, here

$$W_p = \frac{0,15}{0,1s+1}; \quad W_a = 10\frac{0,4s+24,76}{s+25,17}; \quad W_g=1.$$

As a result of the developed H_∞-synthesis procedure execution the optimal H_∞-controller for the robust two-degrees-of-freedom system has been obtained. This process is characterized by the parameter $\gamma=0,1426$. After maximally possible reduction of the obtained controller (from 10th order to 7th respectively) the structure and parameters of the controller can be described by the following quadruple of matrices in the state space

$$\mathbf{A}_s = \begin{bmatrix}
1,032 & 0,086 & 0,013 & -0,109 & -0,009 & 0,037 & 0,008 \\
-0,028 & 0,627 & -0,048 & 0,669 & -0,197 & 0,025 & -0,009 \\
0,044 & 0,044 & 1,007 & -0,052 & -0,021 & 0,035 & 0,007 \\
-0,256 & 0,245 & -0,083 & 0,215 & 0,196 & -0,252 & -0,001 \\
0,333 & 0,344 & 0,185 & -0,321 & 1,033 & 0,166 & 0,049 \\
-0,097 & -0,203 & -0,052 & 0,247 & -0,046 & 0,828 & -0,013 \\
0,193 & 0,376 & 0,106 & -0,072 & 0,165 & 0,398 & 0,769
\end{bmatrix};$$

$$\mathbf{B}_s^T = \begin{bmatrix}
0,324 & 0,246 & -0,272 & -0,517 & 0,621 & -0,287 & 0,7 \\
-23,28 & -8,63 & -20,75 & 108,3 & -111,7 & 51,9 & -107,4
\end{bmatrix};$$

$$\mathbf{C}_s = [0,006 \quad 0,008 \quad 0,003 \quad -0,01 \quad 0,004 \quad 0,002 \quad 0,001]; \quad \mathbf{D}_s = [0,009 \quad -1,659].$$

Results of modeling of the two-degrees-of-freedom system with the synthesized H_∞-controller for the nominal and disturbed systems are represented in Figure 13, Figure 14, and Figure 15.

Practical implementation of the represented results is as follows. Now the inertially stabilized platforms are widely used to stabilize and point sensors, cameras, antennas and weapon systems operated at the aircraft of the different type [1]. The further progress of the above-listed equipment by accuracy and other operating performances is impossible without stabilization of a base, at which they are mounted. This proves the actuality of the inertially stabilized platforms technology development.

FUTURE RESEARCH DIRECTIONS

Nowadays unmanned aerial vehicles are widely used in many industrial areas. Firstly, the research will be developed in the direction of creating algorithms of designing inertial robust platforms for light unmanned aerial vehicles. These algorithms will be based on the robust parametrical optimization using LMI (linear

Figure 13. Results of two-degrees-of-freedom robust platform modeling under constant external disturbance: the horizontal and the vertical channels

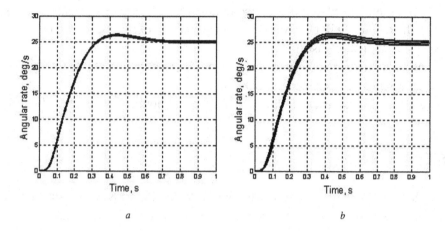

a   *b*

Figure 14. Results of two-degrees-of-freedom robust platform modeling under aerodynamic external disturbance: the horizontal and vertical channels respectively

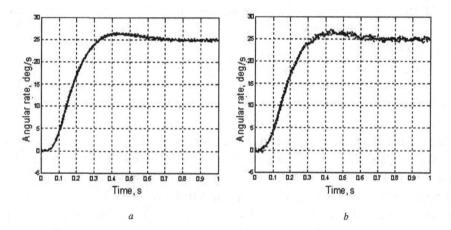

a   *b*

Figure 15. Results of two-degrees-of-freedom robust platform modeling under parametric disturbance such as changing of the inertia of moment: the horizontal and vertical channels respectively

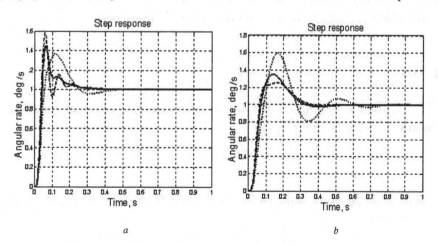

a   *b*

matrix inequalities)-approach. It is planned to create new computer-aided designed procedures, which can be used for the development of stabilization systems operated on the light unmanned aerial vehicles.

Operation of light unmanned aerial vehicles requires using reliable navigation measuring instruments. Secondly, the research will be developed in the direction of the automated choice of structure of redundant measuring navigation instrument based on MEMS-sensors.

Thirdly, it is necessary to widen the area of researches in the direction of maximally possible automation of creating algorithms of information processing.

All the above-stated directions cannot be implemented without elements of artificial intelligence.

CONCLUSION

The analysis of the problem of designing inertially stabilized platforms with instrumentation has been carried out. The problems of synthesis of stabilization systems were researched.

The concept of computer-aided designing robust inertially stabilized platforms with instrumentation was developed. The proposed concept is based on robust parametrical optimization and robust structural synthesis (H_∞-synthesis) taking into consideration elements of artificial intelligence. Such an approach provides an efficient implementation of a set of interactive computer-aided design procedures necessary for the synthesis of the robust controller, analysis of acceptability of designed stabilization system characteristics, and decision making.

Principles of implementation and structure of a system for computer-aided design of robust stabilized platforms are represented. The characteristic of methods, data ware, mathematical supports are described too. The basic principles of forming mathematical models are given. The basic design procedures and their interconnections are researched. The formalized problem statements of computer-aided design of systems of the researched type are represented. The comparative analysis of modern computing facilities and optimization methods acceptable for computer-aided design of robust inertially stabilized platforms has been done. Optimization criteria for the researched problem solution are chosen. Features of robust parametrical optimization including genetic algorithm using are described.

The represented approaches to computer-aided design of robust inertially stabilized platforms are illustrated on an example of designing the two-degrees-of-freedom robust inertially stabilized platform.

It should be noted that developed algorithm can be used both for modernization and creation of new perspective robust inertially stabilized platforms with aviation instrumentation. Obtained results can be useful during the design of robust control systems of the wide class.

REFERENCES

Balandin, D, Kogan, M. (2007) Synthesis of optimal linear-quadratic matrix non-equalities. *Automation and Telemechanics*, 3, 3–18. (in Russian)

Balas, G. J., Chiang, R. Y., Packard, A., & Safonov, M. G. (2010). *Robust control toolbox 3*. User's guide. The Math Works Inc.

Booker, L. (1987). Improving search in genetic algorithms. In *Genetic Algorithms and Simulated Annealing*. Morgan Kaufmann Publishers.

Egupov, I. (2002) Methods of robust, neuro-fuzzy and adaptive control. Moscow, Russia: MSTU named after N.E. Bauman. (in Russian)

Gu, D., Petkov, P., & Konstantinov, M. (2005). *Robust control design with MATLAB*. London, UK: Springer-Verlag.

Hilkert, J. M. (2008). Inertially stabilized platform technology. *IEEE Control Systems Magazine, 28*(1), 26–46. doi:10.1109/MCS.2007.910256

Kochenderfer, M. J., & Wheeler, T. A. (2019). *Algorithms for optimization*. The MIT Press.

Kwakernaak, H. (1993). Robust control and H∞. *Automatica, 29*(2), 255–273. doi:10.1016/0005-1098(93)90122-A

Polyak, B. T., & Shcherbakov, P. S. (2002). *Robust stability and control*. Moscow, Russia: Nauka. (in Russian)

Polyak, B. T., & Shcherbakov, P. S. (2005). Difficult problems of linear control theory. Some approaches to solution. *Automation and Telemechanics*, 5, pp. 7–46. (in Russian)

Pupkov, K. A. (2004). Methods of classical and modern automatic control theory. Vol. 2. In K. A. Pupkov, & N. D. Egupov (Eds.), Statistical dynamics and identification of automatic control systems. Moscow, Russia: MSTU named after N.E. Bauman. (in Russian)

Kim, S., Choi, S., Kim, H., Shin, J., Shim, H., & Kim, H. J. (2018). Robust control of an equipment-added multirotor using disturbance observer. *IEEE Transactions on Control Systems Technology, 26*(4), 1524–1531.

Skogestad, S., & Postlethwaite, I. (2001). *Multivariable feedback control*. New York: John Wiley & Sons.

Solnitsev, R. I. (1991). *Automation of Design of Automatic Control Systems*. Moscow, Russia: Vishaya Shkola. (in Russian)

Sushchenko, O. (2017). Approach to computer-aided design of UAV robust inertial platforms. In *Proceedings of IEEE 4th International Conference on Actual Problems of Unmanned Air Vehicles Developments (APUAVD 2017)*, (pp. 53–57). Kiev, Ukraine. 10.1109/APUAVD.2017.8308775

Sushchenko, O. A. (2015). Design of two-axis robust system for stabilization of information-measuring devices operated at UAVs. In *Proceedings of IEEE 3rd International Conference Actual Problems of Unmanned Aerial Vehicles Developments, (APUAVD-2015)*, (pp. 198–201), Kiev, Ukraine.

Sushchenko, O. A., & Goncharenko, A. V. (2016). Design of robust systems for stabilization of unmanned aerial vehicle equipment. International Journal of Aerospace Engineering. Article ID 6054081.

Tunik, A. A., Rye, H., & Lee, H. C. (2001). Parametric optimization procedure for robust flight control system design. *KSAS International Journal, 2*(2), 95–107.

Tunik, A. A., & Sushchenko, O. A. (2013). Usage of vector parametric optimization for robust stabilization of ground vehicles information-measuring devices. *Proceedings of the National Aviation University*, 4, 23–32. 10.18372/2306-1472.57.5530

Veremey, E. I. (2012). *Introduction in analysis and synthesis of robust control systems*. Retrieved from http//MatLab.exponenta. ru/optimrobust/ book1/ ndex.php

Wu, N., & Yu, J. (2018) Robust controller design of hypersonic vehicles in uncertainty models. In *Proceedings of 3rd International Conference on Electromechanical Control Technology and Transportation (ICECT-2018)*, (p. 288-293). Huazhong, China. 10.5220/0006969302880293

Yang, W. Y., Cao, T. S., Chung, I., & Morris, J. (2005). *Applied numerical methods using MATLAB. New York*. John Wiley & Sons. doi:10.1002/0471705195

Zaliskyi, M., & Solomentsev, O. (2014) Method of sequential estimation of statistical distribution parameters. In *Proceedings of IEEE 3rd International Conference Methods and Systems of Navigation and Motion Control (MSNMC-2014)*, (p. 135-138). Kiev, Ukraine.

ADDITIONAL READING

Anderson, B. D., & Moore, J. B. (1989). *Optimal Control Linear Quadratic Methods*. New Jersey: Prentice Hall.

Burns, R. S. (2001). *Advanced Control Engineering*. Oxford: Butterworth-Heinemann.

Jahn, J. (2011). *Vector Optimization. Theory, Application and Extensions*. Berlin: Springer-Verlag. doi:10.1007/978-3-642-17005-8

Sushchenko, O. A., & Tunik, A. A. (2013). Robust stabilization of UAV operation equipment. In *Proceedings of IEEE 2nd International Conference on Actual Problems of Unmanned Air Vehicles Developments (APUAVD 2013)*, (pp. 176 – 180). Kyiv, Ukraine. 10.1109/APUAVD.2013.6705318

Zhou, K., & Doyle, J. (1999). *Essentials of Robust Control*. New Jersey: Prentice Hall.

KEY TERMS AND DEFINITIONS

Genetic Algorithm: Heuristic algorithm of search for solving optimization problem by means of random variations of parameters using approaches similar to natural selection.

Inertially Stabilized Platform: Electromechanical configurations providing stabilization of a payload in the inertial space.

Robust System: An automated system is called robust if it keeps stability and performance indices in some permissible range in conditions of disturbances influence without using adaptation methods.

Stabilization System: A system, which keeps constant value of a controlled quantity under influence of changed disturbances, is called stabilization system.

System of Computer-aided Design: Automated system realized information technologies of implementation of design functions.

Unmanned Aerial Vehicle: Aircraft without crew on the board is called an unmanned aerial vehicle.

Chapter 11
Correction of the Temperature Component of Error of Piezoelectric Acceleration Sensor

Anatoly Perederko
National Aviation University, Ukraine

ABSTRACT

In the aviation and aerospace industries, systems of artificial intelligence are being intensively implemented. For their effective functioning, it is necessary to provide reliable primary information. Information is collected by primary converters. In various systems of aviation technology, piezoelectric accelerometers are widely used as primary transducers. In particular, they are part of inertial navigation systems. These systems do not need external sources of information and are not affected by external interference. Therefore, they are widely used in aircraft for various purposes. These accelerometers are also used in other aircraft systems. In systems for monitoring the operation of gas turbine engines, systems for registering overloads. However, in conditions of high-temperature differences, shocks, vibrations, and significant accelerations, piezoelectric transducers have inherent flaws that affect the linearity and accuracy in measuring these values. Especially piezoelectric transducers are critical to extreme temperatures.

BACKGROUND

In aviation intelligent systems, primary transducers (sensors) are used, which convert the value of the measured parameters into a proportional electrical signal, since it allows to transmit information over distances and can be easily processed by various mathematical methods. For example, in aviation inertial navigation systems, the parameters of mechanical movement are used: acceleration, speed, movement, are interconnected by certain relations and are mutually definable. Vibration monitoring is performed in the systems for monitoring the operation of aircraft gas turbine engines.

DOI: 10.4018/978-1-7998-1415-3.ch011

The parameters of the mechanical motion listed above: acceleration, speed of movement, movement, are interconnected with each other by certain relations and are mutually definable.

In practice, the measurement of speed and movement by integrating and double integrating the acceleration signal has become widespread. Since the hardware solution to the problem in the opposite direction: differentiation of the movement and speed signal, for determining acceleration, is more problematic than integration since it creates additional noises in the conversion process.

Among the primary transducers used to measure accelerations and vibrations, accelerometers are currently dominant (Sharapov, Musienko, & Sharapova, 2006). Among their diversity, solid-state piezoelectric accelerometers are occupying leading positions. In such accelerometers, a sensitive element that is most commonly used is made from a special composition of piezoceramics, the domain structure of which is subjected to preliminary polarization and has anisotropic properties.

Piezoelectric accelerometers are currently the most common instruments used to measure accelerations and vibrations, which is primarily explained by the simple design, wide frequency, and dynamic ranges, durability characteristics, reliability and stability of parameters. Piezoelectric accelerometers operate on the principle of direct piezoelectric effect: when a force is applied to a piezoelectric element, its geometrical dimensions change, as a result of which, on specially deposited on piezoelectric element electrodes, a charge is generated that is proportional to the applied force. Thus, sensors based on piezoelectric element do not need additional power sources, and the absence of moving parts ensures durability and stability of their work.

ORDEVAL APPROACH

The structural and electrical schematics of the most commonly used in practice piezoelectric sensors of the generator type for measuring acceleration (vibration) are shown in Figure 1.

The structural diagram of such sensors as a sequence of measuring transducers includes: a transducer of the measured mechanical parameter – acceleration a into force F, then piezoelectric element, which converts the force into an electric charge q and an electrical circuit, which is connecting the piezoelectric element with a matching amplifier and converts the charge into voltage φ_0.at the electronic equipment input.

The main disadvantages of piezoelectric transducers made from special piezoceramic materials relate to their thermal properties:

- High pyroelectric coefficient.
- Temperature changes of the piezoelectric coefficients and dielectric permeability.

Figure 1. Structural diagram of piezoelectric sensors

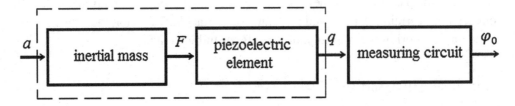

- Temperature hysteresis, which determines the influence of the previous thermal states on the current value of the characteristics.

Also for piezoceramic, there are restrictions on the use by thermal range. If the piezoelectric ceramic material is heated to the Curie point, then the domains will become disordered and the orientation that was obtained as the result of primary polarization will be lost, and the material becomes depolarized. The recommended upper working temperature limit for ceramics is usually located in the middle of the range between $0°$ C and the Curie point. Within the recommended operating temperature range, the temperature corresponding changes in the orientation of the domains are reversible. On the other hand, these changes can create a charge of .displacement and the appearance of an electric field. Also, rapid fluctuations in temperature can generate relatively high voltages, which are also capable of depolarizing the ceramic element.

Therefore, with an external temperature effect on the piezoelectric elements, both the temperature level and the rate of its change relative to the previous value are important.

The temperature value is an important parameter affecting the electrical sensitivity of the sensor since the piezoelectric coefficient depends on it and, possibly, there is also a change in the mechanical parameters determining the coefficient of conversion of mechanical quantities into electrical ones (for example, a change in the mechanical strain). At a constant temperature, this effect on sensitivity theoretically can be ignored.

A change in the temperature of the piezoelectric element entails the emergence of electrical signals even in the absence of external forces caused by vibration or other disturbances. Charges on the electrodes of the piezoelectric element result from the mechanical stresses to which the piezoelectric element is subjected, for example, due to the different expansion of parts and due to the pyroelectric effect.

Materials that are least affected by the temperature influence usually have the least sensitivity when using the direct piezoelectric effect.

Very low-frequency signals (drift), generated by piezoelectric element under the influence of ambient temperature and the pyroelectric effect, are attenuated by the filter of the signal conditioning circuit, which passes only higher frequencies. However, this, in turn, limits the possibility of expanding the bandwidth in the direction of low frequencies.

Therefore, for precision measurements, it is necessary to know the dependence of the sensitivity of the device on temperature.

During the examination of physical processes in the piezoelectric element, it is necessary to consider the anisotropic properties of the latter. Properties such as electrical conductivity, magnetic permeability, thermal conductivity, elasticity, depending on the direction of impact and the structure of the material used.

Let's examine the direct piezoelectric effect: piezoelectric element generates a charge on its electrodes under the action of an external force on it.

The direct piezoelectric effect equation for the vector component of electrical induction in the direction of the residual polarization axis D_z. the piezoelectric element works on the longitudinal piezoelectric effect and the working piezo module is d_{33} (Figure 2) would have the following form:

$$D_z = d_{33}\sigma_{zz} + d_{31}\left(\sigma_{rr} + \sigma_{\theta\theta}\right) + \varepsilon_{33}\varepsilon_z. \tag{1}$$

where:

Figure 2. A piezoelectric element operating on a longitudinal piezoelectric effect

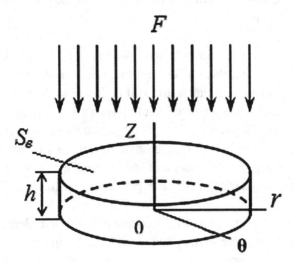

d_{33} and d_{31}- piezo modules (longitudinal and transverse with respect to the axis of residual polarization, respectively);

ε_{33}- dielectrical permeability;

$\sigma_{zz}, \sigma_{rr}, \sigma_{\theta\theta}$.- mechanical stress components by axes Z, r, θ ;

ε_z.– electric field strength.

From formula (1) it is clear that for the correct definition of the piezo module d_{33}. the following conditions must be met:

$\varepsilon_z=0$; $\sigma_{rr}=\sigma_{\theta\theta}=0$; σ_{zz} = const.

Then the piezo module alongside the Z axis (Figure 2):

$$d_{33} = D_z / \tilde{A}_{zz} \cdot, \qquad (2)$$

where $D_z = a/S_e$, $\sigma_{zz} = F/S_0$.

q - charge on piezoelectric element electrodes.

S_e. surface area of the piezoelectric element electrode.

S_0. surface area of piezoelectric element on which external force F acts.

In our case $S_e \approx S_0$.

Then, substituting expressions for D_z and σ_{zz} in (2) we will get:

$$d_{33} = q/F, \text{ or: } q = d_{33}. \qquad (3)$$

From the obtained equation (3) it follows that the generated charge q, which occurs on the piezoelectric element electrodes under the action of the force F, can be influenced by the temperature deviation of the piezoelectric module d_{33}. which is the reaction of the domain structure of the piezoelectric element that occurs when its geometric dimensions change under the action of temperature.

Figure 3. shows the temperature dependence of the piezo module d_{33}.from temperature (Antononenko and Kudzin & Gavshin, 2009). in the temperature range from 20 ° C to 250 ° C (half of the Curie point temperature for the studied piezoceramic material).

The graph in Figure 3, shows that the temperature dependence of the longitudinal relative to the axis of residual polarization of the piezoelectric module d_{33}.is substantial and non-linear.

For a comprehensive consideration of the occurrence of additive error introduced by temperature instability when measuring acceleration (vibration) with piezoelectric element accelerometer, we will analyze the typical measuring tact, which includes: a primary converter, a connecting line, and a charge amplifier.

The equivalent electrical schematic of such a measuring tact is shown in Figure 4.

Charge amplifiers are widely used with sensors whose output signal is an electric charge. Such amplifiers allow the use of connecting cables of great length.

The charge amplifier consists of an operational amplifier with capacitive feedback and converts the charge generated by the sensor into voltage (Levshina,1983). An operational amplifier with a capacitor in the feedback circuit is essentially an electronic integrator integrating the electrical current supplied to its input, which is proportional to the charge generated on the piezoelectric element surfaces covered with electrodes.

The output voltage of the charge amplifier is determined by the formula:

$$\varphi = \frac{q}{C_{oc}} \cdot \frac{j\omega R_{oc} C_{oc}}{1+j\omega R_{oc}C_{oc} + \frac{1}{k}\left[j\omega R_{oc}C_{oc}\left(1+\frac{C_0 + C_k + C_{2E}}{C_{oc}}\right) +1+\frac{R_{oc}}{R_{2E}}\right]}, \cdot \qquad (4)$$

where R_{oc} and C_{oc} amplifier feedback resistance and capacitance.

Figure 3. Temperature dependence of the piezo module d_{33}.of the PZT-19b piezoelectric ceramics

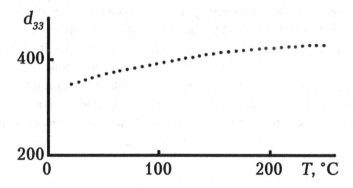

Figure 4. Equivalent Piezo Accelerometer Electrical Circuit with Charge Amplifier

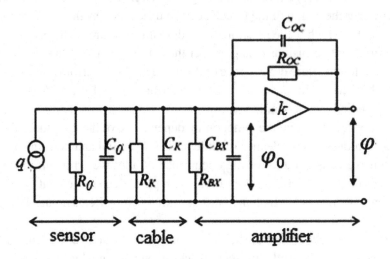

k – Coefficient of amplifier gain.

R_{BX}– Amplifier input resistance.

C_{6x}, C_k, C_0 – The capacitance of the connecting line and the piezoelectric transducer, respectively.

R_o and R_k – Resistance of piezoelectric transducer and connecting line.

In the frequency range $\omega \gg 1/\tau$. where $\tau = R_{oc}C_{oc}$ – is constant of amplifier feedback circuit time, equation (4) takes the form:

$$\varphi = \frac{q}{C_{oc} + \dfrac{C_{oc}}{k} + \dfrac{(C_0 + C_k + C_{2E})}{k}} \cdot \cdot \tag{5}$$

Analyzing equation (5), we can come to the following conclusion: since the resistances and capacitors have very small temperature coefficients of the parameter spread, and the gain of the amplifier is stable in the band of transmitted frequencies, the change in the output voltage φ under the influence of temperature largely depends on the charge q generated by the piezo sensor and also from its capacitance C_0.

Not related to the measured parameter acting on the piezoelectric element, the components of electrical induction generate parasitic signals at the amplifier input, which determines the additive error or zero error of the piezoelectric sensors. Especially this error manifests itself in a changing temperature when measuring parameters of low-frequency accelerations and vibrations.

To take into account the influence of the additive component of the error introduced by the influence of temperature, when measuring the parameters of accelerations and vibrations, it is necessary to have an idea of their dependence. The temperature deviations of the conversion coefficient (see the diagram in Figure 4) may depend on temperature changes of both the physical properties of the converter itself and other components of the circuit.:

- Piezo module.

- Dielectric permeability.
- Isolation resistance.
- The quality of the connection of the piezoelectric element with the substrate and the inertial mass of the accelerometer.

Each of these factors may contribute to the measurement process in the form of an additive component of measurement error caused by temperature instability.

Based on the above, to analyze the relationship between the temperature characteristics of piezoelectric element and coefficient of transmission, we need to examine the transformation function of a piezoelectric sensor, assuming that the sensor contains piezoelectric element operating on a longitudinal piezoelectric effect, that is, the working module is a piezoelectric module d33. In this case, the characteristics of piezoceramic material (d33, ε33 and density - $\tilde{\rho}$. are considered as functions of temperature, and temperature-induced changes of geometric dimensions of other parts, as well as the parameters of the amplifier and cable, are neglected.

Then the coefficient of force-to-voltage conversion, with a direct piezoelectric effect, at the input of the amplifier, as a function of temperature $K_{\varphi 0}(\Delta T)$. can be expressed by the following relation:

$$K_{\varphi_0}\left(\Delta T\right) = \frac{\varphi_0\left(\Delta T\right)}{F} = \frac{q\left(\Delta T\right)R\left(\Delta T\right)\omega}{\sqrt{1 + R^2\left(\Delta T\right)C^2\left(\Delta T\right)\omega^2}}, . \tag{6}$$

Where:

$$q(\Delta T) = d_{33}(\Delta T)\frac{S_5}{S_0} . ;$$

$$C\left(\Delta T\right) = C_0\left(\Delta T\right) + C_: + C_{2E} .$$

$$C_0\left(\Delta T\right) = \mu_{33}\left(\Delta T\right) \bullet \frac{S_5}{h} .;$$

$$\frac{1}{R(\Delta T)} = \frac{1}{R_0(\Delta T)} + \frac{1}{R_:} + \frac{1}{R_{2E}} ; .$$

$$R_0\left(\Delta T\right) = \acute{A}\left(\Delta T\right)\frac{h}{S_5} .$$

Using the conclusions concerning the analysis of the components of equation (5) and equation (3), as well as the fact that the inertial force *F*, which acts on the piezoelectric element as a result of acceleration *a* (Figure 5), is equal to:

$$F = am, \tag{7}$$

where m- inertial mass, and also considering that $\varphi_0 = q/C_0$. the conversion coefficients of the piezoelectric accelerometer on the charge $K_q(T)$ and voltage $K_{\varphi_0}\left(\Delta T\right)$.respectively, as a function of temperature change ΔT have values:

$$\mathrm{K}_q\left(\Delta\mathrm{T}\right) = \mathrm{d}_{33}(\Delta\mathrm{T})\mathrm{m}, . \tag{8}$$

$$\mathrm{K}_{\varphi_0}\left(\Delta\mathrm{T}\right) = \frac{\mathrm{d}_{33}(\Delta\mathrm{T})\mathrm{m}}{\mathrm{C}_0} .. \tag{9}$$

Since in (8) and (9) the mass m does not depend on temperature, we can conclude that:

- The temperature dependence of the conversion coefficient of the piezoelectric accelerometer on the charge $K_q(\Delta T)$.is due to the appearance of a pyroelectric effect and a transverse piezoelectric effect [3] induced by planar temperature stresses in the piezoelectric element.
- The temperature dependence of the conversion coefficient of the piezoelectric accelerometer on the voltage $K_{\varphi_0}\left(\Delta T\right)$.is due to the change in the geometric dimensions and, as a result, the influence on interdependent parameters of capacitance C_0 and dielectric permeability ε_{33} are interconnected by the empirical formula:

$$\frac{\mu_{33}}{\mu_0} = \frac{11,3h}{S_e}C_0, .$$

There

ε_0- Dielectric permeability in vacuum.
h – Piezoelectric element thickness (see Figure 2).

In the future, we will give our calculations relative to the piezoelectric accelerometer with the output by the charge.

A piezoelectric element, changing its geometrical dimensions with a change in temperature, without the influence of an external force F, generates a charge q. The more the linear expansion coefficient of the piezoelectric element differs from the linear expansion coefficients of other elements of the accelerometer design (Figure 5), the more its domain structure will be subjected to mechanical stress and generate an additional charge. This charge is nothing more than as a component of the additive measurement error of a piezo accelerometer caused by temperature effect.

To some extent, the leveling effect of temperature on the occurrence of mechanical stresses in the piezoelectric element can be achieved by using the compensation element shown in the design of the accelerometer in Figure 5a. This element, in the form of a washer, is located between the piezoelectric element and the inertial mass and pulled into a common package by bolt 5. To reduce the dependence of

Figure 5. The design of the piezoelectric accelerometer: 1- substrate, 2- piezoelectric element, 3- compensation element, 4- inertial mass m, 5- bolt, h_0 piezoelectric element height, h_1. compensation element height

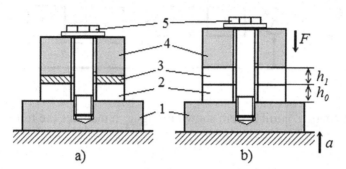

the linear dimensions on temperature changes, this element is made of a material whose linear thermal expansion coefficient is less than that of the piezoelectric material. This solution has apparent simplicity in implementation, but in practice, it is very difficult to select or create composite materials with the necessary value of the coefficient of linear thermal expansion for the manufacture of such an element of compensation for the required piezoceramics brand.

Therefore, a solution has been proposed for compaction of the effect of temperature on a piezoelectric sensor by using an accelerometer as compensation element 3 (Figure 5b), a piezoelectric element - actuator operating on the principle of reverse piezoelectric effect along the polarization axis (mode d_{33}.. The drive is controlled by the automatic control system shown in Figure 6, which operates according to the principle of control by deviation. In the initial state, at normal temperatures ($T=18...22$ C., it is assumed that the height of the piezoelectric element and the height of the compensation element are in a sum equal to h: $h=h_0+h_1$.

The essence of the proposal is that the change in temperature acting on the piezoelectric element causes a change in its geometric dimensions by the value $\Delta h_0(T)$ which leads to the appearance of a constant voltage component at the output of the measuring path due to the appearance of the force $F(T)$ as a reaction of the structure accelerometer. There is observed a so-called zero sensor drift in the absence of influence from the measured parameter.

To eliminate this effect, the automatic control system, analyzing the level of the constant component in the complex signal coming from the piezoelectric sensor, acts on the drive control voltage U, which leads to a change in the drive geometrical size by the value $\Delta h_1(U)$. In this case, the control system does the monitoring so the following condition is met:

$$\Delta h_0\left(T\right) - \Delta h_1\left(U\right) = 0. \tag{10}$$

Then:

$$h = \left(h_0 + \Delta h_0\left(T\right)\right) + \left(h_1 - \Delta h_1\left(U\right)\right)$$

and the next condition is fulfilled: h=const.

Figure 6. Structural diagram of the automatic control system

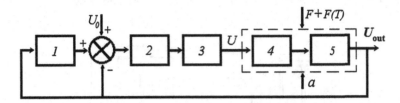

Scheme of the automatic control system shown in Figure 6 includes the following items:

- Phase filter.
- Integrator.
- Power amplifier.
- Piezo element-drive.
- Piezo element-sensor.

The examined scheme regulates by deviation, the principle of which is as follows:

- The regulator is given a signal that the regulated value deviates from the setpoint.
- Deviation controller modifies the control voltage U in such a way as to reduce the deviation $\Delta h_0(T)$.

It should be noted that the specified regulator changes the regulating effect regardless of the reason that caused the regulation error and estimates the error by the level of the constant component in the complex signal. Therefore, additionally introduced a correction for the constant component in the form of U_0. This is the constant component of the accelerometer signal at normal temperature values.

The regulation error in this system is fundamentally not completely removable since the regulating effect is formed only depending on mistake occurrence. The error will be closer to zero, the more accurate the comparison scheme will be implemented.

The scheme works as follows: when the temperature of the medium increases, the geometrical dimensions of the piezoelectric element 5 (see Figure 6) change, which leads to the increase in the value of the constant component in the output signal. The output signal is fed to one input of the adder on a straight line, and the other input of the adder through a phase filter, which shifts it by 180 °. After the adder, the constant component falls on the integrator with the integration time commensurate with the rate of temperature change. The power amplifier controls the piezo element-drive 4 by applying a voltage U to its electrodes, the polarity of which is opposite to the polarity of polarization. Pesoelement-drive fulfilling the control action tends to fulfill the requirements of equation (10).

In the control scheme, it is necessary to note the method for realizing the selection of the constant component from the complex signal, which consists in the fact that the complex signal is delayed by a parallel channel in phase for half a period, and then on the adder it is added to the signal of the main channel. Thus, the variable components arriving at the adder in the antiphase are mutually subtracted. The diagram of the device that implements the specified method is shown in Figure 7 a).

For the delay of the signal on the parallel channel, phase filters of the first order are used. The phase filters of the first order (Tietze & Schenk, 2007), shown in Figure 8, can be used as a wideband phase shifter. By changing the resistance of the resistor R, you can set the phase shift in the range from 0 to $-180°$ without changing the amplitude of the output signal. The magnitude of the phase angle α shift in degrees is calculated by the formula:

$$\alpha = -2arctg\left(\omega RC\right)180 / \pi \tag{11}$$

where $\omega = 2\pi f$.

To ensure the stability of the automatic control system, two series-connected phase detectors are used, the ratings of which are set to a signal phase delay of $-90°$ each.

As a result, two-phase filters, in total, shift the signal by $180°$, and on the adder with transfer coefficients 0.5, the signals are added together. As a result, at the output, we get the constant component with the transfer coefficient equal to 1.

The operation of the device for extracting the constant component from the complex signal of the accelerometer is presented in the oscillograms in Figure 7 b).

Figure 7a. Scheme and oscillograms

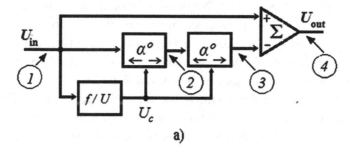

a)

Figure 7b. of the device for extracting the constant component from the complex signal of an accelerometer

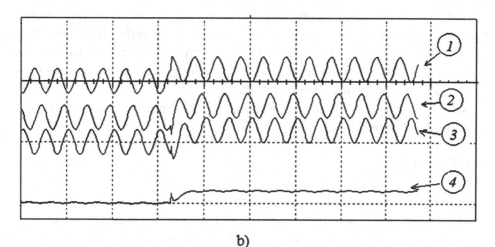

b)

Figure 8. Phase filter of the first order

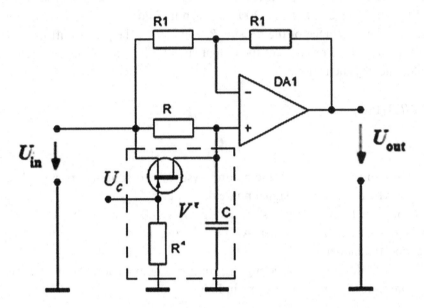

From the formula (11) it is seen that the phase shift also depends on the frequency of the signal of the input phase filter. Therefore, to ensure stable compensation of the temperature component over the entire dynamic range of the piezoelectric accelerometer, and this is a frequency range from 3...10Hz to 10 ... 15 kHz, it is necessary to change the time delay of the phase filter delay circuit. This circuit is implemented on the elements R and C (diagram Figure 8). It is technically more convenient to change the time constant by changing the resistance of the resistor R at C=const.

The dependence of the resistance R from the frequency of the signal f in the dynamic range of 10 ... 10000 Hz to ensure a delay of -90 ° throughout the range is shown in Figure 9.

As can be seen from the graph of Figure 9 the dependence of resistance R from the frequency f is linear and inversely proportional.

To ensure the regulation of R within the specified range, elements highlighted in Figure 8 with a dashed area are introduced into the phase filter circuit. The voltage Uc from the frequency-voltage converter (see Figure 7a) is fed to the control electrode and controls the resistance of the field effect transistor channel V', which is connected in the circuit parallel to the resistor R. Thus, as the signal frequency increases, the voltage Uc at the output of the frequency-voltage converter increases, this causes a proportional decrease in the resistance of the channel of the field effect transistor and, as a result, decreases the total resistance of the time tasking circuit of the phase filter.

Forcing units were not used in the automatic control system, since the effects of temperature on the accelerometer structural elements are rather inertial.

Illustrative Example

The application of the proposed method, we consider the definition of the coordinates of the location of a flying object. The task of determining the coordinates of the location is one of the main when solving problems of navigation of aircraft. At the same time, the constantly growing intensity of airlines in the

Figure 9. The dependence of the value of R from the frequency f in the dynamic range of the accelerometer

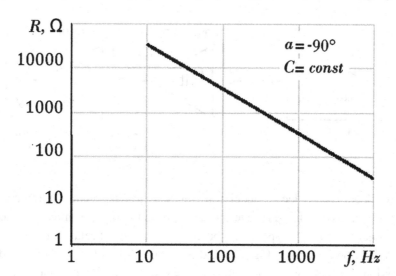

airspace causes a continuous increase in the requirements for the accuracy of determining the navigation parameters. At the same time, in a complex modern jamming environment, inertial navigation systems have recently become one of the main means of navigation on most types of moving objects. The most popular and attractive for unmanned aerial vehicles are strap-down inertial navigation systems. The potential advantages of strap-down inertial navigation systems compared to platform inertial navigation systems include:

- The smaller sizes and power consumption.
- Significant simplification of the mechanical part of the system and its layout and system reliability.
- Reduction of the installation time of the initial parameters.
- The universality of the system, since the definition of navigation parameters, is carried out algorithmically.

Inertial navigation can be considered as the task of determining the position of an object on a flat surface. That is, following the first Newton's law, an object moving along a flat surface, in the absence of a force acting on it, is in a state of rectilinear motion at a constant speed. Knowing the initial conditions, that is, the initial position of the object, its course, and speed, we can determine the current position of the moving object depending on time.

To obtain the motion parameters in the dynamics, accelerometers are used.

For the simplest case of navigation on a flat surface in the Cartesian coordinate system, when the object is at rest, one of these accelerometers is located along the X-axis, the other along the Y-axis.

The acceleration increments from the accelerometer are integrated and provide components of the current speed. The integration of the components of the speed allows obtaining the corresponding increments of the initial values of the x and y coordinates. This determines the current location of the object. The scheme of realization of one channel of inertial navigation on a flat surface is shown in Figure 10.

Figure 10. Scheme of a single channel inertial navigation on a flat surface

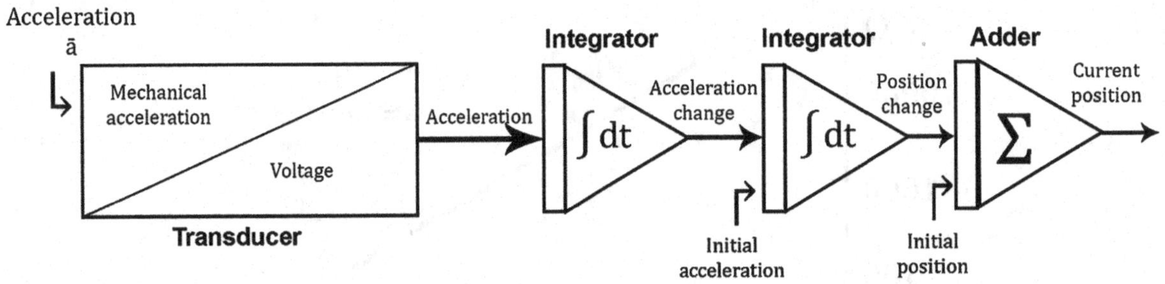

The primary transducer (piezoelectric accelerometer) with a sharp change in temperature, which occurs when the aircraft changes its height, generates an error in the measurement of acceleration. Which leads to an incorrect definition of the speed and position of the object relative to the initial coordinates of the reference. As can be seen from Figure 10, this error will be effectively leveled out at the first stage of the conversion of acceleration a to voltage, because with each subsequent stage of conversion the instrumental error is exacerbated.

FUTURE RESEARCH DIRECTIONS

In the future, as part of the study of the problem considered, it is supposed to work out the option of individual graduation of a specific accelerometer with rapidly changing temperature effects. Graduation data is placed in EPROM (Erasable Programmable Read-Only Memory). In this case, a high-speed thermometer is needed that will directly contact the sensitive element of the accelerometer - a piezoelectric element. According to the temperature data of this thermometer, data from the EPROM table will be extracted and the results of acceleration measurements will be adjusted.

CONCLUSION

Since the information content and reliability of the piezoelectric sensors is characterized by dynamic and frequency ranges, the main and additional errors caused by the influence of influencing factors and taking into account the fact that the most significant and common to all piezoelectric sensors are temperature errors, which for most sensors exceed all other errors, taken together, the method of correction of the additive component of the error caused by changes in those proposed in this paper temperature is relevant.

It allows you to improve the accuracy of measurements performed by piezoelectric accelerometers and to expand their application area about the requirements for the ambient temperature during measurements.

REFERENCES

Aheikin, D. I., Kostina, E. N., & Kuznetsov, N. N. (1965). Sensors for control and regulation. Moscow, Russia. *Mechanical Engineering (New York, N.Y.)*, 496–508.

Antononenko, A. M., Kudzin, A. Y., & Gavshin, M. G. (2009). Influence of the domain structure on the electromechanical properties of ferroelectric ceramics of the TsTs and MNWT [-Moscow, Russia: FTI them. A. F. Ioffe.]. *Physics of the Solid State*, *39*(5). doi:10.1134/1.1129978

Atamanyan, E. G.(1989). Instruments and methods for measuring electrical quantities. Moscow, Russia: Higher. Shk., 134-136.

Bogush, M. V. (2006). *Piezoelectric sensors for extreme operating conditions. T. 3* (pp. 246–248). Rostov-on-Don, Russia: Publishing House SKNTS VS.

Dorf, R., & Bishop, R. (2002). Modern management systems. Per. with Eng. Moscow, Russia: Laboratory of Basic Knowledge, 615-618.

Fraden, J. (1997). *Handbook of modern sensors* (2nd ed., pp. 104–107). Woodbury, NY.

Gorish, A. V., Dudkevich, V. P., & Kupriyanov, M. F. (1999). Piezoelectric instrument-making - T.1. Physics of ferroelectric ceramics. Moscow, Russia: Izdat. forerunners ed. zhur. Radio Engineering.

Ivanov, N. M., Kondakov, E. V., & Miloslavskiy, Y. K. (2016). Opubl Patent RF N° 2584719. Cifrovoj sposob izmerenija parametrov p'ezojelementov [Digital method for measuring the parameters of piezoelectric elements].

Kim, D.(2003). The theory of automatic control. T1. Linear systems / D. Kim. Moscow, Russia: FIZ-MATLIT, 76-93.

Klyuev, V. V. (1978) Instruments and systems for measuring vibration, noise and shock: A handbook. In 2 books. Moscow, Russia: Mechanical Engineering, 328-332.

Levshina, E. S.(1983) Electrical measurements of physical quantities: (Measuring transducers). London, UK: Energoatomizdat. Leningrad separation,132-141.

Milekhin, L. N. (2013) Applied theory of gyroscopes. Moscow, Russia: Higher School, pp. 73-76.

Nesterov, A. A., & Panich, A. A. (2010). Sovremennye problemy materialovedeniy keramicheskih p'ezoelektricheskih materialov: monografiya [Modern Problems of Materials ceramic piezoelectric materials]. P'ezoelektricheskoe priborostroenie. V. 8, Rostov on Don, 163-171.

Osadchiy, E. P. (1979). Design of sensors for measuring mechanical quantities. Moscow, Russia. *Mechanical Engineering (New York, N.Y.)*, 134–146.

Osadchy, E. P. (1979). Designing sensors for measuring mechanical quantities. Moscow, Russia. *Mechanical Engineering (New York, N.Y.)*, 178–182.

Ostrov, K., & Vittenmark, B. (1987). Control systems with computers. Per. with Eng. Moscow, Russia. *WORLD (Oakland, Calif.)*, 234–252.

Serridge, M., & Licht, T. (1987). *Handbook of piezoelectric accelerometers and preamps* (pp. 164–172). Denmark.

Sharapov, V. M., Musienko, M. P., & Sharapova, E. V. (2006). Piezoelectric sensors. Moscow, Russia: Technosphere, 230-245.

Smazhevskaya, E. G., & Feldman, N. B. (1971). P'ezoelektricheskaya keramika [Piezoelectrics ceramic]. Moscow, Russia: Sovetskoe radio, 83-88.

Tietze, U., & Schenk, K. (2007). Semiconductor circuitry. 12th ed. Vol. II: Trans. with him. Moscow, Russia: DMK Press,144-147.

Wang M., & Gross, L. E. (1998). Piezoelectric pseudo-shear multilayer actuator, 84-91.

Yorish, Yu. I. (1963) Vibrometry. Second edition, revised and updated. Moscow, Russia: State. Scientific and Technical Mechanical Engineering Publishing House, 63-67.

Zemlyakov, V. L. (2012). Piezoelectrics and related materials. *Investigations and Applications*, 117–142.

Chapter 12

Research of the Reliability of the Electrical Supply System of Airports and Aerodromes Using Neural Networks

Serhii Mykolaiovych Boiko

Kremenchuk Flight College, Kharkiv National University of Internal Affairs, Ukraine

Yurii Shmelev

Kremenchuk Flight College, Kharkiv National University of Internal Affairs, Ukraine

Viktoriia Chorna

Kremenchuk Flight College, Kharkiv National University of Internal Affairs, Ukraine

Marina Nozhnova

Kremenchuk Flight College, Kharkiv National University of Internal Affairs, Ukraine

ABSTRACT

The system of supplying airports and airfields is subject to high requirements for the degree of reliability. This is due to the existence of a large number of factors that affect the work of airports and airfields. In this regard, the control systems for these complexes must, as soon as possible, adopt the most optimal criteria for the reliability and quality of the solution. This complicates the structure of the electricity supply complex quite a lot and necessitates the use of modern, reliable, and high-precision technologies for the management of these complexes. One of them is artificial intelligence, which allows you to make decisions in a non-standard situation, to give recommendations to the operator to perform actions based on analysis of diagnostic data.

DOI: 10.4018/978-1-7998-1415-3.ch012

INTRODUCTION

Civil aviation airports are key elements of aviation transport. Airport power systems are a complex technical system consisting of a large number of structural elements that are in a state of complex interaction. Elements of the power supply system are characterized by dependence on other technical systems (fuel and energy complex (FEC) and airport facilities) and continuity in time of production processes related to the production, distribution and consumption of electricity.

The considered methods of increasing the reliability of the SES AP form a system of redundancy of the main technical devices to provide electricity to the PE at the airport. The redundancy method, composition and characteristics of technical devices are regulated by normative documents and feasibility study.

Mathematical models of the event tree and the failure tree allow to analytically model and analyze the paths of processes in the system, to determine the interaction characteristics of the model elements and to obtain a simple representation of the obtained reliability characteristics for the structural elements of the SES AP.

In the context of airports and airfields (A&A), which has a very complex and ramified structure, and technological processes are very complex and dependent on many factors, forecasting is a difficult and difficult task, provided that a forecast error of no more than 4% is obtained. Therefore, in the context of AIA, it is advisable to use artificial neural networks (SNMs), which predict significant interconnections between individual factors.

Analysis of studies has shown that multi-layer SNMs have much more ability to solve practical and applied problems .

ANALYSIS OF MATHEMATICAL DESCRIPTION OF NEURAL NETWORKS

ANN concerning the type of activation functions that are part of the structure of the AN is divided (Shibzukhov, 2006):

1) Homogeneous ANNs that consist of one type neurons with a single activation function.
2) Heterogeneous ANNs that consist of neurons with different activation functions.

ANN, depending on the state in which neutrons are, divided into analogue and binary, and depending on the number of neurons that change their status at some time, divided into synchronous ANN when only one neuron changes its state; asynchronous, when its state changes several neutrons (a group of neurons)(Barsky, 2013). Classification of the ANN can be represented as Table 1.

The operation of the NN is based on the function of several variables as the sum of functions of one variable.

The simplest ANN consists of a certain number of ANs, which are grouped into groups that form layers, neurons which are connected by weight links and input signals from other neurons, the previous layers.

As a result of the transformations of the input signals in the neurons of certain layers at the output of the ANN, the signal OUT(Kruglov,2002).

Weights can be written as a matrix that has m lines and n columns:

Table 1. Classification of the ANN

No	Classification Type	Species
1	By the number of layers	Single layer
		Multilayer
2	By way of spreading the signal	Direct Signal Distribution
		Reverse signal propagation
3	According to the architecture of NN	Full networks
		Multilayer network with serial communications
		Weakly connected networks
4	NN with feedback	Hopfield Network
		Hamming's Networks
5	NN with feedback	Loop-to-loop networks
		Layered and fully connected networks
6	Types of activation functions in the neuron	Homogeneous (homogeneous)
		Heterogeneous (different)
7	For the state of neurons (excited, inhibited)	Analogue networks
		Binary networks
8	By the number of neurons that simultaneously change the state	Synchronous
		Asynchronous
9	NN Bidirectional associative memory (BAM)	Feedback Networks
		Modeling of plastic-stable perception
10	NN Architecture of Adaptive-Resonance Theory (ART)	
11	NN architectures ART-1, ART-2, ART-3	
12	Congnitron, neoconnitron	NN modeling of the visual system of the human brain

$$W = \begin{vmatrix} \omega_{11} & \omega_{12} & ... & \omega_{1n} \\ \omega_{21} & \omega_{22} & ... & \omega_{2n} \\ .... \\ \omega_{m1} & \omega_{m2} & ... & \omega_{mn} \end{vmatrix}, \tag{1}$$

where m is the number of inputs, n is the number of ANN neurons.

Then the output vector, whose elements are outputs of OUT signals from the source neurons, are calculated as the matrix product of the matrix-string-input elements on the matrix of weight coefficients:

$$\vec{B} = \vec{X}\vec{W}, \tag{2}$$

where $\vec{B}$ - vector row.

If the desired output from the i-th neuron is denoted by y_i, and the actual output from the i-th neuron is denoted $\hat{y}_i$, then the error for the k-th sample can be calculated using the formula:

$$\Delta E_k = \frac{1}{2}\sum_{i=1}^{n}(y_i - \hat{y}_i)^2, \; i = \overline{1, n} \tag{3}$$

then the total error for the entire election:

$$\Delta E_\Sigma = \sum_{k=1}^{n} E_k, \; k = \overline{1, m}. \tag{4}$$

The combined (total) input to this neuron determines its excitation (activity), that is, the total value of the weight bonds that affect the given neuron determines the real state of the neuron.

For one sample, with the linear function of activating the signal in the neuron, the error can be determined by the formula:

$$E = \frac{1}{2}(y_i - \hat{y})^2 = \frac{1}{2}(y_i - NET)^2. \tag{5}$$

Transforming this expression is obsessed with:

$$E = 0,5(y^2_i - 2y_i\hat{y} + \hat{y}^2) = 0,5\left[y^2_i - 2y_i(NET) + (NET)^2\right].$$

As $\hat{y} = (NET) = (x_1\omega_1 + x_2\omega_2)$ for two weights, then

$$E = 0,5\left[y^2_i - 2y_i(x_1\omega_1 + x_2\omega_2) + (x_1\omega_1 + x_2\omega_2)^2\right]$$
$$= 0,5\left[y^2_i - 2y_ix_1\omega_1 - 2y_ix_2\omega_2 + (x_1\omega_1)^2 + 2x_1\omega_1x_2\omega_2 + (x_2\omega_2)^2\right].$$
$$= 0,5\left[y^2_i - 2y_ix_1\omega_1 - 2y_ix_2\omega_2 + x_1^2\omega_1^2 + 2x_1\omega_1x_2\omega_2 + x_2^2\omega_2^2\right]$$

We find the dependence of the error E from ω_1.

$$E(\omega_1) = 0,5\left[x_1^2\omega_1^2 + (2x_1x_2\omega_2 - 2y_ix_1)\omega_1 + (y_i - 2y_ix_2\omega_2 + \omega_2^2x_2^2)\right]$$
$$= 0,5x_1^2\omega_1^2 + (x_1x_2\omega_2 - y_ix_1)\omega_1 + (0,5y_i - y_ix_2\omega_2 + 0,5\omega_2^2x_2^2)$$

By ticking $0,5x_1^2 = 0$, $x_1x_2\omega_2 - y_ix_1 = b$, $0,5y_i - y_ix_2\omega_2 + 0,5\omega_2^2x_2^2 = c$, we get:

$$E(\omega_1) = a\omega_1^2 + b\omega_1 + c. \tag{6}$$

So the dependence of the error E from ω_1 is parabolic, similarly, the dependence of the error E from ω_2 will also be parabolic. As $a = 0,5x_1^2 > 0$, then the branches of the parabola are directed upwards (in the positive direction of the axis E). This means that the dependency graph $E(\omega_1)$ has a local minimum, which coincides with the vertex of the parabola, which is a dependency graph $E(\omega_1)$.

Since the weight coefficients at the initial stage of the training of the NN gain some random values, the point that characterizes the initial position of the NN in the plane E,ω_1 may be at an arbitrary position and it is unlikely that it will be at the local minimum (Tolkachev, 2016).

Consequently, in the process of learning, the ANN should make correction (change of values) of the weight coefficients so as to minimize the numerical value of the total error (E), that is $E \to E_{min}$, so that this process must pass as soon as possible, that's mean for less number of iterations and epochs. If E will depend on the two variables (ω_1 i ω_2), then in the spatial Cartesian coordinate system (E,ω_1,ω_2) we obtain a volumetric figure of the rotation (paraboloid).

Thatis, weobserve a surface that determines the error value for various combinations of weight coefficients (ω_1 i ω_2), and a weight vector whose origin is at the origin, and the end at a point which is the projection of a local minimum of a paraboloid to a plane, indicates the direction of minimization of the error.

To correct the weights, one can apply the Wadow-Hoff or delta rule, which generally takes the following mathematical record:

$$\Delta \omega_{ij} = \eta \delta_j x_i, \quad i = \overrightarrow{1,n}, \ j = \overrightarrow{1,m}, \tag{7}$$

where $\delta_j = y_j - \hat{y}_j$ - the required output from the neutron j, and $\hat{y}_j$ - the actual (real) output from the j-neutron, x_i - the signal that came from the i-neutron, η - the coefficient of weight coefficient change (the rate of training), $\Delta \omega_{ij}$ - the magnitude of the change in the weight ratio between and j neutron.

In general, the delta rule can be written as:

$$\Delta \omega_{ij} = \eta \delta_j x_i = \eta(y_j - \hat{y}_j)E_V = \eta(y_j - \sum_{j=1}^{m} E_j\omega_{ij})E_V, \quad i = 1,\overline{n}, \ j = 1,\overline{m}. \tag{8}$$

The rate of change in error $\dfrac{\partial E}{\partial \hat{y}_j}$ relative to the change in the real output of the j-neutron is numerically equal:

$$\frac{\partial E}{\partial \hat{y}_i} = \hat{y}_j - y_j = -\delta j, \quad j = \overline{1,m} \tag{9}$$

and the rate of change in output $\hat{y}_j$ from the j-neutron from the change in weight coefficients $\Delta \omega_{ij}$ can be written as:

$$\frac{\partial \hat{y}_j}{\partial \omega_{ij}} = x_i, \quad i = \overrightarrow{1,n}, \ j = \overrightarrow{1,m} \cdot \tag{10}$$

Then the rate of change in the error from the change in weight coefficients can be written:

$$\frac{\partial E}{\partial \omega_{ij}} = \frac{\partial E}{\partial \hat{y}_i} \frac{\partial \hat{y}_j}{\partial \omega_{ij}}, \quad i = \overrightarrow{1,n}, \ j = \overrightarrow{1,m}, \tag{11}$$

$$-\frac{\partial E}{\partial \omega_{ij}} = -\delta_j x_i, \quad i = \overrightarrow{1,n}, \ j = \overrightarrow{1,m}, \tag{12}$$

$$\frac{\partial E}{\partial \omega_{ij}} = \delta_j x_i, \quad i = \overrightarrow{1,n}, \ j = \overrightarrow{1,m}, \tag{13}$$

where $\dfrac{\partial E}{\partial \omega_{ij}}$ - the derivative of the surface error, depending on the change in weight coefficients.

Then the authors can conclude that the changes in weight factors should occur in the direction opposite to the derivative of the error surface, that is, in the direction opposite to the direction of the gradient - in the direction of decrease, rather than increase.

The number of input neurons corresponds to a certain number of features that determine the dimension of the space from which all samples will be selected at the entrance to the ATM. Depending onthe nature of the solution to the problem of the distribution of samples into separate classes, according to certain features, based on the classification, the problems are divided into two types: linear and nonlinear. There are two ways to solve a nonlinear problem with ANN:

1) Use an ANN, which will be based on two or more direct data split into classes.
2) Change the look (content) of the data that is enter work.

Consider the response of the closed neurons to the input and output from the output layer of the ANN to the input data for the NN with two input, two hidden and one neuron of the output layer, using a threshold function with a numerical threshold value T = 0.4 at displacements $\Theta_{1(1)}=1,2$ for one neuron, $\Theta_{1(1)}=1,2$ for the second hidden layer neuron, $\Theta_{1(2)}=-0,2$ for the output layer; with matrixes of output coefficients

$$W_1 = \begin{vmatrix} 1,2 & 0,8 \\ -0,5 & -0,5 \\ -0,5 & -0,5 \end{vmatrix}$$

for a hidden layer,

$$W_2 = \begin{vmatrix} -0,2 \\ 0,7 \\ -0,3 \end{vmatrix}$$

for the output layer, if the input neurons are signaled x_1 and x_2 whose directions are formed by the next tuple:

$\{(1;1), (1;0), (0;1), (0;0)\}$

The results of the calculations presented are presented as Table 2. Input values of the samples are transformed by the weight coefficients of the matrix and displacements and in the adjacent block (AdB) of the hidden layer neurons, and after activating the threshold function in the activation block (AcB) of the hidden layer neurons, we obtain the numerical values of the outputs from the hidden neurons of NET_I and NET_{II}. Then in the neuron of the output layer under the action of weight coefficients of the matrix W_2 and the bias Θ_3 and activation of the threshold function in the AcB of the source neuron, we obtain the initial numerical value of OUT for each sample of the input signals x_1, x_2.

Based on the considered types of ANNs, the question arises of creating a forecast of power consumption, provided that the reliability of power supply for airfields and airports is increased

MODELING THE RELIABILITY OF THE POWER SUPPLY SYSTEM OF AIRPORTS AND AERODROMES

The schematic diagram of the ANN discussed earlier can be illustrated as follows (figure 1): the input neurons of the zero layer denote 1_0, 2_0, the hidden neurons of the first layer denote $1_{(1)}$, $2_{(1)}$, and the output neuron - symbol $1_{(2)}$, the weighting coefficients of the displacements for the hidden layer – $\Theta_{1(1)}$ and $\Theta_{2(1)}$, for the output layer – $\Theta_{1(2)}$, the input signals to 1_0 neurons through $x_1=(1,1,0,0)$, and the input signals to 2_0 neurons – through $x_2=(1,0,1,0)$, we will write the total output signal to the zero layer in the form of a tuple:$\{(x_1,x_2)\} = \{(1,0:1,0), (1,0;0), (0;1,0), (0;0)\}$, and the output signal $OUT=\{0;1;1;0\}$, - and the weight coefficients, respectively, ω_{ij} and γ_{jk}, the matrix which will be:

Table 2. The results of the calculations

No			$\Theta1_{(1)}$	$\Theta2_{(1)}$	$\sum_{i=1;j=1}^{i=2;j=2} {}_{(I)}x_i\omega_{ij} + \Theta_j$	$\sum_{i=1;j=1}^{i=2;j=2} {}_{(II)}x_i\omega_{ij} + \Theta_j$	NETI	NETII	$\Theta1_{(2)}$	ΣI_{II}	OUVT
1	1,0	1,0	1,2	0,8	0,2	-0,2	0	0	-0,2	-0,2	0
2	1,0	0			0,7	0,3	1	0		0,5	1
3	0	1,0	$W_1 = \begin{vmatrix} -0,5 & -0,5 \\ -0,5 & -0,5 \end{vmatrix}$		0,7	0,3	1	0	$W_2 = \begin{bmatrix} 0,7 \\ -0,3 \end{bmatrix}$	0,5	1
4	0	0			1,2	0,8	1	1		0,2	0

$$W_1 = \begin{vmatrix} \omega_{11} & \omega_{12} \\ \omega_{21} & \omega_{22} \end{vmatrix} \quad \text{and} \quad W_2 = \begin{vmatrix} \gamma_{11} \\ \gamma_{21} \end{vmatrix}$$

$$\Sigma_I = \sum_{i=1; j=1}^{i=n; j=m} x_i \omega_{ij} + \Theta_{1j}; \quad i = \overline{1,n}, \ j = \overline{1,m};$$

$$\Sigma_{II} = \sum_{i=1; j=1}^{i=n; j=m} x_i \omega_{ij} + \Theta_{2j}; \quad i = \overline{1,n}, \ j = \overline{1,m};$$

$$\Sigma_{III} = \sum_{k=1; j=1}^{k=s; j=m} Y_j \gamma_{kj} + \Theta_k; \quad k = \overline{1,s}, \ j = \overline{1,m}.$$

In this case, n = 2, m = 2, s = 1, where n is the number of neurons of the zero layer, m is the number of entrances to the hidden layer neurons from the neurons of the zero layer (the neurons of the first hidden layer), s is the number of neurons in the initial layer (another layer); ω_{ij} – weight coefficient from the i-th layer (input) to the j-th hidden layer neuron. Then ω_{12} - weigh the coupling factor from the first neuron of the zero layer (inbound) to the second neuron of the hidden layer (the first layer), similarly to other weighting coefficients. γ_{jk} - the weighting factor of the j-th neuron of the hidden layer (the first layer) to the k th neuron of the output layer (the second layer).

Figure 1. The schematic diagram of the ANN

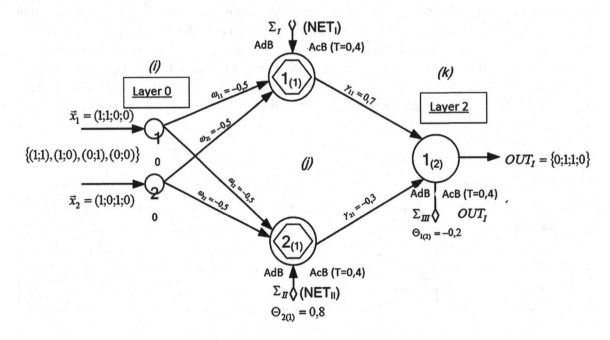

Σ_I- The sum of the weight bonds that belong to the first hidden layer neuron, Σ_{II} is the sum of the weight bonds that belong to the second hidden layer neuron, Σ_{III} - the sum of the weight bonds that are part of the exit neuron (the second layer). $NET_I=Y_I$ is the exit signal from the first hidden layer neuron, after activating the threshold function (T = 0.4 - numeric value of the threshold), NET_{II} - signal from the second neuron of the hidden layer (layer 2), OUT - the numerical value of the output signal from the first the neuron of the output layer (second layer), after activating the threshold function (T = 0.4 - the numeric value of the threshold of the activation function). {(1,0:10), (1,0:0), (0;1,0), (0;0)} - tuple of input signals to a zero layer (input layer), {1,1,0,0} -cortex of output signals from the second layer (the output layer).

$\Theta_{1(1)}$, $\Theta_{2(1)}$, $\Theta_{1(2)}$ - weighted coefficients of displacement of the first and second neurons of the hidden layer (the first layer) and the first neuron of the output layer (second layer). $\vec{x}_1 = (1,0;1,0;0;0)$ - vector of input values to the first neuron (1_0) of the input layer (zero layer), $\vec{x}_2 = (1,0;0;1,0;0)$ - vector of input values to the second neuron (2_0) of the input layer (zero layer).

Schematically depicted ANN in Scheme 4 for different input vectors will be obtained at the output corresponding to each of them, namely (Table 3):

1)   (1,0; 1,0) at the input, and at the output OUT = 0;
2)   At the input (1,0; 0), and at the output OUT = 1;
3)   On the input (0; 1.0), and on the output OUT = 1;
4)   On the input (0; 0), and on the output OUT = 0.

Consequently, each input vector (x_1, x_2) at the output of the ANN corresponds to a certain numeric value OUT.

Threshold function

$$F(NET) = \begin{cases} 0, & if \quad NET < T \\ 1, & if \quad NET \geq T \end{cases}, \tag{14}$$

is an example of a nonlinear activation function.

An example of a linear function is

$$F(NET) = \beta(NET), \tag{15}$$

the area of values of which $F(NET)\in(-\infty;+\infty)$, in the presence of which in AcB, the exit from the neuron will be equal to the value of the entrance to this neuron, if $\beta=1$. Then we can conclude that a multilayered neural network with a linear activation function can solve only those problems that can solve single-layer neural networks. Only incoming and outgoing neurons can be solved. Then we can draw the following conclusion that for multilayer NN it is necessary to use nonlinear activation functions, this is a logistic function

$$F(NET) = \left[1 + \exp\left(-\alpha \cdot NET\right)\right]^{-1}, \tag{16}$$

or other nonlinear functions. The best option is logistic, since it is continuous throughout the definition domain $(-\infty;+\infty)$, differentiated and monotonically increasing for all NETs $(-\infty;+\infty)$ (in the range of values $(0; 1)$).

Consequently, each element and the set of input values x are connected by a weighted value $x_i\omega_{ij}$ with each AN, while in the AdB neuron we find the weight of the inputs to the given neuron that enters the AcB, after activation in which, the output of the output layer will output the NET signal. Output vector $\vec{L}$ of NN, components of which are OUT outputs from the neurons of the output layer, numerically equal to the product of the matrix $\%$ - the vector string of input signals on the matrix W is the matrix of weight coefficients, that is:

$$L = XW, \tag{17}$$

where $\vec{L}$ is the matrix line.

Multilayer NNs have much more possibilities in solving practical and applied problems; they represent a set of layers, such that the output from one layer will be input to the next layer, and the increase of computed capacities (capacities) compared to single-layer NN, is possible only if there is nonlinear activation function between layers.

If there is no nonlinear activation function in multilayer NN, then the calculation of the numerical value of the output layer is found as the product of the input vector X on the first weight matrix W_1, and then on the next weight matrix W_2. That is

$$OUT = (\vec{X}\vec{W}_1)\vec{W}_2, \tag{18}$$

but as a result of the fact that the product of the matrices is associative, then

$$OUT = (\vec{X}\vec{W}_1)\vec{W}_2 = \vec{X}(\vec{W}_1\vec{W}_2). \tag{19}$$

That is, essentially, the transition from a multilayer NN to a single-layered NN, that is, a two-layer NN is equivalent to one hidden layer with a weight matrix, equal to the product of two weight matrices $(W_{703} = W_1W_2)$.

Then

$$OUT = (\vec{X}\vec{W}_1)\vec{W}_2 = \vec{X}(\vec{W}_1\vec{W}_2) = \vec{X}(\vec{W}_{703}). \tag{20}$$

Consequently, any multilayered NN with a linear activation function can be replaced by an equivalent single-layer NM with a weight matrix:

$$W_{703} = W_1W_2. \tag{21}$$

The NMs discussed earlier refer to the direct propagation of the input signal or the network with direct bonds, that is, from the input layer to the output layer of the signal from the NN, that is, from the previous layer to the next and in order. But the possibilities of such NN are limited, they do not have memory, that is, their output is completely determined by the values of input vectors and values of weight coefficients. Such networks are also referred to as NNs without feedback.

NNs that have bundles from the original layers to the inputs are termed networks with reversed bonds. In some NN with backlinks, the value of the output from the network is returned to the input layers, that is, the output will be the input. Therefore, NN with reverse bonds has the property, similar to short-term human memory (Bunakov, 2015).

Illustrative Example

The algorithm of back propagation of the signal in the NN consists of two directions of signal propagation:

1) The direct direction of propagation of the signal from the zero layer to the output layer.
2) The reverse direction of the propagation of the signal from the output layer to the input, passing the error value from the output layer to the input (first) layer, which determines the value of which it is necessary to adjust the weight factors in the learning process of the NN, which is used to ensure that the NN can perform the delivered before it the task of the data that it receives.

Reverse signal propagation mechanism:

1) In the direct propagation of the signal the hidden layer neuron sends signals to each neuron of the output layer.
2) In the reverse propagation of the signal of the hidden layer neurons, receive error signals from each neuron of the output layer.

Figure 2. Figure of NM with feedback

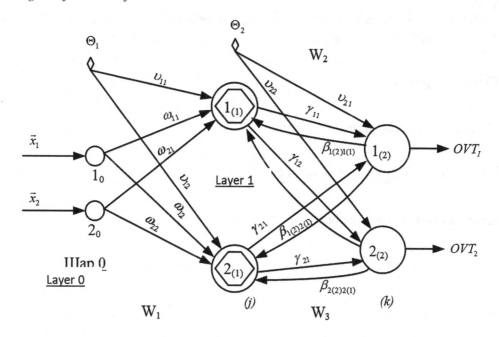

Matrixes of weight coefficients:

$$W_1 = \begin{vmatrix} \omega_{11} & \omega_{12} \\ \omega_{21} & \omega_{22} \end{vmatrix}, \quad W_2 = \begin{vmatrix} \gamma_{11} & \gamma_{12} \\ \gamma_{21} & \gamma_{22} \end{vmatrix}, \quad W_3 = \begin{vmatrix} \beta_{11} & \beta_{12} \\ \beta_{21} & \beta_{22} \end{vmatrix}, \quad W_{\Theta_1} = \begin{vmatrix} \vartheta_{11} \\ \vartheta_{12} \end{vmatrix}, \quad W_{\Theta_2} = \begin{vmatrix} \vartheta_{21} \\ \vartheta_{22} \end{vmatrix},$$

$\beta_{1(2)1(1)}, \beta_{1(2)2(1)}, \beta_{2(1)1(1)}, \beta_{2(2)2(1)}$ - are weight coefficients of feedback between k-m and j-th layer.

When learning NN, every input sample of input signals has its own target output y_k, which corresponds to this input signal x_i, and the actual output from NN is denoted by $\hat{y}_k$, then the error value:

$$\delta_k = y_k - \hat{y}_k.$$

Therefore, the algorithm for the propagation of the NM error signal δ_j is based on the generalization of the delta rule.

The error δ_j corresponds to the error of the neuron of the output layer, but the error of the hidden layer neurons does not directly correlate with the target output values, so the weighted values of the hidden layer neurons need to be corrected in proportion to their effect on the magnitude of the error of the next layer (the output layer for the NN with one hidden layer). That is, the output neuron with greater error more influences the error of the hidden layer neuron, which is associated with this output neuron greater than the weight. Then for a hidden neuron the error can be calculated by the formula:

$$\delta_j = F'(NET_j) \cdot \sum_k \delta_k \beta_{kj}, \tag{22}$$

where k- the index of the layer that sends the error in the opposite direction, β_{kj} is the weighting factor of the feedback.

If the NN has one hidden layer, then the index β corresponds to the original layer. If the NN uses the logistic function of activating the species

$$F(NET) = \left[1 + \exp(-NET_j)\right]^{-1}, \tag{23}$$

Then its derivative has the following entry:

$$F(NET_j) = \exp(-NET_j)\left[1 + \exp(-NET_j)\right]^{-2}. \tag{24}$$

Converting this expression, we obtain:

$$F'(NET_j) = \exp(-NET_j)(1+\exp(-NET_j))^{-1}(1+\exp(-NET_j))^{-1} = \frac{\exp(-NET_j)}{1+\exp(-NET_j)} F(NET_j)$$

$$= \frac{(\exp(-NET_j)+1)-1}{1+\exp(-NET_j)} F(NET_j) = \left(1-\frac{1}{1+\exp(-NET_j)}\right) F(NET_j)$$

$$= F(NET_j)\left[1-\left[1+\exp(-NET_j)\right]^{-1}\right] = F(NET_j)\left[1-F(NET_j)\right]$$

Then the error of the hidden j neuron can be calculated by the expression:

$$\delta_j = F(NET_j)\left[1-F(NET_j)\right]\sum_k \delta_k \beta_{kj}, \tag{25}$$

where k- the index of the layer that sends the error δ_k in the opposite direction, β_{kj} is the weight ratio of the feedback between the neuron of the k-th layer and the j-neuron of the hidden layer, j is the index of the neuron in the hidden layer, $F(NET_j)$ - the logistic function of activating the exit from the j neuron of the hidden layer.

ANN can learn how to solve their tasks. F. Rosenblatt proposed a method of teaching perceptron and proved the statement: "the perceptron can be trained for everything that it can practically realize" (Neural networks, 2008). Learning perceptron can be done both as a teacher and without him. Teaching with a teacher implies the presence of influence from the outside, to assess the behavior of the NN and manage it to make the necessary modifications. Construction of a matrix of weight coefficients that would gradually reduce the error in the output of the neural network (Yurevich, 2014). Teaching without a teacher is carried out through the self-organization of the NN, in order to carry out the necessary changes in weight coefficients, to change the vector of weight coefficients in the field of small errors, is proportional to the error of the output and is equal to zero, if the error is zero. The vector of weight coefficients is modified by the formula:

$$W(t+\Delta T) = W(t) + \eta x^\alpha (\delta^\alpha)^T, \tag{26}$$

where $W(t)$, if $t=0$ - the initial state of the weight coefficients of all neurons (correction), x^α- input image, $\alpha=1,p$, $\delta^\alpha = y^\alpha - \hat{y}^\alpha$ - error vector at the output of the NN, η- subjects of training, $0<\eta<1$, $W(t+\Delta T)$ - (state) of the weight coefficients of all neurons at the moment $(t+\Delta T)$, ΔT - increase of time (after correction).

A set of pairs of vectors of the species (x^α, y^α), where $\alpha=1,p$, which is called a tutorial sample. The NN becomes trained on some training sample (TS), if at the input to the inputs of the NN of each vector x^α on the output each time we get the corresponding vector y^α.

The algorithm for training the neuron has 5 steps (0,1,2,3,4,5), developed by F. Rosenblatt:

Step 0: The random values of the weight coefficients of all the neurons that form the matrix $W(t_0)$ are given.

Step 1: An input sample x^α is submitted to the NN, which will cause formation at the output $\hat{y}^\alpha \neq y^\alpha$ of the NN.

The second step is to calculate the error vector $\delta^\alpha = y^\alpha - \hat{y}^\alpha$ at the output of theNN, which determines the nature of the change in the vector of weight coefficients in the field of small errors.

Step 3: Modification of the vector of weight coefficients by the formula (26).

Step 4: Steps 1,2,3 are repeated for all tutorial vectors. One cycle of sequential input to the NN of the entire sample is called an epoch. The training ends in a few epochs:

1)   If the vector of weight coefficients does not change.
2)   If the absolute error, which is elucidated in all vectors, will be less than a certain small predetermined value.

That is, the HN is trained to get the desired set of outputs for a certain set of inputs to the 0-th layer.

$\{(0;1), (1;1), (1;0), (0;0)\}$ T1=T2=0,5; $\Theta_{1(1)}$=1,4, $\Theta_{2(1)}$=0,6, $\Theta_{1(2)}$=-0,3.

$$W_1 = \begin{vmatrix} 0,3 & -0,4 \\ -0,6 & 0,2 \end{vmatrix}, \quad W_2 = \begin{vmatrix} 0,9 \\ -0,2 \end{vmatrix}.$$

The algorithm of training perceptron is fashionable to use for multilayered perceptrons, taking into account the following basic properties of the perceptron:

1)   Any perceptron may have one or two layers if two layers and weighing characteristics of the first layer are not taught.
2)   Weighing coefficients of any perceptron can be replaced by integers.
3)   During training, after the number of iterations, the following options are possible:
    a)   Perceptron has finished training.
    b)   Stagnates.

The algorithm for training perceptron gives answers as to how many steps it requires for its training and its possible advantages over other methods of learning perceptron.

The multilayer structure of NN is one of the most common, in which each neuron of any layer is bound to the neurons of the previous layer, and the first layer of the neurons is bound to the neurons of the zero layer (input layer) of the NN, and therefore they are called completely connected Nm In multilayer neural networks, the initial values of the neurons of all layers, except for the latter, are unknown, and therefore the perceptron of two and more layers can not be trained only by the magnitude of the error (δ_k) at the input of the NN, and therefore the most suitable variant of training NN is the algorithm of the propagation of the signal from the original layer to previous layers

Figure 3. Figure of NM with feedback

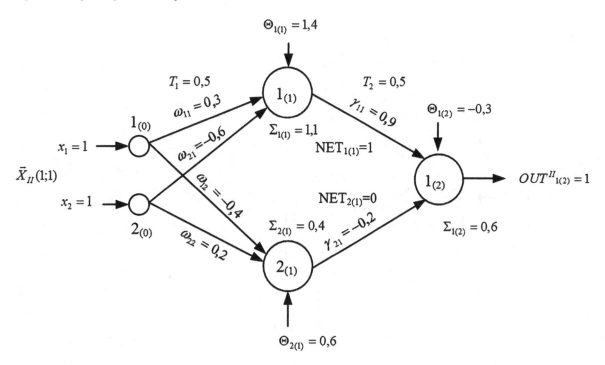

Figure 4. Figure of NM with feedback

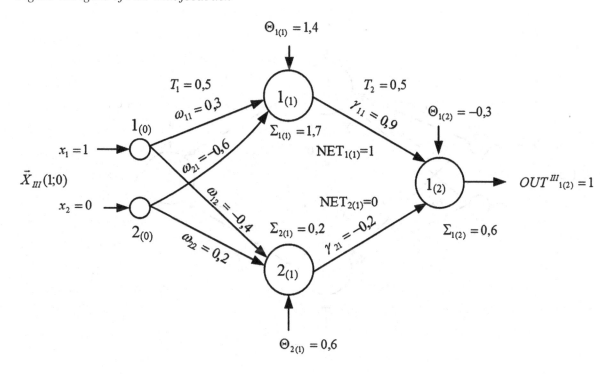

Figure 5. Figure of NM with feedback

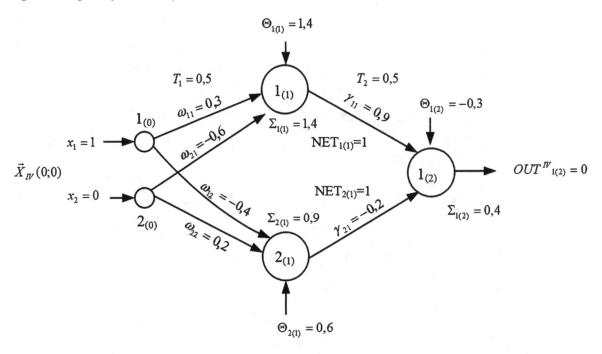

Figure 6. Figure of NM with feedback

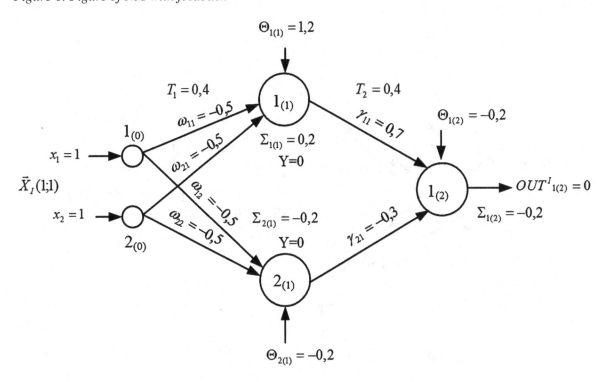

Figure 7. Figure of NM with feedback

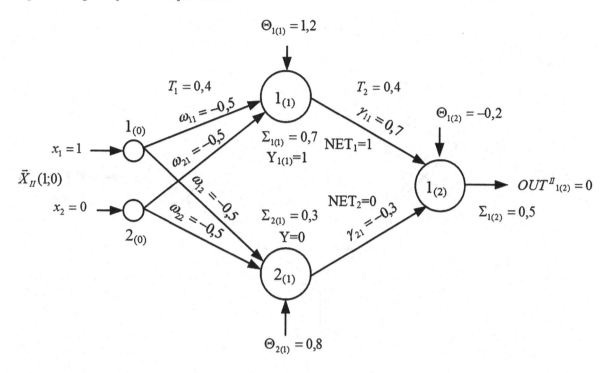

Figure 8. Figure of NM with feedback

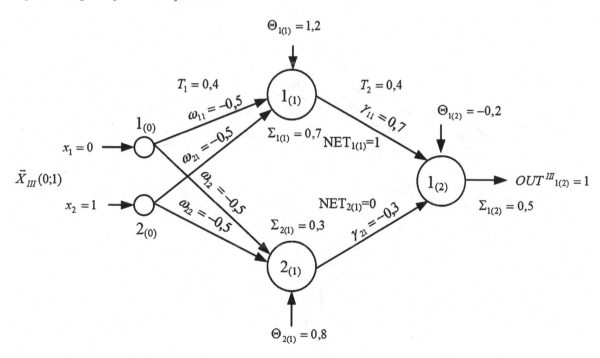

Figure 9. Figure of NM with feedback

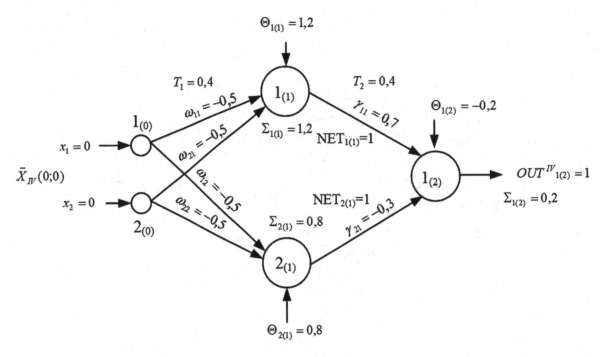

Figure 10. Figure of NM with feedback

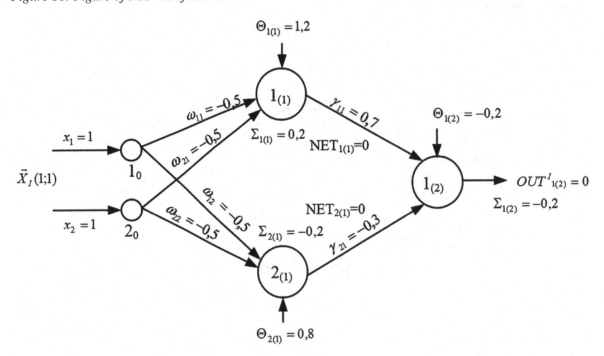

In the NN reverse propagation (NNRP), the main element of the network is a neuron that has a matching block (MB) and an activation block (AB) with an activation function F that has a certain T threshold and outputs a resulting output signal OUT. For the logistic (sigmoidal) activation function $OUT = 1 + \exp(-NET)^{-1}$, for which the first derivative takes place by N = T:

$$\frac{\partial OUT}{\partial NET} = OUT(1 - OUT = OUT - (OUT)^2 . \tag{27}$$

OUT values are in the range (0; 1) for the logistic function, and therefore it provides the necessary non-linearity, which automatically controls the gain of the signal:

1) for weak signals ($NET \rightarrow 0$) the input-output graphic has a strong angle of inclination, which provides a high value of signal amplification;
2) for strong input signals amplification of input signals is reduced.

NN with the propagation of neuro-pulses from the 0-th (input) layer through several hidden layers to the source layer of the neurons is called a "generalized multilayered perceptron", that is, an NN that consists of several consecutively connected layers of the neurons:

1) Input layer (0th layer).
2) Hidden layers: In which each neuron has several inputs connected to the outputs of the neurons of the previous layer and have only one output.
3) Outputs of the neurons of the output layer give the result of the work of the NN - OUT.

The purpose of NN training is to provide the necessary set of output signals from the output layer for a given set of inputs by selecting weight coefficients. That is, for each input vector in the NN there is a corresponding target vector y at the output of the NN, which is called the learning pair (LP) of the vectors [x,y], a certain set of which ensures the learning process of the NN. And a certain set of LP form a teaching set

$$Y = \{(x_1; y_1), (x_2; y_2), (x_3; y_3)...(x_n; y_n)\} . \tag{28}$$

A set of random values of weight coefficients provides a normal course of LP training. The training algorithm for NNRP involves the following sequence of operations:

1) From the teaching set "U", a vector E is applied successively to the input of the NN.
2) The resulting output and NN are compared with the required value by determining the difference δ_k between the target vector of the LP and the real output from the NN.
3) The weighting correction is carried out in order to achieve the minimum value δ_k.
4) The repetition of the points of the algorithm 1,2,3 for each vector of the teaching set until δ_k reaches the desired result, given that the calculations are performed in a layered manner. That is, firstly, the outputs of the previous layers are calculated, and then the results obtained are used as inputs to the next layer, and also calculate the output from the source layer neurons (OUT) that form the source vector of the NN.

From the outgoing NN signals (OUT) subtract the corresponding components of the target vector, find the numerical value of the error (δ_k), which affects the level of correction of the weight factors of the NN.

If, after a sufficient repetition, p. 1,2,3 assumes a value from the permissible interval, then it is assumed that the NN has successfully passed training and the NN is modest to use for recognition, with constant values of weight coefficients.

The propagation of the signal from the input to the input layer is called the forward signal passing (FSP), and from the output layer to the input - the reverse passage of the signal (RPS) in the NN.

Then "forward signal passing" (FSP) can be symbolically written as follows:

$$\% \xrightarrow[i\,F(NET)]{\Sigma \omega x_i} Y \xrightarrow[j\,F(NET)]{\Sigma \omega x_i} \{OUT_k\} = Y_k \,. \tag{29}$$

That is, for an arbitrary layer of NN, the output vector NET can be presented as a product of vectors $\vec{\%}$ and $\vec{W}$, where $\vec{\%}$ - the vector of input signals to the zero layer of NN, $\vec{W}$ - the matrix of weight coefficients between the neurons of the previous and subsequent layers. Applying the activation function F to each vector component of the NET receives the source vector $\vec{Y} = F(\widehat{\vec{XW}})$.

That is, the calculation of the results of the output layer (k) forgets the application of the equation $Y = F(\widehat{\vec{XW}})$ to each layer from the input layer (zero) to the output layer (k).

The reverse signal propagation (RSP) in the NN involves determining the difference between the target value y_k and the output actual (actual y_k) value of the signal and the neuron of the output layer (k) $\delta_{q_k}^{\alpha} = y_{q_k} - \hat{y}_{q_k}$, given that the inner (hidden) layers do not have target values for their output signals, where $\delta_{q_k}^{\alpha}$ - the error (difference) of the output signal from q of the neuron of the output layer (k), $\alpha = \overline{1, p}$. Then the error

$$\delta_{q_k} = F`(NET_{p_j})\delta^{\alpha} = F(NET_{p_j})(1 - F(NET_{p_j})\delta_{q_k}{}^{\alpha} \tag{30}$$

The magnitude of the change in the weight ratio between the neuron p_j and q_k is calculated by the formula:

$$\Delta\omega_{p_j q_k} = \eta\delta_{q_k}OUT_{p_j} = \eta F(NET_{p_j})(1 - F(NET_{p_j})\delta_{q_k}{}^{\alpha}OUT_{p_j}, \tag{31}$$

where OUT_{p_j}- the signal from the neuron p_j, η - the rate of learning.

Then the value of the weight coefficient from the neuron p in the hidden layer "j" to the neuron q in the output layer "k" in the (n + 1) -th step (after correction) is calculated by the formula:

$$\omega_{p_j q_k}(n+1) = \omega_{p_j q_k}(n) + \Delta\omega_{p_j q_k} \tag{32}$$

where $\omega_{p_j q_k}(n)$ - the magnitude of the weight coefficient from the p neuron in the hidden layer "j" to the neuron q in the output layer "k" (to the correction) of the weight coefficients (in step n), $\omega_{p_j q_k}(n+1)$

is the magnitude of the weight coefficient in the step (n + 1) from the p_j neuron to q_k after correction), $\Delta\omega_{p_jq_k}$ - the value for which it is necessary to change (correct) the weight coefficient, which comes from the neuron p_j and enters the neuron q_k.

Each neuron of the latent layer "j" transmits its OUTij signal to the neurons in the output layer "k" by connecting their weight bonds, and during the training, the weight bundles are directed in the opposite direction, directing the value δ_{q_k} from the output layer "ξ" to the hidden layer "j", each weight coefficient is multiplied by the magnitude of the error of the neuron of the output layer "k", from which the weighting factor is obtained. Weights that connect the output layer (k) and the preceding layer (j) are multiplied by the magnitude of the error, between the target and actual values of the output signal from the given neuron of the output layer.

In the hidden layer (j), the product weight coefficients on the error of the output layer, in the block MB find the sum $\sum_q \gamma_{p_{qk}} \delta^\alpha_{q_k}$ of these products and multiply the derivative of the logistic function $F(NET_{p_j})\left[1 - F(NET_{p_j})\right]$. Then the magnitude of the error for the neuron p of the hidden layer is calculated by the expression:

$$\delta_{p_jq_k} = F(NET_{p_j})(1 - F(NET_{p_j})\sum \delta_{q_k}\lambda_{p_jq_k} \tag{33}$$

After calculation δ_{pq_k}, the weighting factors correlating the output layers (k-th) and the preceding (j-th) layer are corrected. That is, for each neuron of the jth hidden layer, the error δ_{pq_k} and the value of the corrected weight coefficients are calculated, until all weighting factors are corrected.

Enter the designation:

1) $C_i(\delta^k_q)$ - The set of errors δ^k_q from the q-neurons of the output layer (k-th), where $q = \overline{1, s}$.

2) $C_j(\delta^j_p)$ is the set of errors from p-th neurons of the j-th layer, where $p = \overline{1, k}$.

3) $W^j_k(\gamma^{q_k}_{p_j})$ is a matrix of weight coefficients of q-th neurons of the kth layer of the th-th neurons of the j-th layer.

Then, in the vector form, the operation of the reciprocal distribution of the error from the q-neurons of the output layer (k-th) to the p-th neurons of the j-th layer can be written as follows:

$$C_j(\delta^j_p) = \left[C_k(\delta^k_p)W^T_k(\gamma^{q_k}_{p_j}) \right]F`(NET_{p_j}), \tag{34}$$

where $F`(NET_{p_j})$ is the derivative of the activation function of the p-th neuron in the j-th (hidden) layer; $\gamma^{q_k}_{p_j}$ - weight factors that come out of q-th neurons of the k-th layer and are included in the p-th neurons of the j-th layer; $W^T_k(\gamma^{q_k}_{p_j})$ - transposed matrix of weight coefficients, which connects the hidden layer neurons (j-th) with the neurons of the output layer (k-th).

If the activation function is a logistic function $F(NET) = \dfrac{1}{1 + 5^{-NET1}}$, then

$$F`(NET_{p_j}) = F(NET_{p_j})(1 - F(NET_{p_j})).$$

Then $!_{j}(\delta_p^j)$ we write in the form:

$$C_j(\delta_{p_j}) = \left[C_k(\delta_{q_k}) W_k^T(\gamma_{p_j}^{q_k}) \right] F(NET_{p_j}) \cdot \left[1 - F(NET_{p_j}) \right]. \tag{35}$$

That is $!_{j}(\delta_{p_j})$, the result of the product of the vector of the initial layer (k-th) $!_{k}(\delta_{q_k})$ on the transpose matrix of weight coefficients $W_k^T(\gamma_{p_j}^{q_k})$, which connects the hidden layer neurons (j-th) with the neurons of the output layer (k-th), and the result of the component multiplied by the derivative of the logistic function of the corresponding (p_j) j-th layer neuron.

To accelerate the learning process, it is enough for each neuron to determine the training bias θ, which provides a change in the position of the initial reference of the logistic function, by providing in the training algorithm an increase in the numerical value of the weight coefficients ($\Delta\omega_i$) to each neuron. Moreover, the additional weighting factors take part in the process of learning NN. There are other methods for accelerating NN training for the algorithm of back propagation of the signal. Methods that accelerate the learning process and ensure its stable stability include the method of momentum and the method of exponential smoothing, each of which has both positive and negative manifestations, depending on what tasks they are offered to solve.

Thus, the "pulse" method involves an increase in the correction of the weight coefficient by a value proportional to the value of the previous change in weight coefficients. That is, the magnitude of each correction of weight coefficients affects the nature of all subsequent corrections. Consider the correction equation at the stages "n" and "n + 1", which can be written symbolically in the following way:

$$\omega_{p_j q_k}(n+1) = \omega_{p_j q_k}(n) + \Delta\omega_{p_j q_k}(n+1), \tag{36}$$

where $\omega_{p_j q_k}(n)$ is the value of the weight coefficients between p-th neurons of the j-th layer and q-neurons of the k-th layer at the stage n (before correction), $\omega_{p_j q_k}(n+1)$ - the importance of the weight coefficients between the neurons of the j-th layer and the q-mi neurons k-th layer at the stage (n + 1) (after correction) $\Delta\omega_{p_j q_k}(n+1)$ is the value for which the numerical value of the weight coefficients between the neurons of the j-th layer and the k-th layer at (n + 1) stage (after correction) has changed.

The magnitude of the change in weight coefficients $\Delta\omega_{p_j q_k}(n+1)$ at (n + 1) stage is calculated by the formula:

$$\Delta\omega_{p_j q_k}(n+1) = \eta\delta_{q_k} OUT_{p_j} + \alpha\Delta\omega_{p_j q_k}(n), \tag{37}$$

where $\Delta\omega_{p_j q_k}(n+1)$ - the value for which the numerical value of the weight coefficients between neurons p_j and neurons q_k changed at (n + 1) stage (after correction); η- coefficient of speed of training NM $(0,01 \leq \eta \leq 0,1)$; δ_{q_k} - the magnitude of the error for the q-th neuron in the output layer k; OUT_{p_j} - the magnitude (output signal) of the p-th neuron in the hidden layer j; α- the pulse coefficient, the value of which is taken in the vicinity of the number 0.9 (approximately 0.9, $\Delta\omega_{p_j q_k}(n)$ - the magnitude of the change in the weight coefficients between the neurons "p" in the hidden layer "j" to the neurons q-them in the initial layer "k", at the stage " n "(before correction).

For the method of exponential smoothing, the equation of correction of weight coefficients at the stages "n" and "n + 1" can be written as follows:

$$\omega_{p_j q_k}(n+1) = \omega_{p_j q_k}(n) + \eta \Delta\omega_{p_j q_k}(n+1), \tag{38}$$

where $\omega_{p_j q_k}(n+1)$ - The value of weight coefficients at the stage (n + 1) (after correction). $\omega_{p_j q_k}(n)$ - The magnitude of the weight coefficients between p-neurons of the hidden layer j and q-neurons of the initial layer k in the p-stage (to the correction); η - coefficient of speed of training NM $(0,01 \leq \eta \leq 0,1)$, affects the average value of change of weight coefficients; $\Delta\omega_{p_j q_k}(n+1)$ - the value to which the corrected numerical values of the weight coefficients between r-th neurons of the k-th layer at (n + 1) stage have changed.

The magnitude on which the weight coefficients $\Delta\omega_{p_j q_k}(n+1)$ are corrected for the (n + 1) stage are calculated by the formula:

$$\Delta\omega_{p_j q_k}(n+1) = \alpha \delta_{q_k} OUT_{p_j} + \alpha \Delta\omega_{p_j q_k}(n), \tag{39}$$

where $\Delta\omega_{p_j q_k}(n+1)$ - the value for which the weight coefficients were corrected on the (n + 1) stage between the neurons j and the neurons of the k-th layer, α- the smoothing factor $(0,01 \leq \alpha \leq 1,0)$; δ_{q_k} - the magnitude of the error for the q-th neuron in the output layer k; OUT_{p_j} - the magnitude of the output signal from the p-th neuron in the hidden layer j; $\Delta\omega_{p_j q_k}(n)$ - the value from which corrected weight factors at the "n" stage between neurons j-th and neurons k-th layer. If $\alpha = 1,0$, then the new correction is not carried out, but the previous one is repeated, and if $0 \leq \alpha \leq 1$, the correction of the weight coefficients is smoothed in proportion to the value α.

Theoretically it is found, according to Kolmogorov's theory, that for simulation of any task a sufficiently multilayer perceptron with 2 intermediate layers is possible, but it is possible that for solution of some problem the simpler and more convenient NN will be with more layers, and for the vast majority of tasks are only one intermediate (hidden) layer, and the two layers are used as a reserve in individual cases (Parwin, 2015). NN with 3 balls apply very little (infrequently).

In the application of NN with the back propagation of the signal, the negative moment is the indefinite long learning process of the network, which is the result of a non-optimal choice of the length of step η, which can cause paralysis of the network or entry into the local minimum. If the size of step Z is very small, then the convergence is rather slow, and if η is very large, then there may be paralysis or constant instability of the neural network. The learning process of the NN must take place on all elements of the training set without passing through the previously studied one. The correction of weighting factors should be calculated on the whole training set and only after a certain number of training cycles, the weighting factors of the equipment have a minimum error on the output of the NN.

To improve and generalize the main algorithm for back propagation of the signal, a large number of methods have been developed in different countries for solving various applications, for example:

1) In Japan, the NEC company applied a reverse signal propagation for visual recognition of letters.
2) Conversion of printed English text into high-quality language.
3) Machine recognition of handwritten English words.
4) To convert images that are eight times better than the input data.

FUTURE RESEARCH DIRECTIONS

It is our belief, that artificial intelligence can be applied not only in aviation but also in other industries. Everywhere, where you need to calculate and make decisions with minimal risk, make decisions in unusual situations, give recommendations to the operator about further actions based on the collection and processing of diagnostic data. Neuroinformatics can be applied in medicine, security systems, the economy, in any industries where reliable communication and security are important. Already at the moment, scientists around the world have achieved incredible results in the recognition and processing of data by artificial intelligence. But this is only the beginning. Further exploration and development will bring great discoveries in the 21st century.

CONCLUSION

When applying the reciprocal distribution algorithm in the learning process, the NN minimizes the error in the training set, and not the error that was obtained from the NN in the processing of completely new observations. NN with a large number of weight factors can solve more complex tasks, at the same time they are more prone to retraining, and with such a number of weight coefficients NN is not flexible enough to solve certain types of tasks. NNs without intermediate (hidden) layers can solve (simulate) only tasks with ordinary linear dependence. At the same time, the NN with a more complex structure gives an output error, but it can be an overload of the network, which can be prevented by using the check-up mechanism (cross-check) to reserve part of the training observations in order not to use them in the learning process. An important indicator of the quality of NN work is the control error. For almost identical values of control errors, it is necessary to choose NN with a simpler model. With multiple experiments over NN, a control error is decisive when choosing a model for a future NN. In order to confirm the correct choice of model, it is tested on a test set of observations, which is used only once. That is, all classes of training on the algorithm of reciprocal propagation foresee the use of the following sets:

1) Educational.
2) Control.
3) Tests that should be representative from the point of view of the content of the task, as well as individually taken, these sets should be representative.

If the training set is not representative, then the NN model will not be qualitative, or completely (unnecessary) not suitable for solving tasks.

The main reasons that worsen the quality of the training set:

1) The choice of historical data law, in which there is no place in the future development of the process, that is, the future does not always resemble the past.
2) Insufficient level of taking into account the properties of NN data (did not take into account all possibilities of the neural network).
3) Insufficiently deeply analyzed the selected all possible properties of the object of the analysis of NN (neural network is learning all that is easier and faster to learn).
4) Imbalance of the set of data of the training set, that is, the unevenness of the selection of data of observations of different types.

REFERENCES

Anderson, E. W. (1994). Cross-category variation in customer satisfaction and retention. *Marketing Letters*, *5*(1), 19–30. doi:10.1007/BF00993955

Barsky, A. (2007). *Logical Neural Networks: Internet University of Information Technologies, Bean.* Laboratory of Knowledge.

Barsky, A. (2013). Logical Neural Networks. Moscow, Russia: Internet University of Information Technologies (INTUIT).

Bunakov, V. (2015). *Neuron physics. Tutorial: monogr*. Moscow, Russia: Publishing House of St. Petersburg University.

Callan, R. (2003). Basic concepts of neural networks. Moscow, Russia: Williams Publishing House.

Chernodub, A., & Dzyuba, D. (2011). Review of methods of neuro management. *Problems of programming*, 2, 79-94.

DeAgostini. (2007). *How is the human body structured? Iss. 25. Neurons*. Moscow, Russia: Author.

Elisov, L., & Ovchenkov, N. (2017). Aviation security as an object of mathematical modeling. *Scientific Herald of the Moscow State Technical University*, *20*(3), 13–20.

Golovinsky, P. (2012). *Mathematical models. Theoretical physics and analysis of complex systems*. Book 2. From nonlinear oscillations to artificial neurons and complex systems.

Haikin, S. (2006). *Neural networks: Comprehensive foundation* (2nd ed.). Moscow, Russia: Williams.

Haikin, S. (2009). *Neural networks: full course*. Moscow, Russia: Williams.

Jesse, R. (2012). *Artificial neural network*. Moscow, Russia: VSD.

Jesse, R. (2012). *Neuron*. Moscow, Russia: VSD.

Kalatskaya, L., Novikov, V., Sadov, V. (2003). *Organization and training of artificial neural networks: Experimental study*. Minsk, Belarus: Publishing House of the BSU.

Khusainov, D., Korenyuk, I., & Gamma, T. (2012). *Mechanisms of rhythmic activity of grape snail neurons*. LAP Lambert Academic Publishing.

Koshkin, R. (2014). Mathematical models of the processes of creation and functioning of search and analytical information systems for civil aviation. *Scientific Herald of the State Research Institute of Civil Aviation*, 5, pp. 39-49.

Kotikova, E. A. (1979). *Catecholaminergic neurons*. Moscow, Russia: Nauka.

Kruglov, V., & Borisov, V. (2002) Artificial neural networks. Theory and practice: monogr. Ed. 2. Moscow, Russia: Hot line – Telecom.

Mandelstam, Yu. (1983). *Neuron and muscle of the insect: monogr*. Moscow, Russia: Nauka.

Marasanov, P. (2015). Evaluation of the reliability of the civil aviation airport power supply system. *Scientific Bulletin of the Moscow State Technical University of Civil Aviation*, *213*(3), 43–49.

Mazurov, M. (2018). Synchronization of relaxation self-oscillating systems, synchronization in neural networks. *Izv. RAS. Ser. Physical*, *82*(1), 83–87.

Neural networks. (2008). Statistica neural networks. *Methodology and technologies of modern data analysis*. Moscow, Russia: Hotline - Telecom.

Ovchenkov, N., & Elisov, L. (2014). Vulnerability assessment of transport infrastructure and vehicles in civil aviation. *Scientific Herald of the Moscow State Technical University,* 204, pp. 65-68.

Parwin, M. (2015). *Concert for neurons and synapses*. Moscow, Russia: Nauka.

Pozin, N. (1970). *Modeling of Neural Structure*. Moscow, Russia: Nauka.

Rashid, T. (2017). *Create a neural network*. St. Petersburg, Russia: LLC Alpha-book.

Russell, J. (2013). *Verbalization of Neural Networks*. Moscow, Russia: VSD.

Russell, J. (2013). *Artificial Neuron*. Moscow, Russia: VSD.

Shibzukhov, Z. (2006). *Constructive methods of teaching sigma-p neural networks*. Moscow, Russia: Nauka.

Tatuzov, A. (2009). *Neural networks in the problems of radar*. Moscow, Russia: Radiotechnic.

Tolkachev, S. (2016). *Neural programming of dialogue systems*. Moscow, Russia: RSUU.

Yasnitsky, L. (2005). *Introduction to artificial intelligence*. Moscow, Russia: Center. *Academia (Caracas)*.

Yurevich, A. (2014). *Neural networks in economics*. Moscow, Russia: Lambert Academic Publishing.

KEY TERMS AND DEFINITIONS

A&A: Airports and Aerodromes.
AcB: Activation Block.
AdB: Adjacent Block.
ANN: Artificial Neural Network.
FSP: Forward Signal Passing.
LP: Learning Pair.
MB: Matching Block.
NN: Neural Network.
OUT: Outgoing Neural Network Signals.
RPS: Reserve Passage of the Signal.

Chapter 13

Artificial Intelligence Methods in Aviation Specialist Training for the Analysis and Transmission of Operational Meteorological Information

Sergiy I. Rudas
National Aviation University, Ukraine

Evgeniya A. Znakovska
 https://orcid.org/0000-0002-9064-6256
National Aviation University, Ukraine

Dmitriy I. Bondarev
 https://orcid.org/0000-0003-1728-6837
National Aviation University, Ukraine

ABSTRACT

The authors present methods for the application of artificial intelligence for operational meteorological information (OPEC). The means of communication for distribution of meteorological data using information technologies are presented. Practical courses for aviation specialists (pilots, air traffic controllers, operators of unmanned aerial vehicles) are considered in which artificial intelligence methods are used: datamining, deep learning, machine learning, using information technologies.

DOI: 10.4018/978-1-7998-1415-3.ch013

BACKGROUND

The measures for centralization and automation carried out within the framework of meteorological authorities led to the development and implementation of automated pre-flight systems information. Flight crew members, operators and other personnel involved in flight operations, may receive pre-flight information, self-instruct, advice and flight documentation through an automated pre-flight information system. Through these systems. Some of these systems are solely for the above purposes, and others serve as an integrated information system, which may not be limited to meteorological part of preflight planning. There are several systems that provide users with coordinated unified access to the AIS and MET pre-flight information. Automated pre-flight information systems can be part of a multi-purpose system (Annex 3, 2016).

The primary means of communication for distributing OPMET information outside the airfield is served by the AFTN network and satellite broadcasting systems within the framework of the artificial intelligence in aeronautical fixed service (AFS). Both the network itself and broadcasting are part of the service.

AFS, which covers telecommunication systems used for international air navigation, except for transmissions on one-way ground-to-air channels. International Databases ICAO's OPMET, which can be reached through the AFTN network, can provide interregional and regional OPMET information exchanges and dissemination (Annex 3, 2016).

The purpose of OPMET information with the help of artificial intelligence is to provide aviation meteorologists and assistants with a basic understanding of the significance each item of information used in pre-flight planning has in the preparation for a flight. Although some re-planning is often carried out in flight (e.g. when considering the acceptance of a different flight level, an alternative airway routing offered by air traffic control or a change of destination), the use made of the meteorological information required for such re-planning is similar to that in pre-flight planning. Flight preparation has three phases: the take-off and climb to cruise altitude; the cruise to the top of descent; and the approach and landing. These phases are not treated separately as they are interdependent, but for explanatory purposes, it is convenient to consider the specific use made of meteorological information in each of the three phases (Doc. 8896, 2011).

It is the pilot's duty to optimize the performance of the aircraft, in order to maximize the economics of the operation while at the same time complying with all the requirements for take-off (including take-off minima) specified by the operator and approved by the State of the Operator and by the authority responsible for the aerodrome. The planning for the take-off and climb-out phase includes calculating, by the pilot, the maximum permissible take-off mass (standard operating mass + passengers + cargo + fuel, etc.) given the constraints at a particular aerodrome. These constraints include runway length, runway slope, climb-out gradients (which ensure clearance of obstacles with one engine failed), aerodrome elevation, and current meteorological conditions, i.e. surface wind (specifically headwind component and limiting tailwind and crosswind components), temperature and pressure. Humidity, although theoretically also affecting aircraft performance, can be neglected as its effect is minimal. Runway contamination (snow or slush covered, wet, icy, etc.) also plays an important role, but is not usually regarded as "meteorological information". Where aircraft take-off mass is not limited by aircraft performance considerations in the prevailing meteorological conditions, temperature has an effect on take-off speeds and on engine power settings and on the possible need to initiate engine and airframe anti-ice procedures.

The list of items that have to be considered in take-off calculations is rendered more manageable by the use of graphs, charts, nomograms, and tables, etc., produced by the operator to assist the pilot or flight operations officer. In many operations, flight planning, particularly for the en-route stage, is carried out by computer. The pilot is able to control at least some of the many variables affecting the take-off performance of the aircraft; one example would be the choice of flap setting, another would be the cargo mass and/or fuel to be uplifted, although clearly, the desire is to maximize the payload consistent with take-off requirements. Any of the various requirements may limit the operation, resulting in a lower payload or fuel uplift than desired, which may result in the need to land en route in order to refuel or, in extreme circumstances, preclude take-off (at a given mass) altogether (Leshchenko, 2013).

Meteorological conditions en route and meteorological conditions at the destination and alternate aerodromes are elements that are superimposed on the initial flight plan based on temperature and wind. Adverse en-route weather conditions may force the choice of a flight level or route segment not conforming with the optimum one given by the flight plan, although such changes are rare with modern jet aircraft. Unfavorable conditions expected at a destination may force a delay in take-off or the preparation of additional flight plan segments to alternate aerodromes (Ermakova, 1987).

During the flight, pilots may wish to optimize aircraft performance by taking advantage of more favorable winds at another flight level. This situation may arise because initially the aircraft was unable to climb to this level due to air traffic control constraints, or it was too heavy to climb to the level with the most favorable tailwinds. As the aircraft mass progressively decreases as fuel is burned off, the pilot may request reclearance to a higher level. The information available to the pilot in considering these matters is greatly enhanced by the increased use of an onboard inertial reference system (IRS), which has the capability of giving instantaneous wind readouts. Many systems also give information on the increased headwind that can be tolerated by going to a higher level so as to take advantage of the decreased fuel consumption normally found at higher flight levels. This is usually referred to as a "wind/altitude trade" (Doc. 8896, 2011).

For landing, there are two basic considerations: the length of the runway and the missed approach capability. The speed flown by aircraft on approach is a function of the stall speed which is determined by the aircraft mass, all other things being equal. The speed on touchdown will be the indicated airspeed flown plus or minus the headwind/tailwind. The presence of a headwind means that the aircraft will land at a lower ground speed and will, therefore, useless distance to stop. The opposite effect is felt with a tailwind. The stopping distance on the runway is also affected by the runway being wet, as brakes are less effective in these conditions. In addition, aircraft have tailwind and crosswind limits, and again these are lower in wet conditions than in dry; typical limits are shown in Figure1 (Doc. 8896, 2011).

For the missed approach possibility the same factors as runway length must be considered, e.g. temperature and pressure altitude. Also, when icing conditions are present, ice formation on the wing and fuselage will adversely affect performance. A chart illustrating the effect of relevant meteorological factors on landing performance, including climb capability for a missed approach procedure, is given in Figure 2 (Doc. 8896, 2011).

MEANS OF DISTRIBUTION OF METEOROLOGICAL DATA

The aeronautical meteorological offices, using appropriate ICAO and WMO communications, are provided with the following meteorological information (Table 1):

Figure 1. Typical wind limit diagram

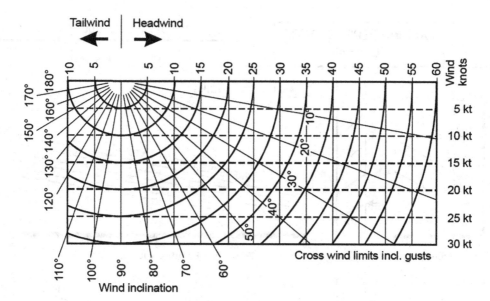

Table 1. Meteorological information

System	Newsletter	Data Type
AFTN/CIDIN (ICAO)	Ground communications	Aerodrome reports (METAR), (SPECI) including, if required, forecasts for landing type TREND,9,18,24 and 30-hour aerodrome forecasts and modified aerodrome forecasts (TAF); SIGMET information, including information on volcanic ash clouds and tropical cyclones; Advisory information on volcanic ash and tropical cyclones. Special air-reports (AIREP), service reports on system operation (runway status reports).
SADIS (ICAO)	Satellite multi-channel communication system (satellite two-way communication system)	Cards in code form. Information in code forms, OPMET alphanumeric data. Advisory information on volcanic ash and tropical cyclones.
GTS (WMO)	Satellite multi-channel communication system, terrestrial circuits, facsimile radio broadcasts	Mainly terrestrial and upper-air reports, analyzes and forecasts in code forms. Volcanic ash and tropical cyclone advisory information, satellite and radar images, airborne reports.
Internet resources	World wide web	A wide range of meteorological data in alphanumeric and graphic form.
Broadcasting systems at the national level	Satellite multi-channel communication system, facsimile broadcasts, facsimile and telex communication, telephone, video text	OPMET alphanumeric data or graphic data, as well as terrestrial and upper-air information, satellite and radar images

Figure 2. Landing performance calculations

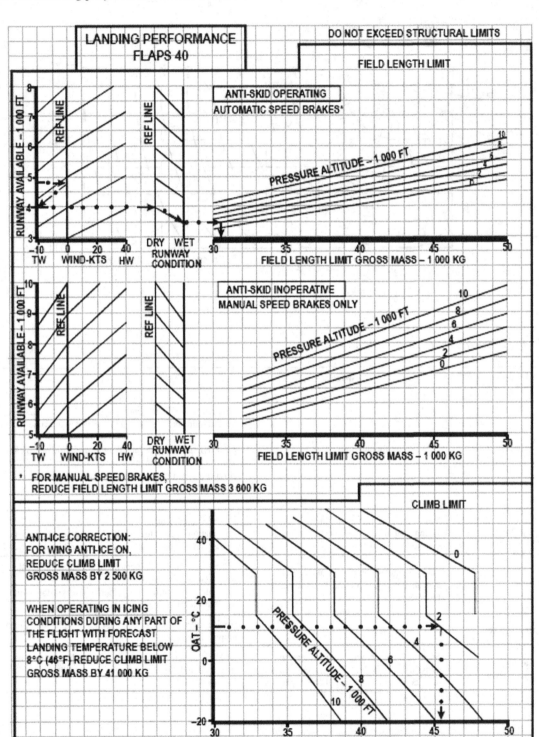

- WAFS data, which include global forecasts of wind, temperature, and humidity at heights, the direction and speed of the maximum wind, the height and temperature of the tropopause, as well as forecasts of special phenomena in digital form.
- Operational meteorological data (OPMET), consisting of regular and special actual weather reports (METAR SPECI), aerodrome weather forecasts (TAF) and area forecasts (GAMET), hazardous weather information (SIGMET, AIRMET), special reports from aircraft (AIREP SPECIAL).
- Advisory data on volcanic ash and tropical cyclones.
- Additional auxiliary and national materials (for example, images obtained by using satellites and radars).

WAFS is an ICAO system providing operational aeronautical meteorological information. This system is based on an organization consisting of the WAFS London and Washington, which provide mutual redundancy so that in the event of a failure of one of the centers, the current materials can continue to be sent using data from another center. One of the responsibilities of ICAO is to ensure that all Contracting States of ICAO have access to all the WAFS data they need through at least one of the aeronautical fixed line (AF) components, i.e. via mutually agreed ground circuits or satellite communications systems (SADIS and ISCS). WAFCs collect all aeronautical meteorological information and send it to states and consumers. In many states, national meteorological services are responsible for making this information available to consumers. Methods for States to transmit this meteorological information to final consumers are different, however, distribution via SADIS and ISCS provides consumers with a standard means of communication that allows them to obtain the required data using AFS (Leshchenko, 2013).

OPMET data are sent using complementary terrestrial and satellite (SADIS) communications. Ground-based communication circuits using advanced AFTN / CIDIN are mainly used to collect data, which are then sent via SADIS. The exchange of many data is also carried out using the Global Telecommunication System (GTS) WMO, operated on the basis of bilateral agreements. GTS consists of terrestrial communications, facsimile broadcasts, and satellite multi-channel communication systems.

OPMET: Data under the exchange lie between aerodromes within the state and at the international level and should be provided with minimal delay:

a) Meteorological reports on the actual weather at the aerodrome should be transmitted no later than 5 minutes after the observations.
b) **Weather Forecasts**: No less than 1 hour before the start of the action.
c) SIGMET, AIRMET information and special reports on weather deterioration at the aerodrome are transmitted immediately.

As an example, we can see different meteorological services that produce some information.

At figure 3 it is shown the website "Flight weather" (Europe-FlightWeather) – the international aviation weather source, like an example it is shown Europe and atmospheric fronts under it.

Figure 4 demonstrates us free meteorological service "Aviation weather center" (Noaa, & National Weather Service). It is shown meteorological data: visibility, wind speed, wind direction, etc.

Figure 5 (Windyty, S. E.) like Figure 1 shows meteorological information.

Figure 3. Open meteorological service "Flight weather"

Figure 4. Open meteorological service "Aviation weather center"

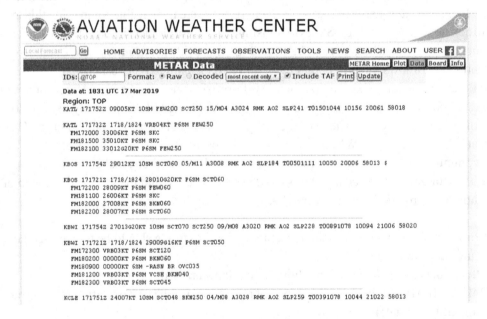

INFORMATION TECHNOLOGIES FOR TRAINING AVIATION SPECIALISTS

Practical courses are conducted for aviation specialists (pilots, air traffic controllers, unmanned aircraft operators), such as Basics of Applied Engineering Technologies, programming of microprocessors and UAV microcontrollers, programming languages, the architecture of computer systems, networks and service platforms. As part of these courses, specialists are trained in the use of various information technologies as well as perform tasks on programming application programs according to the following options:

For air traffic controllers:

Figure 5. Example of another meteorological service

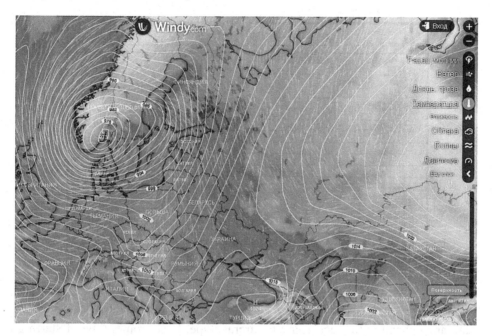

1. To create an application this implements a method of identifying and solving a potential conflict between aircraft during flights on the same tracks. The program should display the current conflict situation and show on screen commands that have to be formulated by the controller in a particular situation.

2. To create an application this implements a method of identifying and solving the potential conflict between aircraft during flights on the same tracks. The program should save the current time and distance remaining between aircraft at the moment of crossing occupied the same track in a single file in the table view.

3. To create an application this implements a method of identifying and solving the potential conflict between aircraft during flights on the same tracks. The program should have an interface of input characteristics of aircraft from the keyboard, calculate Scross according to these characteristics and determine the possibility or impossibility of descending.

4. To create an application that implements fill the middle part of the flight plan (FPL) (Fig. 6) along the route Borispol - Zaporizhzhia/Mokraya.

The result of the program should be checking the correctness of filling the flight plan according to the task option. In the case of the correct filling of the flight plan, all fields are painted in one color, in case of incorrect - in another color.

5. To create an application this implements a method of identifying and solving the potential conflict between aircraft during flights on the same tracks. The program should have an interface of input characteristics of aircraft from the keyboard, calculate actual time and linear interval of divergence of aircraft, identify necessary changes in the speed of aircraft flight.

Figure 6. International flight plan (middle section) (Flight plan, 2019).

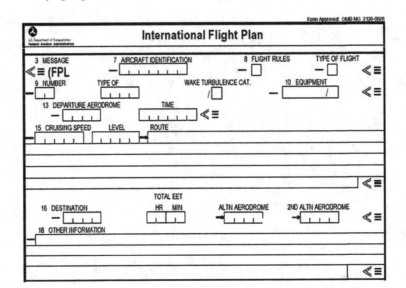

6. To create an application that according to aircraft type, flight route, time of starting maneuver, speed and wind direction, calculates elements of flight during climb/descend to specified flight level. The heading is determined depending on the flight route of the educational training zone. The input data should be read out from separate files that store information about technical characteristics of aircraft and flight route, time of starting maneuver, speed and wind direction. Results of calculations and controller's commands about further actions in a particular situation should be issued on the screen.

7. To create an application that implements the method of identifying and resolving potential conflicts between aircraft during flights on the same tracks. The program should have input interface characteristics of aircraft with the keyboard, Sper regarding the calculation of these characteristics and determine the possibility or impossibility decline.

8. To create an application that implements the method of identifying and resolving potential conflicts between aircraft during flights on tracks that intersect. The program should reflect the current situation of conflict and display on-screen commands to be defined by the controller in a particular situation.

9. To create an application that implements the method of identifying and resolving potential conflicts between aircraft during flights on tracks that intersect. The program should have input interface characteristics of aircraft from the keyboard, to calculate the actual time difference and line spacing of aircraft, identify necessary changes in the speed flight of aircraft.

10. To create an application, which according to the type of aircraft, flight route, time of start maneuver, speed and wind direction, calculates flight elements during on-boron / decrease to the target level flight. Track flight angle determined depending on the flight route learning Trng flax-zone. The input must be read from separate files that store information about the flight performance of aircraft and flight route, start maneuver, speed, and wind direction. The screen should be issued together with the results of calculations as the team manager for further action in a particular situation.

For Radioelectronics and operators of unmanned aerial vehicles
Create an application with GUI under the task variants.

1. There is a billboard with a picture on the turn of the road. An unmanned aerial vehicle (UAV) is approaching the billboard (Fig. 7). The black lines show the boundaries of the image that transfers the UAV to the operator's screen. It is necessary to develop a program with a graphical interface, which shows the operator's screen and the image that transmits the UAV.
2. To create a program that reads the METAR's meteorological aviation code from the file and displays its decoding in Ukrainian or Russian.
3. To create a program that reads the TAF's meteorological aviation code from the file and displays its decoding in Ukrainian or Russian.
4. To create a program that allows the user to create the METAR's meteorological aviation code in the user interface (the interface must have a button saving the code in a file).
5. To create a program that builds on the map the approach (STAR) (Fig. 8)) using data (Table 2):
6. There is a billboard with a picture on the turn of the road. An unmanned aerial vehicle (UAV) is approaching the billboard (Fig. 9). The black lines show the boundaries of the image that transfers the UAV to the operator's screen. It is necessary to develop a program with a graphical interface, which shows the operator's screen and the image that transmits the UAV.
7. To create a program that reads the TAF's meteorological aviation code from the file and displays its decoding in Ukrainian or Russian.
8. To create a program that allows the user to create the TAF's meteorological aviation code in the user interface (the interface must have a button saving the code in a file).
9. To create a program that builds on the map the approach (STAR) (Fig. 8) using data (Table 3):
10. There is a billboard with a picture on the turn of the road. An unmanned aerial vehicle (UAV) is approaching the billboard (Fig. 10). The black lines show the boundaries of the image that transfers the UAV to the operator's screen. It is necessary to develop a program with a graphical interface, which shows the operator's screen and the image that transmits the UAV.

Figure 7. An unmanned aerial vehicle (UAV) is approaching the billboard

Figure 8. STAR (AIS of Ukraine).

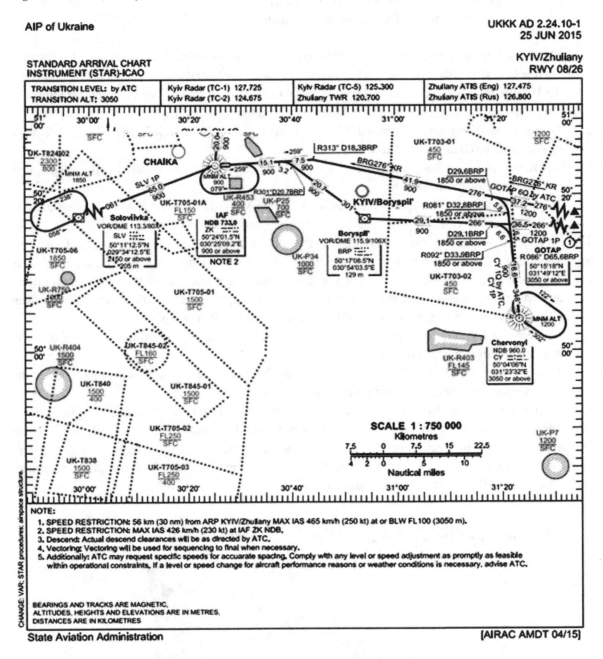

Table 2. Data for building a route

Pointer	Route	Altitude
CY 6A Chervonyi SIX ALPHA	On the road 350° to R094° BRP VOR, turn left onto the track 266° (R086° BRP VOR) on BRP VOR/DME, turn right on R299° BRP VOR, in D22,0 BRP DME turn left onto the track 261° (BRG 261° ZK NDB) for IAF ZK NDB	Cross the CY NDB, BRP VOR/DME to 2150 or higher, IAF ZK NDB to 900 or higher.

Table 3. Data for building a route

Pointer	Route	Altitude
SL 6E Soloviivka SIX ECHO	On road 063° (BRG 063° ZK NDB) to IAF ZK NDB	Cross the SL NDB to 2150 or higher, IAF ZK NDB to 900 or higher.

Figure 9. An unmanned aerial vehicle (UAV) is approaching the billboard

Figure 10. An unmanned aerial vehicle (UAV) is approaching the billboard

Illustrative Example

Illustrative Example shows the realization by specialists of the methodology of solving conflict situations on the same tracks.

First, the theoretical data is studied and made necessary notifications. Then to create the program is performed the several steps that are described below:

1) Firstly create a block-scheme and then is used as a base of the future program.
2) Input given data into the program.
3) Create variables that should be input from the keyboard.
4) Calculated all necessary formulas.
5) Create several cycles to obtain necessary results that are dependent on input data.

6)   Performed steps to output all results, which depend on the given data onto the screen.
7)   To make conclusions which will depend on the result.

Next, is created the block-scheme and mathematical illustration that has built for the given task (Figure 11).

Next, is created the program that calculates Sper (distance between aircraft) and determines the conflict or not situation. It is depending on the input data. Also, the program transfers all input data into system SI and after this start to calculate all values according to formulas. According to the results, comparing classify the situation as a conflict or not.

If the situation is conflict, the air traffic controller gives further instructions to the pilot.

Depending on the situation, ATC can give or not give:

Permission to decrease (climb) may be issued if the condition is performed, otherwise, the controller should use another method:

- Creation of Lateral Spacing.
- Limitation of Velocity of aircraft.
- Changes of Flight Levels of aircraft.

To demonstrate the operation of the application were used data, which were input into the program. Are used the data as an example of a safe situation, when descending or climbing is possible (Figure 12).

Are used the data as an example of a safe situation and obtained that in this case, Sper (Distance between aircraft) is greater than Sbez (Safe distance). That's why we haven't a conflict situation and descending or climbing may be performed.

And now let's consider another variant, and demonstrate the operation of the application in the case of conflict situation (Figure 13).

Are used the data into the program and obtained that in this case Sper is less than Sbez. That's why we have a conflict situation and the descending of the aircraft cannot be performed.

The program can help and simplify the workload of air traffic controllers.

Specialists are involved in the research projects and obtain such copyright certificates for computer programs.

FUTURE RESEARCH DIRECTIONS

Our further research will be aimed at simplifying the understanding of users of Airspace (pilots and other users) not only weather at the aerodrome, but also meteorological codes and maps that describe the weather conditions in the areas of flights and on the route, namely, simplifying the understanding of GAMET codes and tracking maps special weather phenomena, as well as warnings about weather hazards, which are contained in the SIGMET and AIRMET information.

Figure 11. The block-scheme for the task

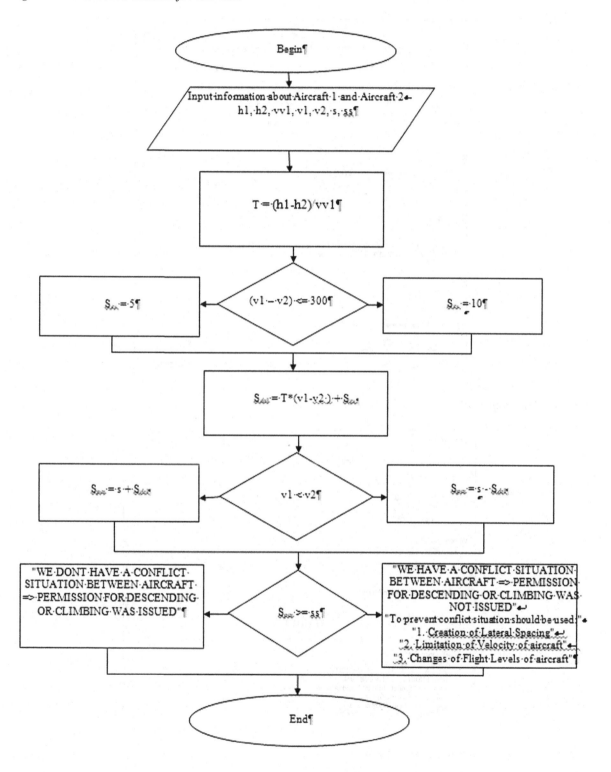

Figure 12. Example of a safe situation

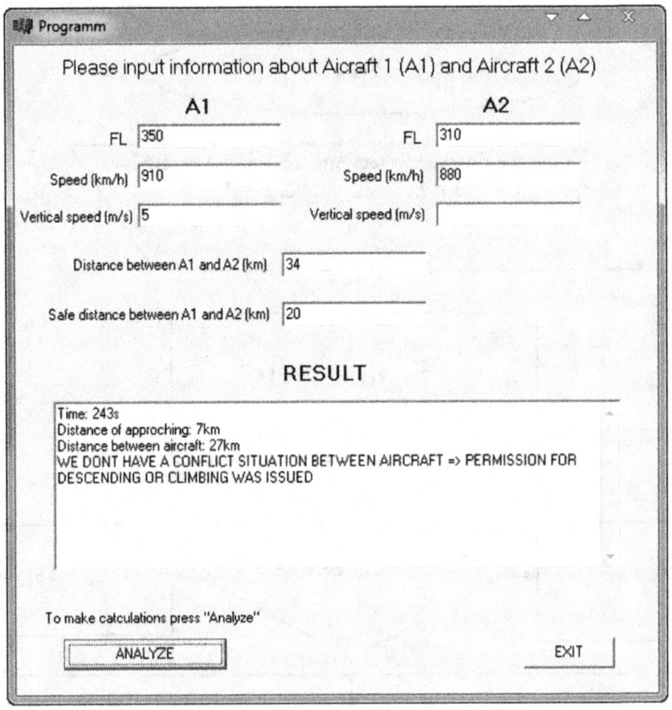

Figure 13. Example of a conflict situation

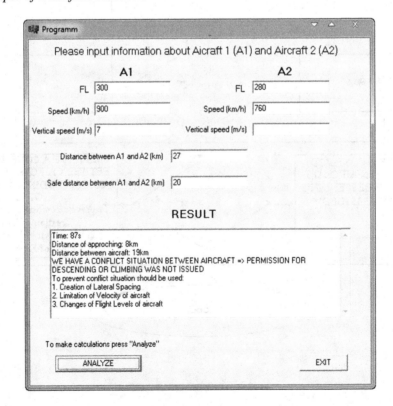

CONCLUSION

Methods for the application of artificial intelligence for Operational meteorological information (OPEC) were presented. Practical courses for aviation specialists (pilots, air traffic controller, operators of unmanned aerial vehicles) are considered in which artificial intelligence methods are used: data mining, deep learning, machine learning, using information technologies.

REFERENCES

AIS of Ukraine. (n.d.). Retrieved from http://www.aisukraine.net/titul_en.php

AN-Conf/13-WP/3. (2009). *The Thirteenth Air Navigation Conference of ICAO.* Montreal, Canada, Oct. 9-19, 2018.

Annex 3. (2016). *Meteorological service for international air navigation. Nineteenth edition.* Montreal, Canada: ICAO, 2016.

Bader, M. J. (1995). Images in weather forecasting. New York, NY: Cambridge University Press.

Doc 10003. (2014). *Manual on the digital exchange of aeronautical meteorological information.* Montreal, Canada: ICAO.

Doc 7192. (2011). *Training Manual. Part F-1: Meteorology for air traffic controllers and pilots.* Montreal, Canada: ICAO.

Doc 7488. (2007). *Manual of the icao standard atmosphere extended to 80 kilometres (262,500 feet).* Montreal, Canada: ICAO.

Doc. 8896. (2011) *Manual of aeronautical meteorological practice. AN/893 Ninth edition.* Canada, Montreal: ICAO.

Doc 9328. (2005) *Manual of runway visual range observing and reporting practices.* Montreal, Canada: ICAO.

Doc 9761. (2003) *Meteorological support for air traffic controllers and pilots.* Montreal, Canada: ICAO.

Doc 9873. (2007) *Manual on the quality management system for the provision of meteorological service to international air navigation.* Montreal, Canada: ICAO.

Doc 9974. (2013) *Flight safety and volcanic ash.* Montreal, Canada: ICAO.

Ermakova, A. I. (1987). *Features of the analysis and assessment of meteorological conditions to ensure the safety of flights on international airlines.* London, UK: Hydrometpublish.

Europe - FlightWeather. (0AD). Retrieved from https://sites.google.com/site/acnetworkweather/home/europe

Flight plan. (2019, April 15). Retrieved from https://en.wikipedia.org/wiki/Flight_plan#/media/File:International_flight_plan.png

Leshchenko, G. P. (2013) Aviation meteorology: Textbook. Kirovograd, Ukraine: Publishing house КЛА НАУ.

Membery, D. A. (1983). Low level wind profiles during the Gulf Shamal. *Weather*, 38.

NOAA, & National Weather Service. (2013, March 15). AWC - Aviation Weather Center. Retrieved from https://www.aviationweather.gov/metar/data

Pearson, G. M. (1988). *Satellite imagery interpretation in synoptic and mesoscale meteorology*. Environment Canada.

Windyty, S. E. (0AD). Windy as forecasted. Retrieved from https://www.windy.com/

KEY TERMS AND DEFINITIONS

Operational Meteorological Information: A statement of observed meteorological conditions related to a specified time and location.

Meteorological Information: The meteorological report, analysis, forecast, and any other statement relating to existing or expected meteorological conditions. aviation specialists.

Operational Flight Plan: The operator's plan for the safe conduct of the flight based on considerations of airplane performance, other operating limitations and relevant expected conditions on the route to be followed and at the aerodromes concerned.

SIGMET Information: Information issued by a meteorological watch office concerning the occurrence or expected occurrence of specified en-route weather and other phenomena in the atmosphere that may affect the safety of aircraft operations.

Chapter 14
Methods for the Synthesis of Optimal Control of Deterministic Compound Dynamical Systems With Branch

Olena Tachinina

National Aviation University, Ukraine

Oleksandr Lysenko

Igor Sikorsky Kyiv Polytechnic Institute, National Technical University of Ukraine, Ukraine

ABSTRACT

This chapter states the result of the development of optimal control methods for deterministic discontinuous systems of optimal control problems for deterministic compound dynamical systems (CDS) with branching paths. The necessary conditions for optimality of the CDS branching paths are formulated in the form convenient for subsequent development of algorithms for the operational synthesis of these paths. The optimality conditions developed by the authors allow both preliminary and in real time (on-line) optimization of the CDS branching paths. The need for an operational synthesis of the CDS branching trajectory is caused by the inaccuracy of prior knowledge of information about the factors affecting CDS movement which are critical for the implementation of the CDS end-use. The developed conditions are universal for solving problems with any finite number of branches of a branching trajectory and are focused on the use of artificially intelligent systems which allow analyzing the structure of optimal control of CDS components as they move along the path branches.

DOI: 10.4018/978-1-7998-1415-3.ch014

BACKGROUND

Currently, the successes in creating precision mechanical objects, wireless telecommunication systems and high-performance compact on-board computers allow designing up-to-date complex technical systems that solve a single technical task without a mechanical connection and having data exchange between the separate components of these objects.

The section describes the unmanned aerial vehicle (UAV) groups that form a mobile sensor network ("flying sensor networks" or "information robots"), as well as reusable aerospace systems of airborne launch type as an alternative to the up-to-date complex technical systems. Theoretically, these control objects can be classified as compound dynamic systems (CDS), i.e., systems consisting of a set of objects (subsystems) with a controlled interaction between them during motion.

The CDS paths in the scientific publications (Lysenko, Tachinina, 2014a) are called branching, on the grounds that they consist of sections of common motion of the CDS subsystems and sections of their individual movement to a target along individual path branches. The CDS functioning efficiency depends on the operational (real-time) optimal choice of spatial coordinates and time moments when the CDS structural transformations occur, as well as on the operational optimal synthesis of control in the intelligent computer by the CDS components as they move along the path branches during the periods of time between the structural transformations.

Therefore, the problem of operational optimization of the CDS branching paths is considered in the scientific community as topical from academic and practical standpoints (Lysenko, Tachinina, 2014b; Lysenko, Tachinina & Chumachenko, 2015). For example, commercial launches into space of a payload by the aerospace systems of airborne launch type, require reliable guarantee of launch under some previously unpredictable change of weather conditions in the launch area. So, the problem of operational correction (optimization) of the branching trajectory of the aerospace system motion, consisting of the aircraft carrier (AC) and orbital stage (OS) paths at the phase of their joint motion in so-called "bundle", common optimization of the AC and OS paths in the process of partition and initial separation with subsequent ascent to the given points of near-earth space, is topical.

The algorithms of the operative correction of the aerospace system trajectory should be programmed in the artificial intelligent system (AIS) of the AC onboard launch system, this makes it possible to correct on-line the branching path in the launch area and this correction will be taken into account when deciding to maneuver.

We emphasize that the task of operational optimization of the branching path of the aerospace system is also of current importance in emergency situations when necessary to safely and quickly separate the OS from the AC.

The precise operational information about victims in the emergency zone under conditions of heavy destruction of infrastructure (fires, earthquakes, tsunamis, tornadoes, etc.) can be obtained by using the sensors placed in the UAV (mobile sensors) that form the "flying sensor network".

The problem of operational optimization of "group behavior" (optimization of the branching path) of mobile sensors in an aggressive environment arising under emergency situation is important.

The algorithm of operational optimization is programmed in the on-board intelligent computer of a telecommunication platform that controls the movement of mobile sensors.

The success of the search and rescue operation is determined primarily by the consistency of "group behavior" of the elements of "flying sensor network" (Kirichek, Paramonov, & Vareldzhyan, 2015; Moiseev, 2017), which should provide latest and precise (timely and reliable) information about victims and required emergency medical aid. The inconsistency of "group behavior" of mobile sensors can lead to a complete failure of the rescue operation. Theoretically, the solution of problem of optimal control of deterministic CDS with a branching motion path is reduced to the task of controlling a deterministic discontinuous system (Ashchepkov, 1987; Bryson & Ho, 1975; Sage &White, 1982).

The analysis of the cited above and other publications (Boudiba & Firsov, 2019; Ledsinger, 2000; Filatyev, 1994; Filatyev & Yanova, 2013) allows us to conclude that today, the theoretical developments obtained by the authors are not studied carefully as applied tasks to a such level that they can be used to solve the problems of designing the motion path of deterministic CDS in real timescale.

This is because the abstract and formal representation of a mathematical model of the CDS motion in the form of a dynamic discontinuous system leads to the dimensions of state and control vectors of the discontinuous system increase in proportion to the number of path branches and it becomes almost impossible to synthesize the algorithm of operational path optimization in the on-board computer.

The applied researches, which would have initially laid down in the problem statement not an abstract view of the discontinuity of the system, but a clear physical view of the character - pattern of the composite system motion along the path branches - would explain the mechanism for constructing optimal conditions for a branching trajectory with an arbitrary pattern, considered all applied aspects and also gave results in the form of the most accessible to a wide range of applied researchers, in terms of the possibility of organizing computational procedures, realizing these conditions, still have not been conducted.

Thus, we have a conflict situation (inconsistency or contradictions): on the one hand, as it stands now, new CDS like "airborne launch type", "flying sensor networks" are designed in hardware and actively continue to be created, and, on the other hand, to control these CDS the control methods are used that do not allow completely use of all technical properties incorporated in these systems.

Therefore, a scientific problem associated with the development of methods for the operational optimization of control process of a deterministic CDS motion is topical.

These methods should formulate optimality conditions for branching paths of CDS motion in the form convenient for real-time synthesis (operational synthesis of deterministic CDS optimal control) and allow solving the problem of operational synthesis of optimal branching paths of deterministic CDS motion using onboard AIS.

METHOD REPRESENTING THE MATHEMATICAL MODEL OF COMPOUND DYNAMIC SYSTEM MOTION

Today, the compound dynamic systems (CDS) are widely used for solving many applied problems in various fields of science and technology.

The CDS is taken to mean a set of objects (subsystems) with consecutive modes of their operation in time, which is described at different time intervals by various differential equations and some finite connections for joining the trajectories of compound parts.

The examples of practical implementation of such CDS are such aerospace systems (AS): «MAKS», «Hotol», «Zanger», «Hermes», «XL-20», «Hope», «Clipper», which are used as an air platform for launching satellites and delivering cargo to near-earth and space orbits.

Modern mechatronics and wireless telecommunications have progressed to the stage where it became possible to create the CDS solving a single technical problem, without mechanical links between the subsystems and with data exchange between components of these systems.

Examples of such systems are wireless sensor networks with mobile sensors and air platforms operate both the sensors carrier and the telecommunications repeater of data collected from the sensors, what are called "flying sensor networks" or "informational robots". The flying sensor networks are a CDS whose elements are: basic unmanned aerial vehicle (UAV); a group of mobile miscellaneous UAVs equipped with multi-sensors and interconnected using a common information and telecommunication network.

The basic UAV is used as an air platform for delivering and initial deployment of drones with multisensors in an emergency zone (EZ), as well as for collecting and processing real-time data and retransmitting received data in real time to the control center (Figure 1, Figure 2, Figure 3).

The trajectories of such CDS have been called branching (Ledsinger, 2000; Filatyev, 1994; Filatyev, Yanova, 2013) paths in contemporary scientific literature, since they consist of sections of the general motion of component parts (subsystems) and sections of their individual motion towards the target along individual path branches.

The application and development of existing and leading-edge CDS leads to the need for operative synthesis of trajectories for systems of this type. This is due to the efficiency of the CDS operation depends on the operative optimal choice of space coordinates and time moments when the CDS structural

Figure 1. Example of "flying sensor networks" use for search and rescue operations in an emergency zone

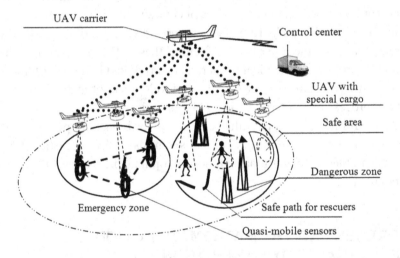

Figure 2. Example of use of airborne and space-based UAVs for transmitting monitoring data

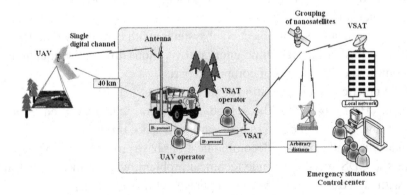

Figure 3. Example of construction a combined mobile sensor network with underwater/surface and flying elements

transformations occur, as well as on the optimal control of the component parts movement along the path branches in time slots between consecutive structural transformations.

Therefore, the scientific problem of the operative optimization of CDS traffic control process is topical. It would allow the on-board computer facilities to solve in real time the task of operative synthesis of optimal branching paths of motion of systems of this kind.

Under the branching trajectory (path), the authors have in mind the group of paths each corresponds to the trajectory of a CDS separate subsystem and is described in the phase space by the points depicting in it the dynamic state of CDS subsystems. The path branch will be called the section of the CDS trajectory between the adjacent points in which structural transformations of the CDS occur.

The structural transformations are the CDS transformations associated with the separation, grouping, start and end of subsystems' and units' motion, compliance with the intermediate phase constraints. These structural transformations can occur both in different and at coincident time moments. Subsystems from the CDS structure, which are considered as whole object, can form units. In the general case, the CDS units and subsystems during their motion can be grouped and divided multiple times.

To solve the task of operational optimization of the CDS traffic control, the task of development of the optimality conditions for CDS branching paths in the form of a real-time synthesis (operative synthesis) of optimal control of these complex systems arises.

To solve this problem, the authors proposed the technique of integrated application of methods for optimal control of discontinuous systems in order to form conditions that allow operative optimization of the CDS branching paths. This method consists of three stages:

- in the first stage there is a transition from the CDS to the discontinuous dynamic system (DDS) with a variable size of state and control vectors;

- in the second stage the optimization of DDS with varying size of state and control vectors is implemented using optimal control methods. In this section, the application of Pontryagin's maximum principle (Boss, 2016; Gabasov et al., 2011; Kim, 2003) will be considered.
- in the third stage, the return to the terms of initial formulation of the task is realized in the form of optimization of the branching path (decomposition of optimal solution obtained for the DDS).

To perform the first stage of the developed methodology and to transit from the CDS to DDS, the authors have designed the method representing a mathematical model of the SDS motion in the form of DDS mathematical model with variable dimensions of state and control vectors at the moments of structural transformations. It consists in the following:

1. Based on the physical considerations of the SDS operation, the diagram of branching path (Figure 4, a) is plotted, the equations of the subsystems (units) motion are compiled along the path branches, the restrictions acting continuously on the subsystems (units) and at the boundary points are written, a criterion is formulated.

Figure 4. Example of a structural diagram of the branching path: a – sequence of structural transformations in time; b – change in size of the state vector and control vector of the DDS; t_i – the time points for the CDS structural transformation; $x \in E^n$ – state vector, $u \in E^m$ – CDS control vector; the arrows show schematically the direction of movement of CDS subsystems and units; the subsystems affecting the movement of other subsystems are marked by «'» symbol.

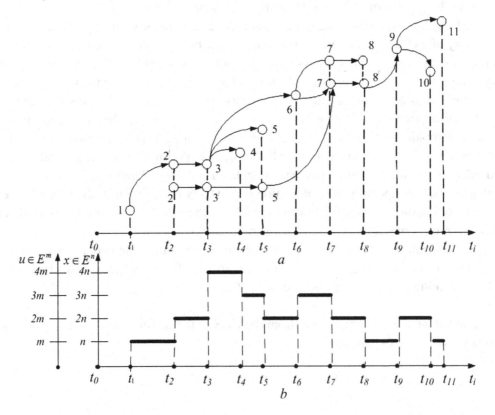

The motion of the CDS along the branches of the branching trajectory is described by a differential system that has the form:

$$\dot{x} = f(x, u; y, v; t), \ t \in [t_0, t_i], \tag{1}$$

where $x \in E^n$ – CDS state vector, $u \in \Omega \subset E^m$ – CDS control vector, Ω – limited set of space E^m; y – phase coordinates, v– vector of control of subsystems from the CDS structure, affecting the movement of other subsystems; t_0, t_i – the time points of start and end of the CDS movement along the corresponding path branches.

The scalar constraints are applied on the paths of subsystems (1):

$$g_i\left(x(t_0), y(t_0), t_0; x(t_i), y(t_i), t_i\right) \begin{cases} = 0, & \overline{i = 1, k_g} \\ \leq 0, & \overline{i = k_g + 1, n_g} \end{cases} \tag{2}$$

$$q_i\left(x(t), u(t), t_0; y(t), v(t); t\right) \begin{cases} = 0, & \overline{i = 1, k_q} \\ \leq 0, & \overline{i = k_q + 1, n_q} \end{cases} \tag{3}$$

where $t \in [t_0, t_i]$.

The criterion that evaluates the efficiency of the CDS operation is described by the expression

$$P = \Pi(\cdot) + \rho_{\sum} \ \rightarrow \min, \tag{4}$$

where $\Pi(\bullet)$: Terminal component of the criterion, which depends on the phase coordinates of subsystems (units) at the time of the CDS structural transformations and time moments.

$\rho_{\sum}$: Integral component of the criterion, consisting of the sum of partial integral components of the type:

$$\rho = \int_{t_0}^{t_f} h(x(t), u(t); y(t), v(t); t) \, dt, \tag{5}$$

corresponding to separate branches of the CDS path.

2. The chronological sequence of time moments for CDS structural transformations is established.

3. The extended state $_iX$ and control $_iU$ ($i = \overline{1, N}$) vectors are introduced in the time intervals between the CDS structural transformations, where $N+1$– the quantity of CDS structural transformations

taking into account the structural transformation associated with the CDS start ($i=0$), consisting of state and control vectors of dynamic subsystems, which move along the path branches in a given time interval.

4. The transition from the CDS to DDS with a variable size of state and control vectors is carried out (Figure 4, b).

The method of transforming the CDS to DDS developed by the authors differs from the existing supplementing definition method (Ashchepkov, 1987; Bryson & Ho, 1975; Sage & White, 1982) wherein the expansion (increase in the size) of the state and control vectors is carried out in time intervals between structural transformations and is equal to the total size of the state and control vectors that are related to those branches of the branching paths that fall into the interval under consideration (Figure 4, b), whereas, when use the supplementing definition method, the dimensions of the state and control vectors of the discontinuity system become equal to the sum of dimensions of the state and control vectors relating to all branches of the branching trajectory.

As a result of implementation of the first stage and application of developed method of transforming the CDS to DDS, we proceed to the second stage of solution of this task –formulation of the optimization problem of the DDS with a variable size of state and control vectors.

$$_i\dot{X} =_i F(_iX,_iU,t), t \in \left[t_{i-1}^+, \quad t_i^-\right], i = \overline{1,N} ; \tag{6}$$

$$Q_{ij}(_iX(t), _iU(t),t)\begin{cases} =0, & j=\overline{1,K_{Q_i}}; \\ \\ \leq 0, & j=\overline{K_{Q_i}+1,N_{Q_i}}; \end{cases} \tag{7}$$

$$G_l(_1X(t_0^+),t_0; _1X(t_1^-), _2X(t_1^+),t_1;\ldots;_N X(t_N^-), t_N)\begin{cases} =0, & l=\overline{1,K_G}; \\ \\ \leq 0, & l=\overline{1,K_G+1,N_G}; \end{cases} \tag{8}$$

$$I = S(_1X(t_0^+),t_0; _1X(t_1^-), _2X(t_1^+),t_1; _2X(t_2^-), _3X(t_2^+),t_2;\ldots$$

$$\ldots; _iX(t_i^-), _{i+1}X(t_i^+),t_i;\ldots;\ldots; _NX(t_N^-), t_N)+ \sum_{i=1}^{N} \int_{t_{i-1}^+}^{t_i^-} \$ (X,U,t)\mathrm{dt} \to \min, \tag{9}$$

where $_iX \in E^{n_{\Sigma^i}}$, $U \in \Omega_i \subset E^{m_{\Sigma^i}}$; $_iX$, $_iU$ – extended vectors of the phase state and control actions, which correspond to i-th time interval between the CDS structural transformations, dimensions n_{Σ_i} i m_{Σ_i}; t_i – time points of the CDS structural transformations; t_i^+, t_i^- – time points to the right and left of t_i; Ω_i – bounded set of space $E^{m_{\Sigma_i}}$; Q_{ij} – restrictions on current values of phase coordinates and controls; G_l – restrictions on the limit values of phase coordinates and moments of its achievement time.

To optimize the discontinuous system with variable size of the state and control vectors (6) - (9), we formulate in the form of the maximum principle (Boss, 2016; Gabasov et al., 2011; Kim, 2003) the necessary conditions for the optimal control of the deterministic CDS moving along a branching path with an arbitrary branching pattern (Figure 4).

Let $_iX(t), _iU(t), t_0, t_1, ..., t_N (t_{i-1} < t_i, t \in [t_0, t_N], i = \overline{1, N})$ – the admissible process of task (6) - (9). For optimal process $_iX(t), _iU(t), t_0, t_1, ..., t_N (t_{i-1} < t_i, t \in [t_0, t_N], i = \overline{1, N})$ the existence of multipliers is required $v_l (l = \overline{0, N_G})$, $\mu_{ij}(t)$ $(i = \overline{1, N}; j = \overline{1, N_{Q_i}})$ $t \in [t_0, t_N]$. It should not be equal at the same time to zero and continuous at $t_0 \leq t \leq, t_N$, $t \neq t_i$ $(i = \overline{1, N})$ decisions $_i\lambda(t)$ of conjugate vector equations

$$_i\dot{\lambda}(t) + \partial H_i(_i\hat{X}(t), _i\hat{U}(t), _i\lambda(t), (t) / \partial _i\hat{X}(t) = 0 \ (i = \overline{1, N}), \tag{10}$$

where "^" means the optimal values of variables and parameters, such that the following conditions are correct:

(1^0) non-negativity and complementary slackness

$v_0 > 0$,

$$v_l \begin{cases} \geq 0, G_l(_1X(t_0^+), t_0; ...;_N X(t_N^-), t_N) = 0, l = \overline{1, K_G}; \\ \geq 0, G_l(_1X(t_0^+), t_0; ...;_N X(t_N^-), t_N) = 0, \\ = 0, G_l(_1X(t_0^+), t_0; ...;_N X(t_N^-), t_N) < 0, \end{cases} l = \overline{K_G + 1, N_G};$$

$$\mu_{ij} \begin{cases} \geq 0, Q_{ij}(_1X(t), _iU(t), t) = 0, j = \overline{1, K}; i = \overline{1, N}; \\ \geq 0, Q_{ij}(_1X(t), _iU(t), t) = 0, \\ = 0, Q_{ij}(_1X(t), _iU(t), t) < 0, \end{cases} j = \overline{K_{Q_i} + 1, N_{Q_i}};$$

(2^0) transversality for conjugate functions and Hamiltonians

$$\left. \frac{\partial S}{\partial _1X(t_0)} \right|_\wedge + _1\lambda(\hat{t}_0^+) = 0;$$

$$\left. \frac{\partial S}{\partial _NX(t_N)} \right|_\wedge - _N\lambda(t_N^-) = 0;$$

$$\left. \frac{\partial S}{\partial t_0} \right|_\wedge - H_1(_1\hat{X}(\hat{t}_0^+), _1\hat{U}(\hat{t}_0^+), _1\lambda(\hat{t}_0^+), \hat{t}_0^+) = 0;$$

$$\left. \frac{\partial S}{\partial t_N} \right|_\wedge - H_N(_N\hat{X}(\hat{t}_N^-), _N\hat{U}(\hat{t}_N^-), _N\lambda(\hat{t}_N^-), \hat{t}_N^-) = 0;$$

(3^0) jump for conjugate functions and Hamiltonians

$$\left.\frac{\partial S}{\partial_i X(\hat{t}_{i-1}^+)}\right|_* + {}_i\lambda(\hat{t}_{i-1}^+) = 0; \qquad (i = \overline{2, N});$$

$$\left.\frac{\partial S}{\partial_i X(t_i^-)}\right|_* - {}_i\lambda(\hat{t}_i^-) = 0; \qquad (i = \overline{1, N-1});$$

$$\left.\frac{\partial S}{\partial t_i}\right|_* + H_i({}_i\hat{X}(\hat{t}_i^-), {}_i\hat{U}(\hat{t}_i^-), {}_i\lambda(\hat{t}_i^-), \hat{t}_i^-) - H_{i+1}({}_{i+1}\hat{X}(\hat{t}_i^+), {}_{i+1}\hat{U}(\hat{t}_i^+), {}_{i+1}\lambda\hat{t}_i^+), \hat{t}_i^+) = 0; \quad (i = \overline{1, N-1});$$

(4^0) minimum of Hamiltonians at the times $t_{i-1} \leq t < t_i$ $(i = \overline{1, N})$

$$H_i({}_i\hat{X}(t), {}_i\hat{U}(t), {}_i\lambda(t), t) = \min_{{}_iU \in \Omega_i} H_i({}_i\hat{X}(t), {}_iU(t), {}_i\lambda(t), t).$$

As mentioned above, the third step in solving the problem of optimizing the CDS branching trajectory is to return to the terms of original formulation of the problem.

The reduction of the conditions for optimal control and the DDS trajectory to CDS optimality, the formal equivalent of which is the DDS, is carried out by decomposing extended state and control vectors, constraints and boundary conditions, auxiliary functions and variables used during the application of optimization method in the reverse sequence of output transformations that led to the transition from the CDS to DDS.

Let formulate, as a consequence, the necessary conditions for the optimal control of the CDS with a simple branching path, taking into account the interaction of the subsystems from the necessary conditions of control optimality (10) for the deterministic CDS moving along a branching path with an arbitrary branching pattern.

Necessary Conditions for the CDS Optimality, Taking into Account the Interaction of Subsystems

The simple branching path with regard for the interaction of subsystems will be considered as the CDS trajectory, consisting of two subsystems and providing no more than one separation.

The chart and timing diagram of a simple branching path with separation are presented in Figure 5.

The problem of optimizing the simplest branching path of the CDS with separation may be written in the following form.

The dynamics of motion of the CDS subsystems along the branching trajectory (Figure 5) is described by the equations:

$$_\beta\dot{x} = {}_\beta f({}_\beta x, {}_\beta u, t), t \in [t_{\beta^*}, t_\beta](\beta = 1, \beta^* = 0; \beta = 11, 12, \beta^* = 1), \tag{11}$$

Figure 5. Chart and timing diagram of a simple branching trajectory with separation

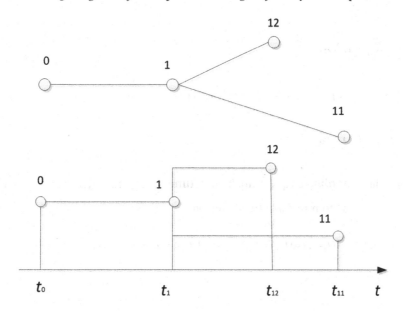

$$_{\beta}x \in E^n, _{\beta}u \in E^{m_{\beta}} \subset \Omega_{\beta}\ (\beta=1,11,12),\ t_0 < t_1 < t_{12} < t_{11}, \tag{12}$$

where $_{\beta}x \in E^n$:Paths of CDS subsystems belonging to the class of piecewise-smooth functions. $_{\beta}u \in \Omega_{\beta} \subset E^{m_{\beta}}$:CDS subsystem control vectors, belonging to the class of piecewise-continuous functions.

On which the restrictions are imposed

$$Q_l^{(1)}(x_1(t), u_1(t), t) \begin{cases} = 0, l = \overline{1, K_Q^{(1)}}; \\ \leq 0, l = \overline{K_Q^{(1)} + 1, N_Q^{(1)}}; \end{cases} \tag{13}$$

$$Q_l^{(11,12)}(_{11,12}x(t), _{11,12}u(t), t) \begin{cases} = 0, l = \overline{1, K_Q^{(11,12)}}; \\ \leq 0, l = \overline{K_Q^{(11,12)} + 1, N_Q^{(11,12)}}; \end{cases} \tag{14}$$

$$Q_l^{(11)}(_{11}x(t), _{11}u(t), t) \begin{cases} = 0, l = \overline{1, K_Q^{(11)}}; \\ \leq 0, l = \overline{K_Q^{(11)} + 1, N_Q^{(11)}}; \end{cases} \tag{15}$$

At the time of separation of subsystems, the conditions should be met

$$_{11}x(t_{11}), t_{11} \begin{cases} = 0, l = \overline{1, K_G}; \\ \\ \leq 0, l = \overline{K_G + 1, N_G}, \end{cases}$$

$$_1x_\tau(t_1) - {}_{11}x_\tau(t_1) = 0, \; {}_1x_\tau(t_1) - {}_{12}x_\tau(t_1) = 0 \; (\tau = \overline{1, n-1}),$$

$$_1x_n(t_1) - {}_{11}x_n(t_1) - {}_{12}x_n(t_1) = 0.$$

Controls $_\beta u(t)$, phase coordinates $_\beta x(t)$, instants of time. t_{β^*}, t_β $(\beta = 1, \beta^* = 0; \beta = 11, 12, \beta^* = 1)$ should be selected in such a way as to minimize the criterion

$$I = S({}_1x(t_0), t_0; {}_1x(t_1), t_1; {}_{11}x(t_{12}), {}_{12}x(t_{12}), t_{12}; {}_{11}x(t_{11}), t_{11}) + I_1 + I_{11} + I_{12} \to \min, \tag{16}$$

where

$$I_1 = \int_{t_0}^{t_1} {}_1(x_1, u_1) dt,$$

$$I_{12} = \int_{t_1}^{t_{12}} {}_{12}(x_{12}, u_{12}; x_{11}, u_{11}) dt, \tag{17}$$

$$I_{11} = \int_{t_1}^{t_{12}} {}_{11}^{12}(x_{12}, u_{12}; x_{11}, u_{11}) dt + \int_{t_{12}}^{t_{11}} {}_{11}(x_{11}, u_{11}) dt.$$

The optimality criterion (16) corresponds to the Bolza formulation, where function $S(\cdot)$ physically reflects the requirements for values of the coordinates of CDS subsystems motion at the time of start and end, as well to the values of instants of time.

The integral members of the criterion express the requirements for character of the CDS subsystems motion along the corresponding path branches.

The mutual influence of subsystems in time interval $[t_{\beta^*}, t_\beta]$ reflected in their motion equations (11) and in particular integral conditions I_β (17).

The assumptions similar to those formulated for the corresponding functions of problem (1) - (9) were stated for all functions under the conditions of problem (11) - (17)

It is now possible to set up the extended criterion

$$\Im = S^* + D + I_1^* + I_{11,12}^* + I_{12}^*, \tag{18}$$

where

$$S^* = v_0 S(\cdot) + \sum_{l=1}^{N_G} {}_l G(\cdot), \tag{19}$$

$$D = \sum_{i=1}^{2} \sum_{\tau=1}^{n-1} \alpha^{(1i)} [{}_1 x_\tau(t_1) - {}_{11} x_\tau(t_1)] + \alpha_n [{}_1 x_n(t_1) - {}_{11} x_n(t_1) - {}_{12} x_n(t_1)],$$

$$I_1^* = \int_{t_0}^{t_1} \left[H_1^*({}_1 x(t), {}_1 u(t), {}_1 \lambda(t), \mu^{(1)}(t), t) - {}_1 \lambda^{\mathrm{T}}(t) {}_1 \dot{x}(t) \right] dt, \tag{20}$$

$$I_{11,12}^* = \int_{t_1}^{t_{12}} \left[H_{11,12}^*({}_{11,12} x(t), {}_{11,12} u(t), {}_{11,12} \lambda(t), \mu^{(11,12)}(t), t) - {}_{11,12} \lambda^{\mathrm{T}}(t) {}_{11,12} \dot{x}(t) \right] dt, \tag{21}$$

$$I_{11}^* = \int_{t_{12}}^{t_{12}} \left[H_{11}^*({}_{11} x(t), {}_{11} u(t), {}_{11} \lambda(t), \mu^{(11)}(t), t) - {}_{11} \lambda^{\mathrm{T}}(t) {}_{11} \dot{x}(t) \right] dt, \tag{22}$$

$$H_1^*(\cdot) = H_1({}_1 x(t), {}_1 u(t), {}_1 \lambda(t), t) + \mu^{(1)}(t)^T Q^{(1)}(x(t), {}_1 u(t), t), \tag{23}$$

$$H_{11,12}^*(\cdot) = \sum_{i=1}^{2} H_{1i}({}_{1i} x(t), {}_{1i} u(t), {}_{1i} \lambda(t), t) + \mu^{(11,12)}(t)^T Q^{(11,12)}({}_{11,12} x(t), {}_{11,12} u(t), t), \tag{24}$$

$$H_{11}^*(\cdot) = H_{11}({}_{11} x(t), {}_{11} u(t), {}_{11} \lambda(t), t) + \mu^{(11)}(t)^T Q^{(11)}({}_{11} x(t), {}_{11} u(t), t), \tag{25}$$

$$H_\beta(\cdot) = v_{0|\ \beta}\ ({}_\beta x(t), {}_\beta u(t), t) + {}_\beta \lambda^{\mathrm{T}}(t) {}_\beta f({}_\beta x(t), {}_\beta u(t), t) \ (\beta = 1, 11, 12), \tag{26}$$

$$\mathcal{V}^{(2)}(t) = col[\mathcal{V}_1^{(2)}(t), \ldots, \mathcal{V}_{N_Q^{(2)}}^{(2)}(t)], \tag{27}$$

$$Q^{(2)}(\cdot) = col[Q_1^{(2)}(\cdot), \ldots, Q_{N_Q^{(2)}}^{(2)}(\cdot)] \ (\beta = 1; 11, 12; 11), \ {}_{11,12} \lambda(t) = col[{}_{11} \lambda(t), {}_{12} \lambda(t)]. \tag{28}$$

We formulate optimality conditions for the CDS with a simple branching path, using the designations introduced in expressions (18) - (28), taking into account the interaction of subsystems (Figure 5).

Assume that ${}_1 x(t), {}_{11} x(t), {}_{12} x(t), {}_1 u(t), {}_{11} u(t), {}_{12} u(t), t_0, t_1, t_{12}, t_{11}$ – admissible processes of tasks (11)–(17). Then to provide their optimality it is necessary that such vector factors exist $v = col(v_0, v_1, \ldots, v_{N_G}), \mu^{(1)}(t), t \in [t_0, t_1], \mu^{(11,12)}(t), t \in [t_1, t_{12}], \mu^{(11)}(t), t \in [t_{12}, t_{11}]$ non-zero at the same time and continuous solutions $_1\lambda(t)$, $t \in [t_0, t_1]$, $_{11}\lambda(t), _{12}\lambda(t)$, $t \in [t_1, t_{12}]$, $_{11}\lambda(t)$, $t \in [t_{12}, t_{11}]$ of conjugate vector equations

$$_1\dot\lambda + \frac{\partial H_1^*}{\partial_1 x}\bigg|_{\wedge} = 0, \; t \in [\hat{t}_0, \hat{t}_1] \tag{29}$$

$$_{11}\dot\lambda + \frac{\partial H_{11,12}^*}{\partial_{11} x}\bigg|_{\wedge} = 0, \; _{12}\dot\lambda + \frac{\partial H_{11,12}^*}{\partial_{12} x}\bigg|_{\wedge} = 0, \; t \in [\hat{t}_1, \hat{t}_{12}] \tag{30}$$

$$_{11}\dot\lambda + \frac{\partial H_{11}^*}{\partial_{11} x}\bigg|_{\wedge} = 0, \; t \in [\hat{t}_{12}, \hat{t}_{11}], \tag{31}$$

such that the conditions are correct:
   (1*) Non-negativity and complementary slackness $v > 0$,

$$v_l \begin{cases} \geq 0, G_l(_1x(t_0), t_0; \ldots; _{11}x(t_{11}), t_{11})|_{\wedge} = 0, l = \overline{1, K_G}, \\ \geq 0, G_l(_1x(t_0), t_0; \ldots; _{11}x(t_{11}), t_{11})|_{\wedge} = 0 \\ = 0, G_l(_1x(t_0), t_0; \ldots; _{11}x(t_{11}), t_{11})|_{\wedge} < 0 \end{cases} l = \overline{K_G + 1, N_G};$$

$$\mu_l^{(\beta)}(t) \begin{cases} \geq 0, Q_l^{(\beta)}(_\beta x(t), _\beta u(t), t)|_{\wedge} = 0, l = \overline{1, K_Q^{(\beta)}}, \\ \geq 0, Q_l^{(\beta)}(_\beta x(t), _\beta u(t), t)|_{\wedge} = 0 \\ = 0, Q_l^{(\beta)}(_\beta x(t), _\beta u(t), t)|_{\wedge} < 0 \end{cases} l = \overline{K_Q^{(\beta)} + 1, N_Q^{(\beta)}};$$

where $\beta = 1; 11, 12; 11$;
   (2*) Transversality for conjugate functions and Hamiltonians

$$\frac{\partial S^*}{\partial_1 x(t_0)}\bigg|_{\wedge} - (-1)_1 \, _1\lambda(\hat{t}_0) = 0; \; \frac{\partial S^*}{\partial t_0}\bigg|_{\wedge} + (-1) H_1^*\big|_{\wedge} = 0,$$

$$\frac{\partial S^*}{\partial_{1i} x(t_{1i})}\bigg|_{\wedge} + (-1)_1 \, _{1i}\lambda(\hat{t}_{1i}) = 0 \; (i = 1, 2),$$

$$\left.\frac{\partial S^*}{\partial t_{11}}\right|_\wedge - (-1)H_{11}^*\big|_\wedge = 0;$$

(3*) Jump for conjugate functions and Hamiltonians

$$\frac{\partial S^*}{\partial_1 x(t_1)}\bigg|_\wedge + (-1)[{}_1\lambda(\hat{t}_1) - {}_{11}\lambda(\hat{t}_1) - {}_{12}\lambda(\hat{t}_1)] = 0, \quad \left.\frac{\partial S^*}{\partial t_1}\right|_\wedge - (-1)(H_1^*\big|_\wedge - H_{11,12}^*\big|_\wedge) = 0,$$

$$\frac{\partial S^*}{\partial_{11} x(t_{12})}\bigg|_\wedge + (-1)[{}_{11}\lambda(\hat{t}_{12}) + (-1)\varepsilon) - {}_{11}\lambda(\hat{t}_{12}) - (-1)\varepsilon)] = 0, \quad \left.\frac{\partial S^*}{\partial t_{12}}\right|_\wedge - (-1)(H_{11,12}^*\big|_\wedge - H_{11}^*\big|_\wedge) = 0;$$

(4*) Minimum of the extended Hamiltonians

$$H_\beta^*({}_\beta \hat{x}(t), {}_\beta \hat{u}(t), {}_\beta\lambda(t), t) = \min_{{}_\beta u(t) \in \Omega_\beta} H_\beta^*({}_\beta x(t), {}_\beta u(t), {}_\beta\lambda(t), t), \quad t \in [t', t''],$$

where $(\beta = 1, t' = t_0, t'' = t_1; \beta = 11, 12, t'' = t_1, t'' = t_1; \beta = 11, t' = t_{12}, t'' = t_{11})$, $\varepsilon \to 0$; ${}_{11}\lambda(\hat{t}_{12} - \varepsilon)$ – the value of the conjugate variable ahead of $t = \hat{t}_{12}$ (to the left of $\hat{t}_{12}$); ${}_{11}\lambda(\hat{t}_{12} + \varepsilon)$ – the value of the conjugate variable following by $t = \hat{t}_{12}$ (to the right of $\hat{t}_{12}$).

Note. The jump condition for the conjugate variable corresponding to *n-th* phase coordinate that means the mass, for mechanical systems has the form

$$\frac{\partial S^*}{\partial_1 x(t_1)}\bigg|_\wedge + (-1)[{}_1\lambda_n(\hat{t}_1) - \zeta_{11}\,{}_{11}\lambda_n(\hat{t}_1) - \zeta_{12}\,{}_{12}\lambda_n(\hat{t}_1)] = 0,$$

$$\zeta_{11} \geq 0, \zeta_{12} \geq 0, \zeta_{11} + \zeta_{12} = 1.$$

The necessary optimal conditions for the control of deterministic CDS with simple branching trajectory of motion, taking into account the interaction of subsystems, formulated in this section, allow real-time optimization of CDS branching paths and real-time correction of their paths in case of occurrence of unexpected factors at the previous stage that are critical to implement the CDS end use.

EXAMPLE

Let us consider an example of the application of formulated necessary optimality conditions (29) - (31) for calculating the reference branching trajectory of the aerospace system (AS) with a hypersonic air launch aircraft. This trajectory should be optimal with minimum time for injection of the orbital aircraft and the air launch aircraft at specified points of the near-earth space. It is the authors' opinion that the reference trajectory is a path which can be taken as a rational first approximation to the optimal trajectory, which is built on more detailed and accurate models of the AS stages motion.

The AS first stage is considered as a hypersonic unmanned air launch aircraft (ALA), and the orbital stage is an unmanned orbital aircraft (OA), which after launching into space, can return to the ground along an aircraft trajectory. The qualitative pattern of the AS flight profile is shown in Figure 6.

It is anticipated that the AS stages move eastward in the equatorial plane, there is no slipping, and the roll angle is zero.

Taking into account the accepted assumptions, the corresponding equations of motion of the ALA + OA center of mass (branch 0–1), ALA (branch 1–12), OA (branch 1–11) were taken as the mathematical model of AS motion along the path branches (Figure 6) in projections on the axis of the trajectory coordinate system (at zero wind) (Bogomaz, Naumenko & Sobolevskaya, 2005; Lozino-Lozinsky, 1998; Lazarev, 2007; Shalygin, Petrova & Sannikov, 2010; Sikharulidze, 1982):

$$m\dot{V} = P\cos\alpha - X_a - mg(r,\omega_E)\sin\theta,$$
$$mV\dot{\theta} = P\sin\alpha + Y_a - mg(r,\omega_E)\cos\theta + 2mV\omega_E + m\frac{V^2}{r}\cos\theta, \tag{32}$$

$$g(r,\acute{E}_E) = \frac{1/4}{r^2} - r\acute{E}_E^2$$

supplemented by kinematic equations

$$\dot{h} = V\sin, , \tag{33}$$

$$\dot{\lambda} = \frac{V}{r}\cos\theta, \tag{34}$$

and mass change equation

Figure 6. Qualitative pattern of the AS flight profile: TE (turbojet engine), RE (ramjet engine) - operating modes of the ALA propulsion system, LRE - orbital aircraft liquid rocket engine

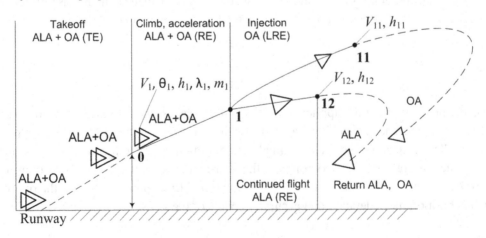

$$\dot{m} = -f, \tag{35}$$

where P– thrust power of AS stages on the path branches, α - angle of attack, θ – path inclination angle, Y_a, X_a– lifting force and drag, ω_E – angular velocity of the Earth rotation, $r=R_E+h$ (R_E– standard radius of the Earth, h– geometric height), μ–product of the gravitational constant by the Earth mass, λ – longitude, f– momentary mass flow rate.

We emphasize that simplification of equations (1) - (5) does not affect the meaning of the problem, consisting in optimizing the branch path. Note that the optimal solution of the trajectory problem obtained for a zero inclination angle of the orbit plane relative to the equator (motion in the equatorial plane) will give an error of no more than 2% of the optimal solution obtained for a non-zero angle (Raikunov, 2011).

For different parts of the branching path, equations (31) - (35) will differ in value of thrust, aerodynamic impact, momentary mass flow rate f and its initial value.

In order to distinguish whether the state vector, control vector, or other parameters for the description of the AS stages along the 0–1 (ALA + OA), 1–12 (ALA), 1–11 (OA) paths, we will mark them with subscript at the left 1, 12 or 11. In cases where the considerations touch on all sections of the trajectory or, when it is clear which particular section of the branching path is involved, we omit the subscript 1, 12, 11.

As a criterion for optimizing the branching path of AS stages, we take the minimum of time for injection the ALA and OA from the climbing – acceleration to the specified end positions

$$I = b_1 \Delta t_1 + b_{12} \Delta t_{12} + b_{11} \Delta t_{11}, \tag{36}$$

where b_1, b_{12}, b_{11} – normalized weight factors ($b_1 = b_{11} = b_{12} = 1$), $\Delta t_1 = t_1 - t_0$; $\Delta t_{1i} = t_{1i} - t_1$ ($i = 1, 2$); t_1, t_{12}, t_{11}– instants of time of structural transformations of the AS trajectory, which moves along a typical branching path with separation (Figure 5).

We assume that ALA + OA start climbing – acceleration at the time $t_0 = 0$ with coordinates $V_1 = 1359,160$ m/s; $\theta_1 = 0$; $h_1 = 28 \cdot 10^3$m; $\lambda_1 = 0$; $m_1 = 294 \cdot 10^3$кg, where: V_1- speed of motion; θ_1 - path inclination angle; h_1- flight altitude; λ_1– longitude, used to replace time as an independent variable; m_1- mass. After climbing – acceleration and separating the stages, OA and ALA should reach speeds and heights $V_{11} = 7843,04$ m/s, $h_{11} = 100 \cdot 10^3$ m and $V_{12} = 3665,824$ m/s, $h_{12} = 45 \cdot 10^3$m at any value $\theta_{1i}, \lambda_{1i}, m_{1i}, (i=1,2)$ at the time of the first achievement of the specified speed and height.

When solving this problem, we will consider as correct such a motion of the AS along the path branches, that the following restrictions on phase coordinates are not exceeded $x = \{V,,,h,»,m\}^T$ and control $u = \{P, \pm\}^T$ of models (31) - (35): on angle of attack $\alpha_{min} \le \alpha \le \alpha_{max}$; thrust $P_{min} \le P \le P_{max}$. The listed set of restrictions corresponds to the physical meaning and is standard for the problems of flight dynamics, in which the task of AS climbing and accelerating is solved (Lazarev, 2007; Sikharulidze, 1982).

The following constraint should also be introduced for the optimization task of the AS branching path:

$$h_{11}(t) - h_{12}(t) \ge A(t), t \in [t_1, t_{12}], \tag{37}$$

It provides safe motion of the OA and ALA after their separation.

Condition (37) requires that the charts of the OA and ALA flight altitudes do not come within a dangerous distance where there is a real threat of collision of the OA and ALA or falling one of them into the wake of another (Okhotsimsky & Sikharulidze, 1990; Shalygin, Petrova & Sannikov, 2010).

The method for solving the problem of AS motion optimal path was developed based on the application of formulated necessary conditions for optimality of the deterministic CDS branching path, taking into account the interaction of the subsystems after separation (29) - (31), presented in the form of the minimum principle and requiring the solution of three two-point boundary problems with meeting the special conditions at the points of CDS structural transformations: conditions for the continuity of all phase coordinates, except for the mass; jump conditions for conjugate variables and Hamiltonians. In this formulation of problem (31) - (37), the equations describing the AS motion along the branches of branched path (Figure 5) take the form:

$$_1\dot{x} =_1 f(_1 x,_1 u,t), t \in [t_0,t_1], \ (_1 x(t_0),t_0) = const, (_1 x(t_1),t_1) = var, \tag{38}$$

$$_{11}\dot{x} =_{11} f(_{11} x,_{11} u,t), t \in [t_1,t_{11}], \ _{11}x(t_1) = const, t_{11} = var, \tag{39}$$

$$_{12}\dot{x} =_{12} f(_{12} x,_{12} u,t), t \in [t_1,t_{12}], \ _{12}x(t_{12}) = const, t_{12} = var, \tag{40}$$

where $_1 x(t) \in E^n$, $_{11}x(t) \in E^n$, $_{12}x(t) \in E^n$ – the state vectors of AS stages; $_1 u(t) \in \Omega_1 \subset E^{m_1}$, $_{11}u(t) \in \Omega_{11} \subset E^{m_{11}}$, $_{12}u(t) \in \Omega_{12} \subset E^{m_{12}}$ –the state vectors.

The phase coordinates of the AS stages are related by the relations at the separation moments

$$_1 x_i(t_1) =_{11} x_i(t_1) =_{12} x_i(t_1), (i = \overline{1,n-1}), \tag{41}$$

$$_1 x_n(t_1) =_{11} x_n(t_1) +_{12} x_n(t_1), \tag{42}$$

$$Q(_{11}x,_{12} x,_{11} u,_{12} u,t) \le 0, t \in [t_1,t_{12}], \tag{43}$$

where the coordinate with subscript n is the mass.

The vector criterion for optimizing the branching path of the AS stages takes the additive form:

$$I = \int_{t_0}^{t_1} {}_1(_1 x,_1 u,t)dt + \int_{t_1}^{t_{12}} [{}_{11}(_{11}x,_{11} u,t) + {}_{12}(_{12}x,_{12} u,t)]dt +$$

$$+ \int_{t_1}^{t_{11}} {}_{11}(_{11}x,_{11} u,t)dt \to \min_{\substack{_1 u(t),t\in[t_0,t_1], _{12}u(t),t\in[t_1,t_{12}], \\ _{11}u(t),t\in[t_1,t_{11}], _1 x(t_1),t_1 t_{11},t_{12}}}. \tag{44}$$

The integral terms of criterion (44) express the requirements for nature of the AS stages motion along the corresponding path branches.

The necessary conditions for optimality of the AS branching trajectory for problem (38) - (44) with regard to (16) - (28) are stated as follows.

Let $_1x(t), _1u(t), t \in [t_0, t_1]$, $_{12}x(t), _{12}u(t), t \in [t_1, t_{12}]$, $_{11}x(t), _{11}u(t), \ t \in [t_1, t_{11}]$, $t_0 < t_1 < t_{12} < t_{11}$ – permissible process.

To provide the optimality of permissible process, functional factor $\mu(t) \geq 0, t \in [t_1, t_{12}]$, continuous solutions $_1\psi(t), \ t \in [t_0, t_1], \ _{11}\psi(t), \ _{12}\psi(t), \ t \in [t_1, t_{12}], \ _{11}\psi(t), \ t \in [t_{12}, t_{11}]$ for differential equations, should be exist

$$_1\dot{\psi} + \frac{\partial H_1}{\partial _1 x}\bigg|_{\wedge} = 0, t \in [t_0, \hat{t}_1],$$ (45)

$$_{11}\dot{\psi} + \frac{\partial H_{11}}{\partial _{11} x}\bigg|_{\wedge} = 0, t \in [\hat{t}_{12}, \hat{t}_{11}],$$ (46)

$$_{12}\dot{\psi} + \frac{\partial H_{12}}{\partial _{12} x}\bigg|_{\wedge} + \mu(t)\frac{\partial Q}{\partial _{12} x}\bigg|_{\wedge} = 0, t \in [\hat{t}_1, \hat{t}_{12}],$$ (47)

$$_{11}\dot{\psi} + \frac{\partial H_{11}}{\partial _{11} x}\bigg|_{\wedge} + \mu(t)\frac{\partial Q}{\partial _{11} x}\bigg|_{\wedge} = 0, t \in [\hat{t}_1, \hat{t}_{12}],$$ (48)

such that the conditions are true:

1) Transversality:

$$H_{11}(_{11}\hat{x}, _{11}\hat{u}, _{11}\psi, \hat{t}_{11}) = 0;$$ (49)

2) Jump:

$$_1\psi(\hat{t}_1) =_{11} \psi(\hat{t}_1) +_{12} \psi(\hat{t}_1),$$ (50)

$$_{11}\psi(\hat{t}_{12} - 0) =_{11} \psi(\hat{t}_{12} + 0),$$ (51)

$$H_1(_1\hat{x}, _1\hat{u}, _1\psi, \hat{t}_1) = H_{12}(_{12}\hat{x}, _{12}\hat{u}, _{12}\psi, \hat{t}_1) + H_{11}(_{11}\hat{x}, _{11}\hat{u}, _{11}\psi, \hat{t}_1) + \mu(\hat{t}_1)Q(_{11}\hat{x}, _{12}\hat{x}, _{11}\hat{u}, _{12}\hat{u}, \hat{t}_1),$$ (52)

$$H_{11}(_{11}\hat{x}, _{11}\hat{u}, _{11}\psi, \hat{t}_{12} - 0) + H_{12}(_{12}\hat{x}, _{12}\hat{u}, _{12}\psi, \hat{t}_{12} - 0) + \mu(\hat{t}_{12} - 0)Q(_{11}\hat{x}, _{12}\hat{x}, _{11}\hat{u}, _{12}\hat{u}, \hat{t}_{12} - 0) -$$
$$-H_{11}(_{11}\hat{x}, _{11}\hat{u}, _{11}\psi, \hat{t}_{12} + 0) = 0;$$

3)   Minimum of the Hamiltonians:

$$H_\beta(_\beta\hat{x}(t),_\beta\hat{u}(t),\ _\beta\psi(t),t) = \min_{_\beta u(t)\in\Omega_\beta} H_\beta(_\beta\hat{x}(t),_\beta u(t),\ _\beta\psi(t),t),$$

$$\beta = 1,\ t \in [\hat{t}_0,\hat{t}_1],\ ;\beta = 11,\ t \in [\hat{t}_{12},\hat{t}_{11}],$$

(53)

$$H_{11}(_{11}\hat{x},\ _{11}\hat{u},\ _{11}\psi,\hat{t}_1) + H_{12}(_{12}\hat{x},\ _{12}\hat{u},\ _{12}\psi,\hat{t}_1) + \mu(\hat{t}_1)Q(_{11}\hat{x},\ _{12}\hat{x},\ _{11}\hat{u},\ _{12}\hat{u},\ \hat{t}_1) =$$

$$= \min_{\substack{_{11}u(t)\in\Omega_{11},\\ _{12}u(t)\in\Omega_{12},\\ t\in[\hat{t}_1,\hat{t}_{12}]}} \begin{bmatrix} H_{11}(_{11}\hat{x},\ _{11}u,\ _{11}\psi,\hat{t}_1) + \\ +H_{12}(_{12}\hat{x},\ _{12}u,\ _{12}\psi,\hat{t}_1) + \\ +\mu(\hat{t}_1)Q(_{11}\hat{x},\ _{12}\hat{x},\ _{11}u,\ _{12}u,\ \hat{t}_1) \end{bmatrix},$$

(54)

where $H_1 = \Phi_1 +_1 \psi^T{}_1 f$, $H_{12} = \Phi_{12} +_{12} \psi^T{}_{12} f$, $H_{11} = \Phi_{11} +_{11} \psi^T{}_{11} f$.

The state and control vectors from equations (38) - (44) will have such structure

$$_\beta x = \left\{V_\beta,\theta_\beta,h_\beta,\lambda_\beta,m_\beta\right\}^T,\quad _\beta u = \left\{\alpha_\beta,P_\beta\right\}^T,\text{ where }\beta-\text{ branch of a branching path }(\beta=1,12,11).\text{ The}$$

restriction (43) takes the form (37).

During the calculations, five options for optimizing the criterion (44) were considered, it corresponds to the criterion (36) in the initial formulation of the problem provided that $\Phi_\beta(_\beta x,_\beta u,t) = b_\beta$, where $\beta-$ branch of a branching path $(\beta = 1,12,11)$.

1.   The main option that requires meeting of all necessary conditions (45) – (54).
2.   The auxiliary option for H (Hamiltonians) that requires meeting of conditions (45) –(49), (52) – (54). Physically, this means that the separation point is optimized only by the time point of separation t_1, for any phase coordinate $_1x(t_1)$, such that the specified conditions are met. Formally, this means that the minimum of expression (44) is searched by all specified controls and parameters except for $_1x(t_1)$.
3.   The auxiliary option of ψ (by conjugate variables) that requires meeting of conditions (45) – (51), (53) – (54). Physically, this means that the separation point is optimized only by the phase coordinate $_1x(t_1)$ for any point in time t_1, such that the specified conditions are met. Formally this means that the minimum of expression (44) is searched by all specified controls and parameters except for t_1.
4.   The alternative option of the OA (by orbital aircraft) that requires meeting of conditions (45), (46) for $t \in [t_1,t_{11}]$, (47), (49), (53), for $^2 = 1$ and $t \in [t_0,t_1]$, $^2 = 11$ and $t \in [t_1,t_{11}]$. Physically, this means that the best conditions are created at the separation point for injection the OA, that ignore the further ALA motion. In other words, the path section $0 - 1 - 11$ is optimized at first, and then, based on the calculated point $(_1x(t_1),t_1)$, ALA motion is optimized along branch $1 - 12$ with a provision that the duration $\Delta t_{12} = t_{12} - t_1$ ($t_1 = const$, $t_{12} = var$) of flight to point $_{12}x(t_{12}) = const$ is minimized and the restriction (43) is met, wherein $_{11}x(\cdot)$ and $_{11}u(\cdot)$ – known functions of time $t \in [t_1,t_{12}]$. Formally, this means that in task (38) – (44) all coordinates $_{11}x(t_1)$ of ALA+OA stage vary at the point of separation except for the mass, as well the time of OA launching into orbit.

5.   The alternative option of the ALA that requires meeting of conditions (40), (41), (43), (44), (46), (49) for $\beta = 1$ and $t \in [t_0, t_1]$, $\beta = 12$ and $t \in [t_1, t_{12}]$, $\beta = 11$ and $t \in [t_{12}, t_{11}]$. Physically, this means that the best conditions are created at the separation point for injection the ALA, that ignore the further OA motion. In other words, the path section $0 - 1 - 12$ is optimized at first, and then, based on the calculated point $({}_{11}x(t_1), t_1)$, OA motion is optimized along branch $1 - 11$ with a provision that the duration $\Delta t = t_{11} - t_1$ ($t_1 = const$, $t_{11} = var$) of flight to point ${}_{11}x(t_{11}) = const$ is minimized and the restriction (38) is met, wherein ${}_{12}x(\cdot)$ and ${}_{12}u(\cdot)$ – known functions of time $t \in [t_1, t_{12}]$.

The method of calculating the AS optimal trajectory will consist of three stages:

- At the first stage, the optimal law of change in the engine thrust of AS stages is analytically justified.
- At the second stage, the models of the AS stages dynamics are reduced, which consists in a sequential transition from the model of the dynamics of AS stages with five dependent variables to a model with one dependent variable.
- At the third stage, the optimal program for changing the angle of attack, phase coordinates and time separation of the AS stages is calculated, it based on a step-by-step approximations of the optimal AS branching path, obtained for the simplest reduced model of the AS dynamics (with one dependent variable) to optimal AS branching path for the initial model (with five dependent variables).

The calculation software consists of five autonomous computing units according to the number of models of the AS stages dynamics.

The calculation starts with the simplest model with one dependent variable and then the result is transmitted for use as a first approximation to the model with two dependent variables, and so on up to the model with five dependent variables (Lysenko, Tachinina & Alekseeva, 2018).

Each unit is divided into several interconnected sub-units: the first iteration generator based on the results of calculating the optimal path for smaller models; the calculator of the Hamiltonian and their minimization of control at given points; approximation of the control in intervals between the minimization points of Hamiltonian; integration of differential equations of the AS stages dynamics; numerical differentiation of Hamiltonians along the motion trajectory of stages and integration of the equations for conjugate variables; procedure of gradient interactive parameter calculation.

The results of calculation of optimal parameters of the paths and controls for all five options of solution the task for a complete mathematical model containing five dependent variables are given in Table 1, Table 2, Table 3, Table 4, Table 5.

The comparative analysis of five options of AS branching trajectory optimization by the main indicator, which is the criterion (36), showed that the main option is most effective, the minimum time by criterion (36) was 302.773 seconds (Figure 7).

The next optimization option is alternative one for the ALA; the minimum time by criterion (36) was 303.136 s, then – the alternative option for the OA; the minimum time by criterion (36) was 308.808 s.

In this case, the optimization option alternative for the ALA essentially ranks below versus the main and alternative options for the OA due to the mass injecting into orbit. The corresponding optimal mass values for these options were: 24.712 kg, 25.903 kg and 26.000 kg.

Table 1. Comparison of optimization options

Options	Branch Path Parameters			
	$\Delta t_1 = \hat{t}_1 - t_0, \text{s}$	$\Delta t_{12} = \hat{t}_{12} - t_1, \text{s}$	$\Delta t_{11} = \hat{t}_{11} - \hat{t}_1, \text{s}$	$\hat{I} = \Delta t_1 + \Delta t_{12} + \Delta t_{11}, \text{s}$
1. Main	46.669	63.164	192.940	302.773
2. Auxiliary for *H*	44.456	65.263	197.029	306.748
3. Auxiliary for ¨	45.457	64.543	196.877	306.877
4. Alternative for OA	47.672	68.606	192.530	308.808
4. Alternative for ALA	41.869	62.980	198.287	303.136

Table 2. Comparison of optimization options

Options	Branch Path Parameters			
	$\Delta t_1 + \Delta t_{12}, \text{s}$	$\Delta t_1 + \Delta t_{11}, \text{s}$	$\alpha_1(t_0 + 0), \text{deg}$	$n_1(t_0 + 0)$
1. Main	109.833	239.609	17.020	1.046
2. Auxiliary for *H*	109.719	241.485	17.443	1.010
3. Auxiliary for ¨	110.000	242.334	17.503	1.004
4. Alternative for OA	116.278	240.020	16.840	1.054
5. Alternative for ALA	104.849	240.156	17.834	0.981

Table 3. Comparison of optimization options

Options	Branch Path Parameters			
	$\hat{V}_1(\hat{t}_1), \text{m/s}$	$\hat{M}_1(\hat{t}_1)$	$\hat{h}_1(\hat{t}_1), \text{km}$	$\theta_1(\hat{t}_1), \text{deg}$
1. Main	2,091.10	6.769	34.217	2.242
2. Auxiliary for *H*	2,050.15	6.636	34.045	2.448
3. Auxiliary for ¨	2,065.82	6.687	34.192	2.397
4. Alternative for OA	2,095.68	6.826	34.420	2.348
5. Alternative for ALA	2,011.95	6.513	34.267	2.273

The results of calculation of optimal branching trajectory by minimum injection time of the orbital aircraft and air launch aircraft to the near-earth space specified points of the aerospace system are shown in Figure 7, Figure 8, Figure 9, Figure 10, Figure 11, Figure 12, Figure 13.

$100 \text{ s} \le t - \hat{t}_1 \le \hat{t}_{11} - \hat{t}_1, \ \hat{t}_1 = 46.669 \text{ c}$):

M – Mach number, n_{xk} – flight path acceleration (tangential G)

Table 4. Comparison of optimization options

Options	Branch Path Parameters			
	$\hat{\alpha}_{11}(t_1+0)$, deg	$\hat{\alpha}_{12}(t_1+0)$, deg	$\Delta t_{n_{.+}}$, s	$\hat{n}_{x_{12}}(t_1+\Delta t_{n_{zk}})$
1. Main	23.314	1.689	2.010	2.821
2. Auxiliary for *H*	16.240	3.763	4.219	2.794
3. Auxiliary for ¨	28.247	4.220	3.823	2.778
4. Alternative for OA	18.587	1.879	5.167	2.718
5. Alternative for ALA	21.166	4.131	1.424	2.792

Table 5. Comparison of optimization options

Options	Branch Path Parameters				
	$\hat{n}_{x_{11}}(t_1+4)$	$\hat{m}_{11}(t_{11})$, kg	$\hat{S}_1(t_1)$, km	$\hat{S}_{12}(t_1)$, km	$\hat{S}_{11}(t_1)$, km
1. Main	1.487	25,903	80.449	257.963	940.807
2. Auxiliary for *H*	1.754	24,946	74.772	258.413	958.930
3. Auxiliary for ¨	1.325	25,025	76.692	257.275	962.048
4. Alternative for OA	1.683	26,000	81.391	274.807	957.946
5. Alternative for ALA	1.611	24,712	69.609	243.623	927.784

The proposed method for calculating the AS optimal trajectory can be used for both preliminary and operational calculation of the AS stages trajectory.

FUTURE LINES OF RESEARCH

Further research should be directed to solving the problem of operational optimization of the CDS motion control process, which would allow existing onboard computer systems to solve real-time the task of operational synthesis of optimal branching paths of such systems. To solve this problem, there is a need to develop the necessary and sufficient conditions for optimality of the CDS motion branching paths in convenient form for real-time synthesis of optimal control of these compound systems. This will allow onboard real-time synthesis and correction of the existing and future CDS optimal trajectories such as aerospace systems, flying sensory networks (information robots), and unmanned aerial vehicle groups.

Figure 7. Flight profiles of AS stages: S - flight range

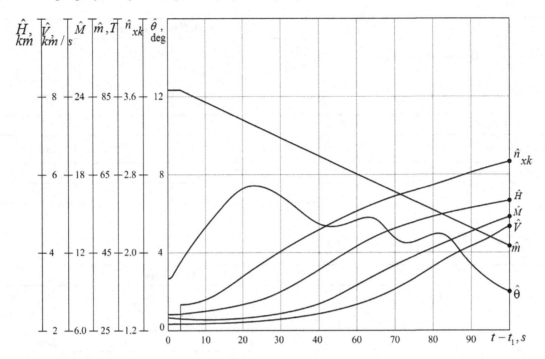

Figure 8. Graphs of optimal phase coordinates for section 0-1 of the AS branching path (ALA+OS flight, $t_0 = 0$ s): M– Mach number, n_{xk}– flight path acceleration (tangential G)

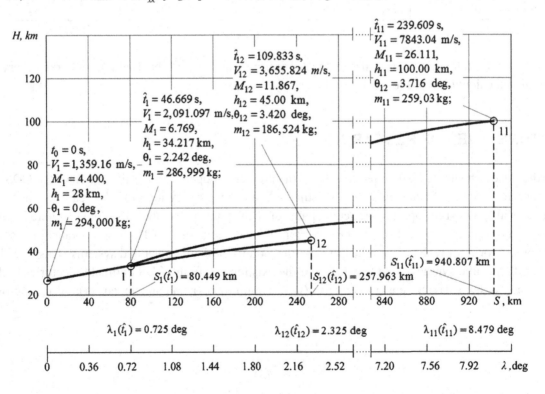

Figure 9. Graphs of optimal phase coordinates for section 1-12 of the AS branching path (ALA flight, $\hat{t}_1 = 46.669$ *s): M– Mach number,* n_{xk} *– flight path acceleration (tangential G)*

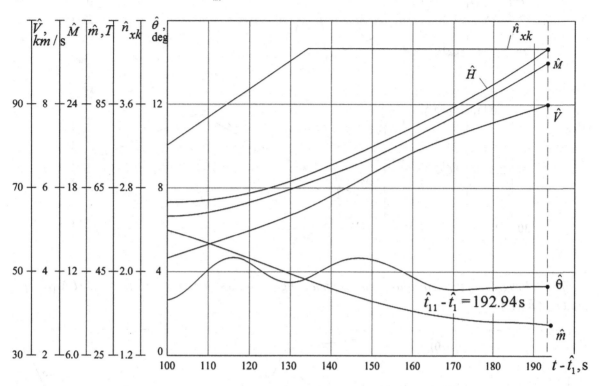

Figure 10. Graphs of optimal phase coordinates for section 1-11 of the AS branching path (OS flight, $100\,s \le t - \hat{t}_1 \le \hat{t}_{11} - \hat{t}_1,\ \hat{t}_1 = 46.669$ *c): M– Mach number,* n_{xk} *– flight path acceleration (tangential G)*

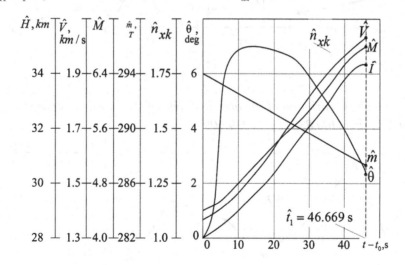

Figure 11. Graphs of optimal phase coordinates for section of the AS branching path (OS flight,

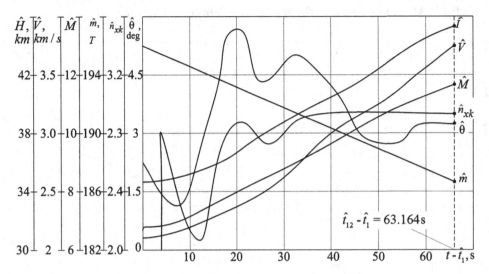

Figure 12. Program of time variation of the optimal angle of attack: for ALA+OA $(\hat{\alpha}_1(t))$, *ALA* $(\hat{\alpha}_{1_9}(t))$ *and OA* $(\hat{\alpha}_{11}(t))$

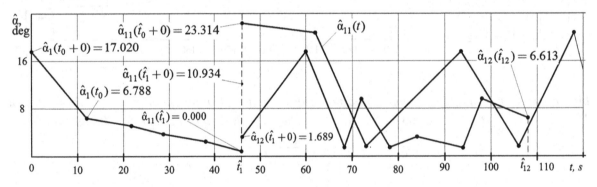

Figure 13. Program of time variation of the optimal angle of attack - continuation of diagram for OA $(\hat{\alpha}_{11}(t))$; $\hat{t}_1 = 46.669$ s, $\hat{t}_{12} = 109.833$ s, $\hat{t}_{11} = 239.609$ s

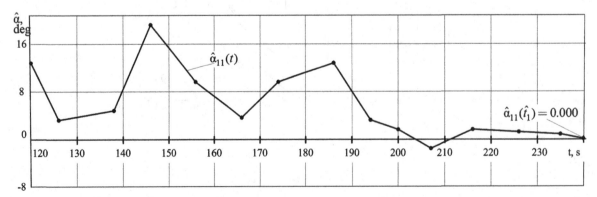

WHAT COMES NEXT?

In the next steps, we plan to consider the application of optimality conditions to improve the algorithms of control system of a group of unmanned aerial vehicles, as well as to improve the algorithms for stabilizing an unmanned aerial vehicle on a given motion trajectory, taking into account possible target changes at any time in a given time slot.

CONCLUSION

In this section, we considered the problem of operational optimization of the control process of CDS motion with branching paths. For the operational synthesis of optimal CDS control, the authors proposed a technique of integrated use of DDS optimal control methods to form conditions that allow the operational optimization of branching paths of the CDS motion. Within the framework of this methodology, the method for transforming the CDS to DDS is proposed with a variable size of control state vectors, which makes it possible to reduce overall the number of computational procedures.

The necessary conditions for the control optimality of CDS moving along the branching trajectory with an arbitrary branching pattern, as well necessary conditions for the control optimality of CDS with the branching trajectory taking into account the interaction of subsystems, are formulated. The necessary optimality conditions developed by the authors allow real-time optimization of the CDS motion branching trajectories and real-time correction of their trajectories in case of affecting factors that were unforeseen at the previous calculation stage and which are critical for implementation of the CDS end use.

The developed conditions offer the advantage of universal application for solving problems with any finite number of path branches and extended coverage of mathematical models of compound systems, which leads to a reduction in computational procedures for control calculating and overcoming the difficulties associated with the indeterminacy of forming of initial conditions and joining of paths. The usability of the form of representing optimality conditions is explained by the use of analytically prepared design ratios that have a clear physical meaning and are focused on the application of standard software.

The example of application of the developed necessary optimality conditions is given. It consists in calculation the reference branching trajectory of aerospace system which is optimal by minimum time for injection of the orbital plane and air launch aircraft to the specified points of near-earth space.

REFERENCES

Ashchepkov, L. T. (1987). *Optimal control of discontinuous systems*. Novosibirsk, Russia: Nauka.

Bogomaz, G., Naumenko, N. & Sobolevskaya, M (2005). Dynamics of launch of spacecraft liquid launch vehicles. Proekt Naukova kniga.

Boss, V. (2016). *Optimal control*. Palmarium Academic Publishing.

Boudiba, O., & Firsov, S. (2019). Designing adaptive PID controller non-sensitive to changes in aerodynamic characteristics of an unmanned aerial vehicle. *Eastern-European Journal of Enterprise Technologies, 1*, 9(97), 69–75.

Bryson, A. E., & Ho, Y.-C. (1975). *Applied optimal control*. New York: John Wiley & Sons.

Filatyev, A. (1994). Optimization of branched trajectories for aerospace transport systems. *19th ICAS Congress, Anaheim, CA*, 325-331.

Filatyev, A., & Yanova, O. (2013).The through optimization of fail-safe branched injection trajectories of launch vehicles in view of aerodynamic load constraints on the basis of the maximum principle. *Progress in Flight Dynamics, GNC and Avionics*, 6, 569-582.

Gabasov, R., Kirillova, F., Paulianok, N., Alsevich, V., Kalinin, A., & Krakhotko, V. (2011). Optimization methods. Minsk, Belarus: Publishing Four Quarters.

Kim, D. P. (2003). *Automatic control theory: Linear systems*. Moscow, Russia: Fyzmatlyt.

Kirichek, R., Paramonov, A. & Vareldzhyan, K. (2015). Optimization of the UAV's motion trajectory in flying sensor networks. *Telecommunication*, 7, 20-25.

Lazarev, Y. N. (2007). *Aerospace trajectory control*. Samara, Russia: Scientific Center of the Russian Academy of Sciences.

Ledsinger, L. A. (2000). Solutions to decomposed branching trajectories with powered flyback using multidisciplinary design optimization. *Thesis,* 162.

Lozino-Lozinsky, G. E. (1998). "Buran" – the way of creation. *Aerospace Engineering and Technology*, 3.

Lysenko, O., & Tachinina, O. (2014a). Mathematical formulation of the problem of optimization of the motion of a group of flying robots on the basis of unmanned aerial vehicles. *Scientific Bulletin Academy of Municipal Administration*, *1*(7), 93–99.

Lysenko, O., & Tachinina, O. (2014b). Method of location of sensors based on the compound dynamic system technology in the area of emergency situation. *International Periodic Scientific Journal SWorld*, *3*(36), 84–89.

Lysenko, O., Tachinina, O., & Alekseeva, I. (2018). Algorithm for operational optimization of two-stage hypersonic unmanned aerial vehicle branching path. *Proceeding of the 2018 IEEE 5th International Conference on Methods and Systems of Navigation and Motion Control (MSNMC)*, 11-16.

Lysenko, O., Tachinina, O., & Chumachenko, S. (2015). The problem statement of branching paths theory for solution of search and rescue task in the area of emergency situation. *Technical Mechanics,* (1), 73–78.

Moiseev, V. (2017). Group application of unmanned aerial vehicles. Kazan, Russia: Editorial and Publishing Center.

Okhotsimsky, D. E., & Sikharulidze, Y. G. (1990). *Basics of space flight mechanics*. Moscow, Russia: Science.

Raikunov, G. G. (2011). *Optimization of the ballistic support of the overflight of the spacecraft system in a circular orbit*. Moscow, Russia: Fyzmatlyt.

Sage, A. P., & White, Ch. C. (1982). Optimum systems control. Moscow, Russia: Radio and Connection.

Shalygin, A. S., Petrova, I. L., & Sannikov, V. A. (2010). *Parametric optimization methods in the flight dynamics of unmanned aerial vehicles*. St. Petersburg, Russia.

Sikharulidze, Y. G. (1982). *Ballistics of aircraft*. Moscow, Russia: Science.

KEY TERMS AND DEFINITIONS

AC: Aircraft Carrier.

AIS: Artificial Intelligent System.

ALA: Air Launch Aircraft.

AS: Aerospace System.

Branching Trajectory (or Path): A group of paths, each corresponds to the trajectory of a separate subsystem of the compound dynamic system and is described in the phase space by the points depicting the dynamic state of CDS subsystems in it.

CDS: Compound Dynamical Systems.

Compound Dynamic Systems: A set of objects (subsystems) operating in time at step-by step mode, which is described at different time intervals by various differential equations and some finite connections for joining the trajectories of compound parts.

DDS: Discontinuous Dynamic System.

EZ: Emergency Zone.

Flying sensor Networks or Informational Robot: A compound dynamic system whose elements are: basic UAV; a group of mobile miscellaneous UAVs equipped with multi-sensors and interconnected using a common information and telecommunication network. The basic UAV is used as an air platform for delivering and initial deployment of drones with multisensors in an emergency zone, as well as for collecting and processing real-time data and retransmitting received data in real time to the control center.

LRE: Liquid Rocket Engine.

OA: Orbital Aircraft.

OS: Orbital Stage.

Path Branch: A section of the compound dynamic system trajectory between the adjacent points in which structural transformations of the CDS occur.

RE: Ramjet Engine.

Structural Transformations: Transformations of compound dynamic system associated with the separation, grouping, start and end of subsystems' and units' motion. These structural transformations can occur both in different and at coincident time moments.

TE: Turbojet Engine.

UAV: Unmanned Aerial Vehicle.

Chapter 15

Intellectual Measuring Complex for Control of Geometrical Parameters of Aviation Details:
Differential–Digital Method of Measurement of Aircraft Parts of Complex Geometric Form

Mariia Kataieva

National Aviation University, Ukraine

Alina Yurchuk

National Aviation University, Ukraine

ABSTRACT

This chapter proposes a new automated method of measuring complex three-dimensional surfaces of aircraft parts in static and dynamic modes. The method allows conducting measurements in closed conditions and at the site of the aircraft disposition. The method consists in the continuous determination of the coordinates of the points of the surface of the detail and their representation in a three-dimensional graphic depiction. New methods of measuring the geometric parameters of parts with the complex spatial surface are suggested. This opens the prospect for the development of new ways of measuring geometric parameters of parts in real-time with high metrological characteristics and computer simulation of the measurement process. The differential-digital method is based on the suggested zero-coordinate principle of the measurement process which involves simultaneous parts availability check, and connects measurement result obtained which provided a reduction in the order of measurement error.

DOI: 10.4018/978-1-7998-1415-3.ch015

BACKGROUND

The rapid development of the aviation industry in modern times is accompanied by a significant increase in the requirements for the reliability of the mechanisms functioning and observance of high accuracy and speed during measuring them. Inaccurate parts manufacturing significantly affects the performance of aircrafts, such as durability, reliability and, to a large extent, depend on the correctness of the choice of shape tolerances and location of the surface.

This, in turn, requires the aviation enterprises to improve the accuracy of the measuring operations, the optimal choice of geometric parameters measuring instruments of the parts and units, the development of new and improved existing methods of measuring the linear-angular dimensions of the parts that would be acceptable in combination with modern computer trichrometers modulation programs and resistant to production conditions.

Therefore, the task of developing the intellectual-measuring complex (IMC) and improving methods of linear-angular measurements of aviation components of complex geometric shape under the conditions of destabilizing factors, which can provide an increased level of reliability of measurement at all stages of the life cycle of a component with a comprehensive change in diagnostic parameters in the event of a malfunction, arises (Ornatsky & Mikhalko & Kataev, 2016; Mulzer & Rote, 2006; Phillips, 2012).

To measure the geometric parameters of aircraft components, it is necessary to develop a measurement algorithm and a procedure for traversing a trajectory (Figure 1).

The principle of operation of the intellectual-measuring complex for the automated process of measuring the aviation components of a complex geometric shape is as follows: when a sensor collides with the surface of the part, the electrical circuit used to detect this moment is locked and a command signal

Figure 1. Algorithm of the process of measurement and positioning

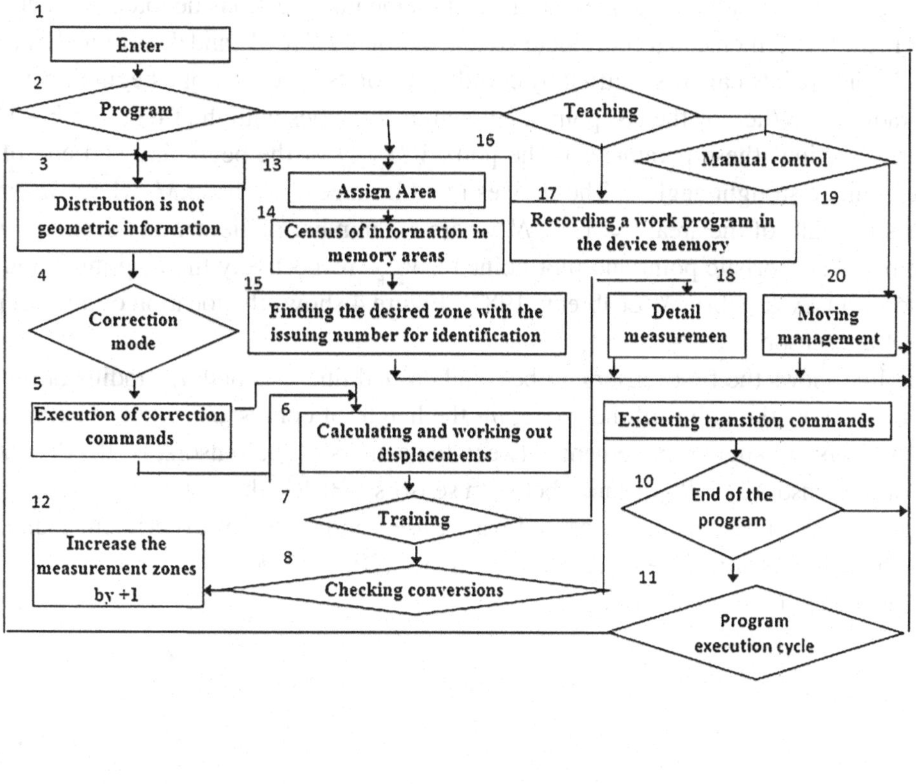

of the IMC is formed. The precision of the reference point of the contact moment of the sensor with the surface of the component depends on the quality of the measuring surface (purity, roughness, oxidation), environmental dustiness, material of the part, curvature radii of the tips and the measuring surface at the point of contact (Kvasnikov & Bilan, 2014).Pre-measured base surfaces of the parts and determine the coordinates of the details in the coordinate system of the IMC. Then, in the measurement mode, in accordance with the developed software mathematical support, the basic data are given the coordinates of the initial measurement point on the surface of the part. The direction of motion and the coordinates of the endpoint with the appropriate step of removing the information, as well as the distance between the plane of the sections. The ICC measurement program provides commands for the automatic control of coordinate drive drives. the formation of sensors through the ADC enters in the synchronization and on the PC where the information is processed, the coordinates of the points of the surface are calculated, a three-dimensional image is constructed. The feedback information for actuating controls enters the IMC device. The periodic and cyclic removal of information by measuring sensors gives the value of the coordinates of the points along the equidistant curve of the surface of the part, as well as the influence of other destabilizing factors.

ORDEVAL APPROACH

To solve the problem a method using artificial intelligence of measuring the complex spatial surfaces of aircraft parts in static and dynamic modes was crafted.

Let's develop a mathematical model and three-dimensional geometric measurement of aircraft parts with complex geometry and on the basis of these models develop a method for measuring geometric parameters. The geometric model may subsequently be used to automatically generate by means of CAD / CAM system which controls a multi-processing CNC center. We will call an elementary section, the section that has constant radius or angle. Number of elementary sections denoted N. Each elementary area viewed in the two-dimensional coordinate system of a detail XdO_dY_d and determined by certain set of parameters. Radius points can be specified by coordinate points $V_i\{x_{vi}, y_{vi}\}$ of a beginning and end of the arc and arc radius R_k. Whereby the last point k an section coincides with the first point $k + 1$ (Figure 2).

For straight sections these parameter is the point $V_i\{x_i, y_i\}$ of the beginning and end of the k segment, and creating a straight angle k. The indices i and k ranges: $i = 1, ..., M$, where M – the number of characteristic points of the area; $k = 1, ..., N$, where N - number of elementary sections, with $M = N + 1$. Random k characteristic point and plot point V_k, V_{k+1} respectively the beginning and end of the section (Chernorutsky & Sibrin & Zhabreev, 1987). Figure 4 shows the location of aircraft parts in the axial plane of the sensors.

The solid line shows the first sensors probes, and dashed line - second. R - radius of the workpiece surface is obtained; r - the radius of the probe; h - the height of combs; β - angular increment between centers of a sensors; δ - angular increment between the centers of the sensors; r - the distance between the centers of the sensors 1-2; R - distance between sensors 1-3; P_1 - the distance between the reference points that lie on the surface of the measured part. Based on numerical experiment found that for an angles $\omega < (15 ... 20)$ it is advisable to take $K = 1$, for $\omega = (20...35)° K = 0,9$, for $\omega > 35° K = 0,8$. Angle the processing area will be equal radius

Figure 2. Breaking the surface of aircraft part into elementary section

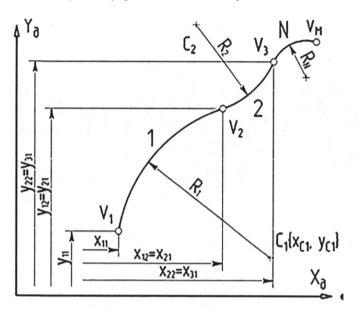

$$\delta = 2\arcsin\left(P / 2\left(R \pm r_{?_i}\right)\right),$$

where "+" sign is taken for the convex surface areas and "-" for concave.

When traversing radial surface (fig.2 a, b) the height of the comb equale

$$h = \pm\sqrt{r^2 + \left(R \pm r\right)^2 - 2 * r\left(R \pm r\right) * \cos\left(\arcsin\left(\frac{\left(R \pm r\right) * \sin\beta / 2}{r}\right) - \frac{\beta}{2}\right)} \mp R,$$

where the upper sign is taken when processing convex surface bottom - concave. When processing line angle (Fig. 3) the height of the ledges will be equal.

To select the position of the reference plane for a part, must be determined three basic points that lie on a straight line (Diaa & ElKott, 2001). Main base point A - surface details contact point. The other two basic points - O and B (Fig. 4).

In determining the coordinates of the base point of sensor details, the distance between the centers of adjacent plates or the distance between the reference points P_1 that lie on the surface of the part must be set. In processing the straight sections are $p_1 = p$. Radial section $p_1 = 2 * R * \sin\left(\dfrac{\beta}{2}\right)$. In determining the coverage angle Ψ of the arc of the circle that forms the surface portion of radius R, you must first calculate the length of the chord section of a radius the expression

$$l_{x_k} = \sqrt{\left(x_{A_{lk}} - x_{A_{nk}}\right)^2 + \left(y_{A_{lk}} - y_{A_{nk}}\right)^2}.$$

Figure 3. Scheme of processing a parts by sensors

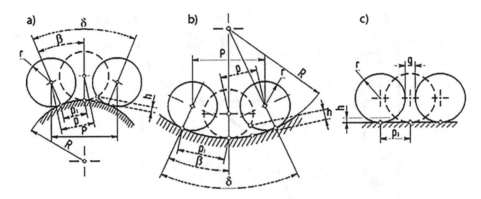

Figure 4. Location of the basic points of the work piece surface

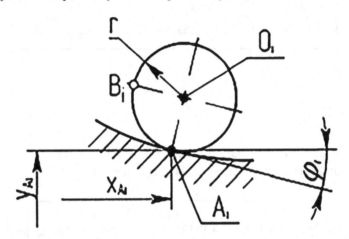

where - $x_{A_{lk}}$ and $y_{A_{lk}}$ - the coordinates of the first point k radius area; $x_{A_{lk}}$ and $y_{A_{lk}}$ respectively - the coordinates of the last point of the site. Then $\varphi_k = 2 * \arcsin\left(\dfrac{l_x}{2R}\right)$.

To further calculation of the base points position wafer surface of aircraft parts in the details coordinate system (Ornatsky & Mikhalko & Kataev, 2016) we need to find angles η between the axis *OX* and a straight line passing through the center of the arc of radius area and point with coordinates (*x*, *y*) lying on this arc and the coordinates of intersection points of each of these direct with radius corners and forming φ (Figure 5).

Bringing the calculations carried out based on conditions of equality interradius distances *P*. For *i*-ht have

Figure 5. Scheme determining the coordinates of base points details

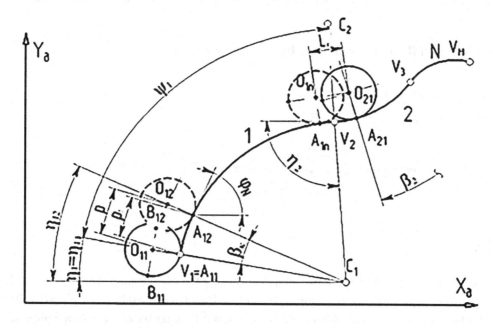

$$\begin{cases} n_i = arctg\left(\dfrac{y_{c_k} - y_i}{x_{c_k} - x_i}\right) npu \left(\dfrac{y_{c_k} - y_i}{x_{c_k} - x_i}\right) > 0; \\[3ex] n_i = 180 + arctg\left(\dfrac{y_{c_k} - y_i}{x_{c_k} - x_i}\right) npu \left(\dfrac{y_{c_k} - y_i}{x_{c_k} - x_i}\right) < 0; \\[2ex] n_i = 0 \, npu \, x_{c_k} - x_i = 0. \end{cases}$$

Before further calculations necessary to determine the exact magnitude of angles above expressions η_{Ak} and η_{Ak+1}, defining the relevant provisions of the first and last points of the k radius surface area.
For concave surface area:

$$\begin{cases} x_{A_i} = x_{C_k} - R_k \cdot cos\eta_i; \\ y_{A_i} = y_{C_k} - R_k \cdot sin\eta_i. \end{cases} \begin{cases} x_{O_i} = x_{A_i} - r_i \cdot cos\eta_i; \\ y_{O_i} = y_{A_i} - r_i \cdot sin\eta_i. \end{cases}$$

For the convex surface areas

$$\begin{cases} x_{A_i} = x_{C_k} - R_k \cdot cos\left(180 - \eta_i\right); \\ y_{A_i} = y_{C_k} + R_k \cdot sin\left(180 - \eta_i\right). \end{cases} \begin{cases} x_{O_i} = x_{A_i} - r_i \cdot cos\left(180 - \eta_i\right); \\ y_{O_i} = y_{A_i} + r_i \cdot sin(180 - \eta_i). \end{cases}$$

After determining the coordinates of the base point A on the k section to calculate the radius for each corner plates φ_i: $\varphi_{i=arctg\left(-\left(x_{C_k}-x_{A_i}\right)/\left(y_{C_k}-y_{A_i}\right)\right)}$.

The coordinates of B_i can be determined by solving the system:

$$\begin{cases} x_{B_{i.kpyr}} = x_{A_i} - \left(r \cdot cos\left(45 - \varphi_i\right)\right) / cos\,45°; \\ y_{B_{i.xpyr}} = y_{A_i} + \left(r \cdot sin\left(45 - \varphi_i\right)\right) / cos\,45°. \end{cases}$$

The coordinates of base points on a cylindrical area:

$$\begin{cases} x_{A_i} = x_{A_{i-1}} + p; \\ y_{A_i} = y_{A_{i-1}}. \end{cases} \qquad \begin{cases} x_{O_i} = x_{A_i}; \\ y_{O_i} = y_{A_i} + r. \end{cases}$$

For canonical origin section(Figure 7) when determining the coordinates of base points follow next sequence of actions. If the drawing taper angle $_k$ not defined clearly, it must be determined by the formula where - the difference between the radii of the current and following basic areas of aviation parts, - the length of the canonical area, defined by the expression:

$$\Delta r_k = y_{2_k} - y_{1_k}, \quad l_{KOH} = x_{2_k} - x_{1_k}$$

If the value of k is negative, the canonical portion is decreasing relative to the first base point, otherwise - upward. Coordinates primary and canonical plot points are taken from the drawing details. The coordinates of base points on the canonical area calculated as follows (Kvasnikov & Bilan, 2014). Where

Figure 6. Scheme determining the coordinates of base points for canonical sites

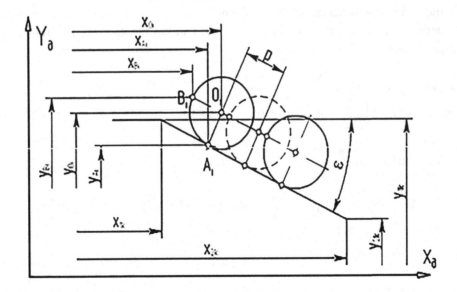

"+" sign is taken for ascending in the direction of creating a bypass circuit, "-" - for diminishing. Angle φ for cylindrical section is 0 and for canonical - the corner ε_k

The systematic increase or decrease of the surface of aircraft parts, fluctuations in its height, waviness and other deviation from flatness can be described by the trend. Usually the surface of complex geometric shapes can be seen as a combination of trend and roughness. Therefore, in many cases, when measuring the workpiece surface, which is characterized by a certain degree of roughness, for declaring it fit or unfit for further use, must clearly identify the presence and characteristics of surfaces trend (Kataeva, 2016). The influence of roughness on the geometric parameters is shown in a very small dimentions (about 0.5 microns). This is true for the trend, but due to longer wavelengths trend impact on the geometric parameters of parts and the likelihood of much greater strain. The result of this approach is shown in Figure 7.

For this recommended (Mulzer & Rote, 2006; Phillips, 2012) hold iteration of random variables derived from measurement details and analyze the trend curve surface for future decisions. To solve the problem of exact mathematical description of the surface trend, consider changes in trend. Based on their thorough analysis determined that reliable description of the trend to use the best method iteration of random variables. Suppose the time series is: $u_1, u_2, \dots$. Find the arithmetic average of each k member of this series

$$V_1 = \frac{1}{k}\sum_{j=1}^{k} u_j, \qquad V_2 = \frac{1}{k}\sum_{j=1}^{k} u_{j+1}, \dots$$

and the composition of these new series of the first iteration $V_1, V_2, \dots$. Then find the arithmetic average member of this new number and composition of the second iteration $W_1, W_2, \dots$

$$W_1 = \frac{1}{k}\sum_{j=1}^{k} V_j, \qquad W_2 = \frac{1}{k}\sum_{j=1}^{k} V_{j+1}, \dots$$

The described process continues until you get the iteration corresponding to the required accuracy trend. With increasing iteration enhanced smoothing time series, it disappears more and more sluggish vibrations. Thus, the described process should end when the roughness disappears and the trend remains. In this case, the latest iteration i of the assessment will be the trend.

Figure 7. The impact of roughness and geometrical parameters for trend details

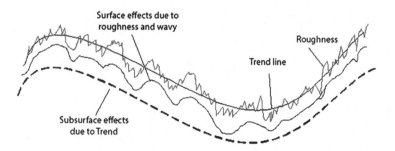

Proved that different iteration procedure can be carried out with different values k. If, for example, for the first iteration is set k_1, for the second – k_2, to third – k_3, the outcome looks like $\dfrac{1}{k_1 k_2 k_3}$. However, the final iteration can be folded so that it exactly describes (that is devoted least time series) polynomial trend, the degree of which does not exceed be any predetermined number p.

It should be noted that if the average member mark time series in finding the arithmetic mean of and use relevant central difference

$$\delta^r u_0 = \delta^{r-1} u_{0+\frac{1}{2}} - \delta^{r-1} u_{0-\frac{1}{2}}, \qquad r = 1, 2, \ldots \tag{1}$$

you can get a formula (for odd k)

$$\frac{1}{k}\left[k\right] u_0 = u_0 + \frac{k^2 - 1^2}{2^2 \cdot 3!} \delta^2 u_0 + \frac{(k^2 - 1^2)(k^2 - 3^2)}{2^4 \cdot 5!} \delta^4 u_0. \tag{2}$$

In order to highlight the trend level o (if it exists) is sufficient to put this formula $\delta^r u_0 = 0$ to all $r > p$. Sharing formula (1) and multiple iterations allows to select various trends needed to solve the problem. In this case highlighted the trend should describe those irregularities that technology should be seen as a significant deviation from flatness.

The question of the best iteration procedure, the choice of numbers k and the total error method requires consideration of some features moving average method. Sharing formula (2) and allow multiple iterations provide a variety of trends needed to solve the problem. In this case highlighted the trend should describe those irregularities that technology should be considered as significant deviations from flatness.

Analysis of the trend (Agterberg & Gaile & Willmott & Reidel & Cliff, 2002; Haining, 2006) showed that should be allocated to recurrent and non-recurrent fluctuation components caused by the influence of destabilizing factors. At the time series is determined by the relation: $u_t = x_1 + x_2 + x_3$, where x_1 - the random component, x_2 - oscillating component (periodical and non), x_3 - trend (systematic component).

Denote T - operator selection trend by moving average. If you use the iterative homogenization, the operator T is linear and hence its effect on time series would be tantamount to his actions on each part: $Tu_t = Tx_1 + Tx_2 + Tx_3$. Suppose now that T - an ideal operator for the allocation of a trend then Tx_1 and Tx_2 can be seen as obstacles encountered in the allocation trend. To these obstacles were minimal, consider each obstacle separately. We assume that the random component x_1 at each t point is an independent random variable ε_t that has a normal distribution with zero expectation and variance D. Then after the first iteration we get

$$j = \left[\frac{k}{2}\right], \quad Tx_1 = \frac{1}{k} \sum \varepsilon_{t+j}, \quad j = -\left[\frac{k}{2}\right],$$

where $\left[\dfrac{k}{2}\right]$ - the whole of magnitude $\dfrac{k}{2}$.

This shows that although the random variables ε_c by definition independent quantities Tx are related to each other, because they contain common members, if the distance between them exceeds k steps. For example if $b-a<k$, the size $Tx_1(a)$ and $Tx_1(b)$ will contain $k+a-b$ common values ε_t and, therefore, will be correlated. Therefore, a number Tx will be smoother than a number x_3 and with increasing multiplicity of iterations, this will increase smoothness. All this leads to a certain oscillatory process. The main characteristics of this process. Due to the random nature of values x_1 and peaks and troughs of the process Tx_1 appear irregularly and therefore there is no value that can be called the period of oscillation. In fact, it is only the distribution period or the average period (Muller, 2004).

The latter can be defined as the average distance between adjacent peaks (or cavity), but this makes it necessary to decide whether to consider a small ripple imposed on the fundamental oscillation. Because most appropriate to determine the average period the average distance between adjacent zeros one sign for this process (Pearson, 2003). Now suppose that as a result of iterative averaging the required obstacle took the form

$$Tx_1 = \sum_j a_j \varepsilon_j,$$

where $\{a_j\}$ - the degree of homogenization. While the range $\left(\dfrac{k}{2}, \dfrac{k}{2}+1\right)$ is zero if

$$V_k = \sum_{j=1}^{k} a_j \varepsilon_j < 0, V_{k+1} = \sum_{j=1}^{k} a_j \varepsilon_{j+1} > 0,$$

top and cavity process. Probability common these conditions is integral in the region bounded by the hyper $V_{(k+1)} = 0$, $V_k = 0$

$$P = P\left(V_k \langle 0, V_{k+1} \rangle 0\right) = \int_{V_{k+1}=0}^{V_k=0} \dots \int f\left(\varepsilon_1, \dots, \varepsilon_{k+1}\right) d\varepsilon_1 \dots d\varepsilon_{k+1}.$$

Zero random process is called crossing point of this process with their expectation. Sign determined by the direction of zero crossing, plus if the intersection is upwards, and negative if the intersection is down. Where $f\left(\varepsilon_1, \dots, \varepsilon_{k+1}\right)$ - joint density values $\varepsilon_1, \dots, \varepsilon_{k+1}$. If these values are normal, it is equal

$P = \dfrac{\theta}{2\pi}$, where $\cos \dfrac{\sum_{j=1}^{k} a_j a_{j+1}}{\sum_{j=1}^{k} a_j^2}$ To a preset formula shows where θ is the angle between the hyper

planes $V_{k+1} = 0$. Since the probability of expectation is the number of positive zeros on the unit length, the average distance between zeros equals $L = \dfrac{1}{p} = \dfrac{2\pi}{\theta}$ Then

$$V_k - V_{k-1} = \sum_{j=1}^{k} a_j \varepsilon_j - \sum_{j=1}^{k} a_j \varepsilon_{j-1} > 0, V_{k+1} - V_k = \sum_{j=1}^{k} a_j \varepsilon_{j+1} - \sum_{j=1}^{k} a_j \varepsilon_{j-1} < 0.$$

Joint probability these conditions will be equal $P_1 = \dfrac{\theta_1}{2\pi}$ where θ - the angle between the hyper $V_k - V_{k-1} = 0$. Therefore

$$\cos\theta_1 = \frac{\left(a_2 - a_1\right)a_1 + \left(a_3 - a_2\right)\left(a_2 - a_1\right) + \ldots a_k\left(a_k - a_{k-1}\right)}{a_1^2\left(a_2 - a_1\right)^2 + \ldots + \left(a_k - a_{k-1}\right)^2 + a_k^2}.$$

The required distance with equal $L_1 = \dfrac{1}{P_1} = \dfrac{2\pi}{\theta_1}$. The dispersion process is equal

$$Tx_1 = D\sum_{j=1}^{k} a_j^2, \tag{3}$$

where D - the variance of x_1.

Since $\displaystyle\sum_{j=1}^{k} a_j = 1$ the sum $S = \displaystyle\sum_{j=1}^{k} a_j^2$ is always <1 and shows as reduced noise variance Tx_1 compared to the variance of the random component tx_1. In general, the value S depends on the multiplicity of iterations q and averaging the number of elements k. As for value k, although its growth Tx_1 also leads to a decrease in the variance, but too much importance to this magnitude can result in smoothing of the trend, which of course is unacceptable. A clear understanding of this issue should be on the analysis of the obstacles Tx_2 and the so-called serial correlation. Consider the simple case (Kataeva, 2015) as a member x_2 of the sinusoid $x_2 = \sin(\gamma + \omega t)$ is described. Then after the first iteration we get

$$Tx_2 = \frac{1}{k}\sum_{j=1}^{k}\sin\left(\gamma + \omega t\right) = \frac{1}{k}\cdot\frac{\sin\dfrac{\omega k}{2}}{\sin\dfrac{\omega}{2}}\cdot\sin\left(\gamma + \omega t\right). \tag{4}$$

Since the resulting value is attributed to the midpoint $t_1 = \dfrac{k+1}{2}$ of the formula (2.16) can be written as (before and after iteration).

$$Tx_2 = \frac{1}{k}\cdot\frac{\sin\dfrac{\omega k}{2}}{\sin\dfrac{\omega}{2}}\cdot\sin\left(\gamma + \omega t_1\right),\; Tx_2 = \left(\frac{1}{k}\cdot\frac{\sin\dfrac{\omega k}{2}}{\sin\dfrac{\omega}{2}}\right)^{q}\cdot\sin\left(\gamma + \omega t_q\right).$$

Thus, to reduce this barrier must take great significance q and k or $k = \dfrac{2m\pi}{\omega}$, $m = 1, 2, \ldots$. Since it was decided to value $q \leq 3$, the value obstacles will k completely depend on the size and frequency of oscillation ω. If ω the interval is small and $k \ll \dfrac{2\pi}{\omega}$ much less averaging period oscillations naturally at slow oscillations, the

$$\left(\frac{1}{k} \cdot \frac{\sin \dfrac{\omega k}{2}}{\sin \dfrac{\omega}{2}} \right)^q \approx \left(\frac{1}{k} \cdot \frac{\dfrac{\omega k}{2}}{\dfrac{\omega}{2}} \right)^q .$$

So, $Tx_2 = x_2$, what can be done next concluded that the value k should be taken so that the interval was significantly higher averaging period of the oscillations to be excluded from the calculation and significantly shorter period of oscillation, which should be left as a trend. In linear angular size of aircraft parts, this means that the averaging interval should be more than a mean surface roughness and waviness less than average period. Thus, with the objective of exact mathematical description of the trend surface aircraft parts with complex geometric form was developed multilevel iteration method of random variables, which allows for non-recurrent and recurrent fluctuation components caused by the influence of destabilizing factors.

The method of selection of useful component of non-stationary random process is highly effective in conditions of a priori uncertainty. The method consists in reproduction is not very original implementation process, and estimates derived in some way. To one of the advantages of the method of detecting abnormal values include: application two threshold criterion for deciding on abnormality mentioned process provides a result in which with increasing magnitude of abnormal values estimation error probability of the first kind α tend to minimum values and estimates the probability of correct detection seeks β - to maximize values (Thompson, 2002).

Along with the advantages of the proposed method of detection of anomalous values presented in (Ornatsky & Kataeva, 2015), found that it has very serious drawbacks, one of which is dependent on the threshold of a constant factor A. The right choice of A will increase the efficiency factor of abnormal values. Modification of the proposed method in the detection of abnormal values A involves the introduction of threshold adaptation coefficient with respect priori fixed value error probability of the first kind α.

The proposed method in this paper requires a single discrete implementation of the investigational non-stationary random process Y_k, which is represented additive model $\lambda_k = a + bt_k + ct_k^2$. Prior information about the test process is that at some time intervals helpful part of the process is a smooth function (Pearson, 2003), that is accurately described by a polynomial of degree not higher than the second. Implementation of the process under study is divided into intervals of random lengths obtained as follows: using the random number uniformly distributed in the range of $(0;1)$, get l-1 numbers $\xi_1^{(1)}, \xi_2^{(1)}, \ldots, \xi_{l-1}^{(1)} \in (0;1)$.

Using the expression $y = t_1 + (t_N - t_1)x$ is carried out mutually unambiguous display range values $(0;1)$ in the range investigated non-stationary random process $(t_1, \ldots, T_N)$, while receiving appropriate partition numbers $\alpha_1^{(1)}, \alpha_2^{(1)}, \ldots, \alpha_{l-1}^{(1)}$ in the range of $(t_1, \ldots, T_N)$ disjoint l intervals

$$\alpha_j^{(1)} = t_1 + \left(t_N - t_1 \right) \xi_j^{(1)}, j = 1, 2, \ldots, l\text{-}1$$

We introduce notation for intervals partition

$$\Delta_1^{(1)} = t_1; \alpha_1^{((1))}, \Delta_2^{(1)} = \left(\alpha_2^{(1)}; \alpha_2^{(1)}, \ldots, \Delta_l^{(1)} = \left(\alpha_{l-1}^{(1)}; t_N\right).\right.$$

Each interval partition $_j^{(1)}(j1, 2, \ldots, l)$ contains at least samples ∇ (minimum length interval partition-ing) source of non-stationary random process from the set $\{t_1, t_2, \ldots, t_N\}$, otherwise random numbers that form the interval partitioning $(\xi_1^{(1)}, \xi_2^{(1)}, \ldots, \xi_{l-1}^{(1)})$ discarded and generated again. The presence of this condition means $l\nabla 0,6N$. For each new evaluation procedure for splitting a segment $(t_1, \ldots, T_N)$ of l random interval length (with the test conditions specified above) is repeated. The result is a temporary N parti-tions interval $(t_1, \ldots, T_N)$.

where the symbol Σ denotes summation for all k-th values that belong to the interval of the partition $\Delta_j^{(i)}$ and $j = \overline{1, l}, i = \overline{1, j}$. At each interval partitioning $\Delta_j^{(i)}$, where $i = \overline{1, p}$, $j = \overline{1, j}$ the method of least squares estimation are $\overline{a}_j^{(i)}$, $\overline{b}_j^{(i)}, \overline{c}_j^{(i)}$, polynomial coefficients approximating $a + bt_k + c_k^2$ as solu-tion of linear equations:

$$\begin{cases} aN + b\sum t_k + c\sum t_k^2 = \sum Y_k, \\ a\sum t_k + b\sum t_k^2 + c\sum t_k^3 = \sum t_k Y_k, \\ a\sum t_k^2 + b\sum t_k^3 + c\sum t_k^4 = \sum t_k^2 Y_k, \end{cases}$$

Indicates where the summation of all -time value belonging interval and splitting.

$$S_{kj}^{(i)} = Y_{kj}^{(i)} - \overline{\lambda}_{kj}^{(i)}, \qquad k = \overline{1, N}. \tag{4}$$

The result of the method is determined to set the interval $(t_1, \ldots, T_N)$ smoothing functions $\overline{\lambda}_{kj}^{(i)}$ where $i = \overline{1, p}$, each of which is a "piecewise quadratic"

$$\overline{\lambda}_{kj}^{(i)} = \begin{cases} \overline{a}_1^{(i)} + \overline{b}_1^{(i)} t_k + \overline{c}_1^{(i)} t_k^2, \\ \overline{a}_2^{(i)} + \overline{b}_2^{(i)} t_k + \overline{c}_2^{(i)} t_k^2, \\ \ldots\ldots\ldots\ldots\ldots\ldots \\ \overline{a}_l^{(i)} + \overline{b}_l^{(i)} t_k + \overline{c}_l^{(i)} t_k^2, \end{cases}$$

where $t_k \in \Delta_l^{(i)}$, $i = \overline{1, p}$ та $j = \overline{1, l}$.

Identify the value of the difference $S_{kj}^{(i)}$ between the original non-stationary random process $Y_{kj}^{(i)}$ and evaluation functions smoothing $\overline{\lambda}_{kj}^{(i)}$. In assessing the difference parameters of the process $S_{kj}^{(i)}$, each interval partitioning $\Delta_j^{(i)}$, using a method of evaluation of stable (Yunusov, 2012), *i*-e parameter estimation expectation *m* and standard deviation $\bar{\sigma}$ conducted at a α^* reduced sample.

For this purpose, each interval partitioning get ranked number values $\Delta_j^{(i)}$ $\bar{m}$ and estimate the expectation and standard deviation $\bar{\sigma}$, which is performed excluding the first and last row rank value. Then expression estimates the expectation $\bar{m}_j^{(i)}$ and standard deviation $\bar{\sigma}_j^{(i)}$ taken as follows (Ferguson, 2009)

$$\bar{m}_j^{(i)} = \frac{1}{r}\sum_{k=2}^{r-1} S_{kj}^{(i)}, \quad \bar{\sigma}_j^{(i)} = \sqrt{\frac{1}{r-1}\sum_{k=2}^{r-1}\left(\bar{m}_j^{(i)} - S_{kj}^{(i)}\right)^2}, \tag{5}$$

where $r\nabla 2$.

Next to each interval partitioning investigational non-stationary random process, set the threshold value where *A* – Factor $0 < A \leq 3$ and $j = \overline{1, l}$. Excess values difference process $S_{kj}^{(i)}$ at every interval partition set threshold (6), fined, that is, if the condition:

$$S_{kj}^{(i)} + \varepsilon_j^{(i)} < Y_{kj}^{(i)} < S_{kj}^{(i)} - \varepsilon_j^{(i)}, \tag{6}$$

then $Y_{kj}^{(i)}$ gets one penalty value.

According to the method of propagation estimates (Ferguson, T. S., 2009; Thompson, W.R., 2002)., the above procedure for determining the fines, repeated times and each repetition of the test condition (3.1) for each value $Y_{kj}^{(i)}$, where $i = \overline{1, p}$; $k = \overline{1, N}$; $j = \overline{1, l}$; *N*- sample size investigational non-stationary random process; *l*- the number of intervals partitioning; *p*- the number of repetitions procedures (4) - (6).Thus, there is an accumulation of a number of penalty values for the elements of the original implementation of the investigational process, that is

$$g_k^{(i)} = \begin{cases} 1, \left|S_{kj}^{(i)} \pm \varepsilon_j^{(i)}\right| > Y_{kj}^{(i)} \\ 0, \left|S_{kj}^{(i)} \pm \varepsilon_j^{(i)}\right| < Y_{kj}^{(i)} \end{cases},$$

where $g_k^{(i)}$ - number of penalty values $k = \overline{1, N}$, $i = \overline{1, p}$; $j = \overline{1, l}$.

Upon completion of processing for all fined initial values determined implementation of the total value of fines and maximum range $max\left(g_k^*\right) = G$. Further test condition: if $g_k^* > 0,7G$, then *k* importance of implementing unsteady input random process Y_k will be treated as anomalous. Condition (6) obtained on the basis of simulation modeling in various models useful and noise components of non-stationary random process. Thresholds are defined in accordance with the expression (5) and using equation (4).

Introduced a series of penalty values g_k obtained in excess of the value $\varepsilon_j^{(i)}$ of each difference threshold process S_k, when the condition (5), the procedure is repeated p times. It is proposed to introduce adaptation threshold appointing fines at a rate fixed value A on a priori error probability of the first kind. To this end, studies dependence of the volume of the sample standard deviation of the value A of a random process L for a variety of stationary processes σ_s with fixed values priori error probability of the first kind α.

The result is a coefficient A estimates depending on the amount of the investigated sample and standard deviation noise component of the process, that is $Af(L,\sigma_s)$. Incoming implementation is stationary centered Gaussian random process. From the analysis of the dependencies for various fixed values of the first kind error α probability with increasing sample size estimation coefficient tends to a constant value and does not depend on the value of standard deviation of a random process. Estimates for sampling factor increases by an average of 5%. The same is negligible factor A depending on the sample size L and standard deviation $\sigma_{\text{ш}}$. When stationary random process represented relyeyev's laws and uniform density probability distribution. Thus, studies show that the value of coefficient estimates A for consideration of distribution laws of random processes depend only on the values given a priori error probability of the first kind α (Nubert, 1997)

$$A(\alpha) = 2,16 - 6,3\alpha + 7,32\alpha^2.$$

(7)

The results of research dependence of $A=f(L,\sigma_w,\alpha)$ can be used in the adaptation threshold (6) instead of a constant coefficient of its value, which is calculated in accordance with (7). Using equation (7) in the evaluation threshold (6) allows the use of the proposed method abnormal values at a fixed value error probability of the first kind α.

Illustrative Examples

The method is fully automated and allows to conduct measurements in closed conditions and at the site of the aircraft disposition. The method consists in the continuous determination of the points coordinates of surface detail and their computer processing and three-dimensional representation of a result (Kataeva, Kvasnikov & Shkvarnitskaya, 2018). We recommend usage of such sensors as: optical, optical fiber and video. By scanning the surface of the detail, these sensors allow the determination of coordinates with high accuracy and speed (more than 10^5 measurement points per 1 second). It is also possible to carry out automatic corrections of temperature and vibration errors, which allows using these sensors in the workshop. A cluster to determine the deviation of the shape and location of the surfaces was entered, which corresponds to the tolerance field according to the spectrum of the color image (De Floriani, Falcidieno, Nagy, Pienovi, 1991).

Differential-digital method is implemented in the following sequence:

1. Installing the detail and based on regular reference points, which are specified in the design documentation.
2. Carry out technical measurements of a limited number of points, which allow the simultaneous combination of a mathematical model (reference model) with a crafted detail.
3. Preliminary calculation of spatial alignment or transfer parameters.

4. Measurement in the automatic mode of optical sensors scanning with not less than 150 - 1100 points on the surface of the detail, depending on the size and field of admission on its geometric parameters (Figure 8).
5. Estimation of the accuracy of the base of the detail.
6. Conduct of control measurements at separate points where there is an abnormal deviation.
7. The color image is converted to numeric data and the distribution of color information is handled.
8. A graphic image is created on clusters and maps are obtained to identify the trend of the surface of the entire detail and for each sector.

The value of the color of the discrete points of the digital image forms the area represented in the form of isolines. As a result, a color image area is formed which corresponds to a deviation from the shape and location of the surfaces and differs from each other in color. Specialized software includes digital model of the reference aircraft detail in form of geometric parameters. The software is based on the mathematical model of the aircraft component. In addition to measuring flat details, the big task is to determine the curvature of the surface and the angle of rotation of the plane of the detail. This dependence is in the form of a three-dimensional array, in which one column occupies the angle of inclination of the plane, the second - the curvature of the detail the third - the coordinate (Detling, Zinchenko, Miroshnichenko, 2006). When scanning on the surface of a detail, a function is used which is based on the mathematical model of the detail process. To scan the surface of the detail, used a function which is based on the mathematical model of the detail process.

The definition of geometric parameters in the Power Inspect program with Artificial Intelligence is similar to the geometric deviations from the shape and location of surfaces, as well as more complex second-order surfaces that include a large number of parameters. The obtained values of the angle of inclination of the plane and the curvature, as well as the value of the coordinates, are compared with the field of admission to this detail and with the spectrum that is represented in the Fig. 9.

Figure 8. Results of the measurement of the surface of the aircraft component with the determination of the error and deviation of the form

Figure 9. Color map of geometric parameters of deviation from the shape and location of the surface

To show the behavior and usability of the adaptive algorithm method of anomalous values for a priori fixed probability of an error of the first kind, we present a step-by-step execution of a sequence of operations in which the method is implemented is as follows:

- Values memorable input values input process, where and, - step rate, *N*- the amount of input implementation.
- Set value error probability of the first kind, the minimum interval partitioning ∇, and the number of partitions.
- Defined evaluation by approximation on each partition intervals not exceeding polynomial of the second degree, where and та.
- Find differences.
- Assessment determined standard deviation for on each of the intervals partition, and given threshold value.
- Where each input value realization compared to a threshold in each of the partitioning interval ls and, if there is an accumulation of penalty values, where - number of penalty values.
- this procedure is repeated -times, bringing the total number of penalties obtained values.
- Where determined maximum value row.
- Test condition: If, then - value of input realization is an abnormal value.

Differential-digital method makes it possible to study the errors, determine the geometric parameters of the detail and obtain the cross sections with the help of automized methods of algorithmic errors corrections, which provides a zero displacement for working conditions of measurement after the training procedure with the obtaining of the statistical result. In the result of research systematic error of coordinate points of the surface details reduced by 0.3 mm.

FUTURE RESEARCH DIRECTIONS

We believe that the differential-digital method of measuring the geometric parameters of complex shapes can be used in other areas, except for aircraft engineering. The practical significance of the results obtained to improve the automation of measurement processes is to develop a differential-digital measurement method using intelligent metering complexes of mechanical quantities based on the proposed method. With the help of the improved method of measuring the linear and angular dimensions of the body parts, it will be possible to easily identify deviations from the shape of the surfaces, determine the size of the step of the detachment, depending on the accuracy of the surface to be controlled, and construct a three-dimensional image. Further testing of the design of the intellectual-measuring complex on the basis of the developed models of the measurement process will enable to identify the relationship between the errors of measurement of geometric parameters with the effect of destabilizing factors and the error of the measuring equipment and prematurely prevented their appearance.

CONCLUSION

The differential-digital method for measuring the geometric parameters of parts with complex spatial surface, which is based on the proposed coordinate-zero principle of measuring the measurement process, and on the elemental bypass of the trajectory of the part with the construction of a three-dimensional image with a minimum measurement error in the conditions of the destabilizing factors, is developed.

The mathematical model of the description of the surface of aviation components during the process of its measurement is developed, which enables the representation of a part in a three-dimensional geometric space of the image with a deviation from the shape and position of the surface and provides the speed of determination and correction of measurement errors. The accuracy and speed of measuring the geometric parameters of parts due to the development of the instrumental part of precision measuring methods have been increased, technical problems of algorithm adaptation and software for specific modifications of automated measuring instruments have been solved.

The methods of experimental research and the determination of abnormal deviations from the shape and location of surfaces when measuring the geometric parameters of a component, which involves the simultaneous verification of the suitability of the part in the production and correctness of the results of measurement, which provided a reduction in the order of measurement error. The developed methods open the perspective for the development of new means of measuring the geometric parameters of the details of factors in real time with increased metrological characteristics and computer simulation of the measurement process.

REFERENCES

Agterberg, F. P., Gaile, G. L., Willmott, C. J., Reidel, D., & Cliff, A. (2002). *Trend surface analysis. Spatial statistics and models. Spatial autocorrelation* (pp. 147–171). London, UK: Pion.

Chernorutsky, G. S., Sibrin, A. P., & Zhabreev, V. S. (1987). *Tracking systems of automatic manipulators*. Moscow, Russia: Science.

De Floriani, L., Falcidieno, B., Nagy, G., & Pienovi, C. (1991). On sorting triangles in a Delaynay tessellation. *Algorithmica*, *6*(1-6), 522–535. doi:10.1007/BF01759057

Detling, V. S., Zinchenko, I. V., & Miroshnichenko, B. M. (2006). Information-measuring system for ensuring the quality of roughness of the surface. *Issue date of the Cherkasy State Technological University, Special Issue*, 135-137.

Elkott, D. F. (2001). Automatic sampling for CMM inspection planning of free-form surfaces. Windsor, Canada.

Ferguson, T. S. C. J. (2009). Proceedings of the 4 Berkeley symposium on mathematical statistics and probability. *Annals of mathematical statistics. 12*, 82–94.

Haining, R. (2006). Trend-surface models with regional and local scales of variation with an application to aerial survey data. *Technometrics, 4*(February), 461–469.

Kataeva, M., Kvasnikov, V., & Shkvarnitskaya, T. (2018). *Development of a differential-digital method for precision measurements of the geometric parameters of aviation details* (pp. 32–35). Vilnius, Lithuania: Science - Future of Lithuania. Transport Engineering and Management.

Kataeva, M. A. (2015). Mathematical construction of the trajectory of the motion of the sensor on the surface of the part. *Measuring and Computer Technology in Technological Processes, 1*, 91-94.

Kataeva, M. O. (2016). Method of describing the trend of a surface of aviation detail of complex geometric shape. *Bulletin of the Engineering Academy of Ukraine, 3*, 186–189.

Kvasnikov, V. P., & Bilan, M. O. (2014). Calculation of tolerances by the method of dimensional analysis, taking into account operational error. *Measuring and Computing Engineering in Technological Processes, 1*, 173-177.

Muller, E. (2004). Koordinen - Messtechnik der Zukunst. *Masehinenbau, 23*, 12–21.

Mulzer, W., & Rote, G. (2006) Minimum-weight triangulation is NP-hard. *Annual Symposium on Computational Geometry, 6*, 86-90.

Nubert, G. P. (1997). *Measuring transducers of non-electric quantities. Introduction to theory, calculation and design*. Moscow, Russia: Bukinist.

Ornatsky, D. P., & Kataeva, M. A. (2015). Development of computerized system for control of curved surfaces using inductive sensors. *Technological Audit and Production Reserves, 1/2*, 83-90.

Ornatsky, D. P., Mikhalko, N. V., & Kataeva, M. O. (2016). Development of a computerized system for controlling curved surfaces using inductive sensors. *Technological Audit and Production Reserves, 1/2(27)*, 83-90.

Pearson, E. S. (2003). Biometrika. *Annals of Mathematical Statistics, 8*, 118–124.

Phillips, G. M. (2012). *Interpolation and Approximation by Polynomials*. Berlin, Germany: Springer.

Thompson, W. R. (2002). The problem of negative estimates of variance components. *Annals of Mathematical Statistics, 12*(2), 214–265.

Yunusov, F. S. (2012). On the problem of mathematical modeling of formation processes of complex surfaces, which are processed on machines with software control *Proceedings of the KAI, 141*, 48-53.

ADDITIONAL READING

Erb, K., & Fisher, P. (1990). Improving the Metrological Characteristics of Measuring Transducers by the Method of Digital Error Compensation Express Information. *Collection Measuring and Measuring Equipment*, *42*, 22–28.

National Instruments. (2016). Measurement and Automation. *Catalog 2016. Austin*. 161. Voltage Output Programmable Sensor Conditioner – Available at: http://elcodis.com/parts/5884374/PGA309.html#datasheet

National Instruments. SCXI Universal Strain Grade Input: SCXI-1540. – Available at: http://www.ni.com/pdf/products/us/4scxisc293_ETC_196.pdf

Chapter 16
Methods for Assessing the Glissade Entrance Quality by the Crew

Yurii Hryshchenko

https://orcid.org/0000-0002-1318-9354

National Aviation University, Ukraine

Viktor Romanenko

National Aviation University, Ukraine

Daria Pipa

National Aviation University, Ukraine

ABSTRACT

This chapter proposes methods for assessing the quality of entry into the glide path using autocorrelation functions and regarding the psycho-physiological state of the human operator. This is due to the possible increase in the psycho-physiological tension of a human operator in special flight situations. The analysis of the pilot's ability to control the trajectory of the aircraft is presented. A method for alerting the crew about the failure in the system for obtaining data on the angle of attack and airspeed to prevent non-coordinated turn is given. The majority of the calculations are based on the method of determining the correlation fields during the flight.

DOI: 10.4018/978-1-7998-1415-3.ch016

INTRODUCTION

Safety issues occupy one of the main places in the air transport system. The quality of maintaining the flight trajectory depends both on the navigation aids and flight navigation equipment, as well as on the quality of the crew training. Most often, there is a deterioration in the quality of piloting in the event of special cases in flight.

The final stage of the landing (landing) depends on the timely entry into the glide path and further maintaining its trajectory.

The literature analysis shows that most authors already considered the influence of the human factor on the glide path approach (Kazak & Budzynska (2009); Gibb & Schvaneveldt (2008); Kashmatov (2008). One of the authors of those books has also adequately investigated the topic. The purpose of the chapter is to research the problem in more detail and propose methods for assessing flight quality with further automation of this process.

This paper examines the accuracy of the aircraft's entry to the glide point by correlation functions and piloting estimation for pre-landing maneuvering, regarding the angle of attack and the flight speed.

Currently, there are deterministic methods for assessing the quality of entry into the glide path under director control mode. However, the pilot's integrated differentiated control movements are stationary. Therefore, we consider it expedient to evaluate the accuracy of entry into the glide path by the correlation functions. But the only accuracy of the glide path entrance is not enough for successful flight. The flight speed must also be observed. We suggest determining the dependence of the velocity on the angle of attack on the contours of the correlation fields. This will allow to identify dangerous configurations of aircraft movement. Thus, according to the above components, we can determine the quality of the glide path entrance.

BACKGROUND

Autocorrelation Functions and Their Use in Assessing Quality of The Approach

The correlation functions between information about the flight trajectory and distortions during the operator tracking operations considered below. This happens due to its psycho-physiological features in a state of high tension.

A comprehensive analysis of the trajectory of the aircraft allows to determine the degree of operator training, his psycho-physiological state, the quality of work of all elements of the aircraft, as well as the reliability of communication in the reception and transmission of commands.

The ideal trajectory control system is the one that provides for the full ergatic compatibility of the operator – aircraft subsystem and continuously processes information about the flight trajectory of the aircraft (AC) without errors and fully performs the specified flight program according to function $I(t)$. In this case, the specified trajectory and information about the real flight trajectory will be the same. Any deviation from a given flight mode is immediately recorded by the onboard equipment, which also notifies the operator. If necessary, the operator can work this difference to zero (Figure 1).

If there is a following situation: the operator uses information I (t), which becomes a subject of distortion due to a combination of certain reasons. In this case, the information management facility will change and become different $-I'(t)$ (Figure 2)

Figure 1. Ideal system   *Figure 2. Non-ideal system*

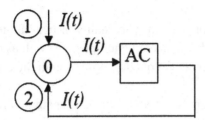   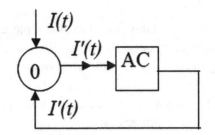

Flight Correlation Function Regarding the Psycho-Physiological State of a Human Operator

Suppose that the pilot, who is in the control loop of the aircraft received information to control the aircraft $-I(t)$ and worked it out, making movements using the controls. As a result of the evolution of control, the information $I(t)$ changes and becomes $I'(t)$. It can be described using the formula:

$$I'(t) = I(t) \cdot (1 + m(t) \cos \Omega t), \qquad (1)$$

where $\Omega = 2\pi f$ – angle speed, f – frequency, $m(t)$ – *amplitude*.

The function I '(t) can take this form based on the experimental fact of the flight parameters amplitude increase (FPAI) due to an increased psycho-physiological tension of the pilot. This is called integro-differentiated motor dynamic stereotype, i.e. the final result of the actions of a human operator during piloting (Gryshchenko, 2004).

If the functions $m(t)$, $I(t)$ are stationary random, $\varphi_i = $ const ($\varphi_i = \Omega_i \tau$), i is the number of the test (landing approach) according to the known category of tests (landings), then the function $I'(t)$ depends only on time t and is fully determined by the result of each test landing.

The correlation function of the flight path $\rho(\tau)$, according to which an airplane should fly without informational distortions for information reception and control (Hryshchenko, Skripets & Tronko, 2015) is

$$\rho(\tau) = I(t) \cdot I(t - \tau) =$$
$$= \lim_{T \to \infty} \left(\frac{1}{T_L}\right) \int_0^{T_L} I(t) \cdot I(t - \tau) dt = \frac{1}{T_L} \int_0^{T_L} I(t) \cdot I(t - \tau) dt, \qquad (2)$$

where τ– the delay time, T_L– the flight time on a particular part of length L, for example, $T_L = T_l$, where T_l is the landing time of the aircraft.

With regard to the flight information distortion the correlation function is defined

$$\rho'(\tau) = \{I(t)[1 + m_i(t) \cos(\Omega_i t + \varphi_i)]\} \cdot \{I(t - \tau)[1 + m_i(t - \tau) \cos[\Omega_i(t - \tau) + \varphi_i]\} =$$
$$= I(t)I(t - \tau) + I(t)I(t - \tau)m_i(t) \cdot \cos(\Omega_i t + \varphi_i) + I(t)I(t - \tau) \cdot \cos[\Omega_i(t - \tau) + \varphi_i] +$$
$$+ I(t)I(t - \tau)m_i(t)m_i(t - \tau)\cos(\Omega_i t + \varphi_i) \cdot \cos[\Omega_i(t - \tau) + \varphi_i]. \qquad (3)$$

Using the inequality of Bunyakovsky-Schwarz, the following can be written:

$$|I(t)I(t-\tau)m_i(t)\cos(\Omega_i t + \varphi_i)| \le |I(t)I(t-\tau)m_i(t)| \cdot \cos(\Omega_i t + \varphi_i). \tag{4}$$

For phenomena that occur long enough, for example, the landing time of the T_l and the period value of the oscillatory process of the aptitude increase $T_i = \dfrac{2\pi}{\Omega_i}$ (according to experimental dependencies [2]) are incommensurable in magnitude

$$T_{L>>}T_{i}.$$

Therefore

$$\cos^2(\Omega_i t + \varphi_i) \approx 0,$$

$$\cos^2[\Omega_i(t-\tau)+\varphi_i] \approx 0.$$

Therefore, the second and the third term of expression (3) can be neglected.

In this way the most significant parameters will remain out of the four components in the correlation function:

$$\rho^{'}(\tau) = \rho(\tau) + I(t)I(t-\tau)m_i(t) \cdot m_i(t-\tau)\cos(\Omega_i t + \varphi_i) \cdot \cos[\Omega_i(t-\tau)+\varphi_i]. \tag{5}$$

Calculating the component:

$$
\begin{aligned}
I(t)I(t-\tau)m_i(t)m_i(t-\tau) \cdot \cos(\Omega_i t + \varphi_i)\cos[\Omega_i(t-\tau)+\varphi_i] = \\
= \frac{1}{2}I(t)I(t-\tau)m_i(t)m_i(t-\tau)\cos\Omega_i\tau + \\
+ \frac{1}{2}I(t)I(t-\tau)m_i(t) \cdot m_i(t-\tau)\cos(2\Omega_i t - \Omega_i\tau + 2\varphi_i] = \\
= \frac{1}{2}I(t)I(t-\tau)m_i(t) \cdot m_i(t-\tau)\cos\Omega_i\tau
\end{aligned}
\tag{6}
$$

In the above expression, the second term is neglected, since it is close to zero

$$2\Omega_i = \frac{4\pi}{T_i} >> \frac{1}{T_?},$$

where T_i is the period of oscillation of the FPAI, T_l is the landing time.

The correlation function is presented in the following form:

$$\rho_{FPAI}(\tau) = \rho(\tau) + \frac{1}{2}I(t) \cdot I(t-\tau)m_i(t) \cdot m_i(t-\tau)\cos\Omega_i\tau. \tag{7}$$

The correlation function of the landing trajectory with the FPAI equals the sum of the correlation function of the landing trajectory without the FPAI and a member dependent on the statistics of the "clean" landing and the statistics of the FPAI.

If the AI is stationary in time $m_i(t)$=const=m, then the correlation function can be represented as:

$$\rho_{FPAI}(\tau) = \rho(\tau) + \rho(\tau)\cos\varphi. \tag{8}$$

If FPAI is not stationary, then:

$$I(t)I(t-\tau)m_i(t)m_i(t-\tau) \leq \rho(\tau)\rho_{FPAI}(\tau). \tag{9}$$

Thus, when landing with FPAI with the random "inclusion", FPAI has the form of an offset correlation function:

$$\rho_{\Sigma FPAI}(\tau) = \rho(\tau) + \rho(\tau)\rho_{FPAI}(\tau)\cos\varphi. \tag{10}$$

Imagine a general correlation function, which is a complex of all the above situations:

$$\rho_{full}(\tau) = \frac{1}{T_L}\int_0^{T_L}[I(t)+I_{FPAI}(t)][I(t-\tau)]+I_{FPAI}(t-\tau)]dt =$$

$$= \frac{1}{T_L}\int_0^{T_L}[I(t)I(t-\tau)+I_{FPAI}(t)\cdot I_{FPAI}(t-\tau)+I(t)I_{FPAI}(t-\tau)+$$

$$+I(t-\tau)I_{FPAI}(t)]dt = \frac{1}{T_L}\int_0^{T_L}I(t)I(t-\tau)dt + \frac{1}{T_L}\int_0^{T_L}I_{FPAI}(t)I_{FPAI}(t-\tau)dt + \tag{11}$$

$$+\frac{1}{T_L}\int_0^{T_L}I(t)I_{FPAI}(t-\tau)dt + +\frac{1}{T_L}\int_0^{T_L}I(t-\tau)I_{FPAI}(t)dt = \rho(\tau) + \rho_{FPAI}(\tau) +$$

$$+\frac{1}{T_L}\int_0^{T_L}I(t)I_{FPAI}(t-\tau)dt + \frac{1}{T_L}\int_0^{T_L}I(t-\tau)I_{FPAI}(t)dt.$$

According to the Bunyakovsky-Schwarz inequality we get:

$$\int_0^{T_L}I(t)\cdot I_l(t-\tau)dt \leq \sqrt{\int_0^{T_L}I^2(t)dt}\sqrt{\int_0^{T_L}I^2_{FPAI}(t-\tau)dt}. \tag{12}$$

With the help of the experimental results (Gryshchenko & Skripets, 2015), we define $I(t-\tau)$

$$I_{FPAI}(t-\tau) = I_{FPAI}\cos(\Omega t + \varphi_{FPAI} - \Omega\tau)$$

then,

$$\sqrt{\int_0^{T_L} I^2{}_{FPAI}(t-\tau)dt} = \sqrt{I^2 \int_0^{T_L} \cos^2(\Omega t + \varphi_{FPAI} - \Omega\tau)dt} =$$

$$= I_{FPAI}\sqrt{\frac{1}{2}\int_0^{T_L}[1+\cos 2(\Omega t + \varphi_{FPAI} - \Omega\tau)]dt} =$$

$$= \frac{I_{FPAI}}{\sqrt{2}}\sqrt{\int_0^{T_L} \cos 2(\Omega t + \varphi_{FPAI} - \Omega\tau)dt} = \tag{13}$$

$$= \frac{I_{FPAI}}{\sqrt{2}}\sqrt{T_L + \frac{1}{2\Omega}\sin(2\Omega t + \varphi_{FPAI} - \Omega\tau)},$$

$$\sin(2\Omega t + \varphi_{FPAI} - \Omega\tau) \le 1,$$

$$T_L \gg \frac{T_{FPAI}}{2\pi},$$

$$\frac{I_{FPAI}}{\sqrt{2}}\sqrt{T_L + \frac{1}{2\Omega}} = \frac{I_{FPAI}}{\sqrt{2}}\sqrt{T_L + \frac{T_{FPAI}}{4\pi}} = I_{FPAI}\sqrt{\frac{T_L}{2}}. \tag{14}$$

Therefore, inequality (12) can be represented using expression (14), thus:

$$\int_0^{T_L} I(t)I_{FPAI}(t-\tau) \le \sqrt{\int_0^{T_L} I^2(t)dt} \cdot I_{FPAI}\sqrt{\frac{T_L}{2}} = \text{const} \tag{15}$$

$$\int_0^{T_L} I_{FPAI}(t)I(t-\tau) \le \sqrt{\int_0^{T_L} I^2{}_{FPAI}(t-\tau)dt} \cdot I\sqrt{\frac{T_L}{2}} = \text{const} \tag{16}$$

Substitute expressions (15) and (16) in (11) and get:

$$\rho_{full}(\tau) = \rho(\tau) + \rho_{FPAI}(\tau) + \text{const}.$$

$$\rho_{FPAI}(\tau) = \frac{1}{T_L}\int_0^{T_L} I_{FPAI}(t)I_{FPAI}(t-\tau)dt =$$

$$= \frac{1}{T_L}\int_0^{T_L} I_{FPAI}\cos(\Omega t + \varphi_{FPAI}) \cdot I_{FPAI}\cos[(\Omega(t-\tau) + \varphi_{FPAI})dt = \tag{17}$$

$$= \frac{I^2{}_{FPAI}}{T_L}\int_0^{T_L} \frac{1}{2}(\cos\Omega\tau + \cos(2\Omega t + 2\varphi_{FPAI} - \Omega\tau)dt =$$

$$= \frac{I^2{}_{FPAI}}{T_L}(\frac{T_L}{2}\cos\Omega\tau + \frac{1}{2\Omega}\sin T_L\Omega\cos(T_L\Omega + 2\varphi_{FPAI})$$

$$\frac{T_L}{2}\cos\Omega\tau \gg \frac{1}{2\Omega}\sin T_L\Omega\cos(T_L\Omega + 2\varphi_{FPAI})$$

$$\rho_{full}(\tau) = \rho(\tau) + \rho_{FPAI}(\tau) + const =$$

$$= \rho(\tau) + \frac{I^2_{FPAI}}{2}\cos\Omega\tau + const$$

$$const < \frac{I^2_{FPAI}}{2}\cdot\cos\Omega\tau \qquad\qquad (18)$$

differs by $\dfrac{1}{2\pi}\dfrac{T_{FPAI}}{T_L} - times$

The correlation functions on the glide path are considered below.

Let the landing length (L_l) be equal to the length (L_g) of the glide path. Let us assume that the calculation $\acute{A}(\c{C})$ taking into account the FPAI on glide path is more accurate.

$$\rho(\chi) = \frac{1}{L_l}\int_0^{L_l}\left[y(x) + y_{FPAI}(x)\right]\left[y(x-\chi)\right]dx =$$

$$= \frac{1}{L_l}\int_0^{L_l}\left[y(x)\cdot y(x-\chi)dx + \frac{1}{L_l}\int_0^{L_l} y_{FPAI}(x)\cdot y(x-\chi)dx\right] +$$

$$+ \frac{1}{L_l}\int_0^{L_l}\left[y(x)\cdot y_{FPAI}(x-\chi)dx + \frac{1}{L_?}\int_0^{L_l} y_{FPAI}(x)\cdot y_{FPAI}(x-\chi)dx\right] = \qquad (19)$$

$$= \rho(\chi) + \rho_{FPAI}(\chi) + \frac{1}{L_l}\int_0^{L_l} y(x)\cdot y_{FPAI}(x-\chi)dx + \int_0^{L_l} y_{FPAI}(x)\cdot y(x-\chi)dx$$

where χ is the deviation from the glide path.

The FPAI is formulated as a function with one spectral component, which is viewed on an experimental curve. However, it should be generalized and recorded in general terms, based on experimental data.

On the basis of the experimental curves, the curve of the autocorrelation function is constructed(Hryshchenko, 2016; Solomentsev & Zaliskyi, 2018), and on its basis, the FPAI function is constructed. Analysis of the ability to control the trajectory of the aircraft with the help of the correlation function

If the correlation function between a given and untimely introduction the aircraft in the glide path by the pilot considered, it would be seen that the function is connected with the psycho-physiological characteristics of the pilot.

Simulation of the human factor action on the pilot by introducing complex failures on the simulator opens up huge opportunities for anti-stress training of crew members (Lapa, 1974; Gryshchenko & Romanenko, 2005).

Imagine a mathematical model of the aircraft control on the correlation functions at the site of entry into the glide path to improve the quality of landing.

In general, the flight path of the aircraft will be described using the function:

$Z = f(x, y).$

This trajectory is determined by the glide path:

$Z = f(x, y) = \text{const}.$

The flight path is determined regarding the pitch angles (υ), roll (γ), inclination from the trajectory (θ) and heading (ψ), as well as the speed (ν) of the aircraft. The coordinates of the flight trajectory dependent on all the listed parameters and are determined by functional expressions:

$Z = F_1(υ, γ, θ, v), \quad Y = F_2(υ, γ, θ, ψ, v), \quad X = F_3(υ, γ, θ, ψ, v).$

The glide path coordinates ($y = \text{const}$):

$Z = F_4(υ, γ, θ, v), ψ = \text{const} \quad Z = F_5(υ, γ, θ, ψ, v), \quad y = \text{const}.$

Let us determine the trajectory of the glide path by a straight line connecting the position of the beacon ($x = L, Z = 0$) and the starting point of the landing ($x = 0, Z_0 = h$). Onthe Figure 3 these points are characterized by a significant angle of the trajectory change α. The real flight trajectory implies a smoothing of the angles, which must be taken into account in future (Hryshchenko, Skripets & Tronko, 2015).

The glide path coordinates are determined by the dependency:

$$Z = Z_0 + xtgα, \tag{20}$$

where Z_0 is the initial coordinate of height, α– the angle between the path line and the X direction; $Z_0 = h$, $tgα = -h/L$, L is the length of the glide path. In these designations, the glide path is:

$$Z(x) = h - \frac{h}{L} \cdot x \tag{21}$$

It should be noted that such a landing trajectory is possible in the complete absence of external environmental influences, normal operation of all aircraft mechanisms, operator's high professionalism without psychological deviations and other factorial overlays (Gryshchenko, 2004).

Landing trajectory is described by the formula (21), we denote as an ideal glide path.

For further analysis of the movement of the vessel during landing (Figure 4), the trajectoryof correlation function is described by equation (22):

Figure 3. Glide path trajectory (L– glide path length,h– initial highs at the moment of landing)

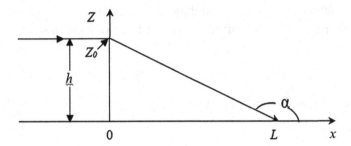

$$Á(\varsigma) = \frac{1}{L} \int\limits_0^l Z(x) \cdot Z(x - \chi) dx \tag{22}$$

whereχ– is the magnitude of the trajectory shift along the x coordinate: $x-\chi$ is the lag value for $\chi > 0$ or the advance value for $\chi < 0$.

The functions $Z(x), Z(x-\chi), Z(x+\chi)$ on the integration interval from 0 to L have the following values:

$$Z(x) = \begin{cases} h - \dfrac{h}{L}x, 0 \leq x \leq L \\ h \leq 0 \\ 0 \geq L \end{cases} \tag{23a}$$

$$Z(x - \chi) = \begin{cases} h, -\infty \leq x \leq \chi^* \\ h - \dfrac{h}{L}(x - \chi), \chi \leq x \leq L + \chi \\ 0, L + \chi \leq x \end{cases} \tag{23b}$$

Figure 4.The trajectory of the glide path delay – 1, the advance glide path approach – 2

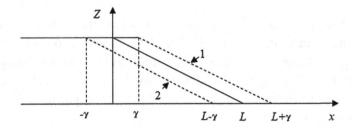

$$Z\left(x+\chi\right)=\begin{cases}h,-\infty\le x\le\chi\\h-\dfrac{h}{L}\left(x+\chi\right),-\chi\le x\le L-\chi\\0,L-\chi\le L\end{cases}\qquad(23c)$$

Based on the conditions determined by the expressions (23a, b, c), the correlation function of delay and advance according to the areas of integration have the following form:

$$\begin{aligned}
\text{Á}\left(\text{Ç}\right)&=\frac{1}{L}\int_{0}^{l}Z\left(x\right)Z\left(x-\chi\right)dx=\frac{1}{L}\int_{0}^{\text{Ç}}\left(h-\frac{h}{L}x\right)\cdot0\,dx+\\
&+\frac{1}{L}\int_{\text{Ç}}^{L-\text{Ç}}\left(h-\frac{h}{L}x\right)\cdot\left[h-\frac{h}{L}\left(x-\chi\right)\right]dx+\frac{1}{L}\int_{L-\chi}^{L}\left(h-\frac{h}{L}x\right)\cdot0\,dx=\\
&=\int_{\text{Ç}}^{L}\left(h-\frac{h}{L}x\right)\cdot\left[h-\frac{h}{L}\left(x-\chi\right)\right]dx=\\
&=\frac{h^{2}}{L}\left(1+\frac{\chi}{L}\right)\left(l-\chi\right)-\frac{h^{2}}{L}\frac{\left(2L+\chi\right)}{2}\left(L^{2}-\chi^{2}\right)+\frac{h^{2}}{3L^{3}}\left(L^{3}-\chi^{3}\right)=\\
&=\frac{h^{2}}{3}-\frac{h^{2}\chi}{2L}+\frac{h^{2}\chi^{3}}{6L^{3}},\chi\ge0
\end{aligned}\qquad(24)$$

The timing correlation function is:

$$\begin{aligned}
\rho\left(-\text{Ç}\right)&=\frac{1}{L}\int_{0}^{l}Z\left(x\right)Z\left(x+\text{Ç}\right)dx=\frac{1}{L}\int_{0}^{L-\text{Ç}}\left(h-\frac{h}{L}x\right)\left[h-\frac{h}{L}\left(x+\text{Ç}\right)\right]dx+\\
&+\frac{1}{L}\int_{L-\text{Ç}}^{L}\left(h-\frac{h}{L}x\right)\cdot0=\frac{1}{L}\int_{0}^{L-\chi}\left(h-\frac{h}{L}x\right)\left[h-\frac{h}{lL}\left(x+\chi\right)\right]dx=\\
&=\frac{h^{2}}{L^{3}}-\frac{h^{2}\chi}{2L}+\frac{h^{2}\chi^{3}}{6L^{3}}\quad8;8\quad\rho\left(\chi\right)=\frac{h^{2}}{3}+\frac{h^{2}\chi}{2L}-\frac{h^{2}\chi^{3}}{6L^{3}},\chi\le0
\end{aligned}\qquad(25)$$

Autocorrelation function: $\rho\left(0\right)=\dfrac{h^{2}}{3}$;

with$\chi=L$

$\rho\left(-L\right)=\rho\left(L\right)=0.$

It should be noted that for arbitrary χ in the range$-L$ to $+L$

$\rho\left(-\chi\right)=\rho\left(+\chi\right).$

The factor overlay leads to a change in the glide path trajectory Z (x) by $\Delta Z_{FA}=Z_{FA}-Z\left(x\right)$ or

$$Z_{FA} = Z(x) + \Delta Z_{FA}(x), \tag{26}$$

where $Z(x)$ is the flight trajectory without factor overlays, $\Delta Z_{FA}(x)$ is the change in the flight path when exposed to factor overlays. This change is determined not only by the influence of technical errors and environmental state on the pilot, but also by his professional level of training and the psychological state. It is necessary to note another important point. Equation (6) is a function of the coordinates of the flight trajectory and thetime, since there is a one-to-one relationship between coordinate x and speed of movement one-to-one $x = x(v, t)$.

$$Z_{FA}(t) = Z(t) + \Delta Z_{FA}^{**}(t). \tag{27}$$

Calculation of the correlation function $Z_{FA}(x)$, defined by the expression (26):

$$\rho_{FO}^{full}(\chi) = \frac{1}{L} \int_0^L Z_{FO}(x) \cdot Z_{FO}(x - \chi) dx =$$
$$= \rho(\chi) + \rho_{FN}^{pure}(\chi) + \frac{1}{L} \int_0^L \left[Z(x) \cdot \Delta Z_{FN}(x - \chi) + \Delta Z_{FN}(x) \cdot Z(x - \chi) \right] dx, \tag{28}$$

where$\rho(\chi)$ is the correlation function of the "pure" flight, $\rho_{FA}(\chi)$ is the correlation function of the factor lining

$$\rho(\chi) = \frac{1}{L} \int_0^L Z(x) \cdot Z(x - \chi) dx \tag{29}$$

$$\rho_{\Phi H}(\chi) = \frac{1}{L} \int_0^L \Delta Z_{FA}(x - \chi) dx \tag{30}$$

The third member of the right side of equation (28) is a correlation function of two independent functions $Z(x)$ and $\Delta Z_{FO}(x)$.

This mixed function is denoted as ρ_{mix}.

$$\rho_{mix} = \frac{1}{L} \int_0^L \left[Z(x) \cdot \Delta Z_{FO}(x - \chi) + \Delta Z_{FO}(x) \cdot Z(x - \chi) \right] dx \tag{31}$$

Most often, there is no correlation between the flight trajectory (including the glide path) ΔZ_{FA} and ρ_{mix}(that is equal or close to zero). For example, take an factor overlay that affects the Z coordinate as harmonic oscillations over time:

$$\Delta Z_{FO}(t) = a_0 \cos(\Omega t) \tag{32a}$$

or

$$\Delta Z_{FO}(t) = a_0 \cos \Omega \frac{x}{v}, h \gg a_0 \tag{32b}$$

$$Z_{FO}(x) = h.$$

where a_0– the amplitude of the factor overlay.

The aircraft flies at a constant height h with a harmonic change in height according to the law defined by the formula (32b).

Then the mixed correlation function ρ_{mix} is calculated by the formula (11) and equal to

$$\rho_{mix} = \frac{h \cdot a_0}{L} \int_0^L \left(\cos \frac{\Omega}{v} x \right) dx = \frac{4h \cdot a}{L} \cdot \frac{\Omega^2}{v^2} \sin \frac{\Omega}{2v} L \cdot \cos \frac{\Omega}{2v} \left(L - \frac{\chi}{2} \right) \tag{33}$$

since $h \ll L$, then $\rho_{mix} \approx 0$.

Note that when $\frac{\Omega}{v} \cdot \chi = 180°$, $\rho_{mix} = 0$, when $\frac{\Omega}{v} \cdot \chi = 0$, amplitude ρ_{mix} is maximum. The mixed correlation function periodically depends on L and χ.

Note that when $\rho_{mix} = 0$, when the amplitude ρcm is maximum. The mixed correlation function periodically depends on L and χ.

$$\rho(L) = h^2, \rho_{FA}(\chi) = \frac{a_0^2}{L} \int_0^L \cos \frac{\Omega}{A} x \cdot \cos \frac{\Omega}{A} (x - \chi) dx =$$

$$= \frac{a_0^2}{2L} \int_0^L \left[\cos \frac{2\Omega}{v} \left(x - \frac{\chi}{2} \right) + \cos \chi \right] dx = \frac{a_0^2}{2L} \int_0^L \left[\cos \frac{2\Omega}{v} \left(x - \frac{\chi}{2} \right) + \cos \frac{\Omega}{v} \chi \right] dx =$$

$$= \frac{a_0^2}{2L} \cdot \frac{v}{2\Omega} \cdot \sin \frac{2\Omega}{v} \left(x - \frac{\chi}{2} \right) \Big|_0^L + \frac{a_0^2}{2L} \cos \frac{\Omega}{v} \chi \cdot L =$$

$$= \frac{a_0^2}{2L} \cdot \cos \left(\frac{\Omega}{v} \cdot \chi \right) + \frac{a_0^2}{2L} \cdot \frac{v}{2\Omega} \left[\sin \frac{2\Omega}{v} \left(L - \frac{\chi}{2} \right) + \left(\frac{2\Omega}{v} \cdot \frac{\chi}{2} \right) \right] \approx$$

$$\approx \frac{a_0^2}{2L} \cdot \cos \left(\frac{\Omega}{v} \cdot \chi \right).$$

Thus, the full correlation function is equal to:

$$\rho_{FO}\left(\chi\right) = h^2 + \frac{a_0^2}{2}\cos\left(\frac{\Omega}{Å}\cdot\chi\right)$$

The value of ρ_{FO} during the horizontal flight with a periodically factor overlay has a periodic dependence on χ. The graph of this dependence is presented on Figure 5.

From (33) it can be seen that the factor overlay at certain stages of the flight can be neglected due to the fact that $h \gg a_0$.

From the example above, the following conclusion can be made.

When flying with an factor overlay and in case of occurrence of the flight trajectory small periodical amplitude oscillations, the full correlation function will be the sum of the flight correlation functions without the factor overlay and the trajectory correlation functions only with the factor overlay.

$$\rho_{FO}^{full}\left(\chi\right) = \rho\left(\chi\right) + \rho_{FO}^{pure}\left(\chi\right)$$

under the condition of a long flight path on a glide path $L > h$, $L \gg a_0$, $h > a_0$.

At low altitudes, when the condition $h > a_0$ is not fulfilled, the height of the flight trajectory is commensurate with the amplitude of the factor overlay oscillations. It is necessary to take into account the correlation function of the displaced process, the trajectory of the ideal flight and the appearance of the factor overlay.

Analysis of Typical for Modern Aviation Glide Path Landing

In the most realistic situation is the aircraft, moving away from the landing approach, tries to return to it (Figure 5). In contrast to the glide path in Fig. 2 in this case, all the trajectories converge at the end of the runway. In the most cases, the direction of the correct glide path is set by radio systems for near navigation (Solomentsev, Herasymenko, Zaliskyi & Cheked, 2016). In case of late entry pilots try to return to the correct glide path. We are considering one of the possible options.

Normal landing $Z = h - \dfrac{h}{L}\cdot x = h\dfrac{L - x}{L} \quad 0 < x \ll L$

Figure 5. Dependence of the correlation function on the magnitude of the delay χ in case of "pure" flight (χ) and with the factor periodic overlay $\rho_{FA}(\chi)$

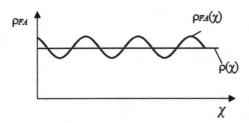

Advance $Z = h\dfrac{L-x}{L+x} - \chi \quad -\chi \ll x \ll L$

Delay $Z = h\dfrac{L-x}{L-x} \quad \chi \ll x \ll L$

$$\rho(\chi) = \frac{1}{L}\int_0^L h\frac{L-x}{L}\cdot h\frac{L-x}{L+x}\,dx =$$

$$= \frac{h^2}{L^2(L+\chi)}\int_0^L (L-x)^2\,dx = \frac{h^2}{L^2(L+\chi)}\int_0^L (L^2 - 2Lx + x^2)\,dx =$$

$$= \frac{h^2}{L^2(L+\chi)}\left[L^3 - L^3 + \frac{1}{3}L^3\right] = \frac{h^2\cdot L}{3(L+\chi)}.$$

$0 < x \ll L$

Having obtained the formula of a normal landing:

$$\rho(\chi) = \frac{h^2\cdot L}{3(L+\chi)},$$

Formula for the delay in theglide path entrance.

$$\rho(+\chi) = \frac{1}{L}\int_0^l h\frac{L-x}{L}\cdot Z_{del}(x)\,dx =$$

$$\frac{1}{L}\int_0^\chi h\frac{L-x}{L}\cdot 0\,dx + \frac{1}{L}\int_\chi^L h\frac{L-x}{L}\cdot h\frac{L-x}{L-\chi}\int_\chi^L (L-x)^2\,dx =$$

$$= \frac{h^2}{L^2(L-\chi)}\int_\chi^L (L^2 - 2Lx + x^2)\,dx =$$

$$= \frac{h^2}{L^2(L-\chi)}\left[L^2(L-\chi) - L(L^2-\chi^2) + \frac{1}{3}(L^3-\chi^3)\right] =$$

$$= \frac{h^2}{L^2}\left[L^2 - L(L+\chi) + \frac{1}{3}(L^2 + L\chi + \chi^2)\right] =$$

$$= \frac{h^2}{L^2}\left[-L\chi + \frac{1}{3}L^2 + \frac{1}{3}L\chi + \frac{1}{3}\chi^2\right] =$$

$$= \frac{h^2}{L^2}\left[\frac{1}{3}L^2 + \frac{2}{3}L\chi + \frac{1}{3}\chi^2\right] = h^2\left[\frac{1}{3} - \frac{2}{3}\frac{\chi}{L} + \frac{1}{3}\left(\frac{\chi}{L}\right)^2\right]$$

$$L^3 - \chi^3 = \left(L^2 + L\chi + \chi^2\right)\left(L - \chi\right)$$

It can be seen from the formula that when the delay χ of the start of landing increases, the correlation function falls.

For example, if the delay χ equals $0.1\,L$, then the probability of an accident also drops by about 10%, which is undesirable.

Now let's compare the entry form to the glide path with delay and advance.

Delay $\rho\left(+\chi\right) = h^2\left[\dfrac{1}{3} - \dfrac{2}{3}\dfrac{\chi}{L} + \dfrac{1}{3}\left(\dfrac{\chi}{L}\right)^2\right] = \dfrac{h^2}{3}\left(1 - \dfrac{\chi}{L}\right)^2$

Advance $\rho\left(-\chi\right) = \dfrac{h^2}{3}\dfrac{1}{1 + \dfrac{\chi}{L}}$

Defining $\dfrac{\rho\left(+\chi\right)}{\rho\left(-\chi\right)}$

$$\dfrac{\rho\left(+\chi\right)}{\rho\left(-\chi\right)} = \dfrac{\left(1 - \dfrac{\chi}{L}\right)}{\dfrac{1}{1 + \dfrac{\chi}{L}}} = \left(1 - \dfrac{\chi}{L}\right)^2\left(1 + \dfrac{\chi}{L}\right) = \left[1 - \left(\dfrac{\chi}{L}\right)^2\right]\left(1 - \dfrac{\chi}{L}\right)$$

From this formula follows that:

$$\rho\left(-\chi\right) > \rho\left(+\chi\right)$$

Thus, as a result of the analysis performed, it was shown that the delay in the entry to the glide path performed by the aircraft is more dangerous for the quality of the approach of the aircraft during the landing, than advance. The lag-ahead correlation function drops faster.

It follows from the above that it is possible to control the trajectory of the aircraft on the glide path using the above formulas and with the help of correlation function without the action of factor overlays and on the glide path with a periodic factor overlay. The glide path correlation function allows one to determine stationary random functions of the flight path.

These results require the implementation of the technology of flight operations to improve the quality of control of the aircraft's trajectory.

Method for estimating the accuracy of entry into the glide path by the correlation function

The deviation of the flight trajectory from the intended course is characterized by a correlation function between a given and realized flight trajectory (Hryshchenk, Romanenko, & Tronko, 2017, Hryshchenko, Romanenko, & Pipa, 2018).

The square of the integral difference of the flight paths (planned and real) in a certain area is equal to

$$\Delta = \int_{x_1}^{x_2} \left[Z_p(x) - Z_r(x) \right]^2 dx \tag{1}$$

where
  $Z_p(x)$ – planned flight trajectory,
  $Z_r(x)$ – real flight trajectory.
  The flight path is on the plane $y = $ const. Z coordinate depends on x, $Z(x)$, that is the height of the flight path in the Cartesian coordinate system.
  x_1, x_2 – the starting and ending points of reference for flight paths horizontally.

Calculating the expression (1) it is obtained:

$$\Delta = \int_{x_1}^{x_2} Z_p^2(x) dx - 2 \int_{x_1}^{x_2} Z_p(x) Z_r(x) dx + \int_{x_1}^{x_2} Z_p^2(x) dx \tag{2}$$

Denoting the components of the equation | 2 |

$$\int_{x_1}^{x_2} Z_p^2(x) dx = L\rho_p$$

$$\int_{x_1}^{x_2} Z_p(x) Z_r(x) dx = L\rho_{pr}$$

$$\int_{x_1}^{x_2} Z_p^2(x) dx = L\rho_r,$$

where $L = x_2 - x_1$, functions $\rho_r, \rho_{pr}, \rho_r$ – are respectively the autocorrelation functions of the planned flight (ρ_{pr}), he correlation function between the planned trajectory and the real trajectory (ρ_r) and ρ_r – is the autocorrelation function of the real flight trajectory. Expression (1) in this case has the form

$$\Delta = L\rho_p - 2L\rho_{pr} + L\rho_r \tag{3}$$

The autocorrelation functions are approximately equal. Consider the case where,

$$\rho_p \approx \rho_r \approx \rho_A, \tag{4}$$

and ρ_A– autocorrelation function of the planned and implemented process. That is, when the correlation function, the planned and realized flight path, differ slightly.

Rewriting the formula (3) with (4)

$$\Delta = 2L\left(\rho_A - \rho_{pr}\right) \tag{5}$$

$$\rho_{pr} = \rho_A - \frac{\Delta}{2L} \tag{6}$$

From (5) it is clear that if the trajectories coincide, then

The value of the integral error per unit length is equal to the difference between the autocorrelation and correlation functions. These three functions for different types of trajectories are calculated below.

The special case of the delay of the landing trajectory is considered.

$$Z_p = h - \frac{h}{L}(x - \chi), \tag{7}$$

where– the delay value;

Then the correlation and autocorrelation functions are equal to:

$$\rho_3 = \frac{1}{L}\int_0^L \left(h - \frac{h}{L}x\right)^2 dx = \frac{h^2}{3} \tag{8.1}$$

$$\rho_{p3} = \frac{1}{L}\int_L^0 \int_0^L \left[\left(h - \frac{h}{L}x\right)\left(h - \frac{h}{L}(x - \chi)\right)\right] dx = \frac{h^2}{3} + \frac{h^2}{2L} \tag{8.2}$$

$$\rho_k = \frac{1}{L}\int_L^0 \left[h - \frac{h}{L}(x - \chi)\right]^2 dx = \frac{h^2}{3} + \frac{h^2}{L} + \frac{h^2}{L^2}\chi^2 \tag{8.3}$$

Substituting the values (8.1, 8.2, 8.3) into the formula (3) and the integral difference between the two paths is found:

$$\Delta = L\frac{h^2}{3} - 2L\frac{h^2}{3} - h^2\chi + \frac{h^2 L}{3} - \frac{h^2}{L}\chi^2 + h^2\chi^2 = \frac{h^2}{L}\chi^2 \tag{9}$$

При $\chi = 0, \Delta = 0, \chi = L, \Delta = h^2 L$ \tag{9.1}

$$\Delta = h^2 L \left(\frac{\chi}{L}\right)^2 \tag{10}$$

From the formula (8) it is seen that with an increase from 0 to L, the value Δ increases
The advance flight path on the glide path is

$$(x+\chi) = h - \frac{h}{L}(x+\chi) \tag{11}$$

Dividing the range $(0, L)$ into two sections $(0, L - \chi)$ and $(0, L + \chi)$. The lead function on the segment is equal to zero. Consequently, the correlation function of the timing is determined by integration only on the interval

$$\rho_{pr} = \rho(+\chi) = \frac{1}{L} \int_0^{-L} \left(h - \frac{h}{L}\chi\right)\left[h - \frac{h}{L}(+x)\right] dx = \frac{h^2}{3} - \frac{h^2}{2L} \tag{12}$$

Comparing expressions (8.2) and (12)

$$\rho(-\chi) - \rho(+\chi) = \frac{h^2}{L}$$

$$\rho(-\chi) > \rho(\chi)^*$$

The autocorrelation function of the advance path is equal to

$$\frac{\Delta}{L} = \frac{1}{3}Z^2 - \frac{2Z^2}{3} + \frac{Z^2}{L}\chi + \frac{Z^2}{3} + \frac{Z^2}{L^2}\chi - \frac{Z^2}{L}\chi + \frac{Z^2}{L^2}\chi^2 = \frac{Z^2}{L^2}\chi^2 \tag{13}$$

Provided that the delay value is much smaller than the glide path length, which is quite realistic, the autocorrelation function of the advance path is equal to

$$\rho_k(+\chi) = \frac{1}{3}h^2 \tag{14}$$

Substituting the values $\rho_{p,}\rho_k(+\chi)$ and $\rho_{pk}(+\chi)$, in equation (3), it is obtained

$$\frac{\Delta'}{L} = \frac{1}{3}h^2 - \frac{2h^2}{3} + \frac{h^2}{L} + \frac{h^2}{3} + \frac{h^2}{L^{23}} - \frac{h^2}{L} + \frac{h^2}{L^{23}}\chi^2 = \frac{h^2}{L^2}\chi^2$$

The dependence $\frac{h^2}{L^2}\chi^2$ on χ is shown on the Figure 7.

Figure 6. The trajectory of the normal glide path – 2, the delay – 3 and the advance glide path – 1 accordingly

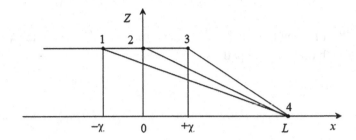

The flight path of the aircraft when it leaves the glide path (Fig. 6) is determined by the expression

$$Z_p(y) = Z - \frac{Z}{L}\chi,$$

where Z – is the flight altitude during the glide path approach, L – is the lateral deviation from the specified point of the glide path (Figure 8).

Improved path autocorrelation function is following:

$$\rho_k(+\chi) = \frac{1}{L}\int_0^L \left[Z - \frac{Z}{L}(x+\chi)\right]^2 dy = \frac{Z^2}{L}\int_0^L \left[1 - \frac{y+\chi}{L}\right]^2 dy = \frac{Z^2}{3} - \frac{Z^2\chi}{L} + \frac{Z^2}{L^{23}}\chi^2$$

Figure 9 can be obtained.

$$\frac{\Delta}{L} = \frac{1}{3}Z^2 - \frac{2Z^2}{3} + \frac{Z^2}{L}\chi + \frac{Z^2}{3} + \frac{Z^2}{L^2}\chi + \frac{Z^2}{L^2}\chi^2 = \frac{Z^2}{L^2}\chi^2$$

In case of deviation from the input line to the glide path, the probability of not reaching the threshold level of the runway increases and, consequently, the probability of an accident increases as well (Gryshchenko, 2003).

When considering the flight of an aircraft along a given trajectory at a constant height, that is, in the plane $Z = $ const, the coordinate Y is a lateral deviation. On a fixed section of the trajectory of a long $L-x_1$, the starting point of reference, x_2 is the final reference point:

$$L = x_2 - x_1.$$

The square of the integral difference of the trajectory of the planned and real flight in a certain area is equal to:

$$\Delta = \int_{x_1}^{x_2} \left[y_p(x) - y_r(x)\right]^2 dx$$

Figure 7. Dependency $\dfrac{h^2}{L^2}\chi^2$ from χ (χ from -46m to 46m)

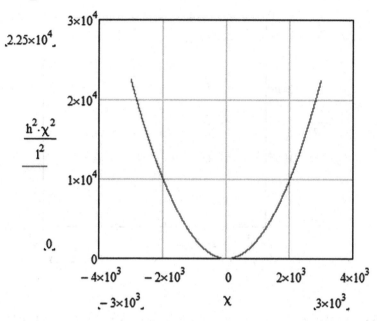

Figure 8. Scheme of entry the glide path in the horizontal plane

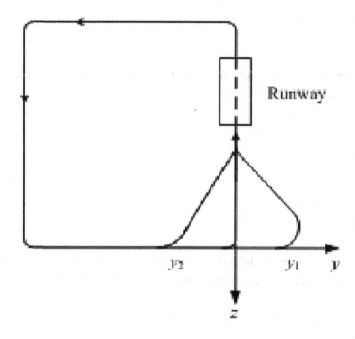

Figure 9. The dependency $\dfrac{Z^2}{L^2}\chi^2$ from χ ($\chi$from–300m to 300m)

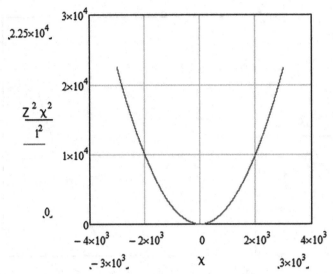

The ratio of the integral difference square of the trajectory Δ to the length L, when the magnitude of the delay χ is determined by the expression:

$$\frac{\Delta}{L} = \frac{1}{3}y^2 + \frac{3y^2}{L}\chi + \frac{2y^2}{L^2}\chi^2 - \frac{y^2}{3L^3}\chi^3.$$

Similarly, an expression can be obtained if, when flying a plane to the zone of its entrance to the glide path, the flight height $z = var$ will vary:

$$\Delta = \int_{y_1}^{y_2} \left[z_p(y) - z_r(y) \right]^2 dx$$

$$\frac{\Delta}{L} = \frac{1}{3}z^2 + \frac{3z^2}{L}\chi + \frac{2z^2}{L^2}\chi^2 - \frac{z^2}{3L^3}\chi^3.$$

Then, for three-dimensional space, lateral deviation y and vertical deviation z, the ratio of the square of the integral difference Δ to the length of the trajectory is written in the form:

$$\frac{\Delta}{L} = \left(\frac{1}{3} - \frac{3\chi}{L} + \frac{2\chi^2}{L^2} - \frac{\chi^3}{3L^3} \right)\left(y^2 + z^2 \right).$$

Illustrative Example

This function Δ / L with specifically defined L and χ will be a paraboloid of rotation (Figure 10-11).

Thus, a formula for determining the allowable entry limits of the glide path in three-dimensional space with lateral deviation y and vertical deviation z and a method for plotting the dependence f (y, z) has been developed (Hryshchenko, Romanenko, 2017). A geometric interpretation of the probability of deviation from a given point at the entrance to the glide path is presented.When χ is more than 1440, we have the following picture (Figure 12-13).

The Problem of Obtaining Incorrect Information on an Angle of Attack and Instrument Rate of the Plane

The accuracy of entry into the glide path, was considered taking into account the influence of negative factors on the flight quality. However, this is not enough. It is necessary to enter with the set speed. Consider the case of incorrect readings of the angle of attack or instrument speed. This may be due to the damage to the sensor that measures the angle of attack or the ingress of foreign objects in the air pressure receiver or in the absence of heating with freezing of moisture in pipelines, etc. Receiving inaccurate information from the systems leads to a deterioration in flight quality by crew members. A system of warning the crew about the failure in the systems for obtaining data on the angle of attack and airspeed and on the execution of a non-coordinated turn is proposed. It is based on a constant calculation of the area of correlation fields (Hryshchenko, Yakimenko & Amelina, 2017). With sufficient speed, uncoordinated turns are not so dangerous. Therefore, it is more important to determine the inconsistency of the readings on the correlation fields of the angle of attack coefficients and instrumental velocity (Hryshchenko, Romanenko, & Amelina, 2017, 2019).

Figure 10. Listing calculation $\Delta / L = f(y, z)$

$L := 12000$ $\chi := 600$

$y := -300..300 =$ $z := -46..46 =$ $y^2 =$ $z^2 =$ $\Delta := z^2 L \left(\frac{\chi}{L}\right)^2 =$ $\frac{\Delta}{L} =$

	y	z	y^2	z^2	Δ	Δ / L
0	-300	-46	$9\cdot10^4$	$2.116\cdot10^3$	$6.348\cdot10^4$	5.29
1	-299	-45	$8.94\cdot10^4$	$2.025\cdot10^3$	$6.075\cdot10^4$	5.063
2	-298	-44	$8.88\cdot10^4$	$1.936\cdot10^3$	$5.808\cdot10^4$	4.84
3	-297	-43	$8.821\cdot10^4$	$1.849\cdot10^3$	$5.547\cdot10^4$	4.623
4	-296	-42	$8.762\cdot10^4$	$1.764\cdot10^3$	$5.292\cdot10^4$	4.41
5	-295	-41	$8.703\cdot10^4$	$1.681\cdot10^3$	$5.043\cdot10^4$	4.203
6	-294	-40	$8.644\cdot10^4$	$1.6\cdot10^3$	$4.8\cdot10^4$	4
7	-293	-39	$8.585\cdot10^4$	$1.521\cdot10^3$	$4.563\cdot10^4$	3.803
8	-292	-38	$8.526\cdot10^4$	$1.444\cdot10^3$	$4.332\cdot10^4$	3.61
9	-291	-37	$8.468\cdot10^4$	$1.369\cdot10^3$	$4.107\cdot10^4$	3.423
10	-290	-36	$8.41\cdot10^4$	$1.296\cdot10^3$	$3.888\cdot10^4$	3.24
11	-289	-35	$8.352\cdot10^4$	$1.225\cdot10^3$	$3.675\cdot10^4$	3.063
12	-288	-34	$8.294\cdot10^4$	$1.156\cdot10^3$	$3.468\cdot10^4$	2.89
13	-287	-33	$8.237\cdot10^4$	$1.089\cdot10^3$	$3.267\cdot10^4$	2.723
14	-286	-32	$8.18\cdot10^4$	$1.024\cdot10^3$	$3.072\cdot10^4$	2.56
15	...	...	...	...	...	...

Figure 11. The graph of dependency f (y, z)

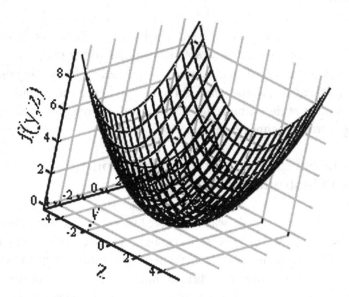

Figure 12. Listing calculation Δ / L = f (y, z), where χ=1450 m

y := -300..300 =   z := 50..-50 =   L := 12000 $\chi := 1450 = 1.45 \times 10^3$

y
-300
-299
-298
-297
-296
-295
-294
-293
-292
-291
-290
-289
-288
-287
-286
...

z
50
49
48
47
46
45
44
43
42
41
40
39
38
37
36
...

$y^2 =$

	0
0	9·10⁴
1	8.94·10⁴
2	8.88·10⁴
3	8.821·10⁴
4	8.762·10⁴
5	8.703·10⁴
6	8.644·10⁴
7	8.585·10⁴
8	8.526·10⁴
9	8.468·10⁴
10	8.41·10⁴
11	8.352·10⁴
12	8.294·10⁴
13	8.237·10⁴
14	8.18·10⁴
15	...

$\Delta := z^2 L \left(\dfrac{\chi}{L}\right)^2 =$

	0
0	4.38·10⁵
1	4.207·10⁵
2	4.037·10⁵
3	3.87·10⁵
4	3.707·10⁵
5	3.548·10⁵
6	3.392·10⁵
7	3.24·10⁵
8	3.091·10⁵
9	2.945·10⁵
10	2.803·10⁵
11	2.665·10⁵
12	2.53·10⁵
13	2.399·10⁵
14	2.271·10⁵
15	...

$z^2 =$

	0
0	2.5·10³
1	2.401·10³
2	2.304·10³
3	2.209·10³
4	2.116·10³
5	2.025·10³
6	1.936·10³
7	1.849·10³
8	1.764·10³
9	1.681·10³
10	1.6·10³
11	1.521·10³
12	1.444·10³
13	1.369·10³
14	1.296·10³
15	...

$$\left(\frac{1}{3} - \frac{3 \cdot x}{L} + \frac{2 \cdot x^2}{L^2} - \frac{x^3}{3L^3}\right)\left(y^2 + z^2\right)$$

$\dfrac{\Delta}{L} =$

	0
0	36.502
1	35.056
2	33.64
3	32.253
4	30.895
5	29.566
6	28.267
7	26.997
8	25.756
9	24.544
10	23.361
11	22.208
12	21.083
13	19.988
14	18.922
15	...

$$f(y,z) := \left(\frac{1}{3} - \frac{3 \cdot x}{L} + \frac{2 \cdot x^2}{L^2} - \frac{x^3}{3L^3}\right)\left(y^2 + z^2\right)$$

Figure 13. The graph of dependency f (y, z), where χ = 1450 m

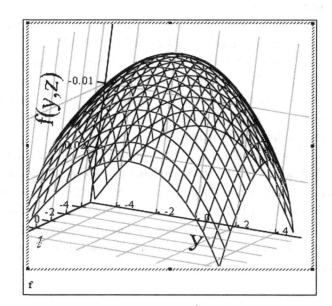

Modern aircraft is equipped with digital avionics systems. Despite the greater reliability of the modern element base, in such systems there are still failures. To eliminate them, just restart the corresponding system. In flight, such actions are provided for individual systems, but this can lead to an increase in the psycho-physiological tension of the human operator. In this case the flight quality may deteriorate during the transition to the steering control mode. Therefore, when developing, it is necessary to have a crew warning system for failures.

It should also be noted that at this stage of the landing approach, the workload of the pilot and crew members increases due to the additional functions of various kinds: change of engine operation modes, communication with ground services, visual monitoring of airspace and others. Simultaneous monitoring and analysis of many parameters and flight factors requires a high concentration of attention from the crew.

During the pre-landing maneuvering of the aircraft by the crew, it is ensured that it enters a given point in the airspace so that the aircraft is on the runway axis with a given height and range. If this problem is solved and the aircraft is in a given small area of airspace, then the parameters of the spatial position of the aircraft and its speed are estimated. When the current parameters meet the specified requirements, a prerequisite for a successful landing is formed, and the crew makes a decision to continue the landing process – entering the glide path using the captured signal of the glide path beacon. If the above conditions are not met, the crew makes a decision about going on the second round.

With a constant engine thrust speed is directly related to the angle of attack. From this it follows that in case of incorrect indications of instrument speed, we can determine this by changing the angle of attack and vice versa, if we proceed from the condition of ordinariness.

The ideal trajectory control system is the one that provides for the full ergatic compatibility of the operator-aircraft subsystem and continuously processes information about the flight trajectory of the aircraft without errors, fully implementing the specified flight program I (t). In this case, the specified trajectory and information about the real flight trajectory will be the same. Any deviation from a given flight mode is immediately recorded by the onboard equipment, which also notifies the operator. If necessary, the operator can work this difference to zero.

In order not to control the aircraft according to the distorted information on the angle of attack and speed, we suggest comparing the correlation fields $\Delta\alpha_j = (\alpha_i-1)/1$ and $\Delta V_j = (V_i-1)/1$, as well as α and γ to obtain information on coordinated reversal, where α is the angle of attack, γ is the angle of heel, V is the instrumental speed and 1 is the minimum allowable error value when reading flight information. This information is transmitted from the sensors to an analog-to-digital converter, and then the areas of correlation fields are compared in the calculator. The pilot takes measures to correct the distorted data for other means of displaying flight information.

Let the correlation field have the values $x_1, x_2...x_n$, which correspond to the values $y_1, y_2...y_n$ (Fig. 1). Combining the set of points with a functional dependence $\varphi(x)$. The voice of it is made on the basis of theoretical considerations or based on how the points are located on the correlation field. To as accurately as possible the process of approximation has to be selected in the selected function $y=\varphi(x, a, b, c,...)$ numerical parameters $a, b, c,...$ To solve this problem, the known method of the least squares is applied.

Below is the formula of the sum of the squares of the difference between the values y_i and the functions $\varphi(x_i, a, b, c,...)$ at the corresponding points. The function of the sum is denoted by $F(a, b, c,...)$.

$$F(a,b,c) = \sum_{i=1}^{n}(y_i - \varphi(x_i, a, b, c))^2$$

For the sum function it is necessary to select numerical parameters $a, b, c, ...$, so that it is the smallest, that is:

$$F(a,b,c...) = \min.$$

In order to determine the minimum of the sum function, one must consider equations with partial derivatives:

$$\frac{\partial F}{\partial a} = 0, \frac{\partial F}{\partial b} = 0, \frac{\partial F}{\partial a} = 0...$$

These equations are represented in the form of a system:

$$\begin{cases} \sum_{i=1}^{n}(y_i - \varphi(x_i, a, b)) \dfrac{\partial\varphi(x_i, a, b, c,...)}{\partial a} = 0 \\[2mm] \sum_{i=1}^{n}(y_i - \varphi(x_i, a, b)) \dfrac{\partial\varphi(x_i, a, b, c,...)}{\partial b} = 0 \\[2mm] \sum_{i=1}^{n}(y_i - \varphi(x_i, a, b)) \dfrac{\partial\varphi(x_i, a, b, c,...)}{\partial c} = 0 \end{cases}$$

The solution of such a system of equations with the substitution of the values x_i and y_i will have numerical parameters a, b, c, ..., which, in the best way, will be minimized by the function of the sum $F(a, b, c, ...)$.

If in the correlation field another function $y = f(x)$ is defined in the same way, then if necessary to calculate the area of the curvilinear figure, between these curves on a certain interval $x \in [x_1, x_2]$.

$$S_{abcd} = \int_{x_1}^{x_2} (\varphi(x) - f(x)) dx$$

If in the correlation field of a function that approximates the given values is not constructed for certain reasons and it is necessary to determine the area of the polygon formed by the given points, then in this case it is necessary to determine the functional dependence of each link of the polygon.

Let the given polygon be convex. Equation of straight lines in sections 1-2, 2-3, ...,n-1-n, 1-n:

$$\frac{x - x_1}{x_2 - x_1} = \frac{y - y_1}{y_2 - y_1}, or\ y_{1-2} = k_1 x + b_1$$

$$\frac{x - x_2}{x_3 - x_2} = \frac{y - y_2}{y_3 - y_2}, or\ y_{2-3} = k_2 x + b_2$$

$$\frac{x - x_{n-1}}{x_n - x_{n-1}} = \frac{y - y_{n-1}}{y_n - y_{n-1}}, or\ y_{n-1-n} = k_n x + b_n$$

$$\frac{x - x_1}{x_n - x_1} = \frac{y - y_1}{y_n - y_1}, or\ y_{1-n} = kx + b$$

The area of the polygon is determined by the formula:

$$S_{1-2-3-n} = \int_{x_1}^{x_2} (y_{1-2} - y_{1-n}) dx + \int_{x_2}^{x_3} (y_{2-3} - y_{1-n}) dx + ... + \int_{x_{n-1}}^{x_n} (y_{n-1-n} - y_{1-n}) dx$$

Reduction of the areas of the parameters indicates a deterioration in flight quality. For each type of aircraft calculations using this methodology on the flight quality should be made.

In accordance with the quantitative indicators of the relative values of the areas, it is proposed to carry out a qualitative assessment of the flight, which assumes the following categories: good flight quality – corresponds to the range of relative values of the areas [0,43-1]; average flight quality – [0,29-0,43]; poor flight quality – [0-0,29] (square bracket – the value enters the interval, the round is not included). Practice shows that most values of the range of relative areas [0,43-1] lie in the range [0,43-0,62], but theoretically it can go beyond its limits and strive for the limit value of *1*.

The Problem of Stress During the Glissade Entry by the Crew

While entering the glide path, the aerodynamic configuration of the aircraft changes, which is associated with the release of flaps and landing gear. When flying in the steering mode, it imposes on the pilot additional actions in controlling the aircraft. With the simultaneous occurrence of failures in the system, crew members may experience increased psycho-physiological tension. Earlier this issue was considered at the site from the end of the fourth turn before landing (Hryshchenko, 2014: Hryshchenko, Romanenko, & Hryshchenko, 2017). By themselves, failures may not be a threat, but under stress, the crew may aggravate the situation. Since the parameters change of the aircraft in this area is ergodic and stationary, the gain (increase in the amplitudes of the parameters) can be determined from the correlation functions.

Calculations are performed by the formula

$$R_i^{(j)} = \frac{1}{n-j+1} \sum_{i=1}^{n-j+} \left(A_i - \frac{1}{n} \sum_{i=1}^{n} A_i \right) \left(A_{i+j-1} - \frac{1}{n} \sum_{i=1}^{n} A_i \right)$$

where n is the number of observations in the time series A_i (parameter amplitude), $j = 1, 2, 3, \ldots L$ represents the argument delay by $0, 1, 2 \ldots (L-1)$.

During the training on a complex airplane simulator, it is necessary to work out exercises that involve entering into a flight situation two or three equipment failures on board simultaneously and determine the gains of each crew member. The block of failures should include failures that do not affect the aerodynamics of the aircraft and the ability to control the aircraft. Experience shows that the uncertainty in assessing the situation during the extreme flight conditions in 70-80% of cases leads to an increase in the amplitude of the flight parameters as a result of increased pilot tension. Thus, the simulator should be used not only for learning specific actions, but also for anti-stress training.

One variant of the application order of trend algorithms for the analysis of the counteraction of pilots overlay factors is considered. Given the numerical data or graphs of headind change (ψ), roll(γ), pitch (ϑ) and vertical velocity (V_y) from the end of the fourth turn to landing, it is necessary to 1) determine the distances from the extremums of these functions to zero 2) calculate the difference between the extremums (without the module) changing each parameter. Results of the values of the amplitudes (A) are taken modulo. Identify the maximum and minimum of each parameter. 3)calculate half-lives (T) corresponding to the maximum and minimum values of each parameter. Then calculations are made according to the following formulas for each parameter:

$$\Delta A = \frac{A_{max} - A_{min}}{A_{min}}$$

After this, the general multichannel picture of the parameters change is obtained:

$$\Delta\Delta A_{\gamma,\Psi,\vartheta,V_y} = \sqrt{\Delta A_\gamma^2 + \Delta A_\Psi^2 + \Delta A_\vartheta^2 + \Delta A_{V_y}^2}$$

Amplitude can be measured and applied to the axis of coordinates when working with numbers in degrees. When working with graphs – in relative units and periods respectively in seconds and conditional relative units.

Erroneous and illogical actions of the crew members during the flight, related in the first place, with changes in mental processes due to the effect on them of factor loads. Receiving spatial signals from the action of factor loads, with their sufficient number, the pilot can fall into the zone of reflected motion. Failure of pilots to actively counteract the factor load can lead to wrong actions in the management of the aircraft (confusion of levers, switches, buttons, etc.). The opposite of this phenomenon is the spatial delay of movements. And because of the training of the entire crew, it is very important to master the technique of spatial delay of movements when operating an aircraft. Also, one should pay attention to another key element of this methodology – flight training logic. It serves as an important basis for the training, as well as the choice of the only correct solution in counteracting unexpected stimulus.

In the ideal case, it is necessary to compare the amplitude value when analyzing flights of the same pilot without failures and with the failures introduced. Thus, by this difference, we can determine the amplitude gain that occurs in the pilot under the influence of factor loads, which on the complex simulator pilots imitated by complex failures.

The System of Warning of Sharp Movements of the Wheel at the Exit to the Second Circle

There are occasions when the crew decides to leave for the second round. In this case, it is important to avoid sharp movements with a steering wheel from yourself to prevent dive.

Alarm works when the following conditions are met:

$$\sqrt{\Delta\alpha_i^2 + V_{yi}^2 + \theta_i^2} > \sqrt{\Delta\alpha_{set}^2 + V_{yset}^2 + \theta_{iset}^2},$$

where

$\Delta\alpha_i$ and $\Delta\alpha_{set}$:The current and set maximum magnitude of the angle of attack, respectively.

V_{yi} and V_{yset} :In respectively, the current and maximum permissible set values of the vertical speed of the aircraft.

θ_i^2 and θ_{iset}^2 :Respectively the current and set values of the pitch angle.

$$V_{dev} < V_{set} + \Delta V,$$

where

V_{dev} :The device speed of the aircraft.

V_{set} :The installed speed of the substation, at which the substation begins to land

ΔV :The speed increase required to trigger the alarm to prevent sharp movements

The magnitude of the angle of attack is calculated by the formula:

$$\Delta\alpha = \frac{\alpha_{max} + \alpha_{min}}{\alpha_{min}}, \alpha_{min} \neq 0,$$

where

$\Delta\alpha$:The magnitude of the angle of attack.

α_{max} :The critical value of the angle of attack, at which flow fluctuates.

α_{min} :The minimum value of the angle of attack amplitude.

Functional scheme of the development device – the warning signal of the prevention of sharp movements presented in Figure 14,

There are three types of signaling: voice ("PULL UPSLIGHTLY" command), visual (illuminated text display on the steering wheel) and physical (small increase of pilot's effort for movement of the wheel). Signals are triggered simultaneously.

The visual alarm lights up the "PULL UPSLIGHTLY" text indicator in yellow, this color is used to indicate boundary modes that require caution.

The developed system of warning of sharp movements of the wheel during going on the second circle maneuver is aimed at improving safety of flights.

Based on the foregoing material, in the future it is planned to build a mathematical model of the ergonomic distribution of the information management system.

CONCLUSION

At low altitudes, it is necessary to take into account the correlation function of the trajectory of the ideal flight with the advent of factorial overlays.

The analysis has shown that the delay with the glide path entrance of the aircraft is more dangerous for the quality of the approach than the advance in entrance of glide path. The lag-ahead correlation function drops faster.

It is shown that it is possible to control the trajectory of the aircraft on the glide path according to the correlation function without the action of factor overlays and with a periodic factor overlay. The correlation function allows to determine the phenomenon of increasing the amplitude parameters of the flight. The results require the implementation of the technology of flight operations to improve the quality of thetrajectory control during the approach.

Figure 14. Functional diagram of the warning signal about the prevention of sharp movem

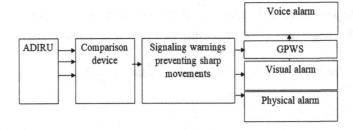

A formula has been developed for determining the allowable entry limits of the glide path in three-dimensional space with lateral deviation y and vertical deviation z and a method for constructing a graph of f (y, z) is presented. A geometric interpretation of the probability of deviation from a given point during the glide path entrance is presented.

In accordance to the quantitative indicators of the relative values of areas, it is proposed to carry out an assessment by flight parameters, which implies the following categories: good flight quality, average flight quality, and poor flight quality. A table has been developed to determine the compliance of this classification.

In the ideal case, it is necessary to compare the amplitude value when analyzing flights of the same pilot without fail and with the failures introduced. Thus, by this difference, we can determine the amplitude gain that occurs in the pilot under the influence of factor loads, which are imitated by means of complex failures on the complex simulator.

The developed system of warning for avoiding the sharp movements of the wheel during the flying on the second circle improves the safety of the flight. There are three types of signaling: voice (the command "PULL UP SLIGHTLY"), visual (illuminated text display on the steering wheel) and physical (small increase of the pilot's efforts for the movement of the wheel). Alarms work simultaneously.

REFERENCES

Gibb, R., Schvaneveldt, R., & Gray, R. (2008). Visual misperception in aviation: glide path performance in a black hole environment. *Human Factors the Journal of the Human Factors and Ergonomics Society*, *50*(4), 699–711. doi:10.1518/001872008X288619 PMID:18767527

Hryschenko, Y. V. (2003). Pair flights as a way of analyzing the dynamic stereotype amplification phenomenon among pilots. *Cybernetics and Computing Engineering. National Academy of Sciences of Ukraine. 140*, 31–34.

Hryschenko, Y. V., & Kurushkina, Y. O. (2004). Using the principle of invariance of control systems in the evaluation of enhanced reflexes. *Cybernetics and Computing Engineering. National Academy of Sciences of Ukraine. 143*, 39–44.

Hryshchenko, Y. V. (2014). Scientific research on the anti-stress preparation of specialists in a quarter century. *Proceedings of the National Aviation University, 1*(58). pp. 53-58.

Hryshchenko, Y. V. (2003). Preparation of pilots for flights in special situations taking into account the phenomenon of dynamic stereotype. *Cybernetics and Computing Engineering. National Academy of Sciences of Ukraine. 139*, 81–85.

Hryshchenko, Y. V. (2016). Reliability problem of ergaticcontrol systems in aviation. Methods and systems of navigation and motion control. *IEEE 4th International Conference,* Kiev, Ukraine. pp. 126-129.

Hryshchenko, Y. V., & Romanenko, V. G. (2017). Quality of flight during approach improvement suggestions. *IEEE 4th International Conference (Oct. 17-19, 2017),* Kiev, Ukraine. pp. 69-72.

Hryshchenko, Y. V., Romanenko, V. G. & Amelina, A. I. (2019). The problem of uncoordinated aircraft turn on small flight speed. *Electronics and Control Systems.* Kiev, Ukraine. NAU. 1(59). pp. 50-57.

Hryshchenko, Y. V., Romanenko, V. G., & Hryshchenko, Y. Y. (2017). Suggestions of the improvement of the quality of flight during landing and missed approach/go around maneuver. *Electronics and control systems*. Kiev, NAU. 2(52). pp. 103-109.

Hryshchenko, Y. V., Romanenko, V. G., & Pipa, D. M. (2018). Evaluation of quality and accuracy of flight path during approach and landing. In *Proceedings IEEE 5th International Conference (Oct. 16-18, 2018)*. Kiev, Ukraine. pp. 191-194.

Hryshchenko, Y. V., Romanenko, V. G., & Pipa, D. M. (2018).Suggestions to the methods for assessing the quality of the glide path entrance. *Electronics and Control Systems,* Kiev, Ukraine. NAU, 2018. 3(57). pp. 41-48.

Hryshchenko, Y. V., Romanenko, V. G., & Polozhevets, A. A. (2005). Mathematical aspects of solving the problems of accounting for a large number of factors during the avionics exploiting. *Cybernetics and Computing Engineering. National Academy of Sciences of Ukraine. 146*, 81–89.

Hryshchenko, Y. V., Romanenko, V. G., Tronko, V. D., & Hryshchenko, Y. Y. (2017). Methods of training of modern aircraft flight crews for inflight abnormal circumstances. *Proceedings of the National Aviation University*, 1. pp. 66-72.

Hryshchenko, Y. V., Skripets, A. V., & Tronko, V. D. (2015). Analysis of the possibility of controlling the trajectory of an airplane's motion by correlation function. *Cybernetics and computing technology: Scientific Journal. House "Academiperiodica" of the National Academy of Sciences of Ukraine. 181.* pp. 35-46.

Hryshchenko, Y. V., Yakimenko, M. S., & Amelina, A. I. (2017). Problems of the impact of failures in avionics systems on the quality of flight. All-Ukrainian scientific and practical conference of young scientists and students "Problems of Navigation and Control of Motion", Section 4, Avionics and Control Systems. Kiev, Ukraine: NAU.

Hryshchenko, Y. V. (2004). Analysis of changes in the dynamic stereotype of pilots in the process of flight training on a complex simulator of an airplane. *Cybernetics and computing. NAN of Ukraine, 142.* pp. 35-40.

Hryshchenko, Y. V., Romanenko, V. G., & Pipa, D. M. (2017). Suggestions to the methods for assessing the quality of the glide path entrance. *Electronics and Control Systems,* Kiev, Ukraine. NAU. 3(57). pp. 41-48.

Hryshchenko, Y. V., Skripets, A. V., & Tronko, V. D. (2015). Autocorrelation functions and their application for assessing. *Proceedings of the National Aviation University, 2(63).* pp. 27-33.

Kashmatov, V. I. (2008). Application of quasi-object control systems by angular position of airplane for the translation of airplane from horizontal flight in a glide slope approach. *Electronics and Control Systems,* Kiev, NAU, 4(18). pp. 79-87.

Kazak, V. M, Budzynska, T. V, & Mischeryakova, V. Yu. (2009). Assessment of the effect of changes in the parameters of the ellipsoid errors on maintaining the landing trajectory of the aircraft. *Science-Intensive Technologies.* Kiev, NAU. 2. pp. 43-45.

Khokhlov, E. M., Hryshchenko, Y. V., & Volodko, O. N. (2016). The method of determining the quality of piloting the contours of the correlation fields of flight parameters in special conditions. *Cybernetics and Computer Science: Science Magazine, 39-51.*

Lapa, V. G. (1974). Mathematical foundations of cybernetics. *Publishing Association «Vishchashkola».* p. 452.

Solomentsev, O. V., Herasymenko, T. S., Zaliskyi, M. Y., & Cheked, I. V. (2016). Statistical data processing procedures for ground navigation equipment. In *Proceedings of IEEE Fourth International Conference Methods and Systems of Navigation and Motion Control, Oct. 18-20, 2016.* 182-185.

Solomentsev, O., & Zaliskyi, M. (2018). Correlated failures analysis in navigation system: Methods and systems of navigation and motion control. *IEEE 5th International Conference,* Kiev, Ukraine. pp. 41-44.

KEY TERMS AND DEFINITIONS

Correlation Field: Is a graphical representation of the source data, auxiliary tool for analyzing sample data.

Correlation Function: Function of time and spatial coordinates, which specifies the correlation in systems with random processes.

Flight Path: The line of movement, the flight course of the aircraft.

Glide Path: The flight path of the aircraft on which it descends just before landing.

Human Factor: Psychological and other characteristics of a person, his capabilities and limitations, determined in the specific conditions of his activity.

Parameter Amplitude: The highest distance from the equilibrium position.

Chapter 17
Basic Analytics of Anti–Failure Avionics

Hanna Polozhevets

https://orcid.org/0000-0003-0986-6559

National Aviation University, Ukraine

Sergiy Derets

National Aviation University, Ukraine

Bogdan Chebukin

Vilnius Gediminas Technical University, Lithuania

ABSTRACT

The chapter analyzes anti-crash avionics. Classifications are given for complex failures of avionics and onboard aviation equipment. The technological failure of the voice informant in an aircraft crash is investigated using process analysis. New technologies to reduce the risks in the elimination of functional failures of avionics are proposed. Considered are the first "computer" accidents and incidents, as well as the causes of errors. Authors present the chronology of the definition of the category of "failure" of radio-electronic equipment and onboard aviation equipment. The complexity of avionics is estimated using process analysis. The methodology of the analysis of technological processes of flights is proposed.

BACKGROUND

The problems of protection from electronic failures first arose in industrial production in the early 70s of the 20th century in the Soviet Union in connection with the transition from relay-contact control systems of industrial machines to electronic - contactless control units. Thousands of presses, flexible automated lines, automated complexes were equipped with such control systems on contactless electronics. In other countries - America, Japan, European countries there was no such transition to contactless electronics. For example, the American press "Blyz" and the Japanese "Kamatsu" had relay contact control systems, and our presses of all efforts (10, 16, 26, 63, 100, 250, 400 tf) and more had a contactless electronic control system.

DOI: 10.4018/978-1-7998-1415-3.ch017

Since industrial machines serviced by teams of instrumentation and automation, electricians, mechanics. Therefore, 80-90% of all failures were in the nature of failures, in particular, electronics to contactless control units. Diagnostics of such failures - suddenly appearance and disappearance stops of industrial machines in the process of work was very difficult. The usual classical instrumentation — a tester, an oscilloscope, frequency counters, and etc. It was impossible to find a failure, to study its cause. Equipment could not be used for a long time.

Thus, the need arose for a new approach to the diagnosis of contactless control units. Therefore, we have created ring semi-automatic devices for dealing with failures of contactless electronics. The effectiveness of such devices (copyright certificate) shows such an example.

At the car factory in the city of Kremenchug, the press with contactless control units was not used due to the failure of the electronics for 1.5 years, the factory did not succeed in diagnostics, the first decision was made - to switch back to the relay-contact system. Diagnostics of the press with a ring semi-automatic machine for 5 minutes showed the cause of the failure of the contactless control units.

This example shows that the fight against failures requires fundamentally new approaches to the creation of any industrial electronics, a complete analysis of the processes occurring in the architecture of the blocks, other elements.

CHRONOLOGY OF THE DEFINITIONS OF THE CATEGORY "FAILURE" OF RADIO-ELECTRONIC EQUIPMENT AND ON-BOARD AVIATION EQUIPMENT

The first scientific definitions of the category's "failures" appeared in the 70s of the twentieth century when creating the fundamentals of operating electronic equipment. "Failure" means a one-time self-eliminating refuse, the duration of which is short compared with the duration of work until an accidental refuse. In this case, the sequence of failures following each other is called an intermittent refuse.

Failures are one of the most complex production phenomena in the practice and theory of the operation of production machines with electronic automated equipment. The problem of failures is one of the most important in the technical and ergonomic operation of avionics, especially in the context of the use of various types of maintenance and repair: status, planned regulations, etc.

Considering that the fundamentals of the theory of failure diagnosis, the theory of design, the operation of "anti-failure" electronics from the 70s of the twentieth century to the present have not actually developed. These definitions of the category of "failures" have survived to the present, and the classification of failures does not actually exist. However, recently, including the year 2019, the following theories have been developed:

- General theory of reliability and safety in flight operations.
- General theory of functional failures (Dianov, 2012).

Such important concepts for failure analysis are introduced: fail-safe, anti-failure equipment, uninformed failures, etc. A process analysis methodology has been developed based on the general theory of statistics for engineering purposes. This creates a broader statistical basis for protecting equipment from refuses and failures.

The First "Computer" Accidents and Incidents: Reason of Errors

In civil aviation, the problem of protection from electronic failures arose in the early 90s of the twentieth century during the transition to electronic avionics, the transition to "flying computers" - flying aviation electronic aircraft complexes.

Before the transition to integrated avionics during the operation of on-board aviation equipment, there were practically no failures. This can be seen from table 1.

Note: The general nature of aircraft failures during operation in 1957-2006 and a comparative analysis of aircraft of eastern and western production according to source (Poltavets, Teymurazov, 2007).

The fact that the transition from electrical to electronic control of an aircraft took place is problematic from the work of expert pilots of the international federation of civil aviation flight unions (Fig. 1, 2). At the same time, pilot experts noted that aircraft electronics became a new risk factor from the beginning. 90s of the 20th century (Fig. 1) and the reason for this is two trends - the hyperbolic increase in the complexity of the flight standard procedures (problems of Standard operating procedure (SOP) - flight operations) and the even steeper hyperbola of the increasing complexity of avionics — (Fig. 2).

Therefore, the main goal of the work is to consider the problem of "failures" of electronics in the structure of on-board avionics of the new generation of aircraft, taking into account these two trends, identifying the basic statistical laws of flight operations, flight technologies, and the construction of basic histograms revealing the manifestation and risks of failure.

The main attention to the "functional failures" of electronic avionics and on-board equipment, their diagnosis during the operation of the aircraft of new generations. Functional failures of avionics are associated with the emergence of new avionics functions.

Among the first "computer" air crashes is considered the third air crash of the "flying computer" of the Airbus A320 in January 1992 in the mountains of Western Europe - Vagesa.

Table 1. General characteristics of failure of "pre-computer" aircrafts B-727 and TU-15 4 (Poltavets, Teymurazov, 2007)

Name of Systems	Type of System Failure	B-727	TU-154
landing gear	Non-release landing gear Folding landing gear Pneumatics destruction Front wheel steering	43 10 3 0	15 4 0 1
	Total system	**56**	**20**
Power unit	Non-localized destruction Engines Starter Destruction Engine shutdown Engine separation from aircraft	13 0 2 2	5 1 0 0
	Total system	**17**	**6**
Flight control system	Flap retention Slat retention	0 2	2 0
	Total system	**2**	**2**
Other systems	For all aircraft systems	5	1
		80	20

Figure 1. Change in complexity of standard operating procedure (SOP)

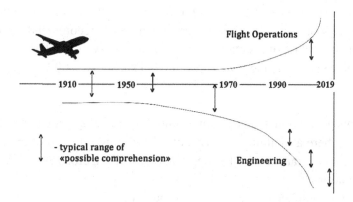

Figure 2. Change in complexity of the avionics

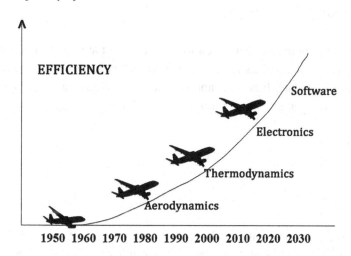

And although, as it turned out later, the ultimate technological cause of this accident was the functional failure of the GPWS (Ground proximity warning system) due to the steep mountain slopes and such accidents occurred earlier, but it was these air crashes that caused the reaction in the world press this way: electronics has been called a new factor of the risk. It was a series of these accidents that led to a comprehensive assessment of avionics, its modification, the emergence of new systems EGPWS (Enhanced-GPWS), TAWS etc., as well as a new function of avionics - an early warning of a collision with the ground.

The main reason for these plane crashes is a false thesis - computer avionics qualitatively simplifies the control of an aircraft - "even a chimpanzee can control an A320 aircraft". To remove this false conceptual premise of Western aviation specialists during the initial operation of the Eastern-produced aircraft IL-96 - 300, TU - 204 with electronic avionics, complex scientific and methodological programs were performed to prevent "computer" aviation accidents.

The results of the work on the programs were presented to flight managers, designers, flight crews, engineering and technical staff, reported on scientific and methodological conferences of civil aviation. For organizational reasons, it was not possible to form such a program to prevent plane crashes on the

AN-70 aircraft. However, after the accident of the Academy of Sciences - 70 February 10, 1995 - according to the report of the Commander of the aircraft AN-70, among other reasons, there was a failure of the electronic display system. Pilots, testers were given urgent recommendations on the use of ring analysis processes, the formation of a stream of operational observations. They made at least 300 comments, and the possible aviation incident was prevented (Tkachenko, 2009.).

So, as results:

1. On aircraft of this generation, technical and technological failures are virtually absent.
2. During the study, there are failures in terms of the warning function of approaching the earth in the systems - automatic takeoff and landing machines - Dangerous Speed Alarm System, GPWS of the first generations.
3. Electronic avionics complicates the piloting of the aircraft.

Classification of Complex Failures of Avionics and Airborne Aviation Equipment

Considering that at present there is no classification of failures of avionics and on-board aviation equipment, we will draw up preliminary classifications of failures that allow for the analysis of failures. Based on the fact that avionics has many functions and structurally many, systems, complexes, it is advisable to allocate for the practice of analyzing and diagnosing failures, there are two types of failures:

*   Functional failures.
*   Technological failures (fig. 3).

We use next definition:

*   Functional failures of avionics and on-board aviation equipment - are partial failures in avionics function, leading to accidents and incidents in complicated situations in flight (weather conditions, actions of risk factors, etc.) and partially disappearing during flight operations under simpler conditions.
*   Technological (systemic) failure of avionics- is called self-propelling and self-vanishing going beyond the allowable parameters of the piloting technique of the avionics system, leading to the appearance of an unwounded position of the aircraft or the creation of complicated piloting conditions for the aircraft.
*   General characteristics of technical failures of aircraft - flying multifunctional devices.

Evaluation of the Complexity of Avionics Using Process Analysis. Methodology of Technology Process Analysis of Flights.

The first "computer" accidents and incidents in the early 90s of the 20th century arose due to incorrect assumptions that the transition from electric to electronic control of an aircraft qualitatively simplifies aircraft flight control, and also due to technological and functional failures of avionics and on-board aviation equipment.

Figure 3. Technological failures

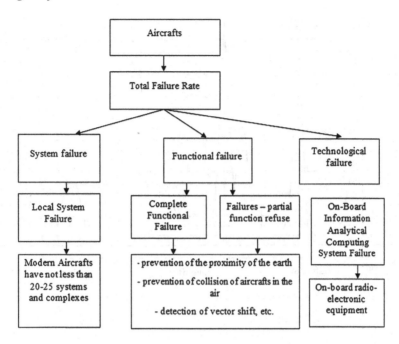

On the implementation of integrated programs to ensure the operation of aircraft with avionics of the first generation on the aircraft IL-96-300, AN-70, TU-204, A320, etc. showed the need to develop a special analytical methodology, a process analysis technology for flights, taking into account the technological complexity of standard flight operations planned by the designer in the manual of the aircraft with first-generation electronic avionics.

At the same time, an analytical apparatus was created for constructing cyclograms of the deployment of flight operations and determining the statistical laws for constructing flight technologies using special deployment histograms.

The results of these studies are shown in the figures: 4, 5. Its classification technology of process analysis of flights and transitions.

Taking into account the extremely short deadlines for the implementation of complex programs for technology process analysis of flights, flight managers set the task of avoiding "computer" accidents with the A320 on planes of eastern production - on IL 96-300. Transitions from histograms to probability distributions were not considered in detail, and only attention was focused on the asymmetry of histograms and their agreement to the log-normal probability distribution.

Such qualitative constraints made it possible in the first phase of the initial operation of the eastern production aircraft to eliminate "computer" accidents. Work was carried out with designers, flight managers, flight crews, etc.

Figure 4. Cyclogram of IL-76

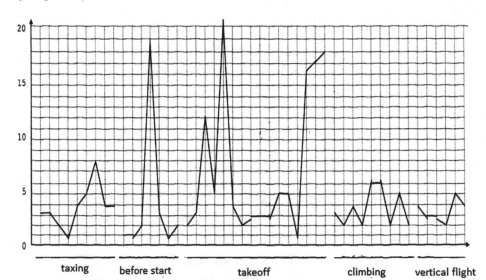

Figure 5. Cyclogram of Boeing 747-400

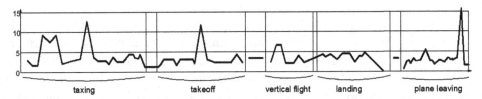

Diagnostics of Failures: Methods of Protection Against Failures at the Design Stage of Avionics

Diagnostics of crashes in flight is a very complicated process, especially for aircraft performing demonstration flights. These difficulties are characterized as follows:

1. Imperfection of the systems warning the pilot about the achievement of the limiting flight parameters by the aircraft, in particular, limiting the angle of attack and overload.
2. Lack of or insufficient information to the crew about the dangerous height and departure from the dangerous height.
3. Uninformed failures of the indication systems of pitch, roll, course in which the corresponding devices are used. In this case, faults like "fading" and "slow drift" are recognized by the flight crew with a great delay (t = 10 sec.), Which leads to emergencies.
4. The discrepancy of the modern level of devices that warn the crew to enter the speed limit, as well as an extensive range of speed limits.

At present, on all aircraft, restrictive systems do not provide for the formation and indication of the necessary nomenclature of permissible angles of attack and overload. Of the ten required values of the angles of attack and signaling of overload, the pilot must maintain 2-3 values at best; the pilot must keep the remaining values in memory and repeatedly read and compare the current angle of attack and overload with the currently acceptable ones during the maneuvering process. This reduces the reserve of attention, complicates piloting and leads to crew errors.

Therefore, on all aircraft, the achievement of permissible angles of attack or overload is indicated by an alarm, which unnecessarily distracts pilots from piloting: permissible angles of attack and overload are operational parameters and should be implemented by pilots without distracting alarms. The indication should be special and ensure the complete safe implementation of the maneuvering characteristics of the aircraft during the flight.

To protect against failures, you need to create special integrated engineering and psychological programs at the design stages. The following recommendations are offered for creating such programs:

- Aircraft design must be included in the structure of the "ring engineering-psychological design", which considers the stages: design, testing, operation of aircraft and feedback to the developers as a single ring process.
- Development of manuals on flight operation as the main document created by designers, suggests the use of statistical processing of process analysis technologies for flights across all systems and complexes of the aircraft, as well as on the functions of avionics.
- The result of the joint work of engineering psychologists, pilots, designers and flight managers are reference books on technologies for process analysis of flights, as the main documents on protection against failures and failures, as well as from "computer disasters", such as the crash of Boeing 737 max October 29, 2018 (near Jacarda) and March 10, 2019 (in Ethiopia).

Analysis of Technological Failures of the Speech Informer in the Air Crash Near Sknilov (06.27.2002)

Currently, the prevention of accidents in terms of the human factor (aviation personnel) is one of the central problems of the global aviation community. International Civil Aviation Organization (ICAO) and its Human factors group are making very significant efforts to address this global problem:

- International conferences on flight safety and human factors were held.
- An ICAO safety and human factors program was established, which consists of amendments to the annexes to the Convention, the ICAO International Circular Standard.

Therefore, the use of experience and knowledge of existing Human factors experts in Ukraine is very limited, although the prevention of ongoing air accidents in civil and military aviation has long required a qualitatively new approach.

Unfortunately, the aircraft designers An-124, An-70, An-140, Yak-42, etc., and their analysis by aeronautical specialists in aerodynamics, flight operations, technical expertise, etc. did not lead to the realization of the leaders and the government of Ukraine to the understanding of the scientific and practical fact that it is necessary to change the methodology of approaches. As an example, let us consider the results of engineering and psychological analysis of an air crash with the Su-27 aircraft in June 2002 (Sknilov).

It should be borne in mind that demonstration flights are one of the most difficult types of flights, when the most difficult aerobatics are performed at air shows. Therefore, such flights can be accompanied by accidents - disasters with a very difficult for analysts and experts of nature and character. An accident in June 2002 brought together an airplane accident and many accidents.

The investigation showed that this accident by experts and commission members is primarily categorized as an accident, where the main cause is the "human factor" (table 2).

Digital speech synthesizer Voice Informer built in the structure of on-board information analytical computing system (Gribov, 2017).

Therefore, specialists of the scientific-methodological center of process analysis, National aviation university of Ukraine, avionics department, flying unions of Ukraine and Russia in the international aviation project "Process analysis of flight safety-2000" - "Priority scientific and methodological program for reaching the zero level of accidents on the human factor (crew) in the global Air Transport Process," a first engineering and psychological study of the causes of possible crew errors using the materials of the investigation and additional engineering-psychological transcript data devices "Tester-U3 series 3" Su-27.

In this case, table of tolerances and control maps were obtained for the angle of attack, speed and overload (Figure 6). The results of information engineering-psychological evaluation of actions and counteraction to sudden and sudden crew in flight showed that the immediate cause of this air accident is not human, but a technical (technological) factor in the form of complex failures of the type "failure". As is known, faults-self-identifiable, discontinuous failures are one of the most complex negative phenomena in the practice and theory of the operation of production machines having electronic automation equipment. Fighting failures is one of the most important problems of technical and ergonomic operation of any complicated machinery, especially in the context of different types of maintenance: "by status", scheduled regulations, etc.

Table 2. New categorization of Aviation Accident 27.07.02 for the main and immediate reasons

Nº	Points	Official View		Engineering-psychological View
		State Aviation Authority Investigation	**Law Enforcement Investigation**	
1.	Main Cause	The main reason: human factor	The main reason: human factor	The main reason was the technical factor.
2.	Immediate cause	"The mistakes of pilot piloting of commander Toponar V.A., made by him in the performance of the unplanned flight task of the aerobatic figure " trunk "instead of the figure "skew loop with a turn "	The crew commander's mistake in the piloting technique due to the lack of skills.	Comprehensive refuse of the "failure" of a three-channel voice informer over all three command-information channels - the angle of attack, the overload rate under the conditions of the partially destructive load of the indication channel.
3.	Applicable methods and classifications, definitions, methodology analysis.	- a method of generalized dichotomy; - a method of selectively applying a fixed tolerance; - hyperbolic classification of crew and crew commander errors; - the method of partial application of the unified regulations; - method of reference points of the trajectory, graphic.	- criminal analysis; - legal classification of crew actions; - a general assessment of the duration of the decisions taken; - method of using the report of the investigation commission.	- the method of factor lists; -informative engineering-psychological assessment; - classification of faults and failures; - fall tolerance rating (tolerance table, control cards).

(comparison of official and engineering psychological approaches)

Figure 6. Map of failures

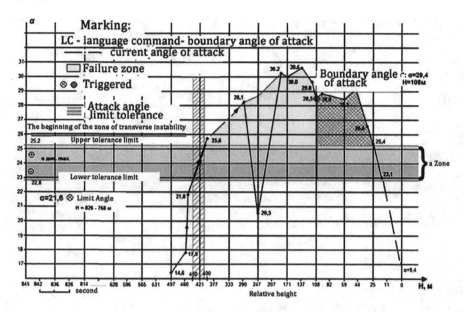

The organization of preventing an accident "with failures" according to the data of foreign and domestic experts on reliability requires the creation of special comprehensive targeted programs - "Safety of flights and human factor".

Functional Failures in the Channel of Warning of the Earth's Approach

Aircraft collision with an Earth or water surface in a controlled flight is the main problem of the global aviation community.

Ways to improve the operation of new aircraft safety systems when dealing with the problem of aircraft collision with the ground (Control flight into terrain - CFIT category) are considered in presentation on theme: "Pilot Guide to Preventing CFIT" (© 2000, 2001 Flight Safety Foundation).

Diagnostics of Failures: Methods of Protection Against Failures at the Stage of Designing Avionics

Diagnosis of flight failures is a very complicated process, especially for aircraft performing demonstration flights. These complexities are described by pilots as follows:

1. Impermeable systems preventing the pilot from reaching the plane of the limiting flight parameters, in particular - the angles of attack and overload. At present, on all planes, restrictive systems do not provide the formation and indication of the necessary nomenclature of permissible angles of attack and overload. Of the 5-10 required values of the angle of attack and overload, at best, it is formed and monitored by 2-3 values; the other values should be kept by the pilot in memory and during the maneuvering process, repeatedly read and compare the current angle of attack and overload with the currently valid ones. This reduces the reserve of attention, complicates piloting and leads to errors.

Figure 7. Controlled flight into terrain statistics

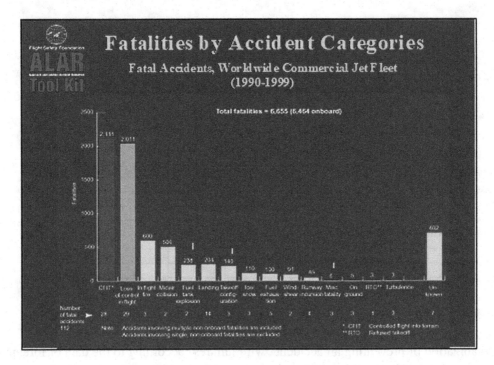

Figure 8. Accident statistics of CFIT

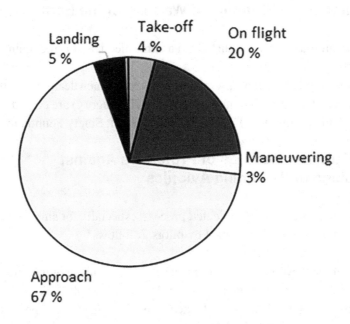

Figure 9. Ground proximity warning system

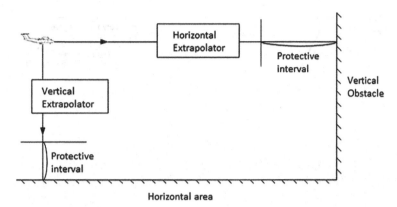

Figure 10. The causes of accidents of aircraft equipped with GPWS systems in mountainous terrain

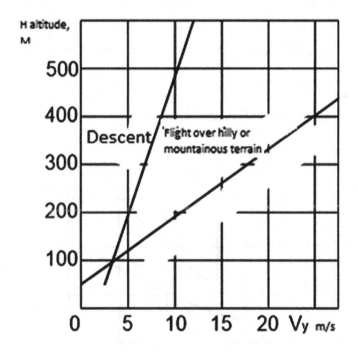

On all planes, reaching the permissible angles of attack or overload is indicated by an alarm, which unnecessarily distract pilots from piloting: the permissible angles of attack and overload are operational parameters and should be implemented by pilots without distracting alarms. The indication of the permissible values of the parameters, both visual and non-visual, should be special and ensure a complete, safe implementation of the maneuverable characteristics of the aircraft in the course of combat maneuvering.

2. Unrecognizable failures of pitch, bank display systems, courses in which devices are used. At the same time, failures such as "fading" or "slow drift" are recognized by the flight crew with a great lateness, which, in complicated meteorological conditions and at night, leads to the emergence of an emergency situation.

Figure 11a. Determination of the maximum available time GPWS before the collision

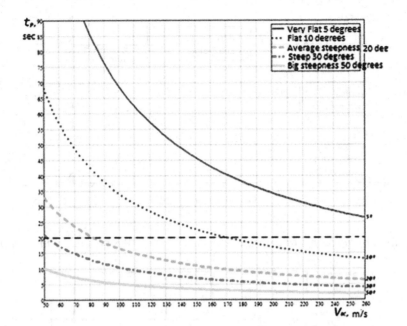

Figure 11b. its engineering and psychological explanation

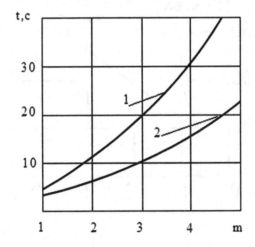

3. There are no aircraft on the planes, or they have absolutely insufficient capabilities of the pilot warning system about dangerous altitudes and dangerous heights.

4. They do not correspond to the modern technical level of the device, warning the pilot of speed limits: from a wide range of speed restrictions, at the most, several values are indicated.

Therefore, in order to protect against failures at the design stages, it is necessary to create special complex engineering-psychological programs.

Our experience in creating these programs for Oriental aircraft makes it possible to draw the following conclusions:

1. Design should be included in the structure of the so-called "circular engineering-psychological design", which deals with the design, testing, operation of aircraft, as a single ring process.
2. When developing as the main document created by the designers, it is necessary to apply the statistical processing of the technology process analysis of flights to all systems and complexes of the aircraft, as well as to the functions of avionics.
3. The implementation of such programs requires the involvement of an experienced flight instructor who can explain to the designers the requirement of technology process analysis of flights handbooks created by engineering psychologists.
4. As a result of the joint work of engineer psychologists, pilots, designers and flight managers, the guides for technology process analysis of flights become the main documents for protection against functional failures and failures, as well as from "computer" disasters.
5. As a prospective analyzer of failures avionics, resulting in the impending position of the aircraft, the analyzer suggested in the work should be recommended.

CONCLUSION

Managing a safety risk involves identifying sources of danger for flight operations, as well as assessing and reducing those risks. This is one of the main requirements of ICAO's international standard. Safety Management.

Functional technical failures, electronic avionics are the central sources of danger in the production of flights leading to aviation incidents such as Loss of control - inflight (LOC) and controlled flight into terrain (CFIT).

The proposed technology process analysis of flights technology is an automatic device and a tool for reducing the risks when eliminating functional failures avionics. The experience of applying technology process analysis of flights to the operation of the aircraft IL-96-300, TU-204, etc. showed a real possibility of preventing "computer" air crashes and reducing the risks.

The estimation of the risks of occurrence of functional and technological failures of avionics should be based on cycloramas and histograms of deployment, expansion of probabilistic distributions and static laws of flight technologies revealed by technology process analysis of flights.

REFERENCES

Asmus, V. F. (2001). Logic: textbook. Ed. 2nd, stereotypical. Moscow, Russia: Editorial URSS.

Dianov, V. N. (2012). Conceptual features of building trouble-free hardware / automatics and telemechanic, 7. pp. 119-138.

Gribov, V. M. (2017). Technical diagnostics of avionics Module 2-2, Kiev, Ukraine.

ICAO Journal. (2012). Doc 9859, AN/474, Safety management manual.

Kondakov, N. I. (1976). Logical dictionary-reference / 2nd ed., Rev. and add. Science, 1976.

Leitmann, G. (Ed.). (1962). *Optimization techniques: with applications to aerospace systems* (Vol. 5). Academic Press.

Lomov, B. F. (1982). Handbook of engineering psychology. Moscow, Russia: Machinery Industry.

Nilsson, N. J. (2014). *Principles of artificial intelligence*. Morgan Kaufmann.

Patton, R. J. (1991). Fault detection and diagnosis in aerospace systems using analytical redundancy. *Computing & Control Engineering Journal, 2*(3), 127–136. doi:10.1049/cce:19910031

Poltavets, V. A., & Teymurazov, R. A. (2007). Safety analysis of aircraft with gas turbine engines for a 50-year period (1957-2006) operation in the USSR civil aviation. Paris, France.

Russell, S. J., & Norvig, P. (2016). *Artificial intelligence: a modern approach*. Malaysia: Pearson Education Limited.

Strelkov, Yu. K. (2001). *Engineering and professional psychology: Proc. Allowance for Stud. Higher Training Institutions. Moscow, Russia: Academy Publishing Center.*

Tkachenko, V. A. (2009). *Flight risk. KIT*. Tersky.

Weiss, G. (Ed.). (1999). *Multiagent systems: A modern approach to distributed artificial intelligence*. Cambridge, MA: MIT Press.

ADDITIONAL READING

SAE ARP4754A. Aircraft Systems (12/01/2010 11/01/1996).

SAE ARP 4761. "Guidelines and Methods for Conducting the Safety Assessment Process on Civil Airborne Systems and Equipment" (March 27, 2014)

KEY TERMS AND DEFINITIONS

Functional Failures of Avionics and On-board Aviation Equipment: Are partial failures in avionics function, leading to accidents and incidents in complicated situations in flight (weather conditions, actions of risk factors, etc.) and partially disappearing during flight operations under simpler conditions.

Technological (Systemic) Failure of Avionics: Self-propelling and self-vanishing going beyond the allowable parameters of the piloting technique of the avionics system, leading to the appearance of an unwounded position of the aircraft or the creation of complicated piloting conditions for the aircraft.

Chapter 18
Perspective Directions of Artificial Intelligence Systems in Aircraft Load Optimization Process

Yelyzaveta Serhiyivna Sahun
 https://orcid.org/0000-0003-4837-4688
Flight Academy, National Aviation University, Ukraine

ABSTRACT

The chapter represents an overview of different approaches towards loading process and load planning. The algorithm and specificities of the current cargo loading process force the scientists to search for new methods of optimizing due to the time, weight, and size constraints of the cargo aircraft and consequently to cut the costs for aircraft load planning and handling procedures. These methods are based on different approaches: mix-integer linear programs, three-dimensional bin packing, knapsack loading algorithms, tabu-search approach, rule-based approach, and heuristics. The perspective direction of aircraft loading process improvement is a combination of multicriteria optimization method and heuristic approach using the expert system.

INTRODUCTION

Goods can be delivered anywhere by the following day through air cargo transport. The short delivery time guarantees fast transportation of commodity, emphasizing the value of temperature sensitive cargo by limiting the impact of non-appropriate conditions. Air cargo loading process plays a significant role in the air company's work and influences the all operations and relationship between air cargo companies, third party organizations and stakeholders. For each flight, air operator, which is responsible for the cargo transportation faces the several challenges simultaneously. The first one is to increase the aircraft's turnaround time. The second one is to solve the problem of over capacity on the air cargo market, which leads to the average load factor's decreasing (IATA Economics, 2016). Subsequently air

DOI: 10.4018/978-1-7998-1415-3.ch018

operators are pushed to cut the handling operation costs, fuel charges and other costs in order to save the business. This situation forces them to face with an aircraft loading problem, which is still the one of the most current explored problems which ties operational research and combinatorial optimization. Load optimizing, indeed leads to an aircraft's turnaround time reduction and consequently brings a notable economic effect. Thus, current program base and previous scientific works related to our problem doesn't provide any decision variables, we search for the optimal method of load planning through the prism of multicriteria optimization and heuristic methods. The main aspect in it is the selection of optimal decision among the possible range, which definitely deals with the peculiarity of artificial intelligence systems and neural networks.

BACKGROUND

It splits to different types of loading sub problems, as a The Bin-Packing Problem (BPP), Container Loading Problem (CLP), Knapsack Problem (KP), and Assignment.

The Bin-Packing Problem (BPP) is the loading problem which deals with packing objects of different sizes into a précised number of similar bins, such that the number of used bins is minimized. The problem is well-known as nondeterministic polynomial time problem (NP-hard) and difficult to solve in practice, especially when dealing with the multi-dimensional cases. (Paquay, Schyns, Limbourg, 2011) mention that the BPP can be one, two or three dimensional and with one or several containers. If there are several containers, they can have the same or different shapes (Dyckho, 1990).

Container Loading Problem (CLP) closely correlates with BPP. Authors search the ways how to optimize a space assignment of cargo inside the container, with the objective to maximize the value of the cargo or aircraft's capacity. The container loading problems can be divided into two types. The first is three-dimensional bin-packing problem. (Bortfeldt & Mack, 2007) note, that the point of the problem, is reducing the number of used containers. The other work (Bortfeldt & Gehring, 2001) defines the second problem as a knapsack problem and its objective is to maximize the capacity of the container while packing to it the cargo which has more value. The CLP focuses on a single container, and has been extended in the literature to handle a variety of different constraints arising from real-world problems. Consider for example the problem of arranging items into an aircraft cargo area such that the barycenter of the loaded plane is as close as possible to an ideal point given by the aircraft's specifications. The position of the barycenter has an impact on the flight performance in terms of safety and efficiency, and even a minor displacement from the ideal barycenter can lead to a high increase of fuel consumption (Trivella & Pisinger, 2017). The other group of loading challenges is a Knapsack Problem (KP). The single-objective knapsack problem consists of choosing a subset of objects from a defined set, maximizing the overall profit, which results from the sum of the individual benefits of the selected objects where a capacity constraint must be fulfilled, i.e., the sum of the weights of the selected objects, must not surpass a given capacity (Martello, Toth, 1990; Kellerer, Pferschy, Pisinger, 2004).

LOAD OPTIMIZATION METHODS AND APPROACHES

Peculiarities of Loading Process

Air cargo transport provides a range of services from point to point and also midpoints in order to move cargo with a help of a *shipper*, a *forwarder*, a *truck transport* (road feeder service), an *airline* (or carrier), and a *consignee*. The shippers' main goal is to send product/item to any place in a world with the lowest price and with a required service level. The forwarder plays a role of a link between the shipper and the career. The road feeder service provides the ground transportation service before and after air flight. The airline provides a chain of services such as: receiving, storage, transfer, loading and unloading cargo, assignment and managing of the compartment's capacity. The consignee gets the shipment. Figure 1 shows the process chain of the air cargo operations.

In order to arrange the handling process, the cargo is usually combined in pallets or containers called unit load devices (ULD) (Figure 2).

Figure 1. Air cargo operations technological chain

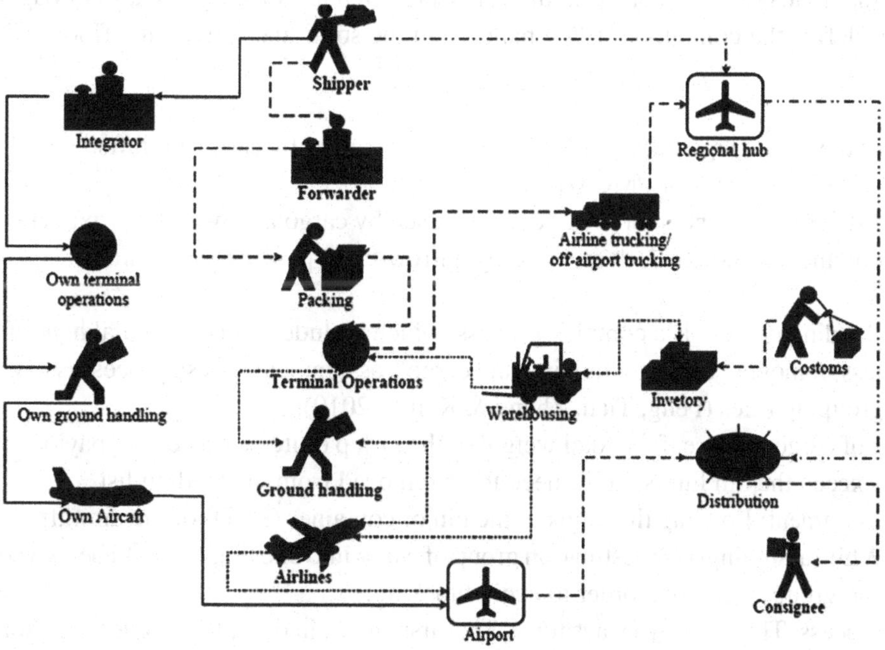

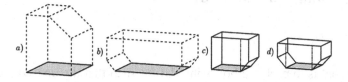

Figure 2. An example of ULD types in natural proportions

Sabine Limbourg in (Limbourg, Schyns, Laporte, 2011), defines ULD as an assembly of components consisting of a container or of a pallet covered with a net, whose purpose is to provide standardized size units for individual pieces of baggage or cargo for rapid loading and unloading. The aircraft loading process of them can differ and depends on the ULD's content and its quantity. Inside the boxes are stacked and are united in such a way to avoid the instability and fragility of the cargo items. Weight constraints inside the ULD allow loading it in appropriate way (Mongeau & Bes, 2003; Souffriau, Demeester, Berghe, & De Causmaecker, 2008). Inside the aircraft the ULDs are placed on designated loading positions and locked into position by latches on the floor. As the aircraft fuselage has a near circular cross section, there exist ULD types with different shapes to efficiently use the aircraft's interior space. Figure 2 shows a few frequently used ULD types (Brandt, 2017).

However, most of the aircrafts need certain ULD types for special loading and assignment. There are also general ULDs, such as aluminum pallets and containers of different formats, and special purpose ULDs for shipping cars, horses, or frozen goods. For transporting bigger cargos, the advantage is given to pallets because they can be packed in containers more easily and their contours can be moved freely. It is also popular to overhang items with the aim of full load of lower deck in width. But in this case each pallet should be covered with a net which need an additional handling effort. From the other hand, such containers possess certain boundaries and walls, so they can stay untied. Thus, they can be loaded only in a position that matches with their contours. They are often used for small cargo or baggage.

Dash lines define the contours of filled pallets and the solid lines point on a floor and walls of the container:

- Pallet for the main cargo deck of PMC type, which is used by cargo aircrafts.
- Pallet for the lower deck of PMC type.
- Half sized AKE containers for lower deck are used by cargo and wide-body passenger aircrafts.
- AKH containers for lower deck are used by narrow-body passenger aircrafts.

Air cargo loading process is a complex process, which includes four individual handling processes: splitting process, removing of cargo, moving and cargo's assignment. These processes have commodity priority and grouping rules (Feng, Tian, Zhang & Kelley, 2010).

Removing of cargo: in case if the total weight of the group of items exceed the payload weight or the total volume exceed the bin limits, items need to be removed from the loading list.

Cargo's assignment: Loading the items to the bin or container (ULD) of the aircraft.

Moving: While assigning a far destination group of cargo into one bin, it should be moved out through near destination group of cargo in order to avoid blocking.

Splitting process: There are two ways for it. The first one is, in order to remove cargo some groups of it should be split in order to transport as much as possible, the second one is, when a big group of items is assigned into précised multiple bins, it should be split into several sub-groups.

Variety of Load Optimization Methods

Mix Integer Linear Programming

Such approach has a goal is to pack the boxes into ULDs such that the unused space is minimized by respecting some basic and specific constraints. Since all the ULDs are identical, it is equivalent to minimize their number. As it solves 3-dimensional Bin Packing Problem-BPP has some general requirements such as the non-overlapping of the boxes and special requirements: uniform distribution of the weight, rotations, stability, fragility etc. The author of this mathematical model considers the 90-rotation since a 180-rotation does not change the space used by the box. So, the space occupied by the box is described more than the box itself. Author proposes four suitable configurations. Some authors (Paquay et al., 2011) propose the mathematical programming for solving the aircraft loading problem. The system determines which container has to be loaded and which has to be left for the next flight. The objective of this method is to minimize fuel consumption and to satisfy stability and safety requirements. The author's approach helped to solve loading problem within 10 minutes using integer linear programming.

The objectives were satisfied as following:

1. Maximization of weight loaded on the aircraft – opposing to volume of cargo.
2. Center of gravity of the aircraft should be as far to the aft as possible - to minimize fuel consumption, but not limiting the benchmark which is imposed by stability constraints (Paquay et al., 2011).

Mathematical formulation includes structural constraints (volume and weight capacities) and total maximal weight per one loading.

Rule Based Optimization

This approach, defined in (Feng et al., 2010) closely correlates with expert systems. The author developed a load planning solution including visualized rule editor, rule engine and an extensible business object interface. Rule Editor is a visualized environment for rule creation. Each rules flow should begin with a Begin Node, then a mixture of conditions, actions and message, and end with End Node or Flow. Rule Engine parse the rule file and execute based on the business object context to change commodity and bin's status. Author considers different constraints for removing and assigning process.

For removing process:

1. Determines the exceed weight and volume of the cargo.
2. Sorts the cargo group with priorities from high to low.
3. According to the sorted order and grouping logic, determines removed cargo to maximize the total weight.

For assigning process, author needs to consider the following constraints:

1. Load commodities with same properties (except weight, pieces and volume) together.
2. Avoid blocking.
3. Minimize move process.
4. Enhance the safety / weight and balance of the aircraft.

423

Different patterns include predetermined ordering and dynamic ordering. The predetermined ordering is achieved by employing some sort criteria, such as box volume, number of boxes, base area, or a certain dimension of a box. Dynamic sorting tends to be based on the usable space remaining after the next placement, volume of remaining boxes of the same type and free-floor space (Zhaoa, Xiaozhou, Bennella, Bektas & Dowsland, 2016).

Tabu Search Approach

Tabu search approach is an example of dynamic ordering.

It involves assigning palletized cargo to specific pallet positions within available aircraft to minimize deviation from preferred aircraft load requirements while satisfying, to the maximum extent possible, the temporal constraints on the pallets. The temporal constraints define the available "window" of acceptable days within which each pallet should reach the destination. Although preferable for a pallet to arrive within its specified time window, the approach considers solutions in which pallets arrive outside the desired window (either early or late). These solutions present decision makers with courses of action from which they can select their preferred option. The temporal constraints in the dynamic airlift problem enable aircraft to conduct multiple trips. The overall goal includes minimizing not only the number of flights required to transport all pallets, but also the total number of aircraft required, while at the same time minimizing aircraft allowable cabin load violations and pallet temporal violations (August, Roesener & Barnes, 2016).

Heuristics

There are lots of approaches which are based on heuristics – the approach that serve as a base of different expert systems. Larsen & Middelsen at (Larsen & Middelsen, 1980) and Amiouny at (Amiouny, Bartholdi, Vande & Zhang, 1992) propose heuristic methods to define a feasible load plan for the aircraft. Larsen & Middelsen implemented an interactive, computer-based procedure for solving a loading problem of containers and pallets into the aircraft. They used benchmarks of ground stability, load limits, position of the cargo, balancing and compartment capacity constraints. The other heuristic approach at (Amiouny et al., 1992) was presented as a one-dimensional loading problem with balancing constraint around the aircraft's midpoint. The requirements are the following:

1) All containers must be loaded.
2) Containers have to be arranged in a one-dimensional hold.

The problem defines the cargo, which must be loaded in a specified sequence with the help of branch and bound method and opinions of loadmasters.

The example of heuristic method can be wall building. The earliest example of it is by George and Robinson (1980). They design a two-section heuristic based on a real industry example of 800 boxes with no more than 20 types. Given the next empty space, initially the whole container, they select the first box of each wall so that each wall has a size that is not too deep or too shallow using the following ranking criteria. First, select the box whose smallest dimension is the largest among all candidates. In case there is a tie in the first criterion, select the box type with the highest quantity. If there is still a tie, select the box type with the longest largest dimension. Once at least one of a certain box type has been packed, that box type is open. Open box types are given preference where the type with largest quantity has the highest ranking. If the length of unfilled container is too short, usually below a certain defined amount, the rest packing no longer follows the wall-building rules (Zhaoa et al., 2016).

Gueret in (Gueret, Jussien, L'homme, Pavageau & Prins, 2003) deals with assignment in military operations such as assignment of troops towards the aircraft cargo compartment. Such loading pattern generation method of two phases:

- Generating a great number of feasible and attractive loadings.
- Selecting a subset from those loadings to cover all the items to be shipped with the final objective of minimizing the number of sorties (Gueret et al., 2003). But this pattern relates to 2-dimensional bin packing problem.

Researchers Thomas, Campbell, Hines, & Racer (1998) also developed heuristic method with branch-and-bound for loading containers into pre-defined positions to save feasibility constraints – determine a feasible packing with minimizing the nonproductive time.

Heuristic methods need to involve high qualified and experienced ground staff in order to try arranging an appropriate loading (satisfy all constraints) by manual experimental methods and error process without maximizing.

The analyzed literature shows that every approach is justified and proved with mathematical and computer -based model. However, the problem of aircraft loading has to be studied further because of its particular constraints, which cannot be satisfied simultaneously with a single algorithm. The range of approaches solve partial tasks connected with separate constraints and does not provide the overall solution that can satisfy the constraints which are in contradiction with each other. The problem needs to be solved with the help of expert systems, which are the part of Artificial Intelligence System (AIS). In order to find such solution, the loading process algorithm was presented and its peculiarities helped to define the base of the load optimization problem solution methods.

The short overview of aircraft loading approaches is performed in Table 1.

Table 1. Aircraft loading approaches

Types of Loading Problem	Focus	Models and Approaches	Objective
BPP	Aircraft loading	Mix integer linear programming	Minimize number of bins packed to the aircraft, maximize the aircraft's capacity minimizing the number of sorties
CLP	Aircraft loading	Nonlinear Mix integer programming	Reduce the number of used containers
KP	Aircraft loading	Mix integer linear programming	Maximize the loaded weights and value of cargo without surpassing a given capacity
Assignment	Cargo Assignment	Rule based optimization Dynamic ordering Tabu-search approach	Minimize deviations from preferred aircraft load requirements Minimize moving process Minimize the number of sorties Minimize the nonproductive time

Aircraft Load Optimization Problem Assumptions

The higher load factor needs more cargo shipments and more earnings. Reduced transfer time makes faster connections and thus higher salaries for the ground service staff. According to IATA information, the average weight load factor in the market varies between 40 to 50 percent (IATA Cargo Strategy, 2018). So, in real-world, load maximization is not the main challenge for the air companies. But the minimization of handling time and rationalization of the aircraft operations are much more difficult tasks to solve.

Cargo shipments need additional handling efforts. Firstly, cargo should be transported with the help of *belt loader assembly* or the forklift inside the terminal. Furthermore, it should be packed in ULDs, fixed with nets and sent further up to its route. Packing procedure is also a challenge in case if the shipment contains heterogeneous items.

While researching such load optimization problems, some aspects should be taken into consideration. These aspects are containers' time of delivery to customers and re-handling procedure. A re-handling means moving an item or cargo bay with the aim of opening the access to another bay, or to change the stowage. Such procedure is considered as a result of ineffective load planning though the overall route of the flight. Permutation can be a synonym of it and means the rearrangement of cargo items or the whole containers towards the cargo compartment. It can be considered as multi-dimensional (note that items cannot be stacked up in a bay) bin packing problem with heterogeneous containers. It is an NP-hard problem (Wilson, Roach, & Ware, 2001).

According to Uwe H. Suhl and Leena M. Suho in (Suhl, Uwe & Suhl, Leena, 1999), solving airline-fleet scheduling problems with mixed-integer programming, they have proposed the following heuristic algorithms:

First-in-first-out (FIFO): The aircraft which arrived first, will leave first.
Last-in-first-out (LIFO): The aircraft which arrived last, will leave first.
Best-first (BF): Choose that flight from available flights which can be flown by an aircraft with the shortest standing time.

The heuristics they used may be used in a load planning algorithm:

First-in-first-out (FIFO): The loaded container which arrived first, will leave first.
Last-in-first-out (LIFO): The loaded container which arrived last, will leave first.

The Best-first (BF) method cannot be used at this research clearly as it ignores the factors of center of gravity and load balancing.

For comparison the author brings an example of the model of Fok and Chun (2004), whose aim as to minimize the cargo loading residual (under-load). A linear program model, similar to the authors' was used.

And the work of Hussein (2012), who also uses LIFO algorithm to solve the task of reloading items, author called the approach as a block relocation problem with weights (BRPW). The problem is encountered in the maritime container shipping industry and other industries where inventory is stored in stacks.

All methods are mentioned previously can't solve the loading problem within time constraints. Pure mathematical programming is very time consuming and provide too much quantity of decision variables. Pure heuristics ignores the balance restrictions as it was mentioned. Therefore, the future step of the problem's solution is a combination of a few previous approaches and development of a new model. This model will contain a formal part – mathematical model and a part of decision making – searching of the feasible alternatives. The reason why the alternatives can be feasible but not optimal is because of uncertain conditions of the model. We have a deal with dynamic ordering as it was mentioned above. So, the stochastic environment which is forming around the loading process forces the author to create a model which would have different variants to choose due to the functions of a rule based expert system and its "if/then logic function.

Load Planning Algorithm Assumptions

1) The first stage is to define which box type to pack next.
2) The containers are homogeneous.
3) Container cannot be rotated.
4) A container can be unloaded if the floor above it is also unloaded.
5) There are two or more legs on the way to the final destination point of the flight.
6) Cargo is stored at the storage and will be moved to the platform/ramp.
7) An aircraft can only carry items that are compatible with it.
8) There is no overlapping inside the container.
9) Avoiding the reloading of cargo on the midpoint.
10) Boxes may only be placed with their edges parallel to the walls of the container.
11) All boxes must be placed within the container.

A subject of the research is a ramp type of loading

The research considers a situation in which the given set of items has to be transported from an estimated airport to three destinations A, B and C. Items have their own characteristics such as:

- Dimensions.
- Weight.
- Type of cargo (according to IATA Dangerous Goods Regulations, 2018).
- Compatibility of cargo.
- Destination (A to B, A to C).

Author also considers having only cargo aircraft with a ramp, not with different placed cargo bays to specify the results. Also, in the work the author looks at the two criteria for sorting the cargo. The first one is to divide the cargo of the same shipment into *as few ULDs as possible*.

The next one is to *avoid mixing different shipments in the same ULD*.

Apart from the type of cargo, some other information should be given:

- Origin and destination; this is given by the IATA three letter code of the airport.
- Scheduled time of arrival.
- Estimated time of departure.

An aircraft rectangular bay is characterized by depth, width and height and its maximal payload. An item is characterized by its dimensions (depth, width, height and weight) and its route (destination).

Model Formalization

Let's consider that the compartment which has to be used to stacked bins consist of a single cargo bay. Our objective function consists of reducing the number of container re-loadings and then minimizing overall time for additional handling efforts.

Therefore, the objective is to reduce time for loading, but with the help of cutting the re loading procedures. The formalization of the model includes the following parameters:

Indices of the mathematical model:

b:Is a bin (b= 1 to n).
t:Is a time range (t= 1 to n_{st}).
s:Are the stacks (s = 1 to n_{st}).

Let XR_{cg} and YR_{cg} stay x-coordinates and y-coordinates towards the aircraft's center of mass after loading.

Output function parameters:

δx_i :Is a center of mass longitudinal deviation of i – bin from the ideal center of mass.
δy_i :Is a center of mass latitudinal deviation of i – bin from the ideal center of mass (Sahun, 2019).
T_{loadV}:Is a loading time of the i – bin to the aircraft.
T_p :Is an aircraft's parking time at the apron (hour).
P :Is an aircraft's parking cost at the apron/hour.
w_i :Is a weight of the i – bin.
C :Is a cargo compartment's capacity.
n :Is a number of containers/bins (integer, >0).
n_{st} :Is a number of stacks (integer, >0).
Nr :Is a maximum number of time ranges that could be needed to reload.
Cargo bins from the compartment (integer, > 0).

Decision variables in mathematical model:

$R_{b,s,t}$ =1 when a cargo bin b is removed from the stack s at the exact time range t.
$P_{b,s,t}$ =1 when a cargo bin b is placed on the stack s at the exact time range t.
$M_{b,t}$ =1 in case if bin b is moved at the exact time range t.

As the objective is to minimize the loading time of set amount of the cargo bins, the *objective function* will be the following:

$$\min \sum_{i=1}^{n} {''}_{load_i} , \tag{1}$$

In this particular case minimization of loading time means reducing number of reloads of cargo during destination points. Let's impose additional criterion for the function:

$$\min \sum_{i=1}^{n} \mathrm{Pr}_i \cdot mN_{rload_i} \cdot X_{i_{(x,y,z.)}}, \tag{2}$$

subject to:

$$XR_{cg} \to 0, \ \delta x_i \geq 0, \ \forall_i \in \{1,...n\}; \tag{3}$$

$$YR_{cg} \to 0, \ \delta x_i \geq 0, \forall_i \in \{1,...n\}, \ w_i \leq C, \ \forall_i \in \{1,...n\}; \tag{4}$$

$$\min \sum_{i=1}^{n} ''_{load_i} \leq T_p; \tag{5}$$

$$P \to \min, \ \forall_i \in \{1,...n\}, \tag{6}$$

where

Pr_i – is a priority due to the delivery time of dt_i bin to its destination;

dt_i – is a delivery time of i – bin $\mathrm{Pr}_i = 1/dt_i$

mN_{rload_i} – is a min number of reloads of the i bin;

$X_i(x, y, z)$ – is a decision variable.

According to the objective, constraints are the following:

Aircraft constraints: constraint (3), (4).
Specific constraints:

$$R_{b,s.t} \leq M_{b,t} \ \forall_b, \ \forall_s, \ \forall_t; \tag{7}$$

$$P_{b,s,t} \leq M_{b,t} \ \forall_b, \ \forall_s, \ \forall_t; \tag{8}$$

$$\sum_{b=1}^{C} M_{b,t} \leq 1 \ \forall_t. \tag{9}$$

Constraint (9) means that not more than one cargo bin should be moved during each time range. This constraint, combined with constraints (7) and (8), includes a supposition that a cargo bin that is placed on a stack should be the same bin which has been removed from a stack in any particular order. Constraint (10) defines that bins should be taken out of the cargo bay in the determined sequence (which depends on an overall route of cargo and an exact arrival midpoint leg of i cargo bin), in order to increase bin number.

$$\sum_{t=1}^{Nr} t \cdot T_{load_{i+1,t}} \geq 1 + \sum_{t=1}^{Nr} t \cdot T_{load_{i,t}}, \ \forall b \text{ from 1 to b}-1. \tag{10}$$

While using the multicriteria optimization method, let the loading time minimization be the main criterion. The objective function was actually build based on this criterion.

The main challenge of multicriteria optimization is the ambiguity of the optimal decision selection. The optimal decision of multicriteria optimization task means one of effective (Pareto effective) decision variables.

Multicriteria tasks appear in cases if a few different contradictory criteria need to be implemented. At the first sight, a possibility of evaluating of the decision with the help of various criteria seems to be unnatural. However, as such tasks appear, a feasible compromise between criteria should be found. The main challenge remain the achievement of a compromise (Pareto front), which consists of Pareto effective points. In this case the multiplicity of effective points remains very extensive and hampers the definite decision selection, and respectively needs imposing of a few «secondary» optimal principles (Muromtsev & Shamkin, 2015).

EXPERT SYSTEM FOR OPTIMIZATION OF AIRCRAFT LOAD PROCESS

The main objective of the research is to find out the opportunities in order to form a system, which will have all features of an expert system (ES). The knowledge base of a future system will contain the information about:

- Cargo information (weight, dimensions, specificity etc.).
- ULD's information (quantity, dimensions).
- Route information (order of a definite leg).
- Information from the load planning system (weight and balance, compartment's dimensions).

The system has an aim to avoid the number of reloading procedures on a route midpoint. Such prototype of ES would let the weight and balance documentation to be arranged in a way to avoid an impact of formed "empty" areas in a cargo compartment, which are left after the first set of unloading procedures at a first midpoint on a further trim and would not go beyond allowed boundaries.

The reasons to use expert systems in decision - making process of the air cargo planning are the following:

- Air cargo transportations is more complicated than passenger transport because it involves more third parties, challenging processes, a combination of weight and dimension constraints, and various prioritized services.
- Air cargo transportations have higher uncertainty than passenger transportations due to it available capacity.
- Cargo capacity forecast is substantially more complicated than passenger aircraft capacity forecast. Passenger aircraft's capacity is constrained with a number of seats, while cargo capacity depends on the used ULD type.

- The load planning conditions are very uncertain and information details can change in the last minute.

However before implementing this approach the author should define which kind of ES would be applicable for the following research. Rule based ES will be the most applicable for the following research. Consequently, the main features and structure of a rule based ES should be considered for the purpose of filling it with a required data.

The components of the expert system are shown in the Figure 3.

The structure of an expert system is illustrated in the Figure 4.

The strength of an expert system arises from its *knowledge base* - an organized collection of facts and heuristic knowledge about the system's domain. The aggregation of knowledge in knowledge bases, from which the inference engine produces conclusions, is the main feature of an expert system (Figure 3). Expert systems are Knowledge-Based Systems (KBS) in which The Knowledge Base (KB) holds information and logical rules for performing inference between facts. The KBS approach is promising as it captures efficient problem-solving of experts, guides the human operator in rapid fault detection, explains the line of reasoning to the human operator, and supports modification and refinement of the process knowledge as experience is gained. The knowledge base of an expert system consists both of factual and heuristic knowledge. *Knowledge representation* is the method used to organize the knowledge in the knowledge base. Knowledge bases must represent concepts as actions to be taken under circumstances, time, causation, correlations, goals, and other higher-level concepts.

Figure 3. Expert system components (adapted from Kaimal, Metkar, & Rakesh (2014))

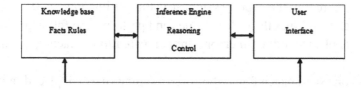

Figure 4. Structure of rule based expert system (adapted from Kaimal et al. (2014))

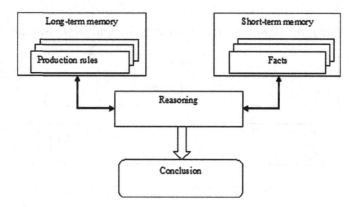

The structures used to store knowledge consist of:

1) **Meta-rule:** Structure - If Complaint Then Hypothesis. Meta-Rules are the supervisory rules to categorize Sub-rules in broad areas or nodes. Meta rules decide what domain rules should be applied next or in what order they should be applied.
2) **Rules:** Structure -If Condition Then Action. Rules are basic components of the Knowledge base. The condition and action parts of the rules are defined as parameters. The condition and action parts can have more than one parameter separated by logical operators. The parameters are specific data variables, which are configurable by the user (Kaimal et al., 2014).

In rule-based expert system, the *domain knowledge* is represented by a set of "if –then" production rules and *data* is represented by a set of *facts* about the current situation.

The inference engine compares each rule stored in the knowledge base with the facts contained in the database (Figure 5).

Taking into account that the future expert system is characterized by greedy constraints relative to the decision making time and program's memory, the inference engine should be realized in such a way for expert system to work faster, and for knowledge base to occupy the minimal memory capacity. This can be implemented by usage of dynamic arrays.

Working Memory (WM) is designed for source and interim facts' storage during the task. Usually, it is deployed in the RAM and reflects the current condition of the subject field by types of facts with certainty factor of their validity.

Kaimal (Kaimal et al., 2014) says, that inference engine (IE) is designed for receiving new facts based on combination of working memory's data and data from the knowledge base. Inference engine plays the most important role in the expert system structure. It implements the algorithms of forward and backward chaining and defines the order of rules and performs the following functions:

Matching: contents of the working memory are compared to the facts and rules contained in the knowledge base.

Selection: When consistent match found the corresponding rules are placed in the conflict set.

Execution: When all matched rules are placed in the conflict set one of the rules is selected for execution.

Figure 5. Inference engine process match – fire procedure (adapted from Kaimal et al. (2014))

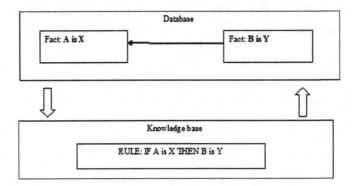

Reasoning subsystem explains how the system received the solution (explanation) of the task (or did not receive) and what kind of knowledge was used during the process, which simplifies testing the system by the expert and increases the user's trust towards the final result. The ability to explain its own functionality is one of the most important feature of an expert system as it causes the following results:

- Increasing the users' trust to the received results.
- Relieving of system's debugging.
- Creation of new patterns' development in the subject area for users.
- Reasoning of final conclusions can serve as a tool of searching the Pareto optimal point in the multiplicity of decisions (Vinogradov, 2013).

According to Kaimal (Kaimal et al., 2014), there are two strategies that can be used by the inference engine to receive a conclusion: forward and backward chaining.

Forward Chaining:
1) Starts from the facts.
2) Apply rules to find all possible conclusions.
3) Corresponds to modus ponens.
4) Data driven.

Backward Chaining:
1) Starts with the desired conclusion(s).
2) Works backwards to find supporting facts.
3) Corresponds to modus tolens.
4) Goal-directed.

Inference Engine Working Principle: The critical alarms that are serving as Meta-rules and subcauses are reflected as rules in the knowledgebase. When the critical alarm appears in the system, the relevant meta-rule will be fired; respectively the rules having the right condition part will get fired in sequence till the final root cause of the caused failure.

Therefore, the ES algorithm will be presented as follows:

1) A data from the knowledge base will be analyzed by the loadmaster.
2) A system will produce decision variables according to the cargo's destination, features and dimensions.

With the help of if-then production rules, cargo will be assigned inside the aircraft cargo bin.

FUTURE RESEARCH DIRECTIONS

So, the main subject of future research direction is the possibility of forming an expert system, which knowledge base contains the information about cargo. The aim is to avoid the reloading procedures on a route midpoint.

Such prototype of an ES would let the weight and balance sheet be arranged in such a way, to prevent an impact of formed "empty" space points, left from the unloading procedures in a first midpoint on a further trim and would not go beyond the required boundaries. The input parameters will be presented as a digital data, processing in real time, from the sensor, which is reflected on a work screen and recording to the ES data base, and also as a reference data, that are placed in a SQL-server, with the possibility of their correction and rewriting. The output parameters are reflected as graphics and numbers on a work screen in real time mode.

CONCLUSION

The following work presents the analysis of the order and specificities of loading process. The research also provides the review, which clarify the different between current load optimization models. The research contains the analysis of an air cargo operations technological chain and also technological issues of various ULD types in natural proportions.

Different load optimization methods and approaches were presented: mix integer linear programming, rule-based optimization, tabu search approach, heuristic methods.

The bottle necks in cargo planning technological chain exist on every step of the load planning procedure. In particular case, author considers bottle necks on the assignment stage. So the point is, to place the cargo in such a way, to foresee it's further motion according to its leg and also to avoid the reloading or unloading procedures. A large amount of scientists were faced with a load optimization problem and solved it with different and contradictory methods with the aim of reducing the fuel costs or optimizing the aircrafts' capacity. But the following work seeks to plan the containers' permutation in order to speed up the handling process, while being under conditions of a flight, consisting of multiple legs. So the author proposed new constraints, which should be satisfied in an optimal load plan. It can be more feasible than optimal, as probably will have multiple decisions and variables.

Some new assumptions were imposed still counting that the problem remains NP-hard. A few criterions were imposed into a new model. Now it includes criteria of loading time minimization which means reducing number of cargo reloads.

The main objective is to define a suitable and robust overall load plan, respective to all the constraints that determined above. Pure mathematical programming cannot provide enough of decision variables and pure heuristics ignores the balance constraints. Consequently, the perspective solution is a combination of multicriteria optimization method and heuristic approach, based on FIFO/LIFO inventory assumptions, using the expert system.

REFERENCES

Amiouny, S. V., Bartholdi, J. J. III, Vande Vate, J. H., & Zhang, J. (1992). Balance loading. *Operations Research, 40*(2), 238–246. doi:10.1287/opre.40.2.238

August G., Roesener, J., & Barnes, W. (2016). An advanced tabu search approach to the dynamic airlift loading problem. *Logistics Research, (1)*, 1–18. doi:. doi:10.100712159-016-0139-6

Bortfeldt, A., & Gehring, H. (2001). A hybrid genetic algorithm for the container loading problem. *European Journal of Operational Research, 131*(1), 143–161. doi:10.1016/S0377-2217(00)00055-2

Bortfeldt, A., & Mack, D. (2007). A heuristic for the three-dimensional strip packing problem. *European Journal of Operational Research, 183*(3), 1267–1279. doi:10.1016/j.ejor.2005.07.031

Brandt, F. (2017). *The air cargo load planning problem.* (Dissertation), University of Karlsruher (KIT), Germany.

Dyckho, H. (1990). A typology of cutting and packing problems. *European Journal of Operational Research, 44,* 145–159. doi:10.1.1.87.4320.

Feng, B., Yanzhi, L., & Shen, Z.-J. M. (2015). Air cargo operations: Literature review and comparison with practices. *Transportation Research Part C, Emerging Technologies, 56,* 263–280. doi:10.1016/j.trc.2015.03.028

Feng, L., Tian, C., Zhang, H., & Kelley, W. (2010). Rule-based optimization approach for airline load planning. *Procedia Computer Science, 1*(1), 1455–1463. doi:10.1016/j.procs.2010.04.161

Fok, K., & Chun, A. (2004). Optimizing air cargo load planning and analysis. In *Proceedings of the International Conference on Computing, Communications, and Control Technologies.* Austin, TX.

George, J. A., & Robinson, D. F. (1980). A heuristic for packing boxes into a container. *Computers & Operations Research, 7*(3), 147–156. doi:10.1016/0305-0548(80)90001-5

Gueret, G., Jussien, N., L'homme, O., Pavageau, C., & Prins, C. (2003). Loading aircraft for military operations. *The Journal of the Operational Research Society, 53*(5), 458–465. doi:10.1057/palgrave.jors.2601551

Hussein, M. I. (2012). *Container handling algorithms and outbound heavy truck movement modeling for seaport container transshipment terminals. (Theses and Dissertations, Paper 56,* University of Wisconsin, Milwaukee, WI).

International Air Transport Association. (2016). Air freight market analysis. *IATA economics.* Retrieved from https://www.iata.org/publications/economics/Reports/freight-monthly-analysis/freight-analysis-dec-2018.pdf

International Air Transport Association. (2018). Dangerous goods regulations. *IATA,* 1–18.

International Air Transport Association. (2018). IATA cargo strategy. *IATA,* 1–31.

Kaimal, L. B., Metkar, A. R., & Rakesh, G. (2014). Self learning real time expert system. Control & Instrumentation Group, C-DAC. *International Journal on Soft Computing* [IJSCAI]. *Artificial Intelligence and Applications, 3*(2), 13–25. doi:10.5121/ijscai.2014.3202

Kellerer, H., Pferschy, U., & Pisinger, D. (2004). *Knapsack problems* (1st ed.). Berlin, Germany: Springer Verlag. doi:10.1007/978-3-540-24777-7

Larsen, O., & Middelsen, G. (1980). An interactive system for the loading of cargo aircraft. *European Journal of Operational Research, 4*(6), 367–373. doi:10.1016/0377-2217(80)90187-3

Limbourg, S., Schyns, M., & Laporte, G. (2011). Automatic aircraft cargo load planning. *The Journal of the Operational Research Society*, 1–13. doi:10.1057/jors.2011.134

Martello, S., & Toth, P. (1990). *Knapsack problems: Algorithms and computer implementations*. Hoboken, NJ: John Wiley & Sons.

Mongeau, M., & Bes, C. (2003). Optimization of aircraft container loading. *IEEE Transactions on Aerospace and Electronic Systems*, *39*(1), 140–150. doi:10.1109/TAES.2003.1188899

Muromtsev, D. Yu., & Shamkin, V. N. (2015). Optimization methods in project decision making. Training manual for master degree students. Tambov State Technical University, Russia, 1–80.

Paquay, C., Schyns, M., & Limbourg, S. (2011). Three dimensional bin packing problem applied to air transport. *Colloque SIL, Dec. 15-16*, 1-6.

Sahun, Ye. (2019). Current status of aircraft load optimization problem. *Proceedings of the National Aviation University*, *1*(78), 35–39.

Souffriau, W., Demeester, P., Vanden Berghe, G., & De Causmaecker, P. (2008). The aircraft weight and balance problem. *Proceedings of ORBEL*, *22*, 44–45.

Suhl, Uwe H., & Leena, M. (1999). Solving airlinefleet scheduling problems with mixed-integer programming. *Operational Research in Industry*, 135–156. doi:. doi:10.1057/9780230372924_7

Thomas, C., Campbell, K., Hines, G., & Racer, M. (1998). Airbus packing at Federal Express. *Interfaces*, *28*(4), 21–30. doi:10.1287/inte.28.4.21

Trivella, A., & Pisinger, D. (2017). *Bin-packing problems with load balancing and stability constraints*. Chicago, IL: INFORMS Transportation and Logistics Society.

Vinogradov, G. P. (2013). *Methods and algorithms of intellectualization of decision-making in ACS productions with continuous and discrete technology*. (Doctor's thesis, Tver, Russia).

Wilson, I. D., Roach, P. A., & Ware, J. A. (2001). Container stowage pre-planning: using search to generate solutions. *Knowledge-Based Systems, (14)*3–4, 137–145.

Zhaoa, X., Bennella, B., Julia, A. T., & Dowsland, K. (2016). A comparative review of 3D container loading algorithms. *International Transactions in Operational Research*, *23*(1-2), 287–320. doi:10.1111/itor.12094

ADDITIONAL READING

Business dictionary. (2019). *Expert systems. Definitions*. Retrieved from http://www.businessdictionary.com/definition/expert-system.html

Techopedia. (2019). *Knowledge based systems. Definition*. Retrieved from https://www.techopedia.com/definition/7969/knowledge-based-system-kbs

YourDictionary. (2019). *Modus tollens. Definition.* Retrieved from https://www.yourdictionary.com/modus-tollens

KEY TERMS

Artificial Intelligence System (AIS): Computer system, that simulate human intelligence processes: learning, reasoning and self-correction.

Bin Packing Problem (BPP): A problem that define items of different volumes which have to be packed into a definite number of bins of finite volume in such a way that minimizes the number of bins used in the aircraft cargo compartment.

Block Relocation Problem with Weights (BRPW): A study, which deals with the problem of block replacement (closely correlates with a Bin Packing Problem).

Container Loading Problem (KLP): A problem that define a loading of the set of rectangular boxes into a rectangular bin of fixed dimensions in order to maximize the volume of the boxes.

Expert System (ES): Artificial intelligence system that converts the knowledge of an expert in a specific subject into a software code.

Knapsack Problem (KP): problem in combinatorial optimization which solve how the chosen set of items, with parameters of weight and a value, should be loaded in a way so that the total weight is less or equal to a set aircraft constraints and the total value is as large as possible.

Knowledge Base (KB): Technology used to store structured and unstructured data used by the computer system.

Knowledge-Based Systems (KBS): Computer system that generates and uses knowledge from different data, sources and information.

Modus Ponens: The rule of logic that says that if a conditional statement ("if–then") is accepted.

Modus Tolens: A valid form of argument in which the consequent of a conditional proposition is denied, thus implying the denial of the antecedent.

Random Access Memory (RAM): A form of computer data storage that stores data and machine code which are currently used.

SQL-server: A relational database management system created by Microsoft.

Unit Load Device (ULD): An assembly of components consisting of a container or of a pallet covered with a net, whose purpose is to provide standardized size units for individual pieces of baggage or cargo for rapid loading and unloading.

Chapter 19
Methods and Techniques of Effective Management of Complexity in Aviation

Maksym Yastrub

https://orcid.org/0000-0002-7434-0310

National Aviation University, Belgium

Mario Boyero Pérez

https://orcid.org/0000-0002-2079-1255

EUROCONTROL, Belgium

Svetlana Kredentsar

National Aviation University, Ukraine

ABSTRACT

This chapter presents the use of enterprise architecture to manage the growing complexity in aviation. Any aviation organization or air traffic management system can be considered as a complex enterprise which involves different stakeholders and uses various systems to execute its business needs. The complexity of such an enterprise makes it quite challenging to introduce any change since it might have an impact on various stakeholders as well as on different systems inside of the enterprise. That is why there is a need for a technique to manage the enterprise and to anticipate, plan, and support the transformation of the enterprise to execute its strategy. This technique can be provided by enterprise architecture, a relatively new discipline that focuses on describing the enterprise current and future states as well as providing a holistic view of it. The authors describe the modern enterprise architecture frameworks and provide an example of an application of one of them (European ATM Architecture framework) to identify and manage changes in aviation.

DOI: 10.4018/978-1-7998-1415-3.ch019

INTRODUCTION

Due to economic globalization and growing demand for air transportation, the level of air traffic has been increasing over the past years. In addition, it is forecasted that the volume of the air traffic will at least double in the next 15 years, with the annual growth of around 4.4 – 4.6 per cent per year (Industry High Level Group, 2017a). On the other hand, the capacity of the current air traffic management (ATM) systems is limited and is approaching, or even has reached, the ceiling to handle more traffic.

Thus, there is a vital need to modernize and enhance the current air traffic management systems to accommodate and support the growing demand for air traffic in a safe and sustainable way. Therefore, a few research and development initiatives were launched worldwide to change air traffic management systems, for example, Single European Sky (SES) initiative and its technological pillar – Single European Sky ATM Research (SESAR) in Europe, Next Generation Air Transportation System (NextGen) in the United States, Collaborative Actions for Renovation of Air Traffic Systems (CARATS) in Japan, etc. These programs and initiatives, in general, are not just single projects but rather a combination of many different projects that contribute to the evolution of the air traffic management systems. (Yastrub & Kredentsar, 2018).

However, any air traffic management system or aviation organization (e.g. aircraft manufacturer, airline, air navigation service provider, airports, network manager, military, etc.) can be considered as a complex enterprise, which involves different stakeholders and uses various systems to execute its business needs. The complexity of such an enterprise makes it quite challenging to introduce any change since it might have an impact on various stakeholders as well as on different systems inside of the enterprise. Besides that, the research programs launched worldwide (e.g. SESAR, NextGen, etc.) consider different aspects of the ATM systems developing various solutions to improve the overall performance. This brings or will bring a number of changes into the current ATM systems that have an impact on different systems and stakeholders. That is why the research programs have used enterprise architecture to describe their ATM systems and to provide a common reference for the research and development activities to ensure the solutions developed by different projects are consistent and interoperable.

In addition, a number of researches have emerged over the past years to address different challenges in aviation using artificial intelligence, machine learning or deep learning, e.g. aircraft design (Tan Wei Min, Sagarna, Gupta, Ong, & Keong Goh, 2017b), trajectory predictability, an efficiency of airport operations (SESAR Joint Undertaking, 2018), etc.

A number of projects have been started in the past years in Europe with the aim to solve ATM-related issues with the help of artificial intelligence, for example:

- Automatic speech recognition to convert speech of air traffic controllers into text and add a new input to an air traffic control system and therefore improve the safety and efficiency of airport operations.
- COPTRA project focused on predicting a trajectory of an aircraft closer to the take-off time or during the flight to predict more precisely when the aircraft will enter a particular part of airspace.
- BigData4ATM project investigating patterns in the behavior of passengers, door-to-door travel time and choice of travel mode using big data and machine learning (SESAR Joint Undertaking, 2018).

The objectives of this chapter are the following:

- Describe what is enterprise architecture and enterprise architecture framework.
- Expose the relationship between artificial intelligence and enterprise architecture.
- Provide an overview of the most relevant enterprise architecture frameworks and their application.
- Look at a practical application of enterprise architecture in aviation – European Air Traffic Management Architecture.
- Give an example of the use of enterprise architecture in aviation based on European Air Traffic Management Architecture.
- Identify the potential elements of the enterprise architecture model where artificial intelligence could be implemented.

BACKGROUND

To understand what enterprise architecture is, it is necessary to give definitions to each of the component of it: enterprise and architecture. According to International Institute of Business Analysis (2015), "An enterprise is a system of one or more organizations and the solutions they use to pursue a shared set of common goals" (p. 15). As can be seen, an enterprise represents a rather complex organizational structure that may involve multiple organizational units, organizations or businesses. As for the "architecture", ISO/IEC/IEEE 42010 (2011) gives a definition to it as following "(system) fundamental concepts or properties of a system in its environment embodied in its elements, relationships, and in the principles of its design and evolution" (p. 2). These two descriptions give an initial idea of what is enterprise architecture: a set of fundamental concepts or properties that allow describing a complex enterprise, its elements and relationships and its design and evolution.

Since enterprise architecture is a relatively new discipline, there is no single definition for it at the moment, however, a number of them were proposed by different scientists and practitioners. The Federation of Enterprise Architecture Professional Organizations (2013) defines the enterprise architecture as "a well-defined practice for conducting enterprise analysis, design, planning, and implementation, using a holistic approach at all times, for the successful development and execution of strategy" (p. 1). According to Marc Lankhorst et al. (2013), the enterprise architecture is a coherent whole of principles, methods and models that are used in the design and realization of an enterprise's organizational structure, business processes, information systems, and infrastructure (p. 3). Scott Bernard in "An Introduction to Enterprise Architecture" (2012) gives the following definition of the enterprise architecture "an analysis and documentation of an enterprise in its current and future states from an integrated strategy, business and technology perspectives" (p. 23).

From the described definitions of enterprise architecture, it is possible to derive the main cornerstones of enterprise architecture:

- Provides a holistic view of an enterprise.
- Allows describing current and future states of an enterprise.
- Supports a decision-making process.

In order to develop and implement enterprise architecture, an enterprise architecture framework is used. The enterprise architecture framework can be described as a set of models, methods, principles, elements and their relationships used to describe the current and future states of the enterprise. Apart from that, the enterprise architecture framework provides a means to communicate and describe the architecture of the enterprise using a common vocabulary and terminology (Cameron & McMilan, 2013, p. 61).

In the past years, different enterprise architecture frameworks have appeared and were adopted by various enterprises across the world, e.g. Department of Defense Architecture Framework (DoDAF), The Open Group Architecture Framework (TOGAF), European Air Traffic Management Architecture (EATMA) framework, etc. Even though those frameworks were developed for different purposes and domains, they all commonly address:

- **Alignment**: Enables better alignment between business and technological or operational and system objectives.
- **Integration**: Enables better interoperability between different components of the enterprise allowing faster and more cost-efficient integration of new components.
- **Value**: An economic value created through reusability of components across the enterprise.
- Change Management: An enterprise-wide process to manage and handle changes in the enterprise.
- **Compliance**: Allows ensuring legal and regulatory compliance of enterprise components (McCarthy, 2006, p. 15).

A few enterprise architecture frameworks, their elements and architecture methods will be described in the next sections of the chapter.

Artificial intelligence (or machine learning) is a transversal technology that can be applied to many enterprises or concepts, but not in the same way for all of them. Therefore, it is important to remark that in order to implement this technology bringing value to the enterprise, it is fundamental to understand where and how it can be applied. In other words, how it fits in the business concept (Mitache, 2019). The implementation would not be the same for a gold mining company as for an aviation enterprise.

Having a good understanding of the enterprise will help to better identify what are the best artificial intelligence use cases that could be implemented to enhance the company. In addition, it will optimize the allocation of the needed resources and will improve the recognition of the impact that it would have on the whole enterprise.

Additionally, more often than desired and since it takes time to complete the model, it contains aged data that has not been updated as it should have been. Besides this, the architecture model changes during the planning and implementation of the changes (Moné, 2019). To solve this issue, artificial intelligence can be used for keeping real-time data in the enterprise architecture model that will lead to making smarter and more realistic decisions.

Even more, artificial intelligence can be useful to ingest all the real-time data contained in the enterprise architecture model to identify any issue or pattern that might not be recognizable by a human being due to a large amount of information that it could contain. Moreover, not limiting the application of machine learning to the identification of issues or patterns, it could be really helpful to identify valuable strategies that could strengthen the enterprise (Crayon, 2018).

Then, it can be concluded that there are at least 3 potential use cases of artificial intelligence working in collaboration with enterprise architecture:

1. Using enterprise architecture to identify the most promising use case for an artificial intelligence implementation within the enterprise.
2. Using artificial intelligence models to keep up-to-date data in the enterprise architecture model.
3. Using artificial intelligence to ingest the vast amount of data contained in the enterprise architecture model to identify issues, patterns or strategies that might not be recognizable by a human being.

The first use case will be applied in the chapter to identify within the enterprise architecture model what are the elements on which artificial intelligence could be potentially applied to enhance the performances of the enterprise.

ENTERPRISE ARCHITECTURE FRAMEWORKS

Department of Defense Architecture Framework

One of the most widespread enterprise architecture frameworks is Department of Defense Architecture Framework (DoDAF). The development of DoDAF started with the creation of the Command, Control, Communications, Computers, Intelligence, Surveillance and Reconnaissance (C^4ISR) architecture framework to develop and define better means and tools to ensure interoperability of C^4ISR capabilities and that they meet the needs of a warfighter (United States Department of Defense, 2007). Later on, the C^4ISR architecture framework has evolved into the DoDAF as it is known now.

The aim of the DoDAF is to provide a fundamental framework to develop and present architecture descriptions ensuring a common denominator for understanding, comparing, and integrating of architectures across various organizations and enterprises. The DoDAF establishes a set of data elements definitions, rules and relationships and a baseline set of products for consistent development of systems, integrated (the one in which elements are uniquely identified and consistently used across all products and views within the architecture), or federated architectures (the one that provides a framework for enterprise architecture development, maintenance, and use) (United States Department of Defense, 2007).

The framework consists of 8 viewpoints (a way to abstract essential information from underlying complexity and present it in a consistent and coherent manner) that are presented on the Figure 1 and detailed in Table 1.

The DoDAF viewpoints allow viewing an enterprise from different perspectives and on in different level of details (e.g. high-level capabilities of the enterprise or resource flows between systems) as well as how different aspects of the enterprise are interconnected (e.g. how business activities are supported by a system). Besides the viewpoint, the framework defines a 6-step architecture development process as a methodology for the development of the architecture and guidance for an architect.

The purpose of DoDAF is to support key processes of the United States Department of Defense, however, it found it use in private and public sectors and become a basis for other enterprise architecture frameworks, for example, British Ministry of Defense Architecture Framework (MoDAF) and North Atlantic Treaty Organization Architecture Framework (NAF) that will be described in more details in the next section.

Figure 1. Department of defense architecture framework viewpoints (Department of Defense, 2010)

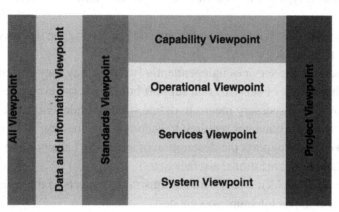

Table 1. DoDAF viewpoint descriptions (Department of Defense, 2010)

Viewpoint	Description
All	Describes overall architecture context related to all viewpoints.
Data and Information	Describes the data relationships and alignment structures in the architecture content for the capability and operational requirements, system engineering processes, and systems and services.
Standards	Articulates the applicable operational, business, technical, and industry policies, standards, guidance, constraints, and forecasts that apply to capability and operational requirements, system engineering processes, and systems and services.
Project	Provides the description of the relationships between operational and capability requirements and the various projects being implemented. It also details dependencies among capability and operational requirements, system engineering processes, systems design, and services design.
Capability	Describes the capability requirements, delivery timing, and the deployed capability.
Operational	Contains the operational scenarios, activities, and requirements that support capabilities.
Services	Provides the design for solutions articulating the Performers, Activities, Services, and their Exchanges, providing for or supporting operational and capability functions.
System	Provides the design for solutions articulating the systems, their composition, interconnectivity, and context providing for or supporting operational and capability functions.

North Atlantic Treaty Organization Architecture Framework

To provide a structured approach for managing the complexity of such large enterprise as North Atlantic Treaty Organization (NATO) while balancing all appropriate user perspectives, NATO has created its own architecture framework called NATO Architecture Framework (NAF). The main aim of NAF is to provide a standard way of developing and describing architectures for both military and business uses. Similarly, to DoDAF, the framework provides a methodology on how to define and run architectures (which is compliant with DoDAF methodology) and viewpoints to facilitate communication of enterprise architecture to different stakeholders (Architecture Capability Team, 2018b).

To manage the complexity, a number of viewpoints of NAF have been developed and defined:

- **Concept (C):** Supports the process of analyzing and optimizing the delivery of enterprise's capabilities in line with its strategy.
- **Service (S):** Describes the services independently from the way they are implemented or used. It helps to establish a library of services that support building service-oriented architectures.
- **Logical (L):** Provides a necessary means to define solution-independent descriptions of the logical nodes, activities and exchanges required to accomplish a mission.
- **Physical Resource (P):** Supports the description of structure, behavior and connectivity of different resources (people, organizations, artefacts, etc.).
- **Architecture Meta-Data (A):** Deals with administrative aspects of the architecture, such as who, when and for whom the architecture was created.

The use of standardized and well-defined viewpoints provides a common language to describe complex real-world concepts and systems. By reducing the complexity, the architecture can be used to support strategic planning, transformation and various analysis (e.g. gap, impact, risk, etc.) to satisfy the needs of stakeholders and decision-making process alongside each of these processes (Architecture Capability Team, 2018b).

The Open Group Architecture Framework

Similarly, to the previous two architecture frameworks, The Open Group Architecture Framework is based on the work of United States Department of Defense – Technical Architecture Framework for Information Management (TAFIM), a reference model for enterprise architecture. TOGAF is a foundational framework which means that it can be used to develop architectures of any kind in context. The purpose of it is to provide tools and methods to assist in production, maintenance and use of enterprise architecture. The key components of TOGAF are a set of best practices, iterative process model and re-usable set of existing architecture assets.

TOGAF consists of 4 generally acceptable subsets of enterprise architecture that it is designated to support. They are:

- **Business Architecture:** Defines key business processes, strategy, organization and governance.
- **Data Architecture:** Provides a description of an organization's logical and physical assets and resources for data management.
- **Application Architecture:** Gives a blueprint for enterprise applications to be deployed, their interactions and relationships to key business processes of the enterprise.
- **Technology Architecture:** Describes software and hardware capabilities that are required to support business, data and application services. It also includes standards, networks, communications, processing, etc.

Besides, TOGAF is composed of the following main parts:

- **Architecture Development Method:** An approach for the development of the enterprise architecture by architects. It is considered to be a core part of TOGAF and consists of iterative phases for architecture development that are described in Fig. 2.
- **Enterprise Continuum:** Provides various reference models (e.g. Technical Reference Model, Standards Information Base, etc.) to illustrate the evolution of the architecture from foundation-based architecture to a target architecture.
- **Architecture Content Framework:** Defines a structure for the architectural content that allows the main deliverable to be defined, structured and presented in a consistent and organized way.
- **Architecture Capability Framework**: Provides descriptions of organization structures, roles, responsibilities, skills, and processes that are required to be deployed to achieve a business capability (The Open Group Standard, 2018c).

Figure 2. Architecture development method cycle (The Open Group Standard, 2018c)

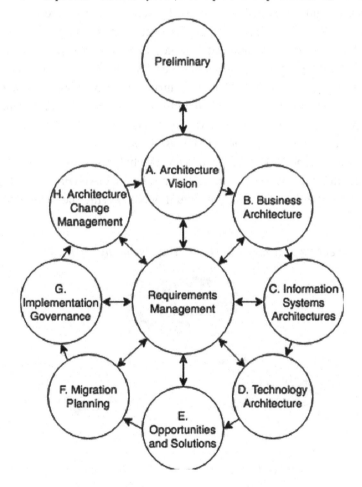

At each iteration of the Architecture Development Method (ADM) cycle, a decision has to be taken regarding the coverage width of the defined enterprise, the level of details, the extent of the time horizon aimed at, and the architectural assets to be used (Lankhorst et al., 2013b).

According to The Open Group (2015), TOGAF is a proven enterprise architecture methodology and framework used by enterprises and organizations to improve their business efficiency and around 80% of companies from Global 50 list and 60% of companies from Fortune 500 list are using TOGAF to define and maintain their enterprise architectures.

European Air Traffic Management Architecture

The SESAR program is a technological pillar of the Single European Sky initiative that was launched to facilitate the fulfilment of high-level goals set by European Commission (e.g. reduction of cost of air navigation service by half, a 3-fold increase in capacity and consecutive decrease of delays, etc.).

The SESAR Joint Undertaking, a public-private partnership established to manage the program, unites more than 100 organization that actively participate in research and development activities under SESAR across Europe. With such a significant number of contributors, there is a strong need to have a common reference and understanding to successfully achieve the changeling goals. That is why the cornerstone of the program is the architecture, the European ATM Architecture (EATMA).

EATMA acts as the main mean to integrate the key elements that are subject to the research and development activities of the programme. Furthermore, EATMA is essential to ensure coherency, consistency and completeness of the content produced during this R&D phase. In addition, EATMA is a key tool for decision making at the planning of the industrialization and deployment of the concepts and technologies developed to achieve the ATM performance targets set by the European Commission, as well supporting the generation of a feasible ATM Master Plan.

EATMA uses its own framework, European ATM Architecture framework that is based on NAF v3.1, however, the architecting process has been inspired by TOGAF (Fig. 3). A top-down approach is followed in order to have the business and conceptual needs driving the technical solutions that cover

Figure 3. EATMA architecting process (SESAR Joint Undertaking, 2017b)

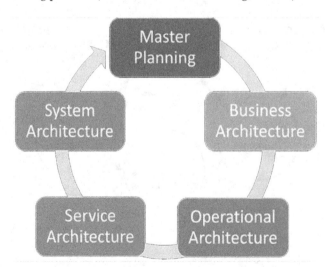

them. Therefore, there are different activities in the process, which cover the various topics of architecting ranging from planning and high-level business architecting to architecting of technical systems.

EATMA framework contains six layers and their purpose is to provide a division of the architecture in various views that together can provide an overall view of the European ATM (Fig. 4). Each layer is composed of a set of elements, which have relationships with elements of the same and other layers.

The list below provides a brief description of each layer composing EATMA:

- **Programme Layer**: Contains the Project and Solution elements, which formalize the SESAR Programme, and the Operational Improvement and Enabler elements, which expose the changes at conceptual and realization level, respectively. This layer provides the project management facet.
- **Capability Layer**: Describes the performance measures as well as the abilities of the European ATM. It can be seen as the strategic/business layer.
- **Operational Layer**: Provides the conceptual view of the enterprise by describing how the actors collaborate within the operational concept. It contains the operational needs.
- **Service Layer**: Provides a link between the operational need and its technical solution by describing services.
- **System Layer**: Is composed of the technical and human resources and describes the internal functional breakdown and the interaction between systems needed for covering an operational need.
- **Standard Layer**: Contains the standards and regulations of the European ATM.

For the purpose of this chapter, not all elements composing the different EATMA layers have been used. Therefore, only the elements necessary for the right understanding of the chapter are presented in the following Table 2.

Besides the elements and the relationships between themselves, EATMA also includes different views. Hereafter a list of the NAF sub-views which are used in the EATMA structure and that will be exploited in the next section.

Figure 4. EATMA layers (SESAR Joint Undertaking, 2017b)

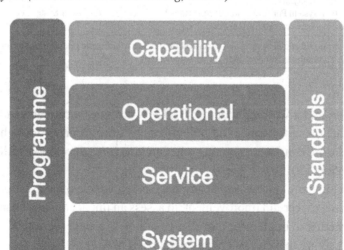

Table 2. Main Architecture Elements per layer in EATMA (SESAR Joint Undertaking, 2017b)

Architecture Layer	Main Architecture Elements
Programme Layer	**SESAR Solution**: Outputs from the SESAR Programme R&I activities which relate to an Operational Improvement (OI) step or a small group of OI steps and its/their associated Enablers, which have been designed, developed and validated in response to validation targets that, when implemented, will deliver business benefits to the European ATM. **Operational Improvement Step**: Operational Improvement Steps (OI Steps) are the means to describe changes in the ATM Operational Environment. **Enabler**: New or modified technical system/infrastructure, human factors element, procedure, standard or regulation necessary to make (or enhance) an operational improvement.
Capability Layer	**Capability:** Capability is the ability of one or more of the enterprise's resources to deliver a specified type of effect or a specified course of action to the enterprise stakeholders. A Capability represents a high-level specification of the enterprise's ability. As such, the whole enterprise can be described via the set of Capabilities that it has. A Capability is a statement of "what" is to be carried out and does not refer to "how" or "by whom" they are carried out. Consequently, capabilities are free from considerations of the physical organization or specific choices of technology. **Measure Category**: A standard unit used to express the size, amount, or degree of something. **Measure**: A certain quantity or degree of something.
Operational Layer	**Node**: A logical entity that performs Activities. **Activity**: A logical process specified independently of how the process is carried out. **Information Exchange:** An information exchange describes the need for actors to deliver and receive information and information products. **Information Element**: A formalized representation of information.
Service Layer	**Service**: The contractual provision of something (a non-physical object), by one, for the use of one or more others. Services involve interactions between providers and consumers, which may be performed in a digital form (data exchanges) or through voice communication or written processes and procedures. **Data Element**: A formalized representation of data.
System Layer	**Capability Configuration**: A Capability Configuration is a combination of Roles and Technical Systems configured to provide a Capability derived from operational and/or business need(s) of a stakeholder type. **Stakeholder**: A stakeholder is an individual, team, or organization (or classes thereof) with interest in, or concerns relative to, an enterprise (e.g. the European ATM). Concerns are those interests, which pertain to the enterprise's development, its operation or any other aspect that is critical or otherwise important to one or more stakeholders. **Technical System**: A collection of Functional Blocks or Functions. **Functional Block**: A logical and cohesive grouping of automated Functions in a Technical System. **Role:** An aspect of a person or organization that enables them to fulfil a particular function. **Function**: An activity which is specified in the context of the resource (human or machine) that performs it **Resource Interaction**: A Resource Interaction is a relationship specifying the need to exchange data between resources such as Capability Configurations and Technical Systems. **Data Element**: *(As defined in the Service layer).* **System Port**: A System Port is an interface provided by a Technical System. A System Port Connector asserts that a connection exists between two System Ports.
Standards Layer	**Standard**: A ratified and peer-reviewed specification that is used to guide or constrain the architecture. **Protocol**: A Protocol is a Standard used for communication.

Besides the structure and a working method, EATMA provides a reference architecture as well. It can be used to accelerate the development of architecture by an enterprise through the reuse of already defined elements and ensure consistency among other architectures developed with the help of EATMA (SESAR Joint Undertaking, 2017b).

European ATM Architecture can be used at any stage of a lifecycle of an aviation enterprise, either to develop a change to achieve a certain improvement, to assess an impact of such change or to facilitate the decision-making on whether to implement that change or not. The next section of this chapter describes an example of an architecture that can be created using EATMA.

Table 3. EATMA Views and brief descriptions (SESAR Joint Undertaking, 2017b)

View	Description
NOV-2: Node View	The NOV-2 is focused on the operational nodes which are logical collections of operational activities. Operational nodes produce or consume information and may represent an operational realization of capabilities. The main features of NOV-2 are the nodes, the links between them, and the characteristics of the information exchanged. Each needline describes the characteristics of the data or information.
NOV-5: Activity View	Describes the operations that are normally conducted in the course of achieving a mission or an operational objective. It describes operational activities (or operational tasks) and flows between activities.
NSV-1: Resource Connectivity View	Links together the Operational View and the System View by depicting which systems and system connections realize which information exchanges. It is based on the definition of Capability Configurations and describes the assets, both technical and human which are required in order to provide a Capability.
NSV-2: Resource Infrastructure View	Defines the connections between individual ports and shows the protocols and hardware spec used for each connection.
NSV-4: Function View	Supports the development of system functional hierarchies and system functions. The intention is to identify components by assigning the sole and exclusive responsibility for a certain system function to a system component. NSV-4 also addresses collaboration between components by describing the system data flows between the system components that realize certain system functions.

EXAMPLE OF APPLICATION OF EUROPEAN AIR TRAFFIC MANAGEMENT ARCHITECTURE

An example will be developed for an Air Navigation Service Provider (ANSP) stakeholder who wants to introduce a new Capability to its enterprise - a Remote Tower Operations Provision.

Using EATMA reference architecture and links between elements, it is possible to identify all the necessary changes needed to implement the Remote Tower Operations Provision in ANSP. The Views and models for this example are created for illustrative purpose only and are not aiming at an accurate description of a SESAR Solution.

An example of a Capability model of an ANSP is described in Figure 5 with a target Capability highlighted with a dashed border.

Also, at this level, an analysis of which is the capability contributing to the feature that is wanted to be enhanced is identified. Therefore, the Remote Tower Operations Provision is also selected as the capability in which artificial intelligence will be applied.

Using the link Capability to SESAR Solution we can derive a SESAR Solution that is capable of delivering improvement or introducing that Capability to an enterprise. For this example, the SESAR Solution #52 '*Remote Tower for two low-density aerodromes*' is selected.

The SESAR Solution #52 addresses the Operational Improvement (OI) Step named *SDM-0205 Remotely Provided Air Traffic Services for Two Low-density Aerodromes* that represents the changes in the ATM operating environment introduced by this Solution. OI Steps are satisfied by Activities from the Operational layer that are used to describe how the implementation of the OI Step will impact the current procedures and what are the new or updated Activities to be performed by the Nodes (Fig. 6).

The Activity View is used to describe how these new or modified Activities are integrated into existing processes and procedures and what are the new information flows introduced by them. An example of such View is presented in Figure 7.

Figure 5. Capability map of air navigation service provider

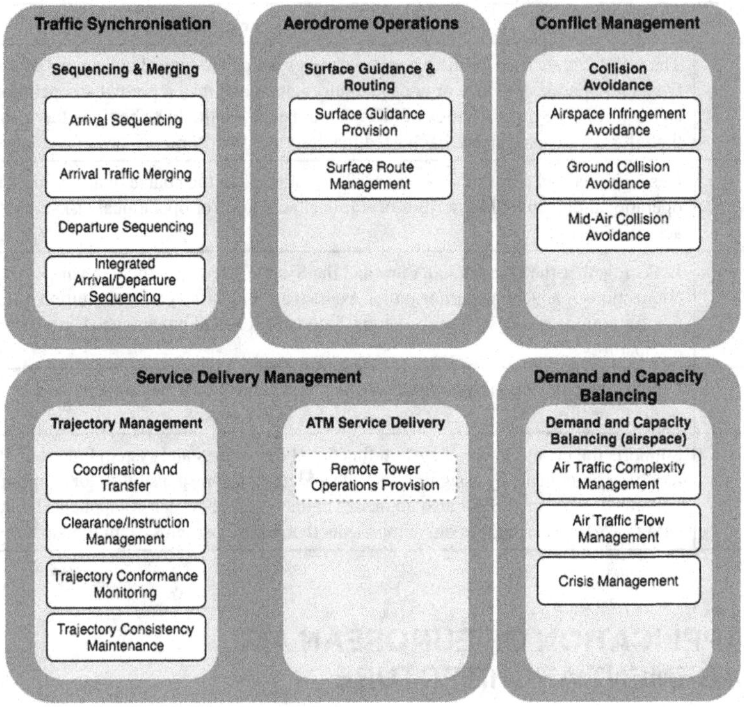

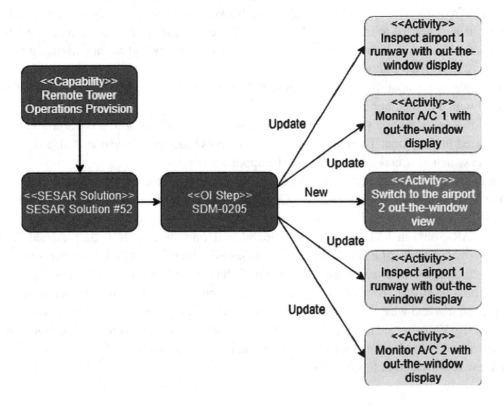

Figure 6. From a capability to the procedural changes

Figure 7. Activity View (NOV-5) of the remote provision of air traffic services to two airports (two arriving aircraft)

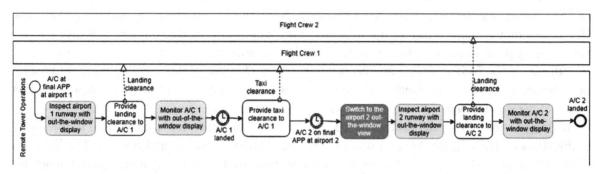

Information flows from Activity Views (NOV-5) are aggregated into Information Exchanges between Nodes on a Node View (NOV-2), which provides a high-level view on the operational architecture of the selected SESAR Solution (Fig. 8). Information Exchange between Nodes is one of the main sources of identification of needs that can be realized through Services.

At this operational level, the activities that can be potentially performed with the help of artificial intelligence are identified: "Inspect airport 1 runway with out-the-window display" and "Inspect airport 2 runway with out-the-window display".

Figure 8. Node View (NOV-2) of the remote provision of air traffic services to two airports

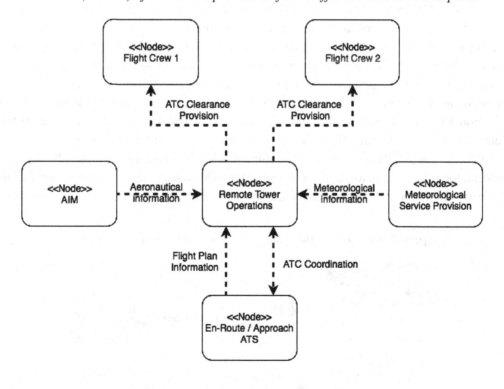

On the created NOV-2, it is possible to see Information Exchanges that can be covered using existing Services from the EATMA reference architecture, for example, the Information Exchange 'Meteorological Information' can be realized via METAR, TAF, SNOWTAM and other meteorological Services. The Information Exchanges that are not covered by Services can be realized through Resource Interactions - legacy point-to-point exchanges between different resources, e.g. Aeronautical Information can be realized via NOTAM Resource Interaction. If there is no Service or Resource Interaction to realize an Information Exchange, then that Information Exchange may lead to the creation of a new Service or update of the existing one.

To describe the system architecture of the implemented Capability, it is necessary to translate Operational Views (NOV-2 and NOV-5) into System Views (NSV-1, NSV-2 and NSV-4). The Resource Connectivity View (NSV-1) uses Capability Configurations to describe what assets (both human and technical) and exchanges are needed to fulfil a Capability. It also captures the infrastructure that is required to achieve the Capability. An example of such View is presented in Fig. 10.

From Fig. 10, it can be seen that Nodes from NOV-2 are realized through Capability Configurations that are presented on NSV-1 (e.g. AIM Node is realized via National AIM Capability Configuration). However, there are no corresponding Nodes for Surveillance Infrastructure Capability Configurations but they are present on NSV-1 to capture the necessary infrastructure.

The changes in the Technical System layer can be highlighted with the help of NSV-1 and NSV-4 Views.

The Remote Tower Centre Capability Configuration depicted on the NSV-1 is an evolution of existing TWR (Tower) Capability Configuration from the EATMA reference architecture. Its internal structure can be seen on another NSV-1 View that is used to describe internal resources (Human Roles and Technical Systems) of a Capability Configuration (Fig. 11). It is also possible to see the required logical and physical interfaces for that Capability Configuration from that NSV-1.

While NSV-1 focuses on the structure of the element and its interfaces, Function View (NSV-4 View) focuses on the behavior aspects of the element (similarly to Activity View (NOV-5) from the Operational layer). Figure 12 describes the internal structure of the Aerodrome ATC Technical System and highlights the new or updated Functional Blocks within it while Figure 13 shows the behavior of these Functional Blocks and interactions of a Human Role with them in a context. Some of the Functions identified during modelling ("Provide zoom and tilt view at airport 1" and "Provide zoom and tilt view at airport 2") potentially could be realized using machine learning to automatically detect outlines of aircraft and focus the remote tower video camera on it. This could help to automate some of the tasks of an air traffic controller and decrease the workload on him or her.

Figure 9. From the operational needs to the system layer through the service layer

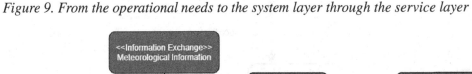
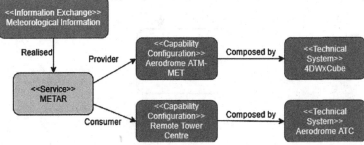

Figure 10. Resource connectivity view (NSV-1) of the remote provision of air traffic services to two airports

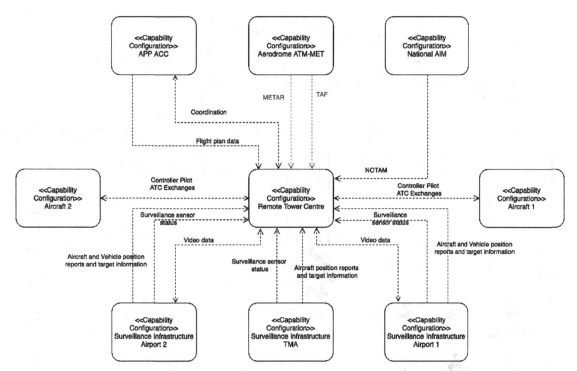

The Resource Infrastructure View (NSV-2) is used to show the underlying communication infrastructure that ensures the physical connectivity of the System layer elements and supports the logical exchanges described on NSV-1 and NSV-4 Views. It shows which technologies (sets of Protocols or Protocol Stacks) are used to support the exchanges and connects System Ports of Capability Configurations or Technical Systems. An example of such View is shown in Figure 14.

As can be seen along this chapter, the architecture is certainly useful for, among other things, supporting at the moment that a decision has to be made by showing all the impacts that this choice has in the enterprise at different levels, from business to the physical assets as systems or human resources, through conceptual processes.

In Figure 15, from the Operational Improvement Step (which was linked to the SESAR Solution #52 and this to the Capability that it is aimed to incorporate to the enterprise) can be easily identified the needed changes in the operational processes. Furthermore, these needed operational modifications are traced to the required changes that have to be done in both, systems and human resources. In addition, these system and human resources changes are aggregated in the form of Enablers, which are also directly linked to the Operational Improvement Step. Therefore, in this example, both the high-level changes (OI Step and Enabler) and the lowest-level modifications can be easily determined.

Along with this example, the use of the enterprise architecting, in which the whole enterprise information is captured, has proven to be an adequate way to identify the elements that potentially can be impacted by the implementation of an artificial intelligence solution. It has been shown, that starting from the capability (Capability layer) to be enhanced or introduced in the enterprise, the conceptual (operational layer) and system (technical layer) elements can be determined easing the process for decision making.

Figure 11. Internal structure view (NSV-1) of remote tower centre capability configuration

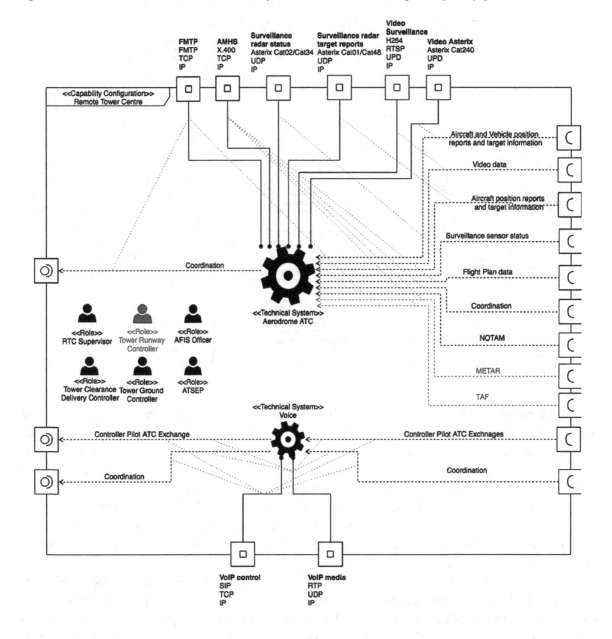

FUTURE RESEARCH DIRECTIONS

The enterprise architecture has proven to be a powerful means to understand and describe an enterprise from many different and complementing facets. Nevertheless, there are some relationships that can be further investigated in the future to enhance even more the utility of architecting. For example, the link of the enterprise architecture with the Cost-Benefit Analysis (CBA) or Performance can be highly valuable. In addition, even if mostly the architecture and requirements are supplemental to each other, there are some requirements that could be interesting to capture in the architecture as the Non-Functional requirements.

Figure 12. Internal structure view (NSV-1) of Aerodrome ATC Technical System

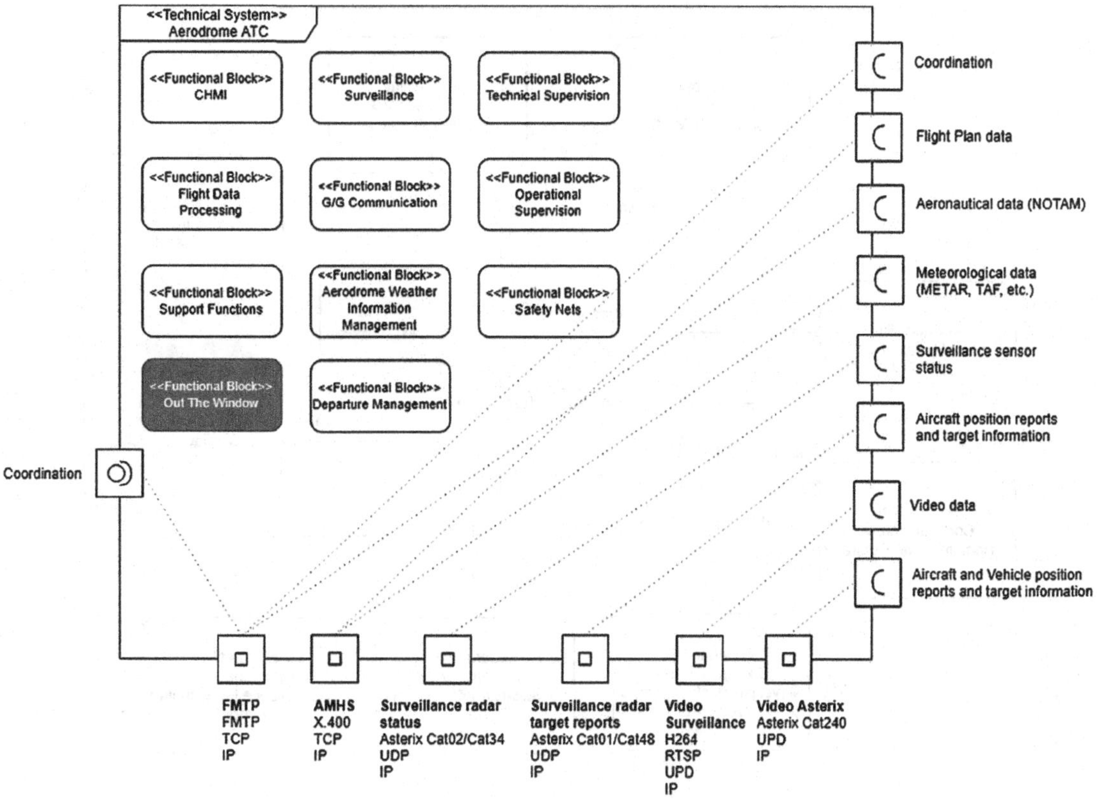

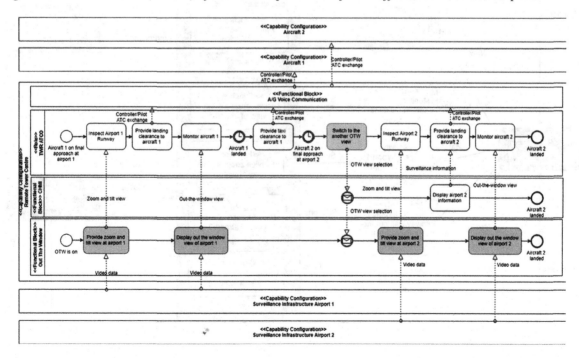

Figure 13. Function view (NSV-4) of the remote provision of air traffic services to two airports

Figure 14. Resource infrastructure view (NSV-2) for the remotely provided air traffic services to two airports

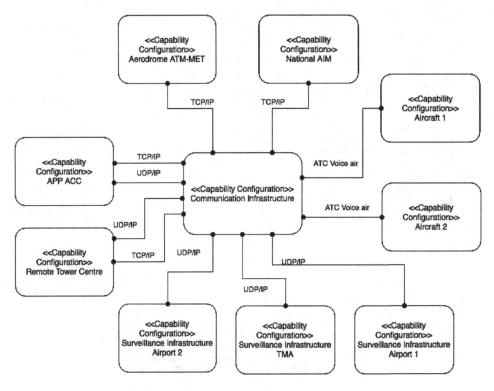

Figure 15. From the operational improvement step to the enablers through the architecture

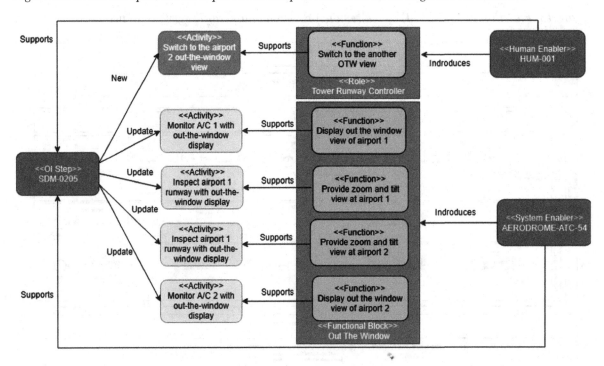

Also, two of the potential use cases regarding artificial intelligence identified in the background section were not addressed in this chapter (keeping real data in the enterprise model and assessment of the model to find occurring issues or potential strategies). Therefore, research about them including the specific detailed mathematical study would be highly beneficial.

Those are the directions proposed to be researched in the future.

CONCLUSION

Enterprise architecture is a powerful tool that is used by various enterprises and organizations around the world for various purposes, e.g. to provide structure and optimize R&D activities, to create a holistic view of an enterprise, to facilitate decision-making, etc. Besides that, architecture plays a central role in planning, managing and implementing changes in the enterprise since it provides an integrated view of it and allows analyzing the impact of the changes. The information provided by enterprise architecture gives a possibility to make more conscious and knowledgeable decisions regarding the future states of the enterprise or organization.

Since the complexity of any aviation organization or enterprise is relatively high, there is a need for a technique and method to help to manage it. This chapter shows that this can be provided by enterprise architecture. The extensive example created with the help of European ATM Architecture framework demonstrates that enterprise architecture can be used to describe all the aspects and layers of an aviation enterprise and help to show the evolution of it in accordance with the selected strategy (for example to implement modern solutions based on artificial intelligence or machine learning in the enterprise). The use of such mature architecture as EATMA helps to develop an architecture of a complex aviation organization in a quick manner and with the necessary level of details.

REFERENCES

Architecture Capability Team. (2018a). *NATO Architecture Framework Version 4.*

Cameron, B. H., & McMilan, E. (2013c). Analyzing the current trends in enterprise architecture frameworks. *Journal of Enterprise Architecture, 9*(1), 61.

Crayon. (2018). *How can machine learning help the enterprise architect?* Retrieved from https://www.crayon.com/en-za/news-and-resources/how-can-machine-learning-help-the-enterprise-architect/

Federation of Enterprise Architecture Professional Organizations. (2013a). Common perspectives on enterprise architecture, *Architecture and Governance Magazine, Issue 9-4.*

Industry High Level Group. (2017a). Aviation benefits 2017, 9–10.

International Institute of Business Analysis. (2015). A guide to the business analysis body of knowledge (version 3.0). Toronto, Ontario.

ISO/IEC/IEEE. (2011). Systems and software engineering – Architecture description (1st edit.). Geneva, Switzerland.

Lankhorst, M. M., Arbab, F., Bekius, S. F., Bonsangue, M., Bosma, H., & Campschroer, J. … Wieringa, R. J. (2013b). Enterprise architecture at work, The Enterprise Engineering Series. Berlin, Germany: Springer-Verlag.

Lesa Moné. (2019). *An enterprise architect's guide to machine learning series: Part 1.* Retrieved from https://blog.leanix.net/en/an-enterprise-architects-guide-to-machine-learning-series-part-1

McCarthy, R. (2006). Toward a unified enterprise architecture framework: An analytical evaluation. *Issues in Information Systems, VII*(2), 15.

Raz Mitache. (2019). *How to get AI right using enterprise architecture.* Retrieved from https://bizzdesign. com/blog/how-to-get-ai-right-using-enterprise-architecture/

SESAR Joint Undertaking. (2017b). *EATMA Guidance Material Version 9.0.*

SESAR Joint Undertaking. (2018). *Artificial intelligence in air traffic management.*

SESAR Joint Undertaking. (2019). *eATM Portal. Research and Development View [Dataset].* Retrieved from https://atmmasterplan.eu/rnd/rd-dashboard

Tan Wei Min, A., Sagarna, R., Gupta, A., Ong, Y.-S., & Keong Goh, C. (2017). Knowledge transfer through machine learning in aircraft design. *IEEE Computational Intelligence Magazine, 12*(4), 48–60. doi:10.1109/MCI.2017.2742781

The Open Group Standard. (2018b). *The TOGAF® Standard, Version 9.2.*

TOGAF® Worldwide. (2015). [Image illustrating the worldwide usage of TOGAF, May 2019]. *TOGAF® Worldwide.* Retrieved from http://web.archive.org/web/20181017230812/http://www.opengroup.org/subjectareas/enterprise/togaf/worldwide

United States Department of Defense. (2007). *DoD Architecture Framework Version 1.5* (Vol. I). Washington, D.C.: Definitions and Guidelines.

United States Department of Defense. (2010). DoD Architecture Framework Version 2.02. Washington D.C.

Yastrub, M. I., & Kredentsar, S. M. (2018). Enterprise architecture as a driver of transformation in air traffic management. In *Proceedings of 2018 IEEE 5th International Conference on Methods and Systems of Navigation and Motion Control (MSNMC 2018)*, 162-165. 10.1109/MSNMC.2018.8576318

KEY TERMS AND DEFINITIONS

Architecture: Fundamental concepts or properties of a system in its environment embodied in its elements, relationships, and in the principles of its design and evolution.

Artificial Intelligence: Artificial intelligence (AI), sometimes called machine intelligence, is intelligence demonstrated by machines, in contrast to the natural intelligence displayed by humans and other animals. In computer science, AI research is defined as the study of "intelligent agents": any device that perceives its environment and takes actions that maximize its chance of successfully achieving its goals.

Enterprise: A system of one or more organizations and the solutions they use to pursue a shared set of common goals.

Enterprise Architecture: A well-defined practice for conducting enterprise analysis, design, planning, and implementation, using a holistic approach at all times, for the successful development and execution of strategy.

Enterprise Architecture Framework: A set of models, methods and principles that can be applied to define and implement enterprise architecture.

Chapter 20
Ensuring the Guaranteed Level of Flight Safety:
The View of the Future

Volodimir Kharchenko
National Aviation University, Ukraine

Oleh Alexeiev
National Aviation University, Ukraine

ABSTRACT

This chapter presents a methodology for the determination of the guaranteed level of flights safety. Purpose of the methodology consists in association in the only complex of tasks of assessment, providing verification of safety aviation activities as a complex hierarchical structure with independent critical elements and also hardware, program, network, and ergatic component which are both means, and subject to safety. The realization of ensuring the guaranteed result consists in realization of management processes so as not to allow transition of infrastructure or its systems to potentially dangerous state and to provide blocking (exception) of the corresponding technical object in case of threat of transition or upon transition to a dangerous state and minimization of consequences of such transition.

BACKGROUND

Considering statistics for the last decade on accident/incident of a question of providing guaranteed flight safety level is the most relevant as shortcomings and problems of functioning aviation activity are explained by lack of general-theoretical basis and the standard scientifically based approaches to safety management system which development has to be guided upon the demand of International Civil Aviation Organization (ICAO) which defines that any region should not have accident/incident frequency level more than twice exceeds universal. These are the following main areas: introduction of an acceptable level of flight safety in the state; mandatory procedures for the development and implementation

DOI: 10.4018/978-1-7998-1415-3.ch020

of safety management system; obligatory procedures for ensuring direct management of the level of the flight safety within acceptable or established level of the enterprise (continuous monitoring and regular assessment of the flight safety, corrective actions required to maintain the agreed flight safety indicators and monitoring, flight information analysis, risk management, etc. Purpose of methodology consists in association in the only complex of tasks of assessment, providing and verification of safety aviation activities as complex hierarchical structure with independent critical elements and also hardware, program, network and ergatic komponen which are both means, and subject to safety. The realization of ensuring the guaranteed result consists in realization of management processes so that not to allow transition of infrastructure or its systems to potentially dangerous state and to provide blocking (exception) of the corresponding technical object in case of threat of transition or upon transition to a dangerous state and minimization of consequences of such transition (ICAO, 2002, 2004).

The Integrated Management System allows for the application of static and dynamic information management principles, which makes it possible to identify restrictions and limitations. This concept of the Integrated Management System operation is based upon the separation of static and dynamic information as well as its graded use in the context of implementation of retroactive, proactive and predictive approaches to the management during the services provision. Retroactive approach is based on the application of corrective measures on the ground of static information received after the fact of a non-conformance or an occurrence has already happened. Distinctive items of static component for this approach are the results of relevant audits (inspections) and investigations of occurrences in the course of which the facts of incompliance with the established requirements are recorded (EUROCONTROL, 2004).

Proactive approach combines application of the static component's information particularly analytic efforts and project evaluation in order to manage or take appropriate corrective measures *before* the fact of non-conformance or an occurrence has already happened. Analytic efforts and project evaluation towards the changes enable the development of preventive measures concerned with the services rendered together with other items of static and dynamic components.

Predictive approach involves application of dynamic information. Such an approach to management processes enables non-conformances identification *under conditions of day-to-day operations* of the Air Navigation system and taking adequate measures for it correction, based on prompt response and predicting the actual state of services if relevant deviations, occurrences, etc.

According to the information received from the results of operation of static and dynamic components of information collection process, routine and periodic analysis, management review, risk assessment and other measures the appropriate corrective measures are being developed. The above measures address the following:

- Development of non-conformances correction strategy.
- Approval of non-conformances correction strategy.
- Allocation of responsibilities on non-conformances correction.
- Execution of non-conformances correction strategy.
- Development of preventive measures which will disable recurrence of non-conformances in the field of services provision.

Corrective measures are planned in line with the procedure for planning of the Integrated Management System operation.

Turning now to a description of the major trends in the matter of rational choice of strategies aimed at ensuring and maintaining a guaranteed result, it is necessary first of all to emphasize that the activity is no unambiguous relationship between the prediction of the results (outcomes) and the problem of decision-making. It has been said that the chaos of possible outcomes facing the decision maker. At the same time, decision should be made, and it should eventually be uniquely. It is important to draw attention to the fact that even the mathematically and the information problems of forecasting and decision-making is usually not the same. (EUROCONTROL, 2004)

At present there are the following principles of rational choice.

Isolationism: Replacement of the i-th participant only its criteria so as to reduce the number of variables that affect the i-th criterion of efficiency, and ideally to reduce it to the criterion of the type that has to be optimized and no matter what the rest of the participants. This method of action is generally accepted in the presence of random factors as efficiency criterion is replaced by its expectation.

Collectivism: Introduction of a single general criterion (general purpose) for the group participants. In this case we speak of coalition and compromise between the parties. The second principle is the formation of rational strategies in the pursuit of good mutual awareness, allows for constant performance criteria form a rational strategy.

The quest for knowledge as the basis of rational choice behavior, of course, is not contrary to the first principle, but rather complements it. For example, the coalition unthinkable substantially without a collective sharing of information, and the extracted individual information reduces the amount of required clotting individual member separate criterion.

The third and very important principle of developing rational behavior consists in the pursuit of *sustainability*, understanding which vary widely. Here, above all it deserves special mention *the principle of guaranteed* **result** calling side operates with the lack of information based on the consideration of the worst possible situations, taking into account available information. It is widely understood principle of guaranteed result can be applied in the selection of rational strategies and the results expected. This principle includes, of course, the usual maximum is used in an antagonistic activity and interaction with the environment, but is not limited to it. (EUROCONTROL, 2004)

It should be emphasized, however, that a reasonable reduction of the number taken into account the values of x1 is reasonable and even an inevitable step. Such a process usually referred to as the method of test, and the 'choice of variables accounted for xx, usually performed by expert procedures.

The most common form of representation of the relationships and interactions of disparate processes and events is a cause-and-effect relationship.

The causal relationship and interaction of processes, events and phenomena in real systems are formed and implemented between objects of different nature. Related technological, informational, administrative, economic, social and other processes are combined in a complex interaction, which is currently not sufficiently precise and easy to use mathematical models. Development of models and methods focused, usually for a specific kind of process and results in a formal apparatus, which is not always convenient to combine disparate processes, objects and phenomena. With the development of common models and methods for solving problems with the use of such models is needed to move from specific and specialized concepts to more general categories of causation. An important and crucial tool models are tangible, imaginable, math.

Isolation in a variety of interactions between objects (processes, events, phenomena) causality is fundamentally difficult.

One option for harmonization of the complex set of developers and users is the availability of the agreement the developer and the user of the universe of objects, processes, events and phenomena used in the synthesis of the complex. Cybernetic sense of purpose related to the behavior of cybernetic systems, presented the process of changing states of the system and the achievement of the desired state of the system. This behavior can be represented by a phase trajectory in the space of states of the system and the set of all possible trajectories of the phase picture.

Model is not a second copy of the original. The model contains or may contain:

- Properties that are available and the original.
- Properties only model.
- Properties for which is not yet known that they belong to the original.

SELECTING RATIONAL STRATEGIES FOR GUARANTEED SAFETY MANAGEMENT

The mathematical models of the iconic, not all designs have a direct interpretation in the model application. Broadly speaking, the development of the formalism of causal systems connected, first of all, with the desire for representation of determinism in the interaction of system components and system actions. The human mind from the experience tends to perceive reality through the causality. It comes down to a causal relationship (Jager, 1986; Hubaev, 1998; 1999; Raz, Michael, 2001).

In assessing the overall meaning of the exchange of information, it should be noted that it should help to reduce the uncertainty in the production process, leaving a narrow variation limits for the selection of operators – in a word, to make the situation more definite.

So, it is advisable to introduce the information sent by one operator to another.

Intuition and experience suggest the reasonableness of collective decisions. One can distinguish three levels of collective action operators $m(m \leq n)$ (we assume that the coalition includes the first m operators):

The exchange of information on activities and process conditions;

The pooling of resources and the subsequent selection of a joint course of action, based on the combined resources.

It is clear that each successive stage creates great opportunities coordination. The possibility of combining in the second or third stage is, in fact, collective rules of conduct, collective strategy. Unification, producing such a strategy, according to tradition will be called coalitions.

A very common type of collective aspirations should be considered joint mixed strategies - distribution laws $\omega_c = x_c$, depending, in general, on the elections x_j operators that are not included in the coalition, and natural uncertainty β. Thus, you can enter ω_e the same as previously defined x_l.

The use of mixed strategies associated with the introduction and averaged criteria coalition:

$$w_i = \int w_i d\omega_c(x_c), i = 1,...m.$$

Regarding the criteria for operators outside the coalition, they are averaging can only be discussed as one of the possible options.

In discussing the possibilities of coalition cannot forget about the additional interactions between the members of the coalition, and between members of coalition and the rest of the operators, although they can be considered as already included x_c, in the future, given their importance, we usually write them separately. Therefore, together with *xc* we consider the vector $Z_c = \{Z_{c1}, ..., Z_{cm}\}$, representing the additional interaction of the coalition as a whole.

$$z_1 = \left\{ \sum\nolimits_{j=1}^{m} \lambda_{ji} z_{ji} - \sum\nolimits_{j=1}^{n} z_{if} , and, u_i = \sum\nolimits_{j=m+1}^{n} \lambda_{ji} z_{ji} \right.$$

But if *i>m*

$$t_i = \left\{ \sum\nolimits_{j=m+1}^{n} \lambda_{ji} z_{ji} - \sum\nolimits_{j=1}^{n} z_{if} , and, v_i = \sum\nolimits_{j=1}^{m} \lambda_{ji} z_{ji} \right.$$

Then the performance criteria can be written as:

$$w_i = f_i(x, \beta) + z_i + u_1, i = 1, ..., m,$$

$$w_i = f_i(x, \beta) + t_i + v_1, i = m+1, ..., n,$$

where the coalition chooses v_1 and t_1 determined by the actions of other operators. If the coalition is exchanged only with additional interactions between its members, the

$$w_i = f_i(x, \beta) + z_i, i = 1, ..., m;$$

$$w_i = f_i(x, \beta) + t_i, i = m+1, ..., n.$$

Where now:

$$z_i = \left\{ \sum\nolimits_{j=1}^{m} \lambda_{ji} z_{ji} - \sum\nolimits_{j=1}^{m} z_{if} , and, t_i = \sum\nolimits_{j=m+1}^{n} \lambda_{ji} z_{ji} - \sum\nolimits_{j=m+1}^{n} z_{ij} \right. .$$

As traditional research has not led to a manageable and unambiguous guidelines, we shall proceed from the inability to complete the formalization of the problem of rational choice, including the choice of coalitions is now necessary to study the process rather particular form, but the study of the processes of rational choice in which it should be possible exhaustive. In addition, analysis of the question of the benefit of joining the union of different species, taking into account possible changes in the mutual awareness of the players. This analysis may be, will reduce the amount considered coalitions and thus make the task of rational choice more transparent. For all these purposes, it is desirable to create a sufficiently flexible formalized description of the behavior of the coalition, similar to reality and yet is relatively simple (Hubaev, 1999).

It seems that one way of formalizing this is the introduction of the common goals of the coalition, reflecting a compromise between the respective characteristics of the operators. Thus, the coalition turns as if to a single operator.

Of course, the efficiency criterion of the coalition can be anything. However, judging from this, it will be difficult to introduce the study of collective action in any foreseeable limits. It is desirable to limit the kind of reasonable compromise criterion on the basis of common sense and the possibilities of mathematical research. Of course, there should be enough space for informal selection criterion performance compromise.

The strategy of ensuring the guaranteed security and reliability of aviation activity determines a set of approaches, principles and measures that ensure the stable functioning of the blood pressure with the specified safety and reliability indicators. Today, administrative, legal, economic, technical methods and tools, as well as methods and means of risk assessment are advocated. Risk is usually assessed as a combination of the probability of occurrence of a dangerous event and its possible consequences. Recently, risk assessment has become increasingly important in security management. At the same time, the basis of safety management are the following principles (Hillson, 1998; ICAO, 1998):

- Absolute safety does not exist - after acceptance protective.
- Actions there is always some residual risk.
- Safety is reached by decrease in risk to established.
- Admissible level, at the same time the residual risk is below admissible level.
- The admissible risk level is established and corrected at all stages of life cycle of the object or process connected with safety.

Here first of all deserves a special mention of the principle of the guaranteed result that is called operates the party at insufficient knowledge to be based on consideration of the worst possible situations taking into account the available information So widely clear principle of the guaranteed result can be applied also at the choice of rational strategy, and at assessment of the expected result. This principle contains, of course, usual maximized used at antagonistic activity and interaction with environment, but to it is not built at all.

The risk management system is a tool for supporting management decisions based on the assessment of the risks associated with traffic safety. In order to provide a given level of security, risk management should be effective and of a systemic nature. The risk management process can not be carried out without identifying the strategic, tactical and operational objectives of the aviation activity.

Safety management – is the main managerial function, which should be considered at a level that is at least adequate in importance to other functions of any airline, whose implementation should be based on a balanced allocation of resources to production tasks and means of protection, to contribute to the establishment of security space. The essence of the postulate is that after the adoption of protective measures residual risk always remains, which is considered as a measure of probability, as well as the severity of harm to the safe functioning of the system and the environment. Damage – is defined as physical damage or harm associated with the deterioration of a person's health or vital functions, which reduces his or her ability to function normally in terms of physiology. Damage can be caused both directly, and indirectly and qualitatively classified by levels as catastrophic, critical, marginal, insignificant.

Model of rational choice of strategies. Suppose there are n experts who seek to improve the performance criterion $\omega_i = (x_1, \ldots, x_n)$, $i = 1, \ldots, n$ by selecting a vector from the set X_i.

The coincidence of interests $\omega_i=\phi_i(\omega)$, where $\phi_i(t)$ the monotonically increase function, and $\omega=const$.

1. The opposite of interests, n=2, $\omega_i=\phi_i(-\omega_1)$ where the monotonically $\phi_i(t)$ increasing function.
2. The situation of the independent interests, $x_i \in X_i$, $\omega_i=\phi_i(\omega)$

Activity setting performance criteria $\omega_i=F_i(x_1,...,x_n,\beta_i)$, $i=1,...,$n, $\beta_i=B_i$– uncertainty. Vector of controlled factors $X=(x_1,...,x_n)$, the i-th expert imposes restrictions $x_i \in P_i$, for example $\phi_i(t) \geq 0$ and referring on this to choose X_i.

When maintaining restrictions and relevant discontinuous performance criteria (1) and (2) is replaced by activities without restrictions P_i any numbers less $\omega_i=f_i(x_i\beta_i)=-\infty$

$$R_i = \inf_{E \in p_i, \beta_i \in B_i} F_i(x_i, \beta_i)$$

The ability to expand strategies by converting the information obtained from *ikj* via

$$Zij\omega_i f_i - \sum\nolimits_{j=1}^{n} Z_{ij} + \sum\nolimits_{j=1}^{n} \lambda_{ij} Z_{ij}$$

efficiency ratio λ_{ij} determine the value for the i-th expert piece of information is transmitted by j operator, with $i \leq m$:

$$Z_i = \sum_{j=1}^{m} \lambda_{ij} Z_{ij} - \sum_{j=1}^{n} Z_{ij} U_i = \sum_{j=m+1}^{n} \lambda_{ij} Z_{ij},$$

with $i < m$:

$$t_i = \sum_{j=m+1}^{n} \lambda_{ij} Z_{ij} - \sum_{j=1}^{n} X_{ij}; \quad \gamma_i = \sum_{j=1}^{m} \lambda_{ij} Z_{ij}.$$

Then,

$$\omega_i = f_i(x\beta) + Z_i + u_i; i = 1...m;$$

$$\omega_i = f_i(x\beta) + t_i + \gamma_i; i = 1...m+1...n.$$

The assessment of the value of risk is determined on the basis of the calculation of two indicators: the possibility of a risk and the value of damage. When calculating the value of the damage, it is advisable to use a fuzzy-logical approach, which is also distinguished by taking into account the independent factors of damage. To assess the value of the damage caused by the occurrence of risks, a fuzzy production damage model has been developed that is filled with elements extracted from an available knowledge data base and it's based on expert opinion. To reduce redundancy, elements at each level should be grouped according to similar characteristics or ranked by degree of significance (ICAO, 1998).

Direct methods for constructing membership functions suggest that the expert determines the rules for setting the values of the membership function $\mu_A(X)$, describing X element.

He chooses these values on the set of elements X in accordance with the following rules.

1. For $\forall x_A x_2 \in X$ the ratio of their membership functions is determined by the inequality $\mu_A(x_1) < \mu_A(x_2)$ when $x2$ more preferably than $x1$, that is, more described by the property A.
2. For $\forall x_A x_2 \in X$ the ratio of their membership functions is determined by the equation $\mu_A(x_1) = \mu_A(x_2)$ and $x2$ identical with respect to property A.

Traditionally, direct methods for determining membership functions are used for quantitatively measurable variables or in cases where opposite values can be distinguished. A subclass of direct methods is direct group methods that assume that a particular object is presented to a certain expert group and each of the experts need to determine whether the given object belongs to a given set. In this case, the number of positive answers divided by the cumulative number of experts determines the value of the function of the object belonging to a given fuzzy set.

Direct methods also include setting the membership function graphically, as well as in the form of a table, a formula.

Analysis of the literature and the results of various studies, as well as practical tools for solving information processing problems showed that it is advisable to use direct methods if there are guarantees that experts rarely make random errors, and they can be queried about the values of the membership function directly. At the same time, distortions arise in any case, for example, a subjective desire to move the estimates of objects to the extreme points of the scale used. In this regard, direct methods should be applied only in cases where expert errors are unlikely.

Indirect methods for determining the object's membership function are applied in cases where there are no measurable properties that define fuzzy sets and are more difficult to implement in practice, but their advantage lies in their persistence with respect to distortions (Kharchenko, Alexeiev, 2017).

In indirect methods, the membership functions must comply with predetermined conditions that can determine the type of information received and the procedures for its conversion. Indirect methods include the statistical method, pairwise comparisons, expert assessments, and others.

The method of statistical data is based on the processing of statistics. The degree to which an object belongs to a fuzzy set is defined as the estimated value of the frequency of use of a linguistic term describing a fuzzy set to characterize this object. At the same time, the use of specialized hint matrices allows obtaining smooth membership functions.

Membership function $\mu_s(X)$ assigns to each element $x \in X$ a number from interval [0, 1], characterizing the degree of belonging of an element X to the set A. Perceiving information, the expert does not use specific numbers, but converts them into terminological concepts - the values of a specific linguistic variable, which is described by the membership function, individual for each expert.

Let analyzing the state of an object for a certain period of time, n times attention is focused on whether fact A occurred. An event consisting of n checks for the presence of fact A is an estimate. Suppose fact A occurred in the checks. The expert determines the frequency $p = k/n$ of fact A and describes it using the words "often", "rarely" and others.

Estimating the frequency p, the expert takes into account his experience, which determines the frequency of occurrence of fact A in the events of the past that seem to him similar to the event in question. It also receives information based on third-party observations of the occurrence of fact A_t, that is, information reflecting public knowledge. Depending on the level of trust in the source of information, it acquires various weight coefficients in order of importance.

The values of the linguistic variable are determined on the scale [0, 1]. Then the belonging of a specific value to a fuzzy set is calculated as the ratio of the number of experiments in which it occurred in a given interval of the scale to the maximum number of experiments for all considered intervals for the considered value. The statistical method is based on the condition that in each interval of the scale used is an equal number of experiments. This condition is not always respected. In practice, an empirical table is constructed, in which experiments can be unevenly distributed over intervals.

The considered method allows obtaining reliable and adequate estimates of the membership functions in the presence of complete statistical information.

The method of constructing the membership function is based on expert assessments. This method of determining membership functions is to use fuzzy numbers, approximately equal to a clear number, and approximate interval estimates, reflecting expert opinions on the analyzed issue. The task is to find the parameters of a certain exponential function, for the solution of which the results of an expert survey are applied.

When constructing the membership function using the parametric method, modified fuzzy terms are constructed based on the existing ones.

The described method of obtaining membership functions was created on the basis of the assumption that an expert, describing the linguistic value of any attribute, can determine three points of the universal scale with minimal cost. Such as: A, B, C, two of which - points B and C – also (or) do not belong to the linguistic meaning being described, A is a point definitely belonging to it.

The sequence of construction of membership functions:

1. On the line $\mu=1$, a point is marked opposite to the value of the carrier $x1$, which definitely belongs to this term (points A and D).
2. On the line $\mu=0$, points are marked opposite the nearest values to the left and right of $x1$ on the carrier (X axis), which definitely do not belong to this term (points B, C, and E).
3. The points marked on the straight lines $\mu=1$ and $\mu=0$ are connected by straight line segments.
4. The shading marks the part of the carrier related to the constructed description.

THE METHOD OF CONSTRUCTING MEMBERSHIP FUNCTIONS USING INTERVAL ESTIMATES

This method is used for the formalized description of selection problems in which there is no distinction between the admissible and the unacceptable, as well as between the ideal and non-ideal states.

The choice of a specific method for constructing membership functions is determined by the class of the problem being solved, the difficulty of acquiring expert and statistical information, the reliability of the data obtained, and the complexity of the algorithms for analyzing and processing information when determining membership functions.

Formation of the production knowledge base describing the influence of the states of the ancestor nodes of the graph on the value of the descendant node specified as a set of production rules.

Obtaining the final damage value of each type of information risk based on the construction of hierarchical systems of fuzzy-logical inference with fuzzy specified input variables based on the fuzzy implication of Larsen and the composition max-prod.

Fuzzy inference is carried out using Larsen's fuzzy implication according to the formula (Hubaev, 1998):

$$\mu_R(x,y) = \mu_A(x)\mu_B(y).$$

As a convolution operation, the multiplication operation is used. Thus, the basis of fuzzy inference is the use of *max-prod composition*

$$\mu_B(y) = \max_{x \in X}\left\{\mu_A(X)\mu_R(x,y)\right\}.$$

The selection of this compositional rule is due to the ease of implementation and great sensitivity to changes in input variables in the premises of fuzzy production rules.

Reduction to clarity (defuzzification) is to convert the fuzzy values of the output variables found in clear. Moreover, all methods for obtaining a clear value of the output variable can be divided into two groups:

- Methods of defuzzification accumulated at the previous stage (from the activated conclusions of all the rules of the base) of the output variable.
- Methods for defuzzifying the output variable without first accumulating the conclusions of the rules.

The first group includes the following defuzzification methods (Hubaev, 1998; 1999):

1. Center of gravity. This defuzzification method can be used only for models based on fuzzy linguistic production rules, in which the sequences are fuzzy statements. A clear value of **y'** output variable is defined as the center of gravity of the obtained membership function and is calculated by the formula:

$$y' = \frac{\displaystyle\int_{y_{\min}}^{y_{\max}} y\mu_B(y)dy}{\displaystyle\int_{y_{\min}}^{y_{\max}} \mu_B(y)dy},$$

where $Y_{\min}$ and Y_{mac} – the bounds of the carrier interval of a fuzzy set of the output variable **y**.

2. Center Square. The clear value of the output variable *y'* is determined by this method from the equation:

$$\underbrace{\int_{y_{min}}^{y} \mu_B(y)dy}_{S_1} = \underbrace{\int_{y}^{y_{max}} \mu_B(y)dy}_{S_2}.$$

3. Maximum membership function. A clear value of the output variable is calculated by the formula:

$$y' = \arg_y \sup(\mu_B(y)),$$

where $\mu_B(y)$– unimodal function, the shape of which can be arbitrary.

4. The first maximum, also called the left maximum. A clear value of *y'* is defined as the smallest value at which the maximum of the total fuzzy set is reached:

$$y' = \min\left\{y_{max} \vdots \mu_B y_{max}) = \max_y \vdots \mu_B(y)\right\}.$$

The rightmost maximum. The clear value of the variable *y* is found as the largest value at which the maximum of the total fuzzy set is reached:

$$y' = max\left\{y_{max} \vdots \mu_B y_{max}) = \max_y \vdots \mu_B(y)\right\}.$$

THE MODEL FOR PROJECT RISK MANAGEMENT

The model for project risk management is based on the use of fuzzy production networks and an unclear cause-and-effect relationship between the antecedent and the consecvet is given in the form of fuzzy products. To build fuzzy risk management models, we will form a set that characterizes risk indicators $-\Omega=\{\Omega_i\}$, $i=1,n$, and the set that characterizes risk response methods $\Psi=\{\Psi_j\}$, $j=1,m$,. When introducing input linguistic variables characterizing risks in the process of aviation activity Θ, we introduce the following sets of terms that determine the levels of risk indicators: m_1- low probability of risk; m_2 - very low probability of risk; m_3 - average probability of risk; m_4- high probability of risk; m_5 - very high probability of risk.

The following algorithm for obtaining expert evaluations for the calculation of risk inherent in the organization and technology of the accounting process is proposed:

1) Conducting an audit procedure: the direction of a written request to an audited person.
2) The creation of an audit team (experts) for the provision of expert assessments and discussion of issues related to the BP.

3) Processing the results of the request by experts - in the auditor's working paper, a special column should be provided to reflect the opinion of the auditor on the degree of impact of the indicator on the likelihood of significant distortions.

4) Passing several evaluation rounds for greater consistency.

5) Calculation of the level of risk.

To determine the relationship (consistency) of the experts' evaluations in the special literature, the coefficient of Kendal's concordance, calculated by the following formulas, is used:

The value of the coefficient varies in the range from 0 to 1. Its equality of the unit means that all experts assigned the objects the same ranges, shows a high degree of consistency in the estimates. The closer the value of the coefficient to zero, the less concurred are expert assessments (ICAO, 2002).

If the calculated coefficient of concordation, according to expert estimates, is close to zero, then the second and subsequent assessment tours must be held to obtain maximum agreement.

In our opinion, the expert estimation method for determining the level of planned audit risk and its components takes place in the audit organizations when the opinion of the majority of employees is taken into account, including the appropriateness of continuing the audit, as the results of the audit organization and the risk of audit activity depend on the findings found. In Figure 1 showed Fuzzy production network for safety management system of aviation activity. Table 1 shows characteristics of the appearance of risk and Type of thermo-sets and action characteristics which are constituents of Fuzzy production network for safety management system of aviation activity.

Figure 1. Fuzzy production network for safety management system of aviation activity

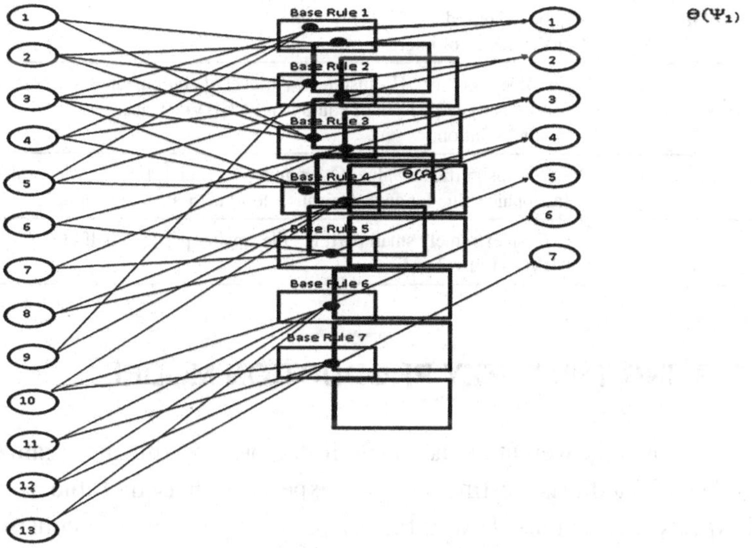

Table 1. Characteristics of the appearance of risk

Definition	Characteristics of the Appearance of Risk
$\Theta(\Omega_1)$:Achieving the goal of security	The goal is not in full compliance with the objectives of the organization.
$\Theta(\Omega_2)$:Administrative	Unskilled management decisions.
$\Theta(\Omega_3)$:Volume, intensity	It is not possible to analyze indetail eachstageofthework, toensuretheinteractionofparticipantsandtheorganizationofworks.
$\Theta(\Omega_4)$:Interaction with the operator	The issue of harmonization of procedures.
$\Theta(\Omega_5)$:Competence of staff	There are no specialists with the necessary competence.
$\Theta(\Omega_6)$:New technology	Problems in using new technologies.
$\Theta(\Omega_7)$:Architectural	Architecture does not provide sustainability of design solutions, which manifests itself in adaptation to possible changes.
$\Theta(\Omega_8)$:Technical	It is difficult to implement the technical requirements for the project.
$\Theta(\Omega_9)$:Policy and standard	Not paying attention to security policies and standards.
$\Theta(\Omega_{10})$:Productivity	The risk of ensuring proper production performance.
$\Theta(\Omega_{11})$:Accessibility	The risk of ensuring the proper availability of production.
$\Theta(\Omega_{12})$:Reliability	The risk of ensuring the proper reliability of production.
$\Theta(\Omega_{13})$:Safety	The risk of ensuring proper production safety.
$\Theta(\Omega_{14})$:Software-hardware level of protection	H - satisfactory, to provide the initial level of protection; Su - sufficient for basic information protection; B - corresponds to the level of confidentiality of information.
$\Theta(\Omega_{15})$:Level of organizational protection	H - poor planning and lack of risk monitoring; C - planning and monitoring of risks is carried out irregularly; B - Timely planning and monitoring of vulnerabilities.
$\Theta(\Omega_{16})$:Level of legal protection	H - incomplete documentation; C - insufficient detailed documentation; In - the documentation is complete and synchronized.
$\Theta(\Omega_{17})$:Motivating sources of threats	$OCHN$ - absent; H - rare manifestation of interest; C - in full interest; In - almost everything, will be interested. OchV - will be of interest.
$\Theta(\Omega_{18})$:Possibility of sources of threats	OnH - does not have; H - insignificant level of equipment; C - average level of equipment; B - a sufficiently large level of equipment; OCHV - IU has significant capabilities of T5. OCH - openinformation.
$\Theta(\Omega_{19})$:Market value of information resource	H - IR is insignificant value; C - commercial secret; B - highly confidential data; Objective - catastrophic value for an organization-level of strategic planning.
$\Theta(\Omega_{20})$:Volume of data in the IP organization	OCH - an extremely small part; H - the smaller part; C - half of the IR; In - a large part of OCHV - a full volume.

RESULTS OF TESTING THE FUZZY PRODUCTION MODEL

For an example of determining the weighting factors of the factors affecting the organization's risk, which generally affects the level of audit risk, estimates of 15 experts were used (Table 3). The verification of the consistency of experts' evaluations showed that there is a high degree of consistency (0.67), so no additional review is required.

The advantage of the method of expert assessments is that its mechanism is easy to implement in the presence of a special graph in the working papers of the auditor, which is marked with the presence and application of the conditions for the formation of accounting information, and the automation of the processing of expert evaluations, in particular using Microsoft Excel spreadsheets.

Table 2. Type of thermo-sets and action characteristics

Definition	Type of Thermo-sets and Action Characteristics
$\Theta(\Psi_1)$:Project plan	H- to conduct additional study of the subject area of the project; V- to conduct the ranking (prioritization) of the requirements with the involvement of external experts; C - to determine the minimum set of functions of the system; B - to attract additional resources of the customer for the analysis of requirements; OV - to conduct an external audit of the system requirements.
$\Theta(\Psi_2)$:Project management	HE - develop a system for increasing motivation of staff; H - to conduct detailed stages and stages of the project; C - appoint a responsible person for quality control of the project; B - define the standard profile used in the project (ISO 9001, TickIT, SEISW-CMM); OV - involve an independent project manager.
$\Theta(\Psi_3)$:Specification	HE - to perform more detailed detailing of processes; H - to introduce less complicated processes; WITH- executereserveplanning, insurance of the most vulnerable zones of the project; B - refuse the implementation of some functions of the system; OV - to develop an alternative scenario for project implementation.
$\Theta(\Psi_4)$:Suggestion	VI - to involve experts-consultants in the field; H - to attract employees from the customer to the project team; C - to conduct an audit of architecture and design decisions; In - use outsourcing of individual parts of the project (with the lack of necessary specialists); OV - to conduct an independent review of the whole
$\Theta(\Psi_5)$:Developed	VI - to conduct a detailed documentation and independent review of the IP code; N - to attract additional employees to the group of developers; C - use proven technology / platform; In - apply advanced prototyping of software solutions; OV - to hold
$\Theta(\Psi_6)$:Testing	VI - to increase the degree of coverage of test scenarios of functionality of IP; N - to develop the most probable test scenarios; C - Develop automated user interface tests; B - Deliver test intermediate versions to end users for their evaluation; OV is to create a safety assessment team
$\Theta(\Psi_7)$:Introduction	HE - to refine (detail) the user's manual; H - to conduct additional training of end users; C - involve experts in quality control of the implemented project; In - create a feedback system; OV - to improve

Path to construct predictive values. Using the experience of previous audits of a particular audited entity can save time and reduce the complexity of work, in particular when calculating the planned audit risk. The auditor can calculate the forecast value of risk with available analytical data from past verifications. This is easily realized through trend analysis and the construction of a trend line of electronic spreadsheets.

To build fuzzy risk management models, we will create a set that characterizes risk indicators $\Omega=\{\Omega_i\}$, $i=1,n$ - and the set that characterizes risk response methods. $\Psi=\{\Psi_j\}$, $j=1,m$.

When introducing input linguistic variables characterizing risks in the process of aviation activities Θ we introduce the following sets of terms that determine the levels of risk indicators: m1- low probability of risk; m2 - very low probability of risk; m_3 - average probability of risk; m_4- high probability of risk; m_5 - very high probability of risk. On the Table 3 presents the results of testing the fuzzy production model that was conducted on the basis of the aviation administration of Ukraine. 15 experts in the field of aviation audit and accident investigation took part in the work.

The coefficient of concordance is 0.665696429 - the greatest coherence of action. During processing, information is given on the damage of the trend line of the 2nd degree polynomial dependence $y = -0,8052x_2 + 1,3442x_2 + 0,2292$, where x is the predicted value of the risk of primary, current or final accounting; y is the predictive value of the technology risk. At the same time, the determination coefficient was RI = 0,996, which indicates a high degree of reliability of the constructed model. In the prediction of the value, for example, the level of initial accountability risk equal to 0.2, in the formula, the auditor may provoke the level of technology efficiency for the future audit. $= 0,8052 \bullet 0,22 + 1,3442 \bullet 0,2 + 0,2292 = 0.465832$.

Table 3. The results of expert

Factor	Expert Evaluation											Sum of Ranks	Deviation From the Average	Quadratic Deviation	Average Rating of Experts
	1	2	3	4	5	6	7	8	9	...	15				
$\Theta(\Omega_1)$	3	2	1	2	1	1	1	2	1		2	25	63	3,969	1,67
$\Theta(\Omega_2)$	10	7	9	9	8	8	7	5	9		8	122	34	1,56	8,13
$\Theta(\Omega_3)$	5	7	9	6	9	8	7	8	8		9	122	34	1,56	8,13
$\Theta(\Omega_4)$	8	9	8	9	8	9	9	8	10		9	133	45	2,025	8,87
$\Theta(\Omega_5)$	8	9	7	8	6	7	10	9	6		9	116	28	784	7,73
$\Theta(\Omega_6)$	7	9	9	8	7	9	8	8	10		8	124	36	1,296	8,27
$\Theta(\Omega_7)$	10	9	9	9	9	9	10	9	10		10	145	57	3,249	9,67
$\Theta(\Omega_8)$	1	2	2	2	1	2	1	2	2		25	-63	25	-63	3,969
$\Theta(\Omega_9)$	1	2	1	1	1	2	1	2	2		1	21	-67	4,489	1,40
$\Theta(\Omega_{10})$	8	6	8	7	10	6	6	8	6		8	115	27	729	7,67
$\Theta(\Omega_{11})$	1	2	2	1	1	2	2	2	1		1	22	-66	4,356	1,47
$\Theta(\Omega_{12})$	8	9	7	8	6	7	10	9	6		9	116	28	784	7,73
$\Theta(\Omega_{13})$	7	9	9	8	7	9	8	8	10		8	124	36	1,296	8,27
$\Theta(\Omega_{14})$	10	9	9	9	9	9	10	9	10		10	145	57	3,249	9,67
$\Theta(\Omega_{15})$	1	2	2	2	1	2	1	2	2		25	-63	25	-63	3,969
$\Theta(\Omega_{16})$	1	2	1	1	1	2	1	2	2		1	21	-67	4,489	1,40
$\Theta(\Omega_{17})$	8	6	8	7	10	6	6	8	6		8	115	27	729	7,67
$\Theta(\Omega_{18})$	8	9	7	8	6	7	10	9	6		9	116	28	784	7,73
$\Theta(\Omega_{19})$	7	9	10	8	7	10	8	8	10		8	124	36	1,296	8,27
$\Theta(\Omega_{20})$	10	9	10	9	10	10	10	9	10		10	145	57	3,249	9,67
$\Theta(\Psi_1)$	1	2	2	2	1	2	1	2	2		25	-63	25	-63	3,969
$\Theta(\Psi_2)$	1	2	1	1	1	2	1	2	2		1	21	-67	4,489	1,40
$\Theta(\Psi_3)$	8	6	8	7	10	6	6	8	6		8	115	27	729	7,67
$\Theta(\Psi_4)$	7	9	9	8	7	9	8	8	10		8	124	36	1,296	8,27
$\Theta(\Psi_5)$	10	9	9	9	9	9	10	9	10		10	145	57	3,249	9,67
$\Theta(\Psi_6)$	1	2	2	2	1	2	1	2	2		25	-63	25	-63	3,969
$\Theta(\Psi_7)$	7	9	9	8	7	9	8	8	10		8	124	36	1,296	8,27

Table 4. Level of risk

Indicator	2016	2017	2018
General risk of accounting technology	0,79	0,7	0,56
The risk of primary accounting	0,85	0,5	0,3
Risk of current accounting	0,82	0,65	0,5
The risk is the total of the accounting	0,67	0,45	0,34

Table 4 presents safety performance results for 2016, 2017, 2018 which have also been tested using the fuzzy production flight management model.

Thus, the risk level of accounting technology for the future audit will be 46.6%.

FUTURE RESEARCH DIRECTIONS

Taking into account the statistics for the last decade on the incident and accident, the issues of provision of guaranteed level of flight safety are the most urgent, since the disadvantages and problems in the functioning of flight safety are due to the lack of the general theoretical basis and generally accepted scientifically sound approaches to the flight safety, the development of which should be guided by the requirement of ICAO, which determines that none the region should not have an incident and accident level more than twice that of the world. These are the following main areas:

- introduction of an acceptable level of flight safety in the state;
- mandatory procedures for the development and implementation of SMS;
- mandatory procedures for ensuring the direct control of the level of the flight safety within an acceptable or established level (continuous monitoring and regular assessment of the flight safety, corrective actions required to maintain agreed flight safety indicators, flight safety monitoring, flight information analysis, risk management of the incident and accident, etc.).

The similarity of the nature of the emergence of risks and the increasing relevance of their reduction to an acceptable level for various critical applications determines the relevance of the establishment of a methodology to ensure and maintain a guaranteed level of safety of future flights.

The purpose of the methodology is to integrate into a single set of tasks the assessment, provision and verification of the security of blood pressure, as a complex hierarchical structure with independent critical elements, as well as hardware, software, network and ergo components, which are both a means and an object of safety.

Implementation of the guaranteed result is to implement the management processes in such a way as to prevent the transition of the infrastructure or its systems to a potentially hazardous state and to ensure the blocking (deletion) of the relevant technical object in the event of a threat of transition or when the transition to a dangerous state and minimization of the consequences of such a transition .

CONCLUSION

Total actual damage due to accidents and other adverse events is determined by the sum of these components as a consequence of the damage of each individual event, and taking into account the real damage for a certain period. To correctly predict the losses need to evaluate two factors determines their value: the average value of the expected losses in the event of an accident or an event and the probability of an accident or event.

In some cases, implemented in the area of guaranteed safety of air navigation services investment project is aimed at reducing the incidence of accidents and events. In this embodiment, a complex calculation that takes into account all types of losses should be performed. Building a risk management system in the area of guaranteed safety and reliability of air navigation services, it is necessary to provide:

- Full and timely implementation of measures aimed at achieving the strategic security objectives.
- Optimal use of resources allocated to the investment; obtaining additional effect due to optimal matching of mutual investment projects implemented, including their location and the time of implementation.
- More efficient use of technical means used and the optimal use of the results of projects implemented in previous periods.

On the basis of the strategic objectives in the field of safety and reliability of air navigation services need to solve a number of the following tasks:

1. Objectives of the formation and perfection of normative-methodical safety and reliability management database.
    1.1. The revision and updating of the existing regulatory framework and its harmonization with international standards.
    1.2. Development of normative-methodical documents aimed at improvement of safety management practices.
    1.3. Development and implementation of risk management practices related to safety, and the development of a safety management system.
    1.4. The development of guidelines and training material for the development and assessment of safety culture.
2. Challenges for the development of technical and technological base.
    2.1. Development and implementation of measures to upgrade the technical technological base associated with safety and reliability of air navigation services.
    2.2. Conducting periodic analysis of efficiency of the use of technology and the results of the ongoing scientific and technical work.
3. Challenges for the development of human resource capacity in the management of safety and reliability of the transportation process.
    3.1. Improving personnel management system relating to safety and reliability.
    3.2. Adaptation of vocational training to the changing technological requirements.
    3.3. Organization of training processes and risk management practices related to safety.
    3.4. Organization of training personnel management practices, risk and reliability in the stages of the life cycle of aviation operations.
4. Challenges for the development of information technology to ensure the safety and reliability of the transportation process.
    4.1. Creation of information decision support systems to ensure the safety and reliability.
    4.2. Implementation of an automated knowledge testing system in terms of safety requirements.
    4.3. Development and implementation of automated management systems, risk and reliability in the stages of the life cycle of the air navigation system.

4.4. Improving automated systems to ensure safety performance monitoring functions of technological safety processes.

4.5. Improvement and development of situational monitoring and control safety and reliability of
air navigation services, taking into account the existing political situation.

4.6. Improve the recording and investigation of accidents and liability allocation rules.

4.7. Development of methods for identifying causal relationships safety violations.

The assessment of the probability of information risk is traditionally based on the use of a statistical approach, of which Bayesian networks are an effective tool. However, the probabilistic approach to
accounting for uncertainty requires a large amount of statistical information presented in the form of
tables of conditional probabilities, which cannot be constructed for the described factors. In this case,
the use of qualitative characteristics or interval values and the approach based on fuzzy logic and the
related theory of possibility is more reasonable.

The assessment of information risk is determined on the basis of the calculation of two indicators:
the possibility of a risk and the amount of damage. When calculating the level of the damage, it is advisable to use a fuzzy-logical approach, which is also distinguished by taking into account the independent
factors of damage.

REFERENCES

Aviation Accident Statistics. (2018). *National Transportation Safety Board*. Retrieved from www.ntsb.
gov/aviation/ aviation.htm

Commission Implementing Regulation (EU). (2013). On the definition of common projects, the es-
tablishment of governance and the identification of incentives supporting the implementation of the
European Air Traffic Management Master Plan No 409/2013. *Official Journal of the European Union,
No. 4.5.2013*, 123/1-123/7.

Commission Implementing Regulation (EU). (2014). On the establishment of the Pilot Common Project
supporting the implementation of the European Air Traffic Management Master Plan No. 716/2014.
Official Journal of the European Union, No. 28.6.2014, 190/19-190/44.

Eddous, M., & Stansfield, R. (1997). Methods of decision-making. *Audit*, 455-459.

EUROCONTROL. (2004). Single European Sky first legislative package (SES I). Retrieved from https://
www.skybrary.aero/index.php/Single_European_Sky_(SES)

Ferat, S. A. (2000). Baesian network approach to the self-organization and learning in intelligent agents.
(Doctoral dissertation), Virginia Polytechnic and State University.

Heckerman, D. (1995). A tutorial on learning with Bayesian netwoks. USA: Technical report MSR -TR-
95-06 - Microsoft Research, Redmond, WA.

Hillson, D. (2003). Using a risk breakdown structure in project management. *Journal of Facilities Management*, 2(1), 77–85. doi:10.1108/14725960410808131

Hubaev, G. N. (1999). Complex systems: Expert methods of comparison. *North Caucasian region.* [H]. *Social Sciences*, 7–24.

Hubaev, G. N. (1999). Statistical methods of planning of extremal experiments with information systems. *Statistics Issues*, 78-83.

International Civil Aviation Organization. (1998). Human factors training manual (1st ed.). Doc. ICAO 9683-AN/950. Canada, Montreal: Author.

International Civil Aviation Organization. (2002). Human factors guidelines for safety audits manual (1st ed.). Doc. ICAO 9806-AN/763. Canada, Montreal: Author.

International Civil Aviation Organization. (2004). *Cross-cultural factors in aviation safety: Human factors digest Nº 16. Circ. ICAO 302-AN/175*. Montreal, Canada: Author.

International Civil Aviation Organization. (2013a). Safety management manual (SMM) (3rd. ed.). Doc. ICAO 9859-AN 474. Canada, Montreal: Author.

International Civil Aviation Organization. (2013b). *State of global aviation safety*. Montreal, Canada: Author.

International Civil Aviation Organization. (2014). *Safety report*. Montreal, Canada: Author.

Jager, R. R. (1986). The sets of the level for the evaluation of the membership of fuzzy subsets. Radio and Communications, 71-78.

Kharchenko, V. P., Alexeiev, O. M., & Babeichuk, D. G. (2010). Method analysis of management decisions making while air navigation functioning in emergency situations. *Proceedings of the National Aviation University*, 19-24. 10.18372/2306-1472.44.1901

Kharchenko, V. P., Alexeiev, O. M., & Kolesnyk, T. A. (2018). Ensuring the guaranteed level of flights safety – the view of the future. *Proceedings of the National Aviation University*, 9-18. 10.18372/2306-1472.4.13492

Kharchenko, V. P, Alexeiev, O. M., Luppo, A. E., & Jurchik, R. I., (2017). The principles to maintain an acceptable of air navigation safety in Ukraine. *Proceedings of Kharkiv Air Force University*, 17-25.

Kulik, N. (2008). *Encyclopedia of aviation safety*. Kiev, Ukraine: Technics.

Raz, T., & Michael, E. (2001). Use and benefits of tools for project risk management. *International Journal of Project Management*, *19*(1), 9–17. doi:10.1016/S0263-7863(99)00036-8

Simonov, C. B. (2003). *Modern concepts of information risk management. Rostov-na-Donu*. Southern Federal University.

Smolyak, S. A. (1992). Risk accounting when setting the discount rate. *Economics and mathematical methods*, 23-27.

The World Bank of Data. (2016). Air transport, registered carrier departures worldwide. Registered carrier departures worldwide are domestic takeoffs and takeoffs abroad of air carriers registered in the country, 8-10.

ubaev, G. N. (1998). Comparison of complex software systems by the criterion of functional completeness. *Software products and systems (SOFTWARE & SYSTEMS)*, 7-9.

Zadeh, L. A. (2002). Toward a perception-based theory of probabilistic reasoning with imprecise probabilities. *Journal of Statistical Planning and Inference*, 230–245.

KEY TERMS AND DEFINITIONS

Air Navigation System: The complex of organizations, personnel, infrastructure, technical equipment, procedures, rules and information that is used to provide of airspace users of safe, regular and efficient air navigation service.

Flight Safety: The state of the aviation system or organization in which the risks associated with aviation activities related to the operation of aircraft or directly providing such operation are reduced to an acceptable level and monitored.

ICAO: International Civil Aviation Organization.

Risk: The probability of a hazard escaping from under control and the severity of the consequences, expressed by the degree of manifestation.

SMS: Safety Management System.

Compilation of References

Abdulov, R. N., Bayramov, A. A., & Hashimov, E. G. (2019). Providing of effective operation of reconnaissance UAVs during analysis and control high dynamic scene. *National Security and Military Sciences*, *5*(3), 27–34.

Adams, D. (2007). *Health monitoring of structural materials and components: methods with applications* . John Wiley & Sons. doi:10.1002/9780470511589

AdditionalREADING

Agterberg, F. P., Gaile, G. L., Willmott, C. J., Reidel, D., & Cliff, A. (2002). *Trend surface analysis. Spatial statistics and models. Spatial autocorrelation* (pp. 147–171). London, UK: Pion.

Aheikin, D. I., Kostina, E. N., & Kuznetsov, N. N. (1965). Sensors for control and regulation. Moscow, Russia. *Mechanical Engineering (New York, N.Y.)*, 496–508.

Ahmed, I., Ikhlef, A., Ng, D. W. K., & Schober, R. (2013). Power allocation for an energy harvesting transmitter with hybrid energy sources. *IEEE Transactions on Vehicular Technology*, *12*, 6255–6267.

AIS of Ukraine. (n.d.). Retrieved from http://www.aisukraine.net/titul_en.php

Alekseev, Yu. S. Yu. M., Zlatkina, V. S., Krivtsov, A. S., Kulik, V. I., Chumachenko, et al. (2012). Designing control systems for rocket and space technology objects: In 3 volumes. Kharkiv, Ukraine: Nat. Aerospace. un-t "Kharkov. Aviation in-t", SPE Hartron-ARKOS, 2012.

Amiouny, S. V., Bartholdi, J. J. III, Vande Vate, J. H., & Zhang, J. (1992). Balance loading. *Operations Research*, *40*(2), 238–246. doi:10.1287/opre.40.2.238

AN-Conf/13-WP/3. (2009). *The Thirteenth Air Navigation Conference of ICAO*. Montreal, Canada, Oct. 9-19, 2018.

Anderson, E. W. (1994). Cross-category variation in customer satisfaction and retention. *Marketing Letters*, *5*(1), 19–30. doi:10.1007/BF00993955

Annex 3. (2016). *Meteorological service for international air navigation. Nineteenth edition*. Montreal, Canada: ICAO, 2016.

Antononenko, A. M., Kudzin, A. Y., & Gavshin, M. G. (2009). Influence of the domain structure on the electromechanical properties of ferroelectric ceramics of the TsTs and MNWT [-Moscow, Russia: FTI them. A. F. Ioffe.]. *Physics of the Solid State*, *39*(5). doi:10.1134/1.1129978

Architecture Capability Team. (2018a). *NATO Architecture Framework Version 4*.

Ashchepkov, L. T. (1987). *Optimal control of discontinuous systems*. Novosibirsk, Russia: Nauka.

Asmus, V. F. (2001). Logic: textbook. Ed. 2nd, stereotypical. Moscow, Russia: Editorial URSS.

Astrom, K. J., Klein, R. E., & Lennartsson, A. (2005). Bicycle dynamics and control. *IEEE Control Systems Magazine*, *25*(4), 26–47. doi:10.1109/MCS.2005.1499389

Atamanyan, E. G.(1989). Instruments and methods for measuring electrical quantities. Moscow, Russia: Higher. Shk., 134-136.

August G., Roesener, J., & Barnes, W. (2016). An advanced tabu search approach to the dynamic airlift loading problem. *Logistics Research, (1)*, 1–18. doi:. doi:10.100712159-016-0139-6

Aviation Accident Statistics. (2017). *National Transportation Safety Board*. Retrieved from www.ntsb.gov/aviation/aviation.htm

Aviation Accident Statistics. (2018). *National Transportation Safety Board*. Retrieved from www.ntsb.gov/aviation/aviation.htm

Aviation Safety Network. (2010). *Final investigation report*. Accident of Sunway Air Carrier (Georgia) IL-76TD Aircraft REG # 4L-GNI at Karachi, Pakistan on Nov. 27, 2010. Retrieved from https://www.reports.aviation-safety.net/2010/20101128-0_IL76_4L-GNI.pdf

Aviation Safety Network. (2018). *ASN data show 2017 was safest year in aviation history*. Retrieved from https://news.aviation-safety.net/2017/12/30/preliminary-asn-data-show-2017-safest-year-aviation-history/

Aviation Safety Network. (2019a). *ASN Wikibase*. Retrieved from https://www.aviation-safety.net/wikibase/

Aviation Safety Network. (2019b). *Aviation Safety Network releases 2018 airliner accident statistics*. Retrieved from https://news.aviation-safety.net/2019/01/01/aviation-safety-network-releases-2018-airliner-accident-statistics/

Avyrom, L. S. (2001). Design reliability of prefabricated buildings and structures. London, UK: Publishing House on Construction.

Babak, V., Kharchenko, V., & Vasylyev, V. (2006). Methods of conflict probability estimation and decision making for air traffic management. *Aviation*, *10*(1), 3–9. doi:10.3846/16487788.2006.9635920

Babich, V., Dovbenko, V., Kuzmych, L., & Dovbenko, T. (2017). Estimation of flexures of the reinforced concrete elements according to the National Ukrainian & European standards. MATEC Web of Conferences, Vol. 116, July 10, 2017, Article number 02005.

Bader, M. J. (1995). Images in weather forecasting. New York, NY: Cambridge University Press.

Bakker, G. J., Kremer, H. J., & Blom, H. A. P. (2001). Probabilistic approaches toward conflict prediction. In G. L. Donohue & A. G. Zellweger (Eds.), *Air Transportation Systems Engineering* (pp. 677–694). American Institute of Aeronautics and Astronautics, Inc.

Balandin, D, Kogan, M. (2007) Synthesis of optimal linear-quadratic matrix non-equalities. *Automation and Telemechanics*, 3, 3–18. (in Russian)

Balas, G. J., Chiang, R. Y., Packard, A., & Safonov, M. G. (2010). *Robust control toolbox 3*. User's guide. The Math Works Inc.

Ball, J. E., Anderson, D. T., & Chan, C. S. (2017). Comprehensive survey of deep learning in remote sensing: Theories, tools, and challenges for the community. *Journal of Applied Remote Sensing, 11*(4), 042609. doi:10.1117/1.JRS.11.042609

Bardish, B., & Burshtinska, H. (2014). Use of vegetation indexes for the identification of terrestrial objects. Modern achievements in geodetic science and production (2), 82-88.

Barlow, R. E., & Proschan, F. (1965). *Mathematical theory of reliability*. New York: John Wiley & Sons.

Barsky, A. (2013). Logical Neural Networks. Moscow, Russia: Internet University of Information Technologies (INTUIT).

Barsky, A. (2007). *Logical Neural Networks: Internet University of Information Technologies, Bean*. Laboratory of Knowledge.

Baxter, G., & Sommerville, I. (2011). Socio-technical systems: From design methods to systems engineering. *Interacting with Computers*, *23*(1), 4–17. doi:10.1016/j.intcom.2010.07.003

Bayramov, A. A., Hashimov, E. G., Hasanov, A. H., Pashayev, A. B., & Sabziev, E. N. (2018). SMART control system of systems for dynamic objects group. In Proceedings of Military Academy after "Георги Стойков Раковски" on National Security (II part. pp. 121-123). Sofia, Military Academy Press, Bulgaria.

Bayramov, A. A., Hashimov, E. H., & Nasibov, Y. A. (2019). Unmanned aerial vehicles application for military GIS tasks solution. In S. Tetiana, Y. Sikirda, N. Rizun, D. Kucherov, & K. Dergachov (Eds.), Automated systems in the aviation and aerospace industries (pp. 273–296). Hershey, PA: IGI Global. doi:10.4018/978-1-5225-7709-6.ch010

Bayramov, A. A., & Hashimov, E. G. (2017). The numerical estimation method of a task success of UAV reconnaissance flight in mountainous battle condition. *Advanced Information Systems*, *1*(2), 70–73. doi:10.20998/2522-9052.2017.2.12

Bayramov, A. A., & Hashimov, E. G. (2018). Assessment of invisible areas and military objects in mountainous terrain. *Defence Science Journal*, *68*(4), 343–346. doi:10.14429/dsj.68.11623

Bellman, R. (1957). *Dynamic programming*. Princeton, NJ: Princeton University Press.

Bellman, R. (1964). *Control processes with adaptation*. Moscow, Russia: Nauka.

Bellman, R. (1966). Dynamic programming. *Science*, *153*(3731), 34–37. doi:10.1126cience.153.3731.34 PMID:17730601

Bertsch, V., Treitz, M., Geldermann, J., & Rentz, O. (2007). Sensitivity analyses in multi-attribute decision support for off-site nuclear emergency and recovery management. *International Journal of Energy Sector Management*, *1*(4), 342–365. doi:10.1108/17506220710836075

Bianco, L., Dell'Olmo, P., & Giordani, S. (2006). Scheduling models for air traffic control in terminal areas. *Journal of Scheduling*, *9*(3), 223–253. doi:10.100710951-006-6779-7

Bicchi, A., & Pallottino, L. (2000). On optimal cooperative conflict resolution for air traffic management systems. *IEEE Transactions on Intelligent Transportation Systems*, *1*(4), 221–232. doi:10.1109/6979.898228

Biondini, F., & Frangopol, D. (Eds.). (2008). *Life-cycle civil engineering* (p. 970). London, UK: Taylor & Francis Group. doi:10.1201/9780203885307

Blin, K., Akian, M., Bonnans, F., Hoffman, E., Martini, C., & Zenghal, K. A. (2000). *A stochastic conflict detection model revisited*. AIAA Guidance, Navigation, and Control Conference and Exhibit. doi:10.2514/6.2000-4270

Blom, H. A. P., Bakker, G. J., Blanker, P. J. G., Daams, J., Everdij, M. H. C., & Klompstra, M. B. (2001). *Accident risk assessment for advanced air traffic management. Air Transportation Systems Engineering* (pp. 463–480). American Institute of Aeronautics and Astronautics, Inc.

Blutner, R., & Hochnadel, E. (2010). Two qubits for C.G. Jung's theory of personality. *Cognitive Systems Research*, *11*(3), 243–259. doi:10.1016/j.cogsys.2009.12.002

Bogomaz, G., Naumenko, N. & Sobolevskaya, M (2005). Dynamics of launch of spacecraft liquid launch vehicles. Proekt Naukova kniga.

Bogush, M. V. (2006). *Piezoelectric sensors for extreme operating conditions. T. 3* (pp. 246–248). Rostov-on-Don, Russia: Publishing House SKNTS VS.

Booker, L. (1987). Improving search in genetic algorithms. In *Genetic Algorithms and Simulated Annealing*. Morgan Kaufmann Publishers.

Bortfeldt, A., & Gehring, H. (2001). A hybrid genetic algorithm for the container loading problem. *European Journal of Operational Research, 131*(1), 143–161. doi:10.1016/S0377-2217(00)00055-2

Bortfeldt, A., & Mack, D. (2007). A heuristic for the three-dimensional strip packing problem. *European Journal of Operational Research, 183*(3), 1267–1279. doi:10.1016/j.ejor.2005.07.031

Boss, V. (2016). *Optimal control.* Palmarium Academic Publishing.

Boudiba, O., & Firsov, S. (2019). Designing adaptive PID controller non-sensitive to changes in aerodynamic characteristics of an unmanned aerial vehicle. *Eastern-European Journal of Enterprise Technologies, 1,* 9(97), 69–75.

Brandt, F. (2017). *The air cargo load planning problem.* (Dissertation), University of Karlsruher (KIT), Germany.

British Standards Institution. (2001). BS EN 13306: Maintenance Terminology, 31 p.

Bryson, A. E., & Ho, Y.-C. (1975). *Applied optimal control.* New York: John Wiley & Sons.

Bunakov, V. (2015). *Neuron physics. Tutorial: monogr.* Moscow, Russia: Publishing House of St. Petersburg University.

Burau, N. (2013). Structural and functional synthesis of systems of diagnostics of structures in operation. In N. Burau, O. Pavlovsky, & D. Shevchuk (Eds.), Bulletin of TNTU. 72, 4. pp. 77-86.

Cafieri, S., & Durand, N. (2014). Aircraft deconfliction with speed regulation: New models from mixed-integer optimization. *Journal of Global Optimization, 58*(4), 613–629. doi:10.100710898-013-0070-1

Callan, R. (2003). Basic concepts of neural networks. Moscow, Russia: Williams Publishing House.

Cameron, B. H., & McMilan, E. (2013c). Analyzing the current trends in enterprise architecture frameworks. *Journal of Enterprise Architecture, 9*(1), 61.

Campbell, R. D., & Bagshaw, M. (2002). *Human performance and limitations in aviation* (3rd ed.). UK: Blackwell Science Ltd.

Carayon, P. (2006). Human factors of complex sociotechnical systems. *Applied Ergonomics, 37*(4), 525–535. doi:10.1016/j.apergo.2006.04.011 PMID:16756937

Carruthers, J. B. (2002). *Wireless infrared communications. Wiley encyclopedia of telecommunications.* Boston, MA: Department of Electrical and Computer Engineering Boston University.

Cetek, C. (2009). Realistic speed change maneuvers for air traffic conflict avoidance and their impact on aircraft economics. *International Journal of Civil Aviation, 1*(1), 62–73.

Chernodub, A., & Dzyuba, D. (2011). Review of methods of neuro management. *Problems of programming, 2,* 79-94.

Chernorutsky, G. S., Sibrin, A. P., & Zhabreev, V. S. (1987). *Tracking systems of automatic manipulators.* Moscow, Russia: Science.

Ching, J. (2009). Equivalence between reliability and factor of safety. *Probabilistic Engineering Mechanics, 24*(2), 159–171. doi:10.1051/matecconf/201711602005

Chybowski, L., & Żółkiewski, S. (2015). *Basic reliability structures of complex technical systems*. Springer International Publishing Switzerland. doi:10.1007/978-3-319-16528-8_31

Cillessen, A. S. N. (2009). Sociometric methods. In W. M. Bukowski, B. Laursen, & K. H. Rubin (Eds.), *Handbook of peer interactions, relationships, and group* (pp. 82–98). London, UK: The Guilford Press.

Ciodaro, T., Deva, D., De Seixas, J., & Damazio, D. (2012). Online particle detection with neural networks based on topological calorimetry information. In *Proceedings of the Journal of Physics: Conference Series*. 10.1088/1742-6596/368/1/012030

Civil Aviation Organization. (1998). Human factors training manual (1st. ed.). Doc. ICAO 9683-AN/950. Canada, Montreal: Author.

Clegg, C. W. (2000). Sociotechnical principles for system design. *Applied Ergonomics*, *31*(5), 463–477. doi:10.1016/S0003-6870(00)00009-0 PMID:11059460

Commission Implementing Regulation (EU). (2013). On the definition of common projects, the establishment of governance and the identification of incentives supporting the implementation of the European Air Traffic Management Master Plan No 409/2013. *Official Journal of the European Union, No. 4.5.2013*, 123/1-123/7.

Commission Implementing Regulation (EU). (2014). On the establishment of the Pilot Common Project supporting the implementation of the European Air Traffic Management Master Plan No. 716/2014. *Official Journal of the European Union, No. 28.6.2014*, 190/19-190/44.

Cordier, M. O., Travé-Massuyes, L., & Pucel, X. (2006, June). Comparing diagnosability in continuous and discrete-event systems. In *Proceedings of the 17th international workshop on principles of diagnosis (DX-06)* (pp. 55-60).

Ćosić, J., Ćurković, P., Kasać, J., & Stepanić, J. (2013). Interpreting development of unmanned aerial vehicles using systems thinking. *Interdisciplinary Description of Complex Systems*, *11*(1), 143–152. doi:10.7906/indecs.11.1.12

Crayon. (2018). *How can machine learning help the enterprise architect?* Retrieved from https://www.crayon.com/en-za/news-and-resources/how-can-machine-learning-help-the-enterprise-architect/

De Floriani, L., Falcidieno, B., Nagy, G., & Pienovi, C. (1991). On sorting triangles in a Delaynay tessellation. *Algorithmica*, *6*(1-6), 522–535. doi:10.1007/BF01759057

DeAgostini. (2007). *How is the human body structured? Iss. 25. Neurons*. Moscow, Russia: Author.

Dergachov, K., & Kulik, A. (2019). Ensuring the safety of UAV flights by means of intellectualization of control systems. In Cases on Modern Computer Systems in Aviation (pp. 287-310). Hershey, PA: IGI Global. doi:10.4018/978-1-5225-7588-7.ch012

Detling, V. S., Zinchenko, I. V., & Miroshnichenko, B. M. (2006). Information-measuring system for ensuring the quality of roughness of the surface. *Issue date of the Cherkasy State Technological University, Special Issue*, 135-137.

Dhillon, B. S. (2006). *Maintainability, maintenance, and reliability for engineers*. New York: Taylor & Francis Group. doi:10.1201/9781420006780

Dianov, V. N. (2012). Conceptual features of building trouble-free hardware / automatics and telemechanic, 7. pp. 119-138.

Doc 10003. (2014). *Manual on the digital exchange of aeronautical meteorological information*. Montreal, Canada: ICAO.

Doc 7192. (2011). *Training Manual. Part F-1: Meteorology for air traffic controllers and pilots*. Montreal, Canada: ICAO.

Doc 7488. (2007). *Manual of the icao standard atmosphere extended to 80 kilometres (262,500 feet)*. Montreal, Canada: ICAO.

Doc 9328. (2005) *Manual of runway visual range observing and reporting practices*. Montreal, Canada: ICAO.

Doc 9761. (2003) *Meteorological support for air traffic controllers and pilots*. Montreal, Canada: ICAO.

Doc 9873. (2007) *Manual on the quality management system for the provision of meteorological service to international air navigation*. Montreal, Canada: ICAO.

Doc 9974. (2013) *Flight safety and volcanic ash*. Montreal, Canada: ICAO.

Doc. 8896. (2011) *Manual of aeronautical meteorological practice. AN/893 Ninth edition*. Canada, Montreal: ICAO.

Dolgikh, S. (2018, November). Spontaneous concept learning with deep autoencoder. *Spontaneous Concept Learning with Deep Autoencoder In International Journal of Computational Intelligence Systems*, *12*(1), 1–12. doi:10.2991/ijcis.2018.25905178

Dorf, R., & Bishop, R. (2002). Modern management systems. Per. with Eng. Moscow, Russia: Laboratory of Basic Knowledge, 615-618.

Dousse, O., Mannersalo, P., & Thiran, P. (2004). Latency of wireless sensor networks with uncoordinated power saving mechanisms. In *Proceedings of the 5th ACM International Symposium on Mobile Ad Hoc Networking and Computing*. (May 24-26, pp. 109-120), Tokyo, Japan. 10.1145/989459.989474

Duanzhang, L., Peng, C., & Nong, C. (2016). A generic trajectory prediction and smoothing algorithm for flight management system. In *Proceedings of the Chinese Guidance, Navigation and Control Conference* (CGNCC) (pp. 1969–1974). IEEE. 10.1109/CGNCC.2016.7829091

Dyckho, H. (1990). A typology of cutting and packing problems. *European Journal of Operational Research, 44*, 145–159. doi:10.1.1.87.4320.

Eaton, C. M., Chong, E. K. P., & Maciejewski, A. A. (2016). Multiple-scenario unmanned aerial system control: A systems engineering approach and review of existing control methods. *Aerospace, 3*(1), 2–26. doi:10.3390/aerospace3010001

Eby, M. S. (1994). A self-organizational approach for resolving air traffic conflicts. *The Lincoln Laboratory Journal, 7*(2), 239–254.

Eby, M. S., & Kelly, W. E. (1999). Free flight separation assurance using distributed algorithms. *Proceedings of 1999 IEEE Aerospace Conference, 2*, 429-441. 10.1109/AERO.1999.793186

Eddous, M., & Stansfield, R. (1997). Methods of decision-making. *Audit*, 455-459.

Egupov, I. (2002) Methods of robust, neuro-fuzzy and adaptive control. Moscow, Russia: MSTU named after N.E. Bauman. (in Russian)

Elisov, L., & Ovchenkov, N. (2017). Aviation security as an object of mathematical modeling. *Scientific Herald of the Moscow State Technical University, 20*(3), 13–20.

Elkott, D. F. (2001). Automatic sampling for CMM inspection planning of free-form surfaces. Windsor, Canada.

Eriksson, D., Frisk, E., & Krysander, M. (2013). A method for quantitative fault diagnosability analysis of stochastic linear descriptor models. *Automatica, 49*(6), 1591–1600. doi:10.1016/j.automatica.2013.02.045

Ermakova, A. I. (1987). *Features of the analysis and assessment of meteorological conditions to ensure the safety of flights on international airlines*. London, UK: Hydrometpublish.

Eurocontrol Experimental Centre. (2011). *User manual for the Base of Aircraft Data (BADA) (Revision 3.9)*. Brétigny-sur-Orge. Author.

EUROCONTROL. (2004). Single European Sky first legislative package (SES I). Retrieved from https://www.skybrary.aero/index.php/Single_European_Sky_(SES)

Europe - FlightWeather. (0AD). Retrieved from https://sites.google.com/site/acnetworkweather/home/europe

European Aviation Safety Agency. (2012). *Decision making for single-pilot helicopter operations*. Koln, Germany: EASA Publ.

European Organization for the Safety of Air Navigation. (2003). *Guidelines for controller training in the handling of unusual/emergency situations* (2nd ed.). Brussels, Belgium: Author.

European Organization for the Safety of Air Navigation. (2004). *ATM training progression and concepts*. Brussels, Belgium: Author.

Federal Aviation Administration. (2008). *Aviation instructor's handbook*. Washington: Author.

Federation of Enterprise Architecture Professional Organizations. (2013a). Common perspectives on enterprise architecture, *Architecture and Governance Magazine, Issue 9-4*.

Feistel, A., Wiczanowski, M., & Stanczak, S. (2007). *Optimization of energy consumption in wireless sensor networks*. Retrieved from https://pdfs.semanticscholar.org/59b3/e2a10ada19151f5730f153661b46812471c5.pdf

Feng, B., Yanzhi, L., & Shen, Z.-J. M. (2015). Air cargo operations: Literature review and comparison with practices. *Transportation Research Part C, Emerging Technologies*, *56*, 263–280. doi:10.1016/j.trc.2015.03.028

Feng, L., Tian, C., Zhang, H., & Kelley, W. (2010). Rule-based optimization approach for airline load planning. *Procedia Computer Science*, *1*(1), 1455–1463. doi:10.1016/j.procs.2010.04.161

Ferat, S. A. (2000). Baesian network approach to the self-organization and learning in intelligent agents. (Doctoral dissertation), Virginia Polytechnic and State University.

Ferguson, T. S. C. J. (2009). Proceedings of the 4 Berkeley symposium on mathematical statistics and probability. *Annals of mathematical statistics. 12*, 82–94.

Filatyev, A., & Yanova, O. (2013).The through optimization of fail-safe branched injection trajectories of launch vehicles in view of aerodynamic load constraints on the basis of the maximum principle. *Progress in Flight Dynamics, GNC and Avionics*, *6*, 569-582.

Filatyev, A. (1994). Optimization of branched trajectories for aerospace transport systems. *19th ICAS Congress, Anaheim, CA*, 325-331.

Flight plan. (2019, April 15). Retrieved from https://en.wikipedia.org/wiki/Flight_plan#/media/File:International_flight_plan.png

Flueler, T. (2006) Decision making for complex socio-technical systems: Robustness from lessons learned in long-term radioactive waste governance. *Environment & Policy*. Springer IATA. Retrieved from https://www.iata.org/publications/Pages/AI-white-paper.aspxInternational

Fok, K., & Chun, A. (2004). Optimizing air cargo load planning and analysis. In *Proceedings of the International Conference on Computing, Communications, and Control Technologies*. Austin, TX.

Fraden, J. (1997). *Handbook of modern sensors* (2nd ed., pp. 104–107). Woodbury, NY.

Frangopol, D. M., Kawatani, M., & Kim, C. W. (Eds.). (2007). Reliability and optimization of structural systems: assessment, design, and life-cycle performance. Taylor & Francis Group, London, UK, p. 269.

Friedman, M., Carterette, E., Wiener, E., & Nagel, D. (2014). *Human factors in aviation* (1st ed.). Massachusetts: Academic Press Publ.

Frolov, K. V. (1992). Problems of safety of complex technical systems.

Frolov, K. V., & Makhutov, N. A. (n.d.). Problems of machine building and machine reliability, 5. pp. 3-11.

Gabasov, R., Kirillova, F., Paulianok, N., Alsevich, V., Kalinin, A., & Krakhotko, V. (2011). Optimization methods. Minsk, Belarus: Publishing Four Quarters.

Galar, D., Sandborn, P., & Kumar, U. (2017). *Maintenance costs and life cycle cost analysis*. Boca Raton, FL: CRC Press. doi:10.1201/9781315154183

George, J. A., & Robinson, D. F. (1980). A heuristic for packing boxes into a container. *Computers & Operations Research*, *7*(3), 147–156. doi:10.1016/0305-0548(80)90001-5

Gertsbakh, I. (2005). *Reliability theory: with applications to preventive maintenance*. New York: Springer. doi:10.1007/978-3-662-04236-6

Ghazi, G., & Botez, R. M. (2019). Identification and validation of an engine performance database model for the flight management system. *Journal of Aerospace Information Systems*, 1-20.

Gianazza, D. (2002). The 21st Digital avionics systems. *Conference Proceedings*. doi:10.1002/9780470774472

Gibb, R., Schvaneveldt, R., & Gray, R. (2008). Visual misperception in aviation: glide path performance in a black hole environment. *Human Factors the Journal of the Human Factors and Ergonomics Society*, *50*(4), 699–711. doi:10.1518/001872008X288619 PMID:18767527

Gnedenko, B. V., Belyayev, Y. K., & Solovyev, A. D. (1969). *Mathematical methods of reliability theory*. New York: Academic.

Golovinsky, P. (2012). *Mathematical models. Theoretical physics and analysis of complex systems*. Book 2. From nonlinear oscillations to artificial neurons and complex systems.

Golovnin, S. M. (2016). *Training pilots and controllers in the virtual environment of piloting and air traffic control*. Retrieved from www.researchgate.net/publication/303859941_Aviation_Training_Pilots_Air_Traffic_Controller

Goncharenko, A. (2017). Aircraft operation depending upon the uncertainty of maintenance alternatives. *Aviation*, *21*(4), 126–131. doi:10.3846/16487788.2017.1415227

Gorish, A. V., Dudkevich, V. P., & Kupriyanov, M. F. (1999). Piezoelectric instrument-making - T.1. Physics of ferroelectric ceramics. Moscow, Russia: Izdat. forerunners ed. zhur. Radio Engineering.

Gribov, V. M. (2017). Technical diagnostics of avionics Module 2-2, Kiev, Ukraine.

Gu, D., Petkov, P., & Konstantinov, M. (2005). *Robust control design with MATLAB*. London, UK: Springer-Verlag.

Gueret, G., Jussien, N., L'homme, O., Pavageau, C., & Prins, C. (2003). Loading aircraft for military operations. *The Journal of the Operational Research Society*, *53*(5), 458–465. doi:10.1057/palgrave.jors.2601551

Haikin, S. (2006). *Neural networks: Comprehensive foundation* (2nd ed.). Moscow, Russia: Williams.

Haikin, S. (2009). *Neural networks: full course*. Moscow, Russia: Williams.

Haining, R. (2006). Trend-surface models with regional and local scales of variation with an application to aerial survey data. *Technometrics*, *4*(February), 461–469.

Hashimov, E. G., Bayramov A. A., & Abdullayev, F. A. (2017). Development of the multirotor unmanned aerial vehicle. *National security and military sciences*. Baku, 3(4), 21-31.

Heckerman, D. (1995). A tutorial on learning with Bayesian netwoks. USA: Technical report MSR -TR-95-06 - Microsoft Research, Redmond, WA.

Herndon, A. A., Cramer, M., Nicholson, T., & Miller, S. (2011). Analysis of advanced flight management systems (FMS), flight management computer (FMC) field observations trials: Area navigation (RNAV) holding patterns. In *Proceedings of the AIAA/IEEE Digital Avionics Systems Conference*, (pp. 4A11–4A117). 10.1109/DASC.2011.6096065

Hilkert, J. M. (2008). Inertially stabilized platform technology. *IEEE Control Systems Magazine*, *28*(1), 26–46. doi:10.1109/MCS.2007.910256

Hillson, D. (2003). Using a risk breakdown structure in project management. *Journal of Facilities Management*, *2*(1), 77–85. doi:10.1108/14725960410808131

Hinton, G., Deng, L., Yu, D., Dahl, G., Mohamed, A.-r., Jaitly, N., & Kingsbury, B. (2012). Deep neural networks for acoustic modeling in speech recognition. *IEEE Signal Processing Magazine*, *29*.

Hlavcheva, D., & Yaloveha, V. (2018). Capsule neural networks. Control. *Navigation and Communication Systems*, *5*(51), 132–135.

Hoover, K. Wildfire Statistics Congressional Research Service. Retrieved from https://fas.org/sgp/crs/misc/IF10244.pdf

Hryschenko, Y. V. (2003). Pair flights as a way of analyzing the dynamic stereotype amplification phenomenon among pilots. *Cybernetics and Computing Engineering. National Academy of Sciences of Ukraine. 140*, 31–34.

Hryschenko, Y. V., & Kurushkina, Y. O. (2004). Using the principle of invariance of control systems in the evaluation of enhanced reflexes. *Cybernetics and Computing Engineering. National Academy of Sciences of Ukraine. 143*, 39–44.

Hryshchenko, Y. V. (2004). Analysis of changes in the dynamic stereotype of pilots in the process of flight training on a complex simulator of an airplane. *Cybernetics and computing. NAN of Ukraine, 142*. pp. 35-40.

Hryshchenko, Y. V. (2014). Scientific research on the anti-stress preparation of specialists in a quarter century. *Proceedings of the National Aviation University, 1*(58). pp. 53-58.

Hryshchenko, Y. V. (2016). Reliability problem of ergaticcontrol systems in aviation. Methods and systems of navigation and motion control. *IEEE 4th International Conference*, Kiev, Ukraine. pp. 126-129.

Hryshchenko, Y. V., & Romanenko, V. G. (2017). Quality of flight during approach improvement suggestions. *IEEE 4th International Conference (Oct. 17-19, 2017)*, Kiev, Ukraine. pp. 69-72.

Hryshchenko, Y. V., Romanenko, V. G. & Amelina, A. I. (2019). The problem of uncoordinated aircraft turn on small flight speed. *Electronics and Control Systems*. Kiev, Ukraine. NAU. 1(59). pp. 50-57.

Hryshchenko, Y. V., Romanenko, V. G., & Hryshchenko, Y. Y. (2017). Suggestions of the improvement of the quality of flight during landing and missed approach/go around maneuver. *Electronics and control systems*. Kiev, NAU. 2(52). pp. 103-109.

Hryshchenko, Y. V., Romanenko, V. G., & Pipa, D. M. (2017). Suggestions to the methods for assessing the quality of the glide path entrance. *Electronics and Control Systems,* Kiev, Ukraine. NAU. 3(57). pp. 41-48.

Hryshchenko, Y. V., Romanenko, V. G., & Pipa, D. M. (2018). Evaluation of quality and accuracy of flight path during approach and landing. In *Proceedings IEEE 5th International Conference (Oct. 16-18, 2018)*. Kiev, Ukraine. pp. 191-194.

Hryshchenko, Y. V., Romanenko, V. G., & Pipa, D. M. (2018).Suggestions to the methods for assessing the quality of the glide path entrance. *Electronics and Control Systems,* Kiev, Ukraine. NAU, 2018. 3(57). pp. 41-48.

Hryshchenko, Y. V., Romanenko, V. G., Tronko, V. D., & Hryshchenko, Y. Y. (2017). Methods of training of modern aircraft flight crews for inflight abnormal circumstances. *Proceedings of the National Aviation University*, 1. pp. 66-72.

Hryshchenko, Y. V., Skripets, A. V., & Tronko, V. D. (2015). Analysis of the possibility of controlling the trajectory of an airplane's motion by correlation function. *Cybernetics and computing technology: Scientific Journal. House "Academiperiodica" of the National Academy of Sciences of Ukraine. 181.* pp. 35-46.

Hryshchenko, Y. V., Skripets, A. V., & Tronko, V. D. (2015). Autocorrelation functions and their application for assessing. *Proceedings of the National Aviation University, 2(63).* pp. 27-33.

Hryshchenko, Y. V., Yakimenko, M. S., & Amelina, A. I. (2017). Problems of the impact of failures in avionics systems on the quality of flight. All-Ukrainian scientific and practical conference of young scientists and students "Problems of Navigation and Control of Motion", Section 4, Avionics and Control Systems. Kiev, Ukraine: NAU.

Hryshchenko, Y. V. (2003). Preparation of pilots for flights in special situations taking into account the phenomenon of dynamic stereotype. *Cybernetics and Computing Engineering. National Academy of Sciences of Ukraine. 139,* 81–85.

Hryshchenko, Y. V. (2016). Reliability problem of ergatic control systems in aviation. In *Proceedings of IEEE 4th International Conference on Methods and Systems of Navigation and Motion Control (MSNMC).* (pp. 126 – 129). Kiev, Ukraine. 10.1109/MSNMC.2016.7783123

Hryshchenko, Y. V., Romanenko, V. G., & Polozhevets, A. A. (2005). Mathematical aspects of solving the problems of accounting for a large number of factors during the avionics exploiting. *Cybernetics and Computing Engineering. National Academy of Sciences of Ukraine. 146,* 81–89.

Hubaev, G. N. (1999). Statistical methods of planning of extremal experiments with information systems. *Statistics Issues,* 78-83.

Hubaev, G. N. (1999). Complex systems: Expert methods of comparison. *North Caucasian region.* [H]. *Social Sciences,* 7–24.

Hubel, D. H., & Wiesel, T. N. (1962). Receptive fields, binocular interaction and functional architecture in the cat's visual cortex. *The Journal of Physiology, 160*(1), 106–154. doi:10.1113/jphysiol.1962.sp006837 PMID:14449617

Hu, J., Prandini, M., & Sastry, S. (2000). Optimal maneuver for multiple aircraft conflict resolution: a braid point of view. *Proceedings of the 39th IEEE Conference on Decision and Control, 4,* 4164-4169. 10.1109/CDC.2000.912369

Hu, J., Prandini, M., & Sastry, S. (2002). Optimal coordinated maneuvers for three dimensional aircraft conflict resolution. *Journal of Guidance, Control, and Dynamics, 25*(5), 888–900. doi:10.2514/2.4982

Hu, L., Tian, Y., Yuan, G., Bai, J., & Zheng, J. (2015). System of systems-oriented flight vehicle conceptual design: Perspectives and progresses. *Chinese Journal of Aeronautics, 28*(3), 617–635. doi:10.1016/j.cja.2015.04.017

Hussein, M. I. (2012). *Container handling algorithms and outbound heavy truck movement modeling for seaport container transshipment terminals. (Theses and Dissertations, Paper 56*, University of Wisconsin, Milwaukee, WI).

Hyunkyung, M., Hayoung, J., & Euiho, S. (2016). An analysis for measure of effectiveness of an unmanned aerial vehicle using simulation. Research & Reviews. *Journal of Engineering and Technology. RRJET, 5*(3), 69–82.

Ibrahim, A., & Alfa, A. (2017). *WSN optimization. Sensors.* doi:10.339017081761

ICAO Journal. (2012). Doc 9859, AN/474, Safety management manual.

ICAO. (2005). Global Navigation Satellite System (GNSS) Manual. The International Civil Aviation Organization, Doc. 9849, Montreal, Canada.

ICAO. (2008) Performance-Based Navigation (PBN) Manual. The International Civil Aviation Organization, Doc. 9613, Montreal, Canada.

İnan, T. T. (2018). The evolution of crew resource management concept in civil aviation. *Journal of Aviation*, 2(1), 45–55. doi:10.30518/jav.409931

Industry High Level Group. (2017a). Aviation benefits 2017, 9–10.

Insurance Information Institute. Facts & Statistics: Wildfires: Wildfire losses in the United States. (2017). Retrieved from https://www.iii.org/fact-statistic/facts-statistics-wildfires

International Air Transport Association. (2016). Air freight market analysis. *IATA economics.* Retrieved from https://www.iata.org/publications/economics/Reports/freight-monthly-analysis/freight-analysis-dec-2018.pdf

International Air Transport Association. (2018). Dangerous goods regulations. *IATA*, 1–18.

International Air Transport Association. (2018). IATA cargo strategy. *IATA*, 1–31.

International Civil Aviation Organization. (1998). Human factors training manual (1st ed.). Doc. ICAO 9683-AN/950. Canada, Montreal: Author.

International Civil Aviation Organization. (1998). Human factors training manual (1st ed.). Doc. ICAO 9683-AN/950. Montreal, Canada: Author.

International Civil Aviation Organization. (2002). Global air navigation plan for CNS/ATM Systems 2002 Doc 9750 AN/963.

International Civil Aviation Organization. (2002). Human factors guidelines for safety audits manual (1st ed.). Doc. ICAO 9806-AN/763. Canada, Montreal: Author.

International Civil Aviation Organization. (2002). Human factors guidelines for safety audits manual (1st ed.). Doc. ICAO 9806-AN/763. Montreal, Canada: Author.

International Civil Aviation Organization. (2004). *Cross-cultural factors in aviation safety: Human factors digest N° 16. Circ. ICAO 302-AN/175.* Montreal, Canada: Author.

International Civil Aviation Organization. (2004). *Cross-cultural factors in aviation safety: Human Factors Digest No. 16. Cir. ICAO 302-AN/175.* Montreal, Canada: Author.

International Civil Aviation Organization. (2004). *Cross-cultural factors in aviation safety: Human Factors Digest No. 16. Circ. ICAO 302-AN/175.* Montreal, Canada: Author.

International Civil Aviation Organization. (2005). *Global air traffic management operational concept. Doc. ICAO 9854.* Montreal, Canada: Author.

International Civil Aviation Organization. (2005). *Global air traffic management operational concept. Doc. ICAO 9854-AN/458.* Montreal, Canada: Author.

International Civil Aviation Organization. (2007). Air traffic management (15th ed.). Doc. ICAO 4444-ATM/501. Canada, Montreal: Author.

International Civil Aviation Organization. (2009). Global performance of the air navigation system (1st ed.). Doc. ICAO 9883. Canada, Montreal: Author.

International Civil Aviation Organization. (2012). Manual on flight and flow information for a collaborative environment (FF-ICE) (1st ed.). Doc. ICAO 9965. Canada, Montreal: Author.

International Civil Aviation Organization. (2013). Safety management manual (SMM) (3rd ed.). Doc. ICAO 9859-AN 474. Canada, Montreal: Author.

International Civil Aviation Organization. (2013a). Safety management manual (SMM) (3rd. ed.). Doc. ICAO 9859-AN 474. Canada, Montreal: Author.

International Civil Aviation Organization. (2013b). *State of global aviation safety*. Montreal, Canada: Author.

International Civil Aviation Organization. (2014). Manual on collaborative decision-making (CDM) (2nd ed.). Doc. ICAO 9971. Canada, Montreal: Author.

International Civil Aviation Organization. (2014). *Safety report*. Montreal, Canada: Author.

International Civil Aviation Organization. (2016). Global Air Navigation Plan 2016-2030 (5th ed.). Doc. ICAO 9750. Canada, Montreal: Author.

International Civil Aviation Organization. (2017). Global Aviation Security Plan (GASP). International Civil Aviation Organization, Canada, Montreal (2017) Global Air Navigation Plan (GANP, Doc 9750 International Civil Aviation Organization. (2013). Safety Management Manual (SMM): Doc. ICAO 9859-AN/474 (3rd ed.). International Civil Aviation Organization, Canada, Montreal (2013).

International Civil Aviation Organization. (2018, October). Potential of Artificial Intelligence (AI) in Air Traffic Management (ATM). In *Thirteenth Air Navigation Conference ICAO,* Montréal, Canada, 9-19.

International Institute of Business Analysis. (2015). A guide to the business analysis body of knowledge (version 3.0). Toronto, Ontario.

ISO/IEC/IEEE. (2011). Systems and software engineering – Architecture description (1st edit.). Geneva, Switzerland.

Ivanov, N. M., Kondakov, E. V., & Miloslavskiy, Y. K. (2016). Opubl Patent RF N° 2584719. Cifrovoj sposob izmerenija parametrov p'ezojelementov [Digital method for measuring the parameters of piezoelectric elements].

Jager, R. R. (1986). The sets of the level for the evaluation of the membership of fuzzy subsets. Radio and Communications, 71-78.

Jamshidi, M. (Ed.). (2009). *Systems of systems engineering. Principles and applications*. Boca Raton, FL: CRC Press Taylor & Francis Group.

Jesse, R. (2012). *Artificial neural network*. Moscow, Russia: VSD.

Jesse, R. (2012). *Neuron*. Moscow, Russia: VSD.

Jones, M. T. (2009). *Artificial intelligence: A systems approach*. Hingham, MA: Jones & Bartlett Learning.

Kaimal, L. B., Metkar, A. R., & Rakesh, G. (2014). Self learning real time expert system. Control & Instrumentation Group, C-DAC. *International Journal on Soft Computing* [IJSCAI]. *Artificial Intelligence and Applications, 3*(2), 13–25. doi:10.5121/ijscai.2014.3202

Kalatskaya, L., Novikov, V., Sadov, V. (2003). *Organization and training of artificial neural networks: Experimental study*. Minsk, Belarus: Publishing House of the BSU.

Kapitza, P. L. (1987). Experiment, theory, practice. In *Experiment, Theory, Practice*. Moscow, Russia: Nauka.

Kaplan, A., & Haenlein, M. (2019). Siri, Siri in my hand, who's the fairest in the land? On the Interpretations, Illustrations, and Implication of Artificial Intelligence. *Business Horizons, 62*(1), 15-26.

Kapur, K. C., & Lamberson, L. R. (1977). *Reliability in Engineering Design*. New York: Wiley.

Kashmatov, V. I. (2008). Application of quasi-object control systems by angular position of airplane for the translation of airplane from horizontal flight in a glide slope approach. *Electronics and Control Systems,* Kiev, NAU, 4(18). pp. 79-87.

Kataeva, M. A. (2015). Mathematical construction of the trajectory of the motion of the sensor on the surface of the part. *Measuring and Computer Technology in Technological Processes, 1*, 91-94.

Kataeva, M. O. (2016). Method of describing the trend of a surface of aviation detail of complex geometric shape. *Bulletin of the Engineering Academy of Ukraine, 3*, 186–189.

Kataeva, M., Kvasnikov, V., & Shkvarnitskaya, T. (2018). *Development of a differential-digital method for precision measurements of the geometric parameters of aviation details* (pp. 32–35). Vilnius, Lithuania: Science - Future of Lithuania. Transport Engineering and Management.

Katipamula, S., & Brambley, M. R. (2005). Methods for fault detection, diagnostics, and prognostics for building systems—A review, part I. *HVAC & R Research, 11*(1), 3–25. doi:10.1080/10789669.2005.10391123

Kazak, V. M, Budzynska, T. V, & Mischeryakova, V. Yu. (2009). Assessment of the effect of changes in the parameters of the ellipsoid errors on maintaining the landing trajectory of the aircraft. *Science-Intensive Technologies*. Kiev, NAU. 2. pp. 43-45.

Keating, C., Fernandez, A. A., Jacobs, D. A., & Kauffmann, P. (2001). A methodology for analysis of complex socio-technical processes. *Business Process Management Journal, 7*(1), 33–50. doi:10.1108/14637150110383926

Kellerer, H., Pferschy, U., & Pisinger, D. (2004). *Knapsack problems* (1st ed.). Berlin, Germany: Springer Verlag. doi:10.1007/978-3-540-24777-7

Kharchenko, V. P, Alexeiev, O. M., Luppo, A. E., & Jurchik, R. I., (2017). The principles to maintain an acceptable of air navigation safety in Ukraine. *Proceedings of Kharkiv Air Force University*, 17-25.

Kharchenko, V., Shmelova, T., & Sikirda, Y. (2012). *Decision-making of operator in air navigation system*: monograph. Kirovograd, Ukraine: KFA of NAU.

Kharchenko, V., Shmelova, T., & Sikirda, Y. (2012). Methodology for analysis of flight situation development using GERT's and Markov's networks: In *Proceedings of the 5th World Congress, Aviation in the XXIst century. Safety in Aviation and Space Technologies*. Kiev, Ukraine.

Kharchenko, V., Shmelova, T., & Zubrytskyi, V. (2017) Decision making in air navigation system as a socio-technical system. In *Proceedings of the International Conference, AVIA 2017*, Kiev, Ukraine.

Kharchenko, V. P., Alexeiev, O. M., & Babeichuk, D. G. (2010). Method analysis of management decisions making while air navigation functioning in emergency situations. *Proceedings of the National Aviation University*, 19-24. 10.18372/2306-1472.44.1901

Kharchenko, V. P., Alexeiev, O. M., & Kolesnyk, T. A. (2018). Ensuring the guaranteed level of flights safety – the view of the future. *Proceedings of the National Aviation University*, 9-18. 10.18372/2306-1472.4.13492

Kharchenko, V., Shmelova, T., & Sikirda, Y. (2011). Graphanalytical models of decision-making by air navigation system's human-operator. *Proceedings of the National Aviation University, 1*, 5–17.

Kharchenko, V., Shmelova, T., & Sikirda, Y. (2011). Methodology for analysis of decision making in air navigation system. *Proceedings of the National Aviation University, 3*, 85–94.

Kharchenko, V., Shmelova, T., & Sikirda, Y. (2012). Modeling of behavioral activity of air navigation system's human-operator in flight. *Proceedings of the National Aviation University, 2*, 5–17.

Kharchenko, V., Shmelova, T., & Sikirda, Y. (2016). *Decision-making in socio-technical systems. Monograph.* Kiev, Russia: National Aviation University.

Kharchenko, V., Shmelova, T., & Sikirda, Y. (2016). *Decision-making in socio-technical systems: Monograph.* Kiev, Ukraine: NAU.

Kharchenko, V., Shmelova, T., & Sikirda, Yu. (2018). Methodology of research and training in air navigation socio-technical system. *Proceedings of the National Aviation University, 1*, 8–23. doi:10.18372/2306-1472.74.12277

Khokhlov, E. M., Hryshchenko, Y. V., & Volodko, O. N. (2016). The method of determining the quality of piloting the contours of the correlation fields of flight parameters in special conditions. *Cybernetics and Computer Science: Science Magazine,* 39-51.

Khusainov, D., Korenyuk, I., & Gamma, T. (2012). *Mechanisms of rhythmic activity of grape snail neurons.* LAP Lambert Academic Publishing.

Kim, D.(2003). The theory of automatic control. T1. Linear systems / D. Kim. Moscow, Russia: FIZMATLIT, 76-93.

Kim, S., Choi, S., Kim, H., Shin, J., Shim, H., & Kim, H. J. (2018). Robust control of an equipment-added multirotor using disturbance observer. *IEEE Transactions on Control Systems Technology, 26*(4), 1524–1531.

Kim, D. P. (2003). *Automatic control theory: Linear systems.* Moscow, Russia: Fyzmatlyt.

Kirichek, R., Paramonov, A. & Vareldzhyan, K. (2015). Optimization of the UAV's motion trajectory in flying sensor networks. *Telecommunication, 7*, 20-25.

Klyuev, V. V. (1978) Instruments and systems for measuring vibration, noise and shock: A handbook. In 2 books. Moscow, Russia: Mechanical Engineering, 328-332.

Klyuev, V. V., Parkhomenko, P. P., & Abramchuk, V. E. (1989). Technical means of diagnosis: a reference book. In V. V. Klyueva (Ed.), *Mashinostroenie.* Moscow, Russia.

Kochenderfer, M. J., & Wheeler, T. A. (2019). *Algorithms for optimization.* The MIT Press.

Kondakov, N. I. (1976). Logical dictionary-reference / 2nd ed., Rev. and add. Science, 1976.

Kosecka, J., Tomlin, C., Pappas, G., & Sastry, S. (1997). Generation of conflict resolution maneuvers for air traffic management. *Proceedings of the 1997 IEEE/RSJ International Conference on Intelligent Robot and Systems, 3*, 1598-1603.

Koshkin, R. (2014). Mathematical models of the processes of creation and functioning of search and analytical information systems for civil aviation. *Scientific Herald of the State Research Institute of Civil Aviation,* 5, pp. 39-49.

Kotikova, E. A. (1979). *Catecholaminergic neurons.* Moscow, Russia: Nauka.

Kruglov, V., & Borisov, V. (2002) Artificial neural networks. Theory and practice: monogr. Ed. 2. Moscow, Russia: Hot line – Telecom.

Ku, M. L., Chen, Y., & Liu, K. J. R. (2015). Data–driven stochastic models and policies for energy harvesting sensor communication. *IEEE J. Sel. Areas Commun.* (33), 1505-1520.

Kuchar, J. K., & Yang, L. C. (2000). A review of conflict detection and resolution modeling methods. *IEEE Transactions on Intelligent Transportation Systems, 1*(4), 179–189. doi:10.1109/6979.898217

Kuchar, J. K., & Yang, L. C. (2000). *A review of conics detection and resolution modelling methods* (pp. 1388–1397). Cambridge, UK: American Institute of Aeronautics and Astronautics.

Kulik, A. (1992). Evaluation of the diagnosability of linear dynamic systems. *Avtomatika i Telemekhanika,* (1), 184-187.

Kulik, A. (2000). Signal-parametric diagnostics of control systems. Kharkiv, Ukraine: National Aerospace University «KhAI»

Kulik, A., ... (2016). Elements of the theory of rational management of objects. Kharkiv, Ukraine. National Aerospace University «KhAI»

Kulik, A., ... (2018). Rational control of objects: theory and applications. Kharkiv, Ukraine. National Aerospace University «KhAI»

Kulik, A., Sirodzha, I. B., & Shevchenko, A. N. (1990). Construction of diagnostic models in the development of diagnostic software for dynamic systems.

Kulik, A. S. (1991). Fault diagnosis in dynamic systems via signal-parametric approach. In *IFAC/IMACS Symp. Baden-Baden* (Vol. 1, pp. 157-162).

Kulik, A., Sirodzha, I. B., & Shevchenko, A. N. (1990). *Construction of diagnostic models in the development of diagnostic software for dynamic systems.* Kharkiv, Ukraine: Part II.

Kulik, N. (2008). *Encyclopedia of aviation safety.* Kiev, Ukraine: Technics.

Kurtov, A., Polikashyn, O., Potikhenskyy, A., & Aleksandrov, V. (2017). Expert assessments. "Delphy" method as technology of administrative decision making. *Collections of scientific works of the Kharkiv National University of Air Force, No. 1*(50), 118–122.

Kutsenko, O., Ilnytska, S., & Konin, V. (2018). Investigation of the residual tropospheric error influence on the coordinate determination accuracy in a satellite landing system. *Aviation, 22*(4), 156–165. doi:10.3846/aviation.2018.7082

Kuzmenko, N. S., & Ostroumov, I. V. (2018). Performance analysis of positioning system by navigational aids in three-dimensional space. In *Proceedings of the International Conference on System Analysis and Intelligent Computing, SAIC 2018,* (pp. 101–104). 10.1109/SAIC.2018.8516790

Kuzmenko, N. S., Ostroumov, I. V., & Marais, K. (2018). An accuracy and availability estimation of aircraft positioning by navigational aids. In *Proceedings of IEEE International Conference on Methods and Systems of Navigation and Motion Control (MSNMC),* (pp. 36 – 40). Kiev, Ukraine. 10.1109/MSNMC.2018.8576276

Kuzmych, L. V. (2018). Approaches for estimating the resource of technical objects and systems [Text] / L. V. Kuzmych / Bulletin of the Zhytomyr State Technological University. Series: Technical Sciences. Vol. 2(82). pp. 204-207. doi: (82) -204-207. doi:10.26642/tn-2018-2

Kuzmych, L. V. (2018). Mechanical effects on the reliability of complex technical systems [Text] / L. V. Kuzmych // Technical sciences and technologies. Vol. 4(14). pp. 28-33. doi:10.251402411-5363-2018-4 (14) -28-33.

Kuzmych, L. V. (2018). Scripts of emergence and development of dangerous states of complex technical systems [Text] / L. V. Kuzmych / Zbirnyk naukovykh prats Viiskovoho instytutu Kyivskoho natsionalnoho universytetu imeni Tarasa Shevchenka. Vol. 62. pp. 35–40.

Kuzmych, L., & Kvasnikov, V. (2017). Study of the durability of reinforced concrete structures of engineering buildings. Advances in Intelligent Systems and Computing. Vol. 543, pp. 659-663. In *Proceedings International Conference on Systems, Control and Information Technologies, SCIT 2016;* Warsaw; Poland; May 20-21, 2016. 10.1007/978-3-319-48923-0_70

Kuzmych, L., Kobylianskyi, O., & Duk, M. (2018, Oct. 1). Current state of tools and methods of control of deformations and mechanical stresses of complex technical systems. *Proc. SPIE 10808, Photonics Applications in Astronomy, Communications, Industry, and High-Energy Physics Experiments.* doi:10.1117/12.2501661

Kuznetsova, V. I., & Barzilovicha, E. Y. (1990). Nadezhnost I effectivnost v tehnike. Moscow, Russia: Mashinostroenie, 8, 248.

Kvasnikov, V. P., & Bilan, M. O. (2014). Calculation of tolerances by the method of dimensional analysis, taking into account operational error. *Measuring and Computing Engineering in Technological Processes, 1,* 173-177.

Kwakernaak, H. (1993). Robust control and H∞. *Automatica, 29*(2), 255–273. doi:10.1016/0005-1098(93)90122-A

Lankhorst, M. M., Arbab, F., Bekius, S. F., Bonsangue, M., Bosma, H., & Campschroer, J. … Wieringa, R. J. (2013b). Enterprise architecture at work, The Enterprise Engineering Series. Berlin, Germany: Springer-Verlag.

Lapa, V. G. (1974). Mathematical foundations of cybernetics. *Publishing Association «Vishchashkola».* p. 452.

Larsen, O., & Middelsen, G. (1980). An interactive system for the loading of cargo aircraft. *European Journal of Operational Research, 4*(6), 367–373. doi:10.1016/0377-2217(80)90187-3

Lazarev, Y. N. (2007). *Aerospace trajectory control.* Samara, Russia: Scientific Center of the Russian Academy of Sciences.

LeCun, Y., Bengio, Y., & Hinton, G. (2015). Deep learning. *Nature, 521*(7553), 436.

Ledsinger, L. A. (2000). Solutions to decomposed branching trajectories with powered flyback using multidisciplinary design optimization. *Thesis,* 162.

Lee, T. H., Adams, G. E., & Gaines, W. M. (1968). *Computer process control: modeling and optimization.* New York.

Lefebvre, B. (2008, January-June). Intentional-reflective model of the agent. *Reflexion Processes and Control, 1, Vol. 1,* 69–78.

Lefebvre, B., & Adams-Webber, J. (2002, January-June). Functions of fast reflexion in bipolar choice. *Reflexion Processes and Control, 1, Vol. 1,* 29–40.

Lefebvre, V. A. (2001). Algebra of Conscience (2nd enlarged ed.). Dordrecht, The Netherlands: Kluwer Publishers.

Leitmann, G. (Ed.). (1962). *Optimization techniques: with applications to aerospace systems* (Vol. 5). Academic Press.

Lesa Moné. (2019). *An enterprise architect's guide to machine learning series: Part 1.* Retrieved from https://blog.leanix.net/en/an-enterprise-architects-guide-to-machine-learning-series-part-1

Leshchenko, G. P. (2013) Aviation meteorology: Textbook. Kirovograd, Ukraine: Publishing house КЛА НАУ.

Levin, B. R. (1978). *Theory of reliability of radio engineering systems.* Moscow, Russia: Radio. (in Russian)

Levshina, E. S.(1983) Electrical measurements of physical quantities: (Measuring transducers). London, UK: Energoatomizdat. Leningrad separation,132-141.

Leychenko, S., Malishevskiy, A., & Mikhalic, N. (2006). *Human factors in aviation: monograph in two books. Book 1st*. Kirovograd, Ukraine: YMEKS.

Limbourg, S., Schyns, M., & Laporte, G. (2011). Automatic aircraft cargo load planning. *The Journal of the Operational Research Society*, 1–13. doi:10.1057/jors.2011.134

Liubchenko, N., Podorozhniak, A., & Bondarchuk, V. (2017). Neural network method of intellectual processing of multispectral images. *Advanced Information Systems*, 1(2), 39–44. doi:10.20998/2522-9052.2017.2.07

Lomov, B. F. (1982). Handbook of engineering psychology. Moscow, Russia: Machinery Industry.

Lozino-Lozinsky, G. E. (1998). "Buran" – the way of creation. *Aerospace Engineering and Technology*, 3.

Lubbers, B., Mildner, S., Oonincx, P., & Scheele, A. (2015). A study on the accuracy of GPS positioning during jamming. In *Proceedings of the 2015 International Association of Institutes Navigation World Congress (IAIN)*, (pp. 1–6). IEEE. 10.1109/IAIN.2015.7352258

Luftner, T., Kropl, C., Hagelauer, R., Huemer, M., Weigel, R., & Hausner, J. (2003). Wireless infrared communications with edge position modulation for mobile devices. *IEEE Wireless Communications*, 10(2), 15–21. doi:10.1109/MWC.2003.1196398

Lyapunov, A. M. (1892). The general problem of motion stability. *Annals of Mathematics Studies*, 17.

Lysenko, O., & Tachinina, O. (2014b). Method of location of sensors based on the compound dynamic system technology in the area of emergency situation. *International Periodic Scientific Journal SWorld, 3*(36), 84–89.

Lysenko, O., Tachinina, O., & Chumachenko, S. (2015). The problem statement of branching paths theory for solution of search and rescue task in the area of emergency situation. *Technical Mechanics,* (1), 73–78.

Lysenko, O., & Tachinina, O. (2014a). Mathematical formulation of the problem of optimization of the motion of a group of flying robots on the basis of unmanned aerial vehicles. *Scientific Bulletin Academy of Municipal Administration, 1*(7), 93–99.

Lysenko, O., Tachinina, O., & Alekseeva, I. (2018). Algorithm for operational optimization of two-stage hypersonic unmanned aerial vehicle branching path. *Proceeding of the 2018 IEEE 5th International Conference on Methods and Systems of Navigation and Motion Control (MSNMC)*, 11-16.

Ma, J., Sheridan, R. P., Liaw, A., Dahl, G. E., & Svetnik, V. (2015). Deep neural nets as a method for quantitative structure–activity relationships. *Journal of Chemical Information and Modeling, 55*(2), 263–274. doi:10.1021/ci500747n PMID:25635324

Makarov, R., Nidziy, N., & Shishkin, G. (2000). *Psychological foundations of didactics in flight education: Monograph*. Moscow, Russia: MAPCHAK.

Makhutov, N. A., & Reznikov, D. O. (2011). A comparative assessment of specification-based and risk- management-based approaches to the security assessment of complex technical systems. *Journal of Machinery Manufacture and Reliability, 40*(6), 579–584. doi:10.3103/S1052618811060124

Malyshevskiy, A. V. (2010). Improving the management and planning in the field of air transport by means of socionics selection of aviation personnel. *Scientific Bulletin of the Moscow State Technical University of Civil Aviation. Ser. Aeromechanics and Strength, no. 1(151)*, 150–157.

Mandelstam, Yu. (1983). *Neuron and muscle of the insect: monogr.* Moscow, Russia: Nauka.

Manita, L. (2016). Optimization problems for WSNs: Trade-off between synchronization errors and energy consumption. *Journal of Physics: Conference Series.* pp. 1-6.

Marasanov, P. (2015). Evaluation of the reliability of the civil aviation airport power supply system. *Scientific Bulletin of the Moscow State Technical University of Civil Aviation, 213*(3), 43–49.

Martello, S., & Toth, P. (1990). *Knapsack problems: Algorithms and computer implementations.* Hoboken, NJ: John Wiley & Sons.

Mazurov, M. (2018). Synchronization of relaxation self-oscillating systems, synchronization in neural networks. *Izv. RAS. Ser. Physical, 82*(1), 83–87.

McCarthy, R. (2006). Toward a unified enterprise architecture framework: An analytical evaluation. *Issues in Information Systems, VII*(2), 15.

McFeeters, S. K. (1996). The use of the Normalized Difference Water Index (NDWI) in the delineation of open water features. *International Journal of Remote Sensing, 17*(7), 1425–1432. doi:10.1080/01431169608948714

Membery, D. A. (1983). Low level wind profiles during the Gulf Shamal. *Weather*, 38.

Merzlyak, M. N., Gitelson, A. A., Chivkunova, O. B., & Rakitin, V. Y. (1999). Non-destructive optical detection of pigment changes during leaf senescence and fruit ripening. *Physiologia Plantarum, 106*(1), 135–141. doi:10.1034/j.1399-3054.1999.106119.x

Mescon, M., Albert, M., & Khedouri, F. (2008). *Management. Monograph.* Moscow, Russia: Vilyams.

Mikolov, T., Deoras, A., Povey, D., Burget, L., & Černocký, J. (2011). Strategies for training large scale neural network language models. Paper presented at the 2011 IEEE Workshop on Automatic Speech Recognition & Understanding. 10.1109/ASRU.2011.6163930

Milekhin, L. N. (2013) Applied theory of gyroscopes. Moscow, Russia: Higher School, pp. 73-76.

Ministry of Civil Aviation of the USSR. (1985). *Manual on the production of flights in the civil aviation of the USSR (with changes and additions).* Moscow, Russia: Air Transport Publ.

Mishkin, E., & Braun, L. (Eds.). (1961). *Adaptive control systems.* McGraw-Hill.

Moiseev, V. (2017). Group application of unmanned aerial vehicles. Kazan, Russia: Editorial and Publishing Center.

Mongeau, M., & Bes, C. (2003). Optimization of aircraft container loading. *IEEE Transactions on Aerospace and Electronic Systems, 39*(1), 140–150. doi:10.1109/TAES.2003.1188899

Mukhometzianov, R., & Carrillo, J. (2018). CapsNet comparative performance evaluation for image classification. *arXiv preprint arXiv:1805.11195.*

Muller, E. (2004). Koordinen - Messtechnik der Zukunst. *Masehinenbau, 23*, 12–21.

Mulzer, W., & Rote, G. (2006) Minimum-weight triangulation is NP-hard. *Annual Symposium on Computational Geometry, 6*, 86-90.

Mumford, E. (2006). The story of socio-technical design: Reflections on its successes, failures, and potential. *Information Systems Journal, 16*(4), 317–342. doi:10.1111/j.1365-2575.2006.00221.x

Muñoz-Marrón, D. (2018). Human factors in aviation: CRM (Crew Resource Management). *Papeles del Psicólogo / Psychologist Papers, vol. 39(3)*, 191-199. doi:. doi:10.23923/pap.psicol2018.2870

Muromtsev, D. Yu., & Shamkin, V. N. (2015). Optimization methods in project decision making. Training manual for master degree students. Tambov State Technical University, Russia, 1–80.

Murrieta-Mendoza, A., & Botez, R. (2015). *Aircraft vertical route optimization deterministic algorithm for a flight management system* (No. 2015-01-2541). SAE Technical Paper.

Nakagawa, T. (2005). *Maintenance theory of reliability*. Springer.

Nanduri, A., & Sherry, L. (2016). *Anomaly detection in aircraft data using Recurrent Neural Networks (RNN). In 2016 Integrated Communications Navigation and Surveillance (ICNS) (pp. 5C2-1)*. IEEE.

Nesterov, A. A., & Panich, A. A. (2010). Sovremennye problemy materialovedeniy keramicheskih p'ezoelektricheskih materialov: monografiya [Modern Problems of Materials ceramic piezoelectric materials]. P'ezoelektricheskoe priborostroenie. V. 8, Rostov on Don, 163-171.

Neural networks. (2008). Statistica neural networks. *Methodology and technologies of modern data analysis*. Moscow, Russia: Hotline - Telecom.

Nilsson, N. J. (2014). *Principles of artificial intelligence*. Morgan Kaufmann.

NOAA, & National Weather Service. (2013, March 15). AWC - Aviation Weather Center. Retrieved from https://www.aviationweather.gov/metar/data

Nubert, G. P. (1997). *Measuring transducers of non-electric quantities. Introduction to theory, calculation and design*. Moscow, Russia: Bukinist.

Okhotsimsky, D. E., & Sikharulidze, Y. G. (1990). *Basics of space flight mechanics*. Moscow, Russia: Science.

Ornatsky, D. P., & Kataeva, M. A. (2015). Development of computerized system for control of curved surfaces using inductive sensors. *Technological Audit and Production Reserves, 1/2*, 83-90.

Ornatsky, D. P., Mikhalko, N. V., & Kataeva, M. O. (2016). Development of a computerized system for controlling curved surfaces using inductive sensors. *Technological Audit and Production Reserves, 1/2(27)*, 83-90.

Osadchiy, E. P. (1979). Design of sensors for measuring mechanical quantities. Moscow, Russia. *Mechanical Engineering (New York, N.Y.)*, 134–146.

Osadchy, E. P. (1979). Designing sensors for measuring mechanical quantities. Moscow, Russia. *Mechanical Engineering (New York, N.Y.)*, 178–182.

Ostroumov, I. V., & Kuzmenko, N. S. (2018). Accuracy assessment of aircraft positioning by multiple radio navigational AIDS. *Telecommunications and Radio Engineering, 77(8)*, 705–715. doi:10.1615/TelecomRadEng.v77.i8.40

Ostroumov, I. V., & Kuzmenko, N. S. (2018). An area navigation (RNAV) system performance monitoring and alerting. In *Proceedings of the International Conference on System Analysis and Intelligent Computing, SAIC 2018*, (pp. 211–214). 10.1109/SAIC.2018.8516750

Ostroumov, I. V., Kuzmenko, N. S., & Marais, K. (2018). Optimal pair of navigational aids selection. In *Proceedings of the International Conference on Methods and Systems of Navigation and Motion Control, MSNMC 2018*, (pp. 90–93). doi:10.1109/MSNMC.2018.8576293

Ostrov, K., & Vittenmark, B. (1987). Control systems with computers. Per. with Eng. Moscow, Russia. *WORLD (Oakland, Calif.)*, 234–252.

Ovchenkov, N., & Elisov, L. (2014). Vulnerability assessment of transport infrastructure and vehicles in civil aviation. *Scientific Herald of the Moscow State Technical University, 204*, pp. 65-68.

Padhy, N. P. (2005). *Artificial intelligence and intelligent systems*. Oxford, UK: Oxford University Press.

Paielli, R. A., & Erzberger, H. (1997). Conflict probability estimation for free flight. *Journal of Guidance, Control, and Dynamics, 20*(3), 588–596. doi:10.2514/2.4081

Paquay, C., Schyns, M., & Limbourg, S. (2011). Three dimensional bin packing problem applied to air transport. *Colloque SIL, Dec. 15-16*, 1-6.

Parwin, M. (2015). *Concert for neurons and synapses*. Moscow, Russia: Nauka.

Pashayev, A. B., & Sabziev, E. N. (2013). Description of fluid flow as a system of systems. In *Proceedings of 8th International Conference on System of Systems Engineering*, June 2-6, pp. 28-33, Maui, Hawaii.

Pasmore, W. (2001). Action research in the workplace: The socio-technical perspective. In *P. Reason & H. Bradbury (2006) Handbook of action research: Concise paperback* (pp. 38–48). London, UK: Sage.

Patton, R. J. (1991). Fault detection and diagnosis in aerospace systems using analytical redundancy. *Computing & Control Engineering Journal, 2*(3), 127–136. doi:10.1049/cce:19910031

Pearson, E. S. (2003). Biometrika. *Annals of Mathematical Statistics, 8*, 118–124.

Pearson, G. M. (1988). *Satellite imagery interpretation in synoptic and mesoscale meteorology*. Environment Canada.

Phillips, G. M. (2012). *Interpolation and Approximation by Polynomials*. Berlin, Germany: Springer.

Podorozhniak, A., Lubchenko, N., Balenko, O., & Zhuikov, D. (2018). Neural network approach for multispectral image processing. In *Proceedings 2018 14th International Conference on Advanced Trends in Radioelectronics, Telecommunications, and Computer Engineering (TCSET)*. 10.1109/TCSET.2018.8336357

Poltavets, V. A., & Teymurazov, R. A. (2007). Safety analysis of aircraft with gas turbine engines for a 50-year period (1957-2006) operation in the USSR civil aviation. Paris, France.

Polyak, B. T., & Shcherbakov, P. S. (2005). Difficult problems of linear control theory. Some approaches to solution. *Automation and Telemechanics, 5*, pp. 7–46. (in Russian)

Polyak, B. T., & Shcherbakov, P. S. (2002). *Robust stability and control*. Moscow, Russia: Nauka. (in Russian)

Pontryagin, L. S. (1986). *Generalization of numbers*. Moscow, Russia: Science.

Pozin, N. (1970). *Modeling of Neural Structure*. Moscow, Russia: Nauka.

Prabhakar, B., Uysal-Biyikoglu, E., & Gamak, A. E. (2000). Energy-efficient transmission over a wireless link via lazy packet scheduling. In Proceedings IEEE InfoCom, (Apr. vol. 1, pp. 386-394).

Prandini, M., Lygeros, J., & Sastry, S. S. (2000). A probabilistic approach to aircraft conflict detection. *IEEE Transactions on Intelligent Transportation Systems, 1*(4), 199–220. doi:10.1109/6979.898224

Pupkov, K. A. (2004). Methods of classical and modern automatic control theory. Vol. 2. In K. A. Pupkov, & N. D. Egupov (Eds.), Statistical dynamics and identification of automatic control systems. Moscow, Russia: MSTU named after N.E. Bauman. (in Russian)

Raikunov, G. G. (2011). *Optimization of the ballistic support of the overflight of the spacecraft system in a circular orbit*. Moscow, Russia: Fyzmatlyt.

Rashid, T. (2017). *Create a neural network*. St. Petersburg, Russia: LLC Alpha-book.

Rausand, M. (2004). *System reliability theory: models, statistical methods and applications*. New York: John Wiley & Sons.

Raz Mitache. (2019). *How to get AI right using enterprise architecture*. Retrieved from https://bizzdesign.com/blog/how-to-get-ai-right-using-enterprise-architecture/

Raz, T., & Michael, E. (2001). Use and benefits of tools for project risk management. *International Journal of Project Management, 19*(1), 9–17. doi:10.1016/S0263-7863(99)00036-8

Reason, J. (1990). *Human error*. Cambridge, UK: Cambridge University Press. doi:10.1017/CBO9781139062367

Rizun, N., & Shmelova, T. (2016). Decision-making models of the human-operator as an element of the socio-technical systems. In R. Batko & A. Szopa (Eds.), *Strategic imperatives and core competencies in the era of robotics and artificial intelligence. Manuscript* (pp. 167–204). Hershey, PA: IGI Global. doi:10.4018/978-1-5225-1656-9.ch009

Ruhnow, W. B., & Goemaat, M. L. (1982). VOR/DME automated station selection algorithm. *Navigation, 29*(4), 289–299. doi:10.1002/j.2161-4296.1982.tb00814.x

Rusell, S., & Norvig, P. (2003). *Artificial intelligence: A modern approach* (2nd ed.). Upper Saddle River, NJ: Prentice Hall.

Russell, J. (2013). *Artificial Neuron*. Moscow, Russia: VSD.

Russell, J. (2013). *Verbalization of Neural Networks*. Moscow, Russia: VSD.

Russell, S. J., & Norvig, P. (2016). *Artificial intelligence: a modern approach*. Malaysia: Pearson Education Limited.

Sabour, S., Frosst, N., & Hinton, G. E. (2017). Dynamic routing between capsules. Paper presented at the Advances in Neural Information Processing Systems.

Safety Management Manual. (2013), Doc 9859, AN/474, 251 p.

Sage, A. P., & White, Ch. C. (1982). Optimum systems control. Moscow, Russia: Radio and Connection.

Sahun, Ye. (2019). Current status of aircraft load optimization problem. *Proceedings of the National Aviation University, 1*(78), 35–39.

Sainath, T. N., Mohamed, A.-R., Kingsbury, B., & Ramabhadran, B. (2013). Deep convolutional neural networks for LVCSR. In *Proceedings 2013 IEEE International Conference on Acoustics, Speech, and Signal Processing*. 10.1109/ICASSP.2013.6639347

Salyga, V. I., Sirodga, I. B., Kulik, A. S., & Obruchev, V. L. (1989). Synthesis of fault-tolerant dynamic control systems with fault identification. PROBL. CONTROL INF. *THEORY, 18*(1), 43–54.

Schrijver, A. (1998). *Theory of linear and integer programming*. John Wiley & Sons.

Serridge, M., & Licht, T. (1987). *Handbook of piezoelectric accelerometers and preamps* (pp. 164–172). Denmark.

SESAR Joint Undertaking. (2017b). *EATMA Guidance Material Version 9.0.*

SESAR Joint Undertaking. (2018). *Artificial intelligence in air traffic management.*

SESAR Joint Undertaking. (2019). *eATM Portal. Research and Development View [Dataset]*. Retrieved from https://atmmasterplan.eu/rnd/rd-dashboard

Shalygin, A. S., Petrova, I. L., & Sannikov, V. A. (2010). *Parametric optimization methods in the flight dynamics of unmanned aerial vehicles*. St. Petersburg, Russia.

Sharapov, V. M., Musienko, M. P., & Sharapova, E. V. (2006). Piezoelectric sensors. Moscow, Russia: Technosphere, 230-245.

Shibzukhov, Z. (2006). *Constructive methods of teaching sigma-p neural networks*. Moscow, Russia: Nauka.

Shi, J., & Zhou, S. (2009). Quality control and improvement for multistage systems: A survey. *IIE Transactions, 41*(9), 744–753. doi:10.1080/07408170902966344

Shmelova, T., & Konovalova, A. (2017). Decision making of ATCO in emergency situation aircraft decompression. In *Proceedings of the International Conference AVIA -2017*. Kiev, Ukraine.

Shmelova, T., & Shyshakov, V. (2016). System analysis of internal system and external system factors of the non-deterministic subsystem "crew-aircraft". *Bulletin of the Engineering Academy of Ukraine, no. 4*, 48-54.

Shmelova, T., & Sikirda, Y. (2016). Calculation the scenarios of the flight situation development using GERT's and Markov's networks. In *Proceedings of the Conference "Management of high-speed moving objects training operators of complex systems"*. Kirovograd, Ukraine.

Shmelova, T., & Sikirda, Y. (2017). Assessments of the influence of organizational factors on flight safety in air traffic control. *Proceedings of the Kharkov National University of Air Forces*.

Shmelova, T., Bondarev, D., & Znakovska, Y. (2016). Modelling of the decision making by UAV's operator in emergency situations. In *Proceedings of the IEEE 4th International Conference on Methods and Systems of Navigation and Motion Control (MSNMC-2016)*. Kiev, Ukraine.

Shmelova, T., Danilenko, O., & Boldysheva, T. (2015). Expert estimation of human factor using Reason model's components. In *Proceedings of the XII International conference AVIA -2015*. Kiev, Ukraine.

Shmelova, T., Dolhov, D., & Muliar, P. (2015) Mathematical models of operators' activities in aeronavigation socio-technical system. In Proceedings of the Conference "Techniques and technology. Science yesterday, today, tomorrow". Warsaw, Poland.

Shmelova, T., Mirskova, M., & Krepak, D. (2017). Decision making in emergency situation of air navigation system's operator. In *Proceedings of the International Conference AVIA -2017*. Kiev, Ukraine.

Shmelova, T., Shyshakov, V., & Shostak, O. (2015). Deterministic models of the operations of the aircraft crew in the case of an unusual situation in flight. *Collegiate of Scientific Works of the Kharkiv National University of the Air Forces, no. 2(19)*, 33-37.

Shmelova, T., Sikirda, Y., & Kasatkin, M. (2019). Applied artificial intelligence for air navigation sociotechnical system development. In *Proceedings of the 15th International Conference on ICT in Education, Research, and Industrial Applications. Integration, Harmonization, and Knowledge Transfer*. Ukraine, pp. 470-475.

Shmelova, T., Sikirda, Y., & Kovalyov, Y. (2013). Models of personality. Behaviour and activity of air navigation system's human-operator: In *Proceedings of the Conference "Management of high-speed moving objects training operators of complex systems "*. Kirovograd, Ukraine.

Shmelova, T., Sikirda, Y., & Sunduchkov, K. (2013). Socio-technical analysis of air navigation system. *Science and technology of the Air Forces of the Armed Forces of Ukraine, 4*(13), 34–39.

Shmelova, T., Sikirda, Y., Rizun, N., Abdel-Badeeh, M., Salem, K. Yu. (2017). Socio-technical decision support in air navigation systems: Emerging research and opportunities. Publisher of Progressive Information Science and Technology Research, Pennsylvania.

Shmelova, T., Sikirda, Yu., Scarponi, C., & Chialastri, A. (2018). Deterministic and stochastic models of decision making in air navigation socio-technical system. In *Proceedings of the 14th International Conference "ICT in Education, Research and Industrial Applications. Integration, Harmonization and Knowledge Transfer" (ICTERI 2018)*, vol. 2: Workshops, part 3: 4th international workshop on theory of reliability and Markov modelling for information technologies (TheRMIT 2018). Kiev, Russia. Taras Shevchenko National University of Kyiv.

Shmelova, T., Sterenharz, A., & Sikirda, Y. (2018). The diagnostics of emotional state deformation in the professional activity of operators in the air navigation system. 8th World Congress. Aviation in the XXIst century. Safety in Aviation and Space Technologies NAU.

Shmelova, T., & Sikirda, Y. (2012). Analysis of human-operator's decision-making in air navigation system. *Radioelectronic and Computer Systems*, 7(59), 319–324.

Shmelova, T., Sikirda, Y., & Jafarzade, T. (2013). Models of flight situations development while decision-making by air navigation system's human-operator. *Information Processing Systems*, 8(115), 136–142.

Shmelova, T., & Sikirda, Yu. (2018). Models of decision-making operators of socio-technical system. In T. Shmelova, Yu. Sikirda, N. Rizun, A.-B. M. Salem, & Yu. Kovalyov (Eds.), *Socio-technical decision support in air navigation systems: emerging research and opportunities. Manuscript* (pp. 21–48). Hershey, PA: IGI Global. doi:10.4018/978-1-5225-3108-1.ch002

Shovengerdt, R. A. (2010). Remote sensing. Models and methods of image processing. Moscow, Russia: Technosphere, 560.

Shvets, V., Ilnytska, S., & Kutsenko, O. (2019). Application of computer modelling in adaptive compensation of interferences on global navigation satellite systems. In T. Shmelova, Y. Sikirda, N. Rizun, & D. Kucherov (Eds.), *Cases on modern computer systems in aviation* (pp. 339–380). Hershey, PA: IGI Global; doi:10.4018/978-1-5225-7588-7.ch014

Sikharulidze, Y. G. (1982). *Ballistics of aircraft*. Moscow, Russia: Science.

Sikirda, Yu., & Shmelova, T. (2018). Socionics and sociometry diagnosting of air navigation system's operator. In T. Shmelova, Yu. Sikirda, N. Rizun, A.-B. M. Salem, & Yu. Kovalyov (Eds.), *Socio-technical decision support in air navigation systems: emerging research and opportunities: Manuscript* (pp. 70–90). Hershey, PA: IGI Global. doi:10.4018/978-1-5225-3108-1.ch004

Sikirda, Yu., Shmelova, T., & Tkachenko, D. (2018). Evaluation of influence of the organizational risk factors on flight safety in air traffic control. *Proceedings of the National Aviation University, no. 2(75)*, 26-34. 10.1109/MSNMC.2018.8576317

Simonov, C. B. (2003). *Modern concepts of information risk management. Rostov-na-Donu*. Southern Federal University.

Sivagami, A., Pavai, K., & Sridharan, D. (2010). Latency optimized data aggregation timing model for wireless sensor networks. *IJCSI International Journal of Computer Science Issues, 7*(3), no. 6.

Sivaram, M., Porkodi, V., Mohammed, A. S., & Manikandan, V. (2019). Detection of accurate facial detection using hybrid deep convolutional recurrent neural network. *ICTACT Journal on Soft Computing*, 09(02), 1844–1850. doi:10.21917/ijsc.2019.0256

Skogestad, S., & Postlethwaite, I. (2001). *Multivariable feedback control*. New York: John Wiley & Sons.

Smazhevskaya, E. G., & Feldman, N. B. (1971). P'ezoelektricheskaya keramika [Piezoelectrics ceramic]. Moscow, Russia: Sovetskoe radio, 83-88.

Smith, D. J. (2005). *Reliability, maintainability and risk. Practical methods for engineers.* London, UK: Elsevier.

Smolyak, S. A. (1992). Risk accounting when setting the discount rate. *Economics and mathematical methods*, 23-27.

Solnitsev, R. I. (1991). *Automation of Design of Automatic Control Systems.* Moscow, Russia: Vishaya Shkola. (in Russian)

Solomentsev, O. V., Herasymenko, T. S., Zaliskyi, M. Y., & Cheked, I. V. (2016). Statistical data processing procedures for ground navigation equipment. In *Proceedings of IEEE Fourth International Conference Methods and Systems of Navigation and Motion Control, Oct. 18-20, 2016.* 182-185.

Solomentsev, O., & Zaliskyi, M. (2018). Correlated failures analysis in navigation system: Methods and systems of navigation and motion control. *IEEE 5th International Conference,* Kiev, Ukraine. pp. 41-44.

Solomentsev, O., Kuzmin, V., Zaliskyi, M., Zuiev, O., & Kaminskyi, Y. (2018). Statistical data processing in radio engineering devices operation system. In *Proceedings of the International Conference on Advanced Trends in Radioelectronics, Telecommunications and Computer Engineering,* TCSET 2018, (pp. 757–760). 10.1109/TCSET.2018.8336310

Solomentsev, O. V., Melkumyan, V. H., Zaliskyi, M. Yu., & Asanov, M. M. (2015). UAV operation system designing. In *Proceedings of IEEE 3rd International Conference on Actual Problems of Unmanned Air Vehicles Developments (APUAVD),* (pp. 95–98). Kiev, Ukraine. 10.1109/APUAVD.2015.7346570

Solomentsev, O., Zaliskyi, M., Herasymenko, T., Kozhokhina, O., & Petrova, Yu. (2018). Data processing in case of radio equipment reliability parameters monitoring. In *Proceedings of IEEE International Conference on Advances in Wireless and Optical Communications (RTUWO 2018).* (pp. 219–222). Riga, Latvia. 10.1109/RTUWO.2018.8587882

Solomentsev, O., Zaliskyi, M., Kozhokhina, O., & Herasymenko, T. (2017). Efficiency of data processing for UAV operation system. In *Proceedings of IEEE 4th International Conference on Actual Problems of UAV Developments (APUAVD).* (pp. 27–31). Kiev, Ukraine. 10.1109/APUAVD.2017.8308769

Solomentsev, O., Zaliskyi, M., Nemyrovets, Yu., & Asanov, M. (2015). Signal processing in case of radio equipment technical state deterioration. In *Proceedings of Signal Processing Symposium 2015 (SPS 2015).* (pp. 1–5). Debe, Poland. 10.1109/SPS.2015.7168312

Solomentsev, O., Zaliskyi, M., & Zuiev, O. (2016). Estimation of quality parameters in the radio flight support operational system. *Aviation, 20*(3), 123–128. doi:10.3846/16487788.2016.1227541

Souffriau, W., Demeester, P., Vanden Berghe, G., & De Causmaecker, P. (2008). The aircraft weight and balance problem. *Proceedings of ORBEL, 22,* 44–45.

State Aviation Administration of Ukraine. (2005). *Regulations on the flight safety management system in aviation transport.* Kiev, Russia: Author.

State Aviation Administration of Ukraine. (2017). *Report of the implementation of flight safety oversight functions in the ATMS in Ukraine 2016.* Kiev, Russia: Author.

Strelkov, Yu. K. (2001). *Engineering and professional psychology: Proc. Allowance for Stud. Higher Training Institutions.* Moscow, Russia: Academy Publishing Center.

Suhl, Uwe H., & Leena, M. (1999). Solving airlinefleet scheduling problems with mixed-integer programming. *Operational Research in Industry,* 135–156. doi:. doi:10.1057/9780230372924_7

Sushchenko, O. A. (2015). Design of two-axis robust system for stabilization of information-measuring devices operated at UAVs. In *Proceedings of IEEE 3rd International Conference Actual Problems of Unmanned Aerial Vehicles Developments, (APUAVD-2015),* (pp. 198–201), Kiev, Ukraine.

Sushchenko, O. A., & Goncharenko, A. V. (2016). Design of robust systems for stabilization of unmanned aerial vehicle equipment. International Journal of Aerospace Engineering. Article ID 6054081.

Sushchenko, O. (2017). Approach to computer-aided design of UAV robust inertial platforms. In *Proceedings of IEEE 4th International Conference on Actual Problems of Unmanned Air Vehicles Developments (APUAVD 2017)*, (pp. 53–57). Kiev, Ukraine. 10.1109/APUAVD.2017.8308775

Sushchenko, O. A., & Golitsyn, V. O. (2016). Data processing system for altitude navigation sensor. In *Proceedings of IEEE 4th International Conference on Methods and Systems of Navigation and Motion Control (MSNMC)*. (pp. 84–87). Kiev, Ukraine. 10.1109/MSNMC.2016.7783112

Systems engineering guide for systems of systems. (2010). *Essentials*. Washington, D.C.: Office of the Director, Defense Research and Engineering, Director of Systems Engineering.

Tala't, M., Yu, Ch.-M., Ku, M.-L., & Feng, K.-T. (2017). On hybrid energy utilization in wireless sensor networks. *Energies*, *10*(12), 1940. doi:10.3390/en10121940

Tan Wei Min, A., Sagarna, R., Gupta, A., Ong, Y.-S., & Keong Goh, C. (2017). Knowledge transfer through machine learning in aircraft design. *IEEE Computational Intelligence Magazine, 12*(4), 48–60. doi:10.1109/MCI.2017.2742781

Tanaka, S., Okita, T., & Müller, P. C. (1987). Failure detection in linear discrete dynamical systems and its detection-ability. *IFAC Proceedings, 20*(5), 75-80. 10.1016/S1474-6670(17)55355-0

Tartakovsky, A., Nikiforov, I., & Basseville, M. (2015). *Sequential analysis: hypothesis testing and changepoint detection*. New York: Taylor & Francis Group.

Tatuzov, A. (2009). *Neural networks in the problems of radar*. Moscow, Russia: Radiotechnic.

The Departmant of Forestry and Fire Protection of California. Public Information Map. Retrieved from http://cdfdata.fire.ca.gov/pub/cdf/images/incidentfile2277_4287.pdf

The Department of Forestry and Fire Protection of California. Top 20 Largest California Wildfires. Retrieved from https://www.fire.ca.gov/communications/downloads/fact_sheets/Top20_Acres.pdf

The Open Group Standard. (2018b). *The TOGAF® Standard, Version 9.2.*

The World Bank of Data. (2016). Air transport, registered carrier departures worldwide. Registered carrier departures worldwide are domestic takeoffs and takeoffs abroad of air carriers registered in the country, 8-10.

Thomas, C., Campbell, K., Hines, G., & Racer, M. (1998). Airbus packing at Federal Express. *Interfaces, 28*(4), 21–30. doi:10.1287/inte.28.4.21

Thompson, W. R. (2002). The problem of negative estimates of variance components. *Annals of Mathematical Statistics*, *12*(2), 214–265.

Tietze, U., & Schenk, K. (2007). Semiconductor circuitry. 12th ed. Vol. II: Trans. with him. Moscow, Russia: DMK Press,144-147.

Tkachenko, V. A. (2009). *Flight risk. KIT*. Tersky.

TOGAF® Worldwide. (2015). [Image illustrating the worldwide usage of TOGAF, May 2019]. *TOGAF® Worldwide*. Retrieved from http://web.archive.org/web/20181017230812/http://www.opengroup.org/subjectareas/enterprise/togaf/worldwide

Tolkachev, S. (2016). *Neural programming of dialogue systems*. Moscow, Russia: RSUU.

Torres, S. (2012). Swarm theory applied to air traffic flow management. *Procedia Computer Science, 12*, 463–470. doi:10.1016/j.procs.2012.09.105

Transmission of audio and/or video and related signals using infra-red radiation. (2015). *State standard specification IEC 61603-1-2014*. Part 1. Moscow, Russia.

Trivella, A., & Pisinger, D. (2017). *Bin-packing problems with load balancing and stability constraints*. Chicago, IL: INFORMS Transportation and Logistics Society.

Tsypkin, Y. Z., & Nikolic, Z. J. (1971). *Adaptation and learning in automatic systems* (Vol. 73). New York: Academic Press.

Tunik, A. A., & Sushchenko, O. A. (2013). Usage of vector parametric optimization for robust stabilization of ground vehicles information-measuring devices. *Proceedings of the National Aviation University, 4*, 23–32. 10.18372/2306-1472.57.5530

Tunik, A. A., Rye, H., & Lee, H. C. (2001). Parametric optimization procedure for robust flight control system design. *KSAS International Journal, 2*(2), 95–107.

U. S. Forest Service. Managing fire (2019). Retrieved from https://www.fs.fed.us/science-technology/fire

U.S. Geological Survey. (2019). Retrieved from https://earthexplorer.usgs.gov/

ubaev, G. N. (1998). Comparison of complex software systems by the criterion of functional completeness. *Software products and systems (SOFTWARE & SYSTEMS)*, 7-9.

United States Department of Defense. (2007). *DoD Architecture Framework Version 1.5* (Vol. I). Washington, D.C.: Definitions and Guidelines.

United States Department of Defense. (2010). DoD Architecture Framework Version 2.02. Washington D.C.

Vasyliev, D.V. (2013). Matematychna model kerovanoho rukhu litaka dlia doslidzhennia protsesiv aeronavihatsiinoho obsluhovuvannia polotiv [Mathematical model of controlled aircraft motion for analysis of air navigation service processes]. *Systemy ozbroiennia i viiskova tekhnika, 23*, 63-67.

Vasyliev, D. (2016). Method of multi-objective resolution of two-aircraft conflict in three-dimensional space based on dynamic programming. *Proceedings of the National Aviation University, 68*(3), 35–45. doi:10.18372/2306-1472.68.10907

Velychko, O., Kolomiyets, L., Hordiyenko, T., Karpenko, S., & Haber, A. (2015). *Group expert assessment and experts' competence: monograph*. Odesa, Ukraine: Aprel Publ.

Veremey, E. I. (2012). *Introduction in analysis and synthesis of robust control systems*. Retrieved from http//MatLab. exponenta. ru/optimrobust/ book1/ ndex.php

Villarroel, J., & Rodrigues, L. (2016). Optimal control framework for cruise economy mode of flight management systems. *Journal of Guidance, Control, and Dynamics, 39*(5), 1022–1033. doi:10.2514/1.G001373

Vinogradov, G. P. (2013). *Methods and algorithms of intellectualization of decision-making in ACS productions with continuous and discrete technology*. (Doctor's thesis, Tver, Russia).

Wang M., & Gross, L. E. (1998). Piezoelectric pseudo-shear multilayer actuator, 84-91.

Weiss, G. (Ed.). (1999). *Multiagent systems: A modern approach to distributed artificial intelligence*. Cambridge, MA: MIT Press.

Westhoff, D., Girao, J., & Sarma, A. (2017). *Security solution for wireless sensor networks. General papers*. Retrieved from https://www.nec.com/en/global/techrep/journal/g06/n03/pdf/t060322.pdf

WHITEPAPER. (2013). *Low latency systems.* Ittiam. Systems Pvt. Ltd. February 2013. Retrieved from https://www.ittiam.com/wp-content/uploads/2017/12/WP005_low-latency-systems.pdf

Wiegmann, D. A., & Shappell, S. A. (2017). *A human error approach to aviation accident analysis: The human factors analysis and classification system.* Routledge. doi:10.4324/9781315263878

Wilamowski, B. M., & Irwin, J. D. (2017). *Intelligent systems. The industrial electronics handbook.* Boca Raton, FL: CRC Press.

Wilson, I. D., Roach, P. A., & Ware, J. A. (2001). Container stowage pre-planning: using search to generate solutions. *Knowledge-Based Systems, (14)*3–4, 137–145.

Windyty, S. E. (0AD). Windy as forecasted. Retrieved from https://www.windy.com/

Wu, N., & Yu, J. (2018) Robust controller design of hypersonic vehicles in uncertainty models. In *Proceedings of 3rd International Conference on Electromechanical Control Technology and Transportation (ICECT-2018)*, (p. 288-293). Huazhong, China. 10.5220/0006969302880293

Wu, S., Clements-Croome, D., Fairey, V., Albany, B., Sidhu, J., Desmond, D., & Neale, K. (2006). Reliability in the whole life cycle of building systems. *Eng. Constr. Architectural Manage, 13*(2), 136–153. doi:10.1108/09699980610659607

Yablonsky, S. V. (1988). Some problems of reliability and control of control systems. *Mathematical Problems of Cybernetics, 1.* pp. 5-25.

Yang, W. Y., Cao, T. S., Chung, I., & Morris, J. (2005). *Applied numerical methods using MATLAB. New York.* John Wiley & Sons. doi:10.1002/0471705195

Yanoff, M., & Duker, J. S. (2009). Ophthalmology. 3rd Edition. MOSBY Elsevier. p. 54.

Yasnitsky, L. (2005). *Introduction to artificial intelligence.* Moscow, Russia: Center. *Academia (Caracas).*

Yastrub, M. I., & Kredentsar, S. M. (2018). Enterprise architecture as a driver of transformation in air traffic management. In *Proceedings of 2018 IEEE 5th International Conference on Methods and Systems of Navigation and Motion Control (MSNMC 2018)*, 162-165. 10.1109/MSNMC.2018.8576318

Yorish, Yu. I. (1963) Vibrometry. Second edition, revised and updated. Moscow, Russia: State. Scientific and Technical Mechanical Engineering Publishing House, 63-67.

Yunusov, F. S. (2012). On the problem of mathematical modeling of formation processes of complex surfaces, which are processed on machines with software control *Proceedings of the KAI, 141*, 48-53.

Yurevich, A. (2014). *Neural networks in economics.* Moscow, Russia: Lambert Academic Publishing.

Yu, Y., Prasanna, V. K., & Krishnamachari, B. (2006). Energy minimization for real–time data gathering in wireless sensor networks. *IEEE Transactions on Wireless Communications, 5*(11), 3088–3096. doi:10.1109/TWC.2006.04709

Zadeh, L. A. (1963). On the definition of adaptivity. *Proceedings of the IEEE, 51*(3), 469–470. doi:10.1109/PROC.1963.1852

Zadeh, L. A. (2002). Toward a perception-based theory of probabilistic reasoning with imprecise probabilities. *Journal of Statistical Planning and Inference,* 230–245.

Zaliskyi, M., & Solomentsev, O. (2014) Method of sequential estimation of statistical distribution parameters. In *Proceedings of IEEE 3rd International Conference Methods and Systems of Navigation and Motion Control (MSNMC-2014)*, (p. 135-138). Kiev, Ukraine.

Zaliskyi, M., Solomentsev, O., Kozhokhina, O., & Herasymenko, T. (2017). Reliability parameters estimation for radio-electronic equipment in case of change-point. *In Proceedings of Signal Processing Symposium 2017 (SPSympo 2017).* (pp. 1–4). Jachranka Village, Poland. 10.1109/SPS.2017.8053676

Zeghal, K. (1998). A review of different approaches based on force fields for airborne conflict resolution. *Proceedings of AIAA Guidance, Navigation, and Control Conference and Exhibit,* 818-827. 10.2514/6.1998-4240

Zemlyakov, V. L. (2012). Piezoelectrics and related materials. *Investigations and Applications,* 117–142.

Zhaoa, X., Bennella, B., Julia, A. T., & Dowsland, K. (2016). A comparative review of 3D container loading algorithms. *International Transactions in Operational Research,* 23(1-2), 287–320. doi:10.1111/itor.12094

Zimmermann, H. J. (2006). *Fuzzy set theory & its application* (4th ed.). USA: Softcove, Kluwer Academic Press Publ.

Zwillinger, D. (1997). *Handbook of differential equations* (3rd ed.). Boston, MA: Academic Press.

About the Contributors

Tetiana Shmelova, Doctor of Science, Professor of Department of Air Navigation Systems in National Aviation University (Ukraine). Areas of Scientific Interests: Artificial Intelligence; Mathematical Models of Decision-Making by operators in the Air Navigation System, especially in Emergency; Decision Support Systems; Research of Air Navigation system as Socio-technical system; Professional Effectiveness using Aviation Sociometry and Socionics; Management and Marketing in aviation; Information technology and Informatics of Decision-Making in the Air Navigation System; Problems of Human Factors in aviation. Author of more than 300 scientific articles, books, methodical manuals, copyright certificates, guides, monographs in fields of aviation, economics, mathematics, the theory of system. Teaching courses: Decision Making; Mathematical Programming; Effectiveness of Air Traffic Management; Effectiveness of Unmanned Systems; Artificial Intelligence.

Yuliya Sikirda, Associate Professor of Management, Economy, Law and Tourism Department at the Flight Academy of National Aviation University. In 2001 has received a Master degree on speciality "Air Traffic Services" specialization "Air Traffic Control". In 2004 has got a diploma of PhD in Technical Sciences on speciality "Automated Control Systems and Advanced Information Technologies". In 2006 has received an attestat of Associate Professor of Management and Economy Department. Since 2014 is the member of the specialized scientific council for the defence of dissertations for PhD in Technical Sciences on the speciality "Navigation and Traffic Control". The author and co-author of about 180 scientific works, including three manuscripts and more than 40 articles in specialized scientific publications, three copyright certificates on computer programs, above 80 teaching and methodological works. Areas of scientific interests: increasing of the decision-making efficiency by the human-operator of the Air Navigation System in flight emergencies; assessment of the influence of aviation enterprises' management environment factors on the activity of the Socio-technical System's operator.

Arnold Sterenharz, Managing Director of ECM Space Technologies GmbH, Dr. Ing. Arnold Sterenharz, will represent the company in the consortium. Relevant skills: Dr.-Ing., Master of Business and Administration. Experience in space technologies and partnerships with research companies in the field of GIS and EO systems, focusing on space tasks and services; administrative manager in 16 European educational and research projects: 7 TEMPUS projects, 3 Erasmus+ projects, an Erasmus Mundus project and 5 FP7 research projects.

* * *

Azad Bayramov, ScD, professor Azad Bayramov has completed his MSc in Azerbaijan State University, Physical faculty, Baku, Azerbaijan in 1975. He has defended his D.'s dissertation in 2000. He is currently serving as a professor in War College of Armed Forces of the Azerbaijan Republic. His research areas include development of technical systems, development of new detection methods of military objects and armoured fighting vehicles, mathematical modelling of combat activities. His contribution to the current study include quantitative assessment of invisible areas and military objects, modelling of UAV flights, comparison data.

Serhii Boiko, PhD, Head of the Department of Air Transport and Electricity and Control in Kremenchuk Flight Colledge of Kharkiv National University of Internal Affairs (Ukraine). Areas of Scientific Interests: Systems of Power Saving and Power Consumption of Industrial Enterprises; Complexes of Distributed Generation Sources; Electric Power Supply Systems for Aircraft; Power Supply Systems for Airfields and Airports. Author of more than 100 scientific articles, 3 articles in Scopus, 21 patents of Ukraine on a utility model, 6 monographs in fields of aviation, power industry and avionics.

Viktoriia Chorna, PhD, Lecturer, Head of the Cyclic Commission in Kremenchuk Flight Colledge of Kharkiv National University of Internal Affairs (Ukraine). Areas of Scientific Interests: Energy Management; Power Supply Systems for Airfields and Airports. Author of more than 100 scientific articles, articles in Scopus, patents of Ukraine on a utility model, methodical manuals, monographs in fields of aviation, power industry and energy management.

Konstantin Dergachov is Head of the Department of Aircraft Control Systems of the National Aerospace University KhAI, Kharkov, Ukraine.

Serge Dolgikh has graduated from National Nuclear University in Moscow having published works on the topology of super-manifolds in the string theory. He went on to study Telecommunications Management and Engineering at Coventry University, United Kingdom, having graduated with a Masters Degree in Operational Telecommunications. Serge has worked in the industry on a number of advanced projects in the domains of telecommunications and information technology maintaining interests in fundamental and applied science. His current research interests include machine learning and Artificial Intelligence, including unsupervised learning and self-learning systems, artificial evolutionary systems, information theory, and self-organizing systems. He has a number of publications in international peer-reviewed editions in these fields.

Elshan Hashimov, PhD, professor, colonel Elshan Hashimov has completed his MSc in Kharkov Military University, Ukraine in 1998. He has defended his PhD's dissertation in 2000. He is currently serving as the head of Department of Adjunction and Military Sciences of War College of Armed Forces of the Azerbaijan Republic. His research areas include quantitative analysis of mountainous terrain by Geography Information Systems and photogrammetry technologies. His contribution to the current study include application GIS and photogrammetry technologies for research of invisible areas and military objects.

Yurii Hryshchenko, Ph.D., Associate Professor of Avionics Department in National Aviation University (Ukraine). Thesis in the field of pilots training for flights in special cases (1995). Research interests: flight safety and reliability of technical and ergatic systems. Author of 43 scientific articles. Teaching courses: Dependability of Avionics Systems, Technical Diagnostics of Avionics, Engineering Psychology, Ergonomics and Human Factor in Aviation, Fundamentals of Avionics Operation, Computer-Integrated Complexes of Flight and Navigation Equipment.

Mariia Kataieva is a Candidate of Technical Sciences, Lecturer in Computer Aided Electrical Systems and Technologies, National Aviation University. Has two degrees in economics, as well as standardization, certification and metrology. Defended dissertation on the topic "Methods of measuring geometric parameters of aviation parts" in the specialty "Devices and methods of measuring mechanical quantities". Has more than 30 scientific publications in domestic and foreign scientific publications.

Volodymyr Kharchenko, Doctor Sc.(Engineering), Professor, Vice-Rector on Scientific Work of the National Aviation University, Kyiv, Ukraine. Editor-in-Chief of the scientific journal Proceedings of the National Aviation University,member of the editorial board of the journal Vilnius Gediminas Technical University. Winner of the State Prize of Ukraine in Science and Technology, Honoured Worker of Science and Technology of Ukraine. Education: Kyiv Institute of Civil Aviation Engineers, Kyiv, Ukraine. Research area: management of complex socio-technical systems, air navigation systems and automatic decision-making systems aimed at avoidance conflict situations, space information technology design, air navigation services in Ukraine provided by CNS/ATM systems. Publications: 570. Born in 1943 in the city of Oratov, Vinnytsia region. Education is higher, in 1967 he graduated from Kiev Institute of Civil Aviation Engineers (now - National Aviation University), specializing in "Technical Operation of Aeronautical Radio Equipment of Airports". He began his career in 1966 as radio engineering of the 2nd form of the Krasnoyarsk Joint Aircraft Division. In 1967-1968 he was an engineer of the Krasnoyarsk Combined Airborne Division. 1968-1969 - engineer of the Kyiv Combined Airborne Division of the Ukrainian Civil Aviation Administration. From 1969 to 1973 - postgraduate student at the Kyiv Institute of Civil Aviation Engineers. In 1973-1977 he was a junior researcher at the Kyiv Institute of Civil Aviation Engineers. 1977-1984 - Senior Researcher at the Kyiv Institute of Civil Aviation Engineers. From 1984 to 1987, he was an associate professor at the Radiation Department of the Kyiv Institute of Civil Aviation Engineers. 1987-1994 - Associate Professor of the Department of Aviation Radioelectronic Systems of the Kyiv Institute of Civil Aviation Engineers. In 1994-1999 he was a professor at the Chair of Aviation Electronic Electronic Systems at the Kyiv Institute of Civil Aviation Engineers. 1999-2000 - Professor of Aeronautical Systems Department, Kyiv Institute of Civil Aviation Engineers. September-October 2000 - Professor of the Department of Information and Diagnostic Systems of the Kyiv Institute of Civil Aviation Engineers. From 2000 to 2001 he was the head of the department of aeronautical systems of the Kyiv Institute of Civil Aviation Engineers. April-June 2011 - Head of the Department of Air Navigation Systems of the National Aviation University. Since December 2001 - Vice-Rector on Scientific Work of the National Aviation University.

Heorhii Kuchuk was born on March 12, 1956. Education: graduated from Moscow State University in 1977, Master of Mathematics; Completed his PhD thesis in 1993, PhD "Operation of weapons and military equipment"; Completed his doctoral thesis in 2013, Full Doctor "Information Technology". Position Professor of the Department "Computer Engineering and Programming" National Technical University "Kharkiv Polytechnic Institute"

Anatolii Kulik, Professor of the Department of Aircraft Control Systems of the National Aerospace University KhAI, Kharkov, Ukraine.

Nataliia Kuzmenko is a senior researcher of the National Aviation University of Ukraine. She obtained her Ph.D. degree of Engineering in Navigation and Traffic Control in 2017 from the National Aviation University of Ukraine. Nataliia is certified aviation security instructor by ICAO (ASTP/Basic, ASTP/Instructors). She has had a traineeship in Tool Development for support to CAA/NSA at Regulatory Division within the Directorate Single Sky (Eurocontrol, Brussels). Current research projects include aviation safety, collision detection, and avoidance, artificial intelligence, Remotely Piloted Aerial Systems, video stream object detection and recognition, kernel density estimation, neural networks.

Lyudmyla Kuzmych is a Postdoctoral Researcher at the Department of the Computerized Electrotechnical Systems and Technology of the National Aviation University, Kyiv, Ukraine. The field of researches is the analysis of the reliability of complex technical structures; methods and instruments of the measurement strains and deformations of the engineering structures.

Volodymyr Kvasnikov is an expert in the field of electronics. Ph.D. Engaged in engineering and scientific work. Professor, Doctor of Sciences, Honored Metrologist of Ukraine, Head of the Department of the Computerized Electrotechnical Systems and Technology at the National Aviation University.

Oleksandr Lysenko, Doctor of Science, Professor of Department of Telecommunication in National technical university of Ukraine "Igor Sikorsky Kyiv Polytechnic Institute" (Ukraine). Areas of Scientific Interests: Digital control systems, Optimal control systems, Operations research, Expert systems, Artificial intelligence, Reusable aerospace systems, Mobile sensor networks based on UAV and telecommunication aeroplatforms. Author of more than 320 scientific articles, books, methodical manuals, copyright certificates, guides, monographs in fields of aviation, mathematics, the theory of system. Teaching courses: Theory of Probability and Mathematical Statistics, Digital Control and Information Processing Systems, System Analysis, The Theory of Decision Making, Operations Research, Mathematical Methods of Modeling and Optimization.

Marina Nozhnova is a lecturer in Kremenchuk Flight Colledge of Kharkiv National University of Internal Affairs (Ukraine). Graduated Flight Academy of National Aviation University (2013, Ukraine). Areas of Scientific Interests: Artificial Intelligence; Information technology and Informatics of Decision-Making in the Air Navigation System; Power Supply Systems for Airfields and Airports.

Ivan Ostroumov has been a faculty of Air Navigation Systems Department of the National Aviation University of Ukraine since September 2007. He obtained his Ph.D. degree of Engineering in Navigation and Traffic Control in 2009 from National Aviation University of Ukraine. Since then, he has been a research scientist and associated professor for National Aviation University. Since 2016 he has also served as navigation instructor at "Aviation Company Ukrainian Helicopters". In 2017/2018 has being a Fulbright scholar in the school of Aeronautics and Astronautics at Purdue. Also, he took part in several projects, including Supporting SESAR on GNSS Vulnerability Assessment by performing Space Weather Analysis (Navigation department, EUROCONTROL, Brussel) and E-learning course development (Institute of Air Navigation Services, EUROCONTROL, Luxembourg). His research theme is advanced methods for Alternative Positioning, Navigation, and Timing. Current research projects include Methods and Algorithms of positioning by multiple navigational aids, Availability and Accuracy estimation of navigation.

Mario Boyero Pérez is an experienced engineer on European R&D programmes of ATM and U-space, mainly contributing to architecting solutions and the development of the framework of the European Air Traffic Management Architecture. Also supporting the development of the Virtual Centre concept as system engineer. With a background in Telecommunication Engineering having completed the studies in Spain and Germany.

Andrii Podorozhniak was born on December 17, 1965. Education: Kharkiv higher military aviation school of radio electronics, department of radio electronics, Degree in Radio Electronics (1983–1988); Kharkiv Military University, department of space systems, Degree in Space Systems (1992–1995); Ivan Kozhedub Kharkiv Air Force University, PhD in Radio Engineering and Television Systems (2003–2006). Occupation: Associate Professor of the Department "Computer Engineering and Programming", National Technical University "Kharkiv Polytechnic Institute".

Victor Romanenko, Ph.D., Associate Professor of Avionics Department in National Aviation University (Ukraine). Thesis in the field of aircraft drive systems (1996). Research interests: flight safety, theory of automatic control systems, cavitation in fluid systems. Author of 62 scientific articles. Teaching courses: Automatic Control Theory, Fundamentals of Information Theory and Coding, Life Cycle Main Stages and Management of Avionics, Metrology, Standardization and Certification.

Yelyzaveta Sahun was born 1992 the 27th of February in Kirovohrad (now Kropivnitsky). Received the certificate of secondary education in May of 2009. The same year began to study at the State Flight Academy of Ukraine (now Flight Academy of National Aviation University). Graduated in 2014 from the Faculty of Management and Economy (Specialty Air Transport) as a manager-economist. The same year started to work at Plant Hydrosila, JSC – gear pumps designing and manufacturing as a price economist. From September 2017 till nowadays is a post graduate student of Flight Academy of National Aviation University, specialty Air Transport.

Yurii Shmelev, PhD, Deputy Director of the College of Educational Work in Kremenchuk Flight Colledge of Kharkiv National University of Internal Affairs (Ukraine). Areas of Scientific Interests: Development and Design of Aviation Training Complexes; Complexes of Avionics; Power Supply Systems for Airfields and Airports. Author of scientific articles, articles in Scopus, patents of Ukraine on a utility model, books, methodical manuals, guides, monographs in fields of aviation, power industry and avionics.

Olha Sushchenko received her M.Sc. degree in gyroscopic devices and units from Kyiv Polytechnic Institute, Ukraine, in 1980, the Ph.D. degree in instrument-making from Research University "Hydraulic Device", USSA, in 1991 and the Sc.D. degree in automation of design works from National Aviation University, Ukraine, in 2015. In 1980, she was a researcher at Kyiv Automatic Plant, Ukraine and in 2000 a faculty member at National Aviation University, Ukraine. Currently, she is a professor in the Department of Aerospace Control Systems at National Aviation University, Ukraine. She has published about 200 refereed journal and conference papers. Her research interest covers control systems, and inertial navigation. Prof. O. Sushchenko is a member of IFANG and IEEE.

Olena Tachinina, Doctor of Science, Professor of Department of Automation and Energy Management in National Aviation University (Ukraine). Areas of Scientific Interests: Digital control systems, Optimal control systems, Artificial intelligence, Reusable aerospace systems, Mobile sensor networks based on UAV and telecommunication aeroplatforms. Author of more than 120 scientific articles, books, methodical manuals, copyright certificates, guides, monographs in fields of aviation, mathematics, the theory of system. Teaching courses: Digital Control and Information Processing Systems, Mathematical Methods of Modeling and Optimization, Information Theory, Control of Complex Robotic Systems, Resource Management Automation.

Dmytro Tkachenko, Master of Air Traffic Services. Postgraduate student of the Air Traffic Services Faculty of the Flight Academy of the National Aviation University, Kropivnitsky, Ukraine. Education: Air Traffic Services Faculty of the Flight Academy of the National Aviation University, Kropivnitsky, Ukraine (2012). Workplace: Lviv Regional Branch of UkSATSE. Position: Air Traffic Controller. Research area: risk factors in Air Navigation Socio-technical System. Publications: 10.

Denys Vasyliev, PhD (Navigation and Air Traffic Control). Deputy Director of the Training and Certification Center, Ukrainian State Air Traffic Services Enterprise (UkSATSE), Boryspil, Ukraine. Research interests: situation analysis and decision-making in air traffic management. Author of 45 scientific papers.

Volodymyr Vasyliev, Professor, Doctor of Science (Navigation and Air Traffic Control). Head of the Department of Aviation Radio-electronic Complexes, National Aviation University, Kyiv, Ukraine. Research interests: system theory, identification and prediction, data processing, decision- making support systems in air navigation and air traffic management. Author of more than 150 scientific papers, methodical manuals and patents. Teaching courses: Mathematical Methods of Optimization; Methods of Mathematical Modeling; Computerized Data Processing Systems and Air Traffic Control.

Maksym Yastrub, PhD student at National Aviation University (with specialisation in Aviation Transport) and background in Air Traffic Management R&D programmes, mainly contributing to the development of the CNS and Airport architectures and enhancement and development of European ATM Architecture framework. The scientific interests are related to the use of enterprise architecture or system engineering techniques in aviation.

Index

A

A and A 280, 305
AC 235, 324, 351, 373
AcB 285, 287-288, 305
Adaptive Approach 179
AdB 285, 288, 305
AI (Artificial Intelligence) 34
air cargo 419, 421-422, 430, 434
air navigation service provider 89, 439, 449-450
Air Navigation Socio-technical System 1, 34
Air Navigation System (ANS) 2, 90
air traffic 1, 4, 8, 12, 15, 17, 26-27, 30, 66-67, 70, 75-76, 78, 80, 86, 89-90, 149-150, 184, 306-308, 312, 318, 321, 438-439, 441, 446, 449, 451-453, 455-456
Air Traffic Control (ATC) 75, 90
Air Traffic Management (ATM) 12, 27, 67, 89-90, 439
air traffic management system 26, 90, 438-439
aircraft 1, 4, 8, 15-17, 23-24, 26-27, 34-39, 42, 48, 62, 65, 67-68, 70, 72-73, 81, 85-86, 90, 136, 147, 149-150, 180-181, 186-187, 189, 193, 196, 232, 258, 262-263, 274, 276, 307-308, 312, 318, 322, 324, 337-338, 344, 349, 351-356, 359, 363, 366-368, 372-374, 378-379, 383-384, 386, 390, 395-400, 403-404, 406-411, 413, 415, 417-420, 422-428, 430, 433, 437, 439, 451-452, 479
AIS 2, 90, 307, 316, 324-325, 351, 425, 437
ALA 338-340, 343, 346-348, 351
architecture 139-141, 144, 194, 196, 198-199, 210, 312, 405, 438-449, 451-454, 456-459
artificial intelligence 1-2, 4-5, 9-10, 30, 34, 38-39, 90, 149-151, 154, 177, 180-181, 185, 189-190, 231, 234, 240-241, 246-247, 250, 253, 257, 260, 263, 279, 302, 306-307, 321, 354, 367, 419-420, 425, 437, 439, 441-442, 449, 451, 453, 457-458
Artificial Intelligence System (AIS) 425, 437
Artificial Neuron Network 147
automated systems 3, 23, 149
automatic control system 41, 65, 240, 271-274

autonomous fly vehicles 36, 62, 65
aviation specialists 306, 312, 321-322, 407
avionics 180-184, 189-190, 395, 404-410, 412-413, 417-418

B

Bin Packing Problem (BPP) 437
Block Relocation Problem with Weights (BRPW) 426, 437
branching trajectory 323-325, 327, 329-330, 332-333, 337, 341, 343-344, 349, 351

C

capsule neural networks 139-140, 143-145
CDS 323-328, 330-331, 340, 351
changepoint 148, 152-154, 165, 167-168, 179
charge amplifier 267-268
Collaborative Decision Making (CDM) 4, 9, 17, 26-27, 66, 90
compound dynamic systems 324-325, 351
conflict detection 12
Container Loading Problem (KLP) 437
control system 26, 36-39, 41-42, 65, 151, 155, 165, 179-180, 193, 231, 239-240, 246, 252, 255, 271-274, 349, 373, 395, 404
convolutional neural networks 138-140, 145
correlation field 396-397, 403
correlation function 374-376, 378-379, 381-384, 386-389, 400, 403

D

DDS 328, 330, 332, 349, 351
Decision Making (DM) 67, 90
Decision Support System (DSS) 3, 81, 90
deep learning 134-135, 138-139, 145, 147, 306, 321, 439

deformation 17, 58, 218, 223
destabilizing effects 39, 41-43, 45-46, 48-49, 52, 57, 61, 65
diagnosability 43-46, 65
diagnostic variables 148-149, 151, 153, 165-166, 177
differential-digital method 352, 366, 368-369
DoDAF 441-443
dynamic programming 16, 210

E

Earth remote sensing 136, 147
enterprise architecture 438-444, 446, 454, 457, 459
enterprise architecture framework 441, 459
European Air Traffic Management Architecture 441, 446, 449
Expected Aircraft's Operating Conditions 34
Expert Systems (ES) 17, 34
EZ 326, 351

F

Flight Emergency (FLEM) 68, 90
Flight management system 180-181
Flight n Flow Information for a Collaborative Environment (FF-ICE) 9, 90
flight path 184-185, 344, 346-347, 374, 379, 382, 384, 386-390, 403
flight safety 5, 66-67, 75-76, 78-80, 86, 413, 460-461, 475, 479
flying sensor networks 324-326, 351
FSP 298, 305
functional failures 404, 406, 408, 413, 417-418

G

genetic algorithm 246-247, 250-253, 260, 262
GERT-network (Graphical Evaluation and Review Technique) 34
glide path 372-373, 378-382, 384, 386, 389-393, 395, 398, 400-401, 403
guaranteed flight safety level 460

H

human factor 5, 8-9, 30, 67, 181, 190, 373, 378, 403, 411-413
human-operator (H-O) 2, 90

I

ICAO 2-3, 8, 10, 12, 17, 27, 67-70, 76, 90, 149, 185, 188-189, 307-308, 311, 411, 417, 460-461, 465, 467, 471, 475, 479
Index Image 147
inertial navigation system 185
inertially stabilized platform 239-240, 257, 260, 262
information technologies 2, 5, 30, 262, 306, 312, 321
Informational Robot 351
intellectualization 151, 154, 177, 179
Intelligent Automated System (IAS) 81, 90
intelligent computer 38, 324
irradiated power 195, 211

J

joint decision-making 69

K

Knapsack Problem (KP) 420, 437
Knowledge Base (KB) 431, 437
Knowledge-Based Systems (KBS) 431, 437

L

lifecycle 1, 448
load planning 419-420, 423, 426-427, 434
LP 297, 305
LRE 338, 351

M

maintenance 2, 76, 86, 149-150, 165-166, 179, 215, 223, 405, 412, 442, 444
Markov network 17, 34
mathematical model 151-153, 185, 214, 220-222, 241, 244, 325, 328, 338, 343, 354, 367, 369, 379, 400, 423, 427
MB 297, 299, 305
measurement error 269-270, 352, 369
meteorological information 149, 306-308, 311, 321-322, 452
Modus Ponens 437
Modus Tolens 437
monitoring 5-6, 28-29, 70, 135, 145, 149, 151, 157, 164-166, 169, 175, 177, 179, 188-189, 209, 222-223, 229, 263, 271, 326, 395, 461

multi-criteria approach 27, 66, 71, 86
multicriteria optimization 419-420, 430, 434
Multispectral Image 136, 147

N

NAF 442-444, 446-447
navigation 1-2, 4-5, 30, 34, 66-67, 69-70, 75, 81, 89-90,
149-150, 180-181, 184-190, 260, 263, 274-276,
307, 373, 384, 439, 446, 449-450, 461, 476, 479
network graph 27-28, 71, 73
neural network 1-2, 17, 82-84, 86, 134, 139-145, 252,
287, 291, 301, 305
non-professional factors 5-6, 34, 81

O

OA 338-340, 343, 348, 351
on-board aviation equipment 405-406, 408, 418
Operating State 65
operation system 148-151, 154-155, 177, 179
Operational Flight Plan 322
Operational Meteorological Information 306, 321-322
optimal control 323, 325, 327, 331-332, 345, 349
OS 149, 151, 156-158, 160-165, 177, 324, 346-348, 351
out 38, 42, 46, 50, 52, 57, 63, 67, 72, 76, 78, 80-81,
151, 156-158, 165-166, 169-170, 194-198, 206,
210-211, 223, 233-234, 240, 250-251, 255, 260,
276, 280-281, 285, 287-288, 291, 297-299, 301,
305, 307-308, 311, 330, 332, 356, 360, 363, 366,
374-375, 397-398, 401, 407, 409, 422, 429-430,
465, 469

P

Parameter Amplitude 398, 403
Path Branch 327, 351
perceptron 23, 291-292, 297, 301
piezo module 265-267
power supply 50, 206, 280, 285
professional factors 5, 7, 34, 81, 85-86

R

radioelectronic equipment 148-150, 163, 169, 177
Random Access Memory (RAM) 437
Rational Control Object 65
RE 338-339, 351, 418, 428
reconnaissance flight 203
recoverability 46-47, 53, 65
reflection 34, 38, 137, 216

reliability 5, 24, 39, 148-149, 153, 165, 177, 179,
214-217, 219-220, 222-224, 228, 231, 233, 264,
276, 279-280, 285, 353, 373, 395, 413, 465, 468,
473, 476
reliability parameters 148-149, 153, 165, 177
Remotely Piloted Aircraft 23
residual resource 214-215, 222-224, 229
resource 39, 69, 214-215, 218-225, 228-229, 442,
452-453, 456
risk 1-2, 15-17, 25, 30, 48, 66, 75-76, 78-80, 84, 86,
175-176, 302, 406-407, 417-418, 444, 461, 465-
466, 469-473, 475-477, 479
Robust System 247, 262
RPS 298, 305
rule-based optimization 434

S

safety management system 460-461, 471, 479
safety risk 417
sensors 39, 48-50, 52, 54, 56-58, 180-181, 183, 185,
193, 195, 197, 199, 206, 209-211, 232-233, 240,
255, 258, 263-264, 267-268, 276, 324-326, 354,
356, 366, 396
SIGMET Information 322
signal-noise ratio 193, 195, 211
Single European Sky ATM Research 439
SMART 30, 193-194, 196, 198-200, 202, 204, 206,
209-211
SMS 479
socio-technical systems 3-4, 34-35
spectral indices 134-136, 145
SQL-server 434, 437
stabilization system 232, 235-237, 245, 250, 252,
255-257, 260, 262
statistical data processing 151, 179
structural transformations 324, 326-330, 339-340, 351
synchronization 27, 29, 66, 69-70, 354
System of Computer-aided Design 262
System-Wide Information Sharing and Management
(SWIM) 90

T

TE 338, 351
technical condition 56, 149-150, 154, 165-167, 169-
173, 175-177, 179, 214, 218, 222, 224
technology process analysis of flights 408-409, 417
temperature effects 276
TOGAF 441, 444-446

training 4, 6, 10, 15-16, 22, 29, 34, 69, 72-73, 81, 85, 140-141, 143-145, 149, 165, 283, 291-292, 297-303, 306, 312, 368, 373, 378, 382, 398-399

U

UAV 17, 19-21, 23-24, 26, 30, 85, 193-198, 204-205, 207, 211, 312, 315, 317, 324, 326, 351
Unexpected Aircraft's Operating Conditions 35
Unit Load Device (ULD) 437
unmanned aerial vehicle 194, 262, 315, 317, 324, 326, 345, 349, 351

V

veracity 175-177, 179

W

wildfire 134-135

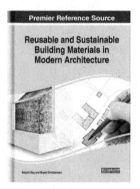
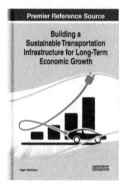
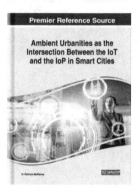
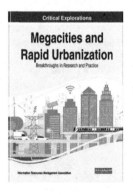
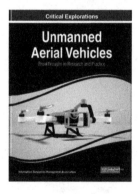

Ensure Quality Research is Introduced to the Academic Community

Become an IGI Global Reviewer for Authored Book Projects

Premier Reference Source

Emerging GIS Applications for Emergency and Disaster Management

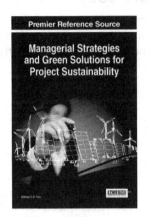

Premier Reference Source

Managerial Strategies and Green Solutions for Project Sustainability

Premier Reference Source

Comparative Approaches to Using R and Python for Statistical Data Analysis

Premier Reference Source

Solutions for High-Touch Communications in a High-Tech World

The overall success of an authored book project is dependent on quality and timely reviews.

In this competitive age of scholarly publishing, constructive and timely feedback significantly expedites the turnaround time of manuscripts from submission to acceptance, allowing the publication and discovery of forward-thinking research at a much more expeditious rate. Several IGI Global authored book projects are currently seeking highly-qualified experts in the field to fill vacancies on their respective editorial review boards:

Applications and Inquiries may be sent to:
development@igi-global.com

Applicants must have a doctorate (or an equivalent degree) as well as publishing and reviewing experience. Reviewers are asked to complete the open-ended evaluation questions with as much detail as possible in a timely, collegial, and constructive manner. All reviewers' tenures run for one-year terms on the editorial review boards and are expected to complete at least three reviews per term. Upon successful completion of this term, reviewers can be considered for an additional term.

If you have a colleague that may be interested in this opportunity, we encourage you to share this information with them.

Printed in the United States
By Bookmasters